Engineering A Level: Compulsory Units for AS and A Level Engineering

Mike Tooley
formerly Vice Principal
Brooklands College of Further and Higher Education

OXFORD AMSTERDAM BOSTON LONDON NEW YORK PARIS
SAN DIEGO SAN FRANCISCO SINGAPORE SYDNEY TOKYO

Newnes
An imprint of Butterworth-Heinemann
Linacre House, Jordan Hill, Oxford OX2 8DP
30 Corporate Drive, Burlington, MA 01803

First published 2005

British Library Cataloguing in Publication Data
A catalogue record for this book is available from the British Library

ISBN 0 7506 66927

For information on all Newnes publications
visit our website at www.newnespress.com

Typeset by the author
Printed and bound in Great Britain

Contents

Preface		*vii*
Unit 1	Engineering Materials, Processes and Techniques	1
Unit 2	The Role of the Engineer	127
Unit 3	Principles of Design, Planning and Prototyping	175
Unit 4	Applied Engineering Systems	261
Unit 5	The Engineering Environment	349
Unit 6	Applied Design, Planning and Prototyping	391
Index		409

Preface

Welcome to the challenging and exciting world of engineering! This book is designed to help you succeed on a course leading to an AS or A2 level qualification in Engineering. It contains all of the essential underpinning knowledge required of a student who may never have studied engineering before and who wishes to explore the subject for the first time.

About you

Have you got what it takes to be an engineer? The GCE advanced level (AS/A2) courses in Engineering will help you find out and still keep your options open. Successful completion of the course also qualifies for UCAS points and thus could lead you into further studies at degree level.

Engineering is an immensely diverse field but, to put it simply, engineering, in whatever area that you choose, is about thinking *and* doing. The 'thinking' that an engineer does is both logical and systematic. The 'doing' that an engineer does can be anything from building a bridge to testing a space vehicle. In either case, the essential engineering skills are the same.

You do not need to have studied engineering before starting the advanced level course. All that is required to successfully complete the course is an enquiring mind, an interest in engineering, and the ability to explore new ideas in a systematic way. You also need to be able to express your ideas and communicate these in a clear and logical way to other people. If, on the other hand, you have already completed a GCSE or BTEC First Award in Engineering, you will be on home territory and things should fall into place fairly easily.

As you study the GCE Engineering (at both AS and A2 levels) you will be learning in a practical environment as well as in a classroom. This will help you to put into practice the things that you learn in a formal class situation. You will also discover that engineering is fun—it's not just about learning a whole lot of meaningless facts and figures!

About the AS and A Level in Engineering

The GCE in Engineering will help you to build and apply knowledge in a wide variety of engineering contexts. The course can be taken along with GCE Maths or any of the science GCEs.

When taken on a stand alone basis, the three AS units will provide you with a valuable insight into engineering. The three AS units also provide progression on to the three A2 units, allowing you to further develop your understanding of engineering as well as an appreciation of the environment in which engineers work.

Taken as a whole, the course will help you to:

- understand the nature and demands of different types of engineering
- develop an understanding of engineering technologies and the science that underpins them
- apply your understanding of engineering to the design, development, production and manufacture of engineered products or engineering services
- understand the context in which engineers work, including the environmental impact of engineering activities and the legislation that impacts on the design, production and utilisation of engineered products and engineering services.

It is important to be aware that you will be required to design and manufacture engineered products at both the AS and A2 levels. You will also be required to carry out investigations and evaluations of engineered products or engineering services. These practical aspects of the course form a major part of the assessment of the relevant units.

The three units that make up the AS course cover *Engineering Materials, Processes and Techniques*, *The Role of the Engineer*, and *Principles of Design, Planning and Prototyping*. The three units that make up the A2 course are *Applied Engineering Systems*, *The Engineering Environment*, and *Applied Design, Planning and Prototyping*. The links between these units are shown in the diagram below.

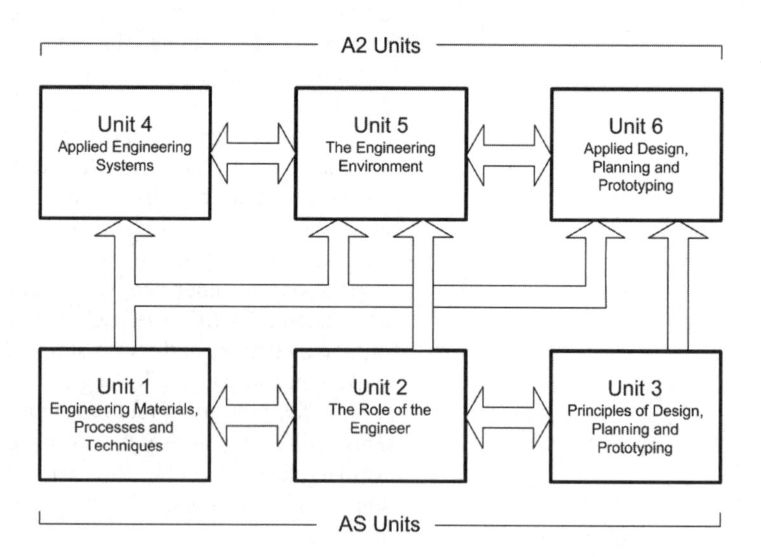

Relationship between the units that make up the GCE Engineering course

How to use this book

This book covers all six units that make up the AS and A2 Engineering GCE course. One chapter is devoted to each unit and each of these chapters contain text, 'test your knowledge' questions, activities, and review questions.

The 'test your knowledge' questions are interspersed with the text throughout the book. These questions allow you to check your understanding of the preceding text. They also provide you with an opportunity to reflect on what you have learned and consolidate this in manageable chunks.

Most 'test you knowledge' questions can be answered in only a few minutes and the necessary information can be gleaned from the surrounding text. Activities, on the other hand, require a significantly greater amount of time to complete. Furthermore, they often require additional library or resource area research coupled with access to computing and other information technology resources.

As you work through this book, you will undertake a programme of activities as directed by your teacher or lecturer. Don't expect to complete *all* of the activities in this book—your teacher or lecturer will ensure that those activities that you do undertake relate to the resources available to you and that they can be completed within the timescale of the course. Activities also make excellent vehicles for improving your skills and for gathering the evidence that can be used to demonstrate that you are competent in core skills.

To help you with Unit 5 (*The Engineering Environment*) and Unit 6 (*Applied Design, Planning and Prototyping*) I have provided you with samples of investigations and product developments undertaken by two A2 students at a mythical North Downs College. These samples of students' work are designed to give you an idea of how to tackle the assessment of these two units. Sufficient detail has been included for you to get a feel of what students are actually required to do and, more importantly, how the students have used the skills and knowledge that they acquired as a result of studying the earlier units.

The 'review questions' presented at the end of each chapter are designed to provide you with an opportunity to test your understanding of the unit to which they relate. These questions can be used for revision or as a means of generating a 'checklist' of topics that you should be familiar with. Here again, your tutor may suggest that you answer specific questions which relate to the context in which you are studying the course.

Finally, here are a few general points worth noting:

- Allow regular time for reading—get into the habit of setting aside an hour, or two, at the weekend to take a second look at the topics that you have covered during the week.
- Make notes and file these away neatly for future reference—lists of facts, definitions and formulae are particularly useful for revision!
- Look out for the inter-relationship between subjects and units —you will find many ideas and a number of themes that crop up in different places and in different units. These can often help to reinforce your understanding.
- Don't be afraid to put your new ideas into practice. Remember that engineering is about thinking *and* doing—so get out there and *do* it!
- Lastly, I hope that you will find some useful support material at the book's companion website which you will find at: http://books.elsevier.com/companions/0750666927

Good luck with your AS and A2 Engineering studies!

Mike Tooley

Unit 1 Engineering materials, processes and techniques

Summary

Welcome to the fascinating world of engineering materials! All branches of engineering are concerned with the use and processing of materials and so this makes an excellent starting point for your study of engineering. The correct choice and processing of materials can often be crucial. For example, Civil Engineers need to ensure that the materials from which they build their roads, bridges and dams are not only able to withstand the forces that will be applied to them but also that they will not deteriorate with time. Think about what might happen if the materials chosen were not appropriate for the job. For example, if a suspension bridge was built from a material that is not sufficiently strong it might collapse under a heavy load. Even on a much smaller scale knowledge of materials is vitally important. An Electronic Engineer designing a complex flight control system must ensure that the materials chosen do not prematurely deteriorate in the harsh environment in which the system will operate. Failure of the system could be equally catastrophic, particularly if the failure just happened to occur at a critical point in flight, such as landing or take-off. A study of materials is, therefore, an essential part of every engineer's portfolio and for this reason alone, is worthy of our attention right at the outset.

1.1 Materials

We start by taking a look at the main classes of materials used in engineering as well as the terminology that we apply to them. This theme is further developed when we look at the ferrous and non-ferrous metals that are widely used. We shall also study non-metals, including ceramics, polymers (plastics), elastomers and composite materials. We shall introduce some of the most commonly used processes and techniques used for joining and improving the properties and appearance of a wide range of engineering materials.

Classes of material

What do you already know?

For most engineering applications the most important criteria for the selection of materials is that they do the job properly (i.e. that they perform according to the specification that has been agreed) and that they do it as cheaply as possible. Whether a material does its job depends on its *properties*, which are a measure of how the material reacts to the various influences to which it is exposed. For example, loads, atmospheric environment, electromotive forces, heat, light, chemicals, and so on. To aid our understanding of the different types of materials and their properties I have divided them into four categories; metals, polymers, ceramics, and composites. Strictly speaking, composites are not really a separate group since they are made up from the other categories of material. However, because they display unique properties and are a very important engineering group, they are treated separately here.

Metals

You will be familiar with metals such as aluminium, iron, and copper, in a wide variety of everyday applications: i.e. aluminium saucepans, copper water pipes and iron stoves. Metals can be mixed with other elements (often other metals) to form an *alloy.* Metal alloys are used to provide improved properties because they are often stronger or tougher than the parent pure metal. Other improvements can be made to metal alloys by *heat-treating* them as part of the manufacturing process. Thus steel is an alloy of iron and carbon and small quantities of other elements. If, after alloying, the steel is quickly cooled by quenching in oil or brine, a very hard steel can be produced. Much more will be said later about alloys and the heat-treatment of metals.

Polymers

Polymers are characterized by their ability to be (initially at least) moulded into shape. They are chemical materials and often have long and unattractive chemical names. There is considerable incentive to seek more convenient names and abbreviations for everyday use. Thus you will be familiar with PVC (polyvinyl chloride) and PTFE (polytetrafluoroethylene).

Polymers are made from molecules which join together to form long chains in a process known as *polymerization.* There are essentially three major types of polymer. *Thermoplastics*, which have the ability to be remoulded and reheated after manufacture. *Thermosetting plastics*, which once manufactured remain in their original moulded form and cannot be re-worked. *Elastomers or rubbers*, which often have very large elastic strains, elastic bands and car tyres are two familiar forms of rubber.

Ceramics

This class of material is again a chemical compound, formed from oxides such as silica (sand), sodium and calcium, as well as silicates such as clay. Glass is an example of a ceramic material, with its main constituent being silica. The oxides and silicates mentioned above have very high melting temperatures and on their own are very difficult to form. They are usually made more manageable by mixing them in powder form, with water, and then hardening them by heating. Ceramics include, brick, earthenware, pots, clay, glasses and refractory (furnace) materials. Ceramics are usually hard and brittle, good electrical and thermal insulators and have good resistance to chemical attack.

Another view

In whatever branch of engineering you decide to study or work, you will be concerned with the use of a variety of different materials. An understanding of the characteristics and properties of these materials will not only allow you to select the right material for a particular application but it will also help you to understand the processes and treatments that can be applied to it.

Composites

A composite is a material with two or more distinct constituents. These separate constituents act together to give the necessary strength and stiffness to the composite material. The most common example of a composite material today is that of fibre reinforcement of a resin matrix but the term can also be applied to other materials such as metal-skinned honeycomb panels. The property that these materials have in common is that they are light, stiff and strong and, what's more, they can be extremely tough.

Reinforced concrete is another example of a composite material that is invaluable in engineering. The steel and concrete retain their individual identities in the finished structure. However, because they

Activity 1.1

How much do you already know about different materials and their properties? Test your knowledge by trying the exercise set out below. In attempting to tackle this task you might find it helpful to explore the objects that exist within your own home, and ask yourself why they are made from those particular materials. Complete the table by using a grading scheme such as: excellent, good, fair, poor; or high, above average, below average, low, or some other similar scheme.

Material properties	Class of material			
	Metals	Polymers	Ceramics	Composites
Density				
Stiffness				
Strength				
Toughness				
Shock resistance				
Hardness				
Thermal conductivity				
Electrical conductivity				
Corrosion resistance				
Melting temperature				

Test your knowledge 1.1

(a) Name the THREE main classes of polymer materials

(b) Name TWO examples of composite materials.

work together, the steel carries the tension loads and the concrete carries the compression loads. Furthermore, although not considered as a separate class of material, some *natural materials* exist in the form of a composite. The best known examples are wood, mollusc shells and bone. Wood is an interesting example of a natural fibre composite; the longitudinal hollow cells of wood are made up of layers of spirally wound cellulose fibres with varying spiral angle, bonded together with lignin during the growing of the tree.

The previous classification of materials is rather crude and many important subdivisions exist within each category. The natural materials, except for those mentioned above under composites, have been deliberately left out. The study of materials such as wool and cotton is better placed in a course concerned with the textile industry. Here, we will be concentrating on the engineering application of materials and will only mention naturally occurring materials where appropriate.

Finally, what you may have discovered from Activity 1.1 was how difficult it is to make generalisations about the properties of different classes of materials. We shall now continue this theme by examining each class of material in a little more detail but, before we do, I suggest that you take a look at Activity 1.2.

Activity 1.2

As you progress through Unit 1 you will find it extremely useful to have data on a variety of materials that are commonly used in engineering. Sources of data include reference books and text books as well as several on-line databases accessible via the Internet. You are advised to become familiar with several of these sources. However, in order to provide you with a simplified set of data, I have created the matSdata database specifically for A-level engineering students. The database will provide you with data and datasheets on a wide range of commonly used engineering materials. Furthermore, unlike other databases, matSdata allows you to maintain your own personal copy of the database. This means that you can modifying existing records and adding your own data as you progress through the course.

You can download a copy of the matSdata database from www.key2study.com/matsdata or from the URL shown in the Preface. To get started you should first download the database and then extract the files in order to make your own personal copy on a floppy disk or on a USB memory stick. Next your should run the matSdata program (matSdata.exe) and read the Help file. Finally, it's worth printing out datasheets for each of the following materials that you will be meeting in the next section of this book:

1. Cartridge brass
2. Carbon steel
3. Polyvinyl chloride (PVC)

Another view

Despite what you might think, composite materials are not new. However, this class of materials is extremely interesting to engineers because, unlike conventional materials, we can control the properties of a composite material during its manufacture. This leads to some exciting possibilities—for example, a material that is very strong in one direction but relatively weak in another.

Ferrous metals

As previously stated, ferrous metals are based upon the metal iron. For engineering purposes iron is usually associated with various amounts of the non-metal carbon. When the amount of carbon present is less than 1.8% we call the material steel. The figure of 1.8% is the theoretical maximum. In practice there is no advantage in increasing the amount of carbon present above 1.4%. We are only going to consider the plain carbon steels. Alloy steels are beyond the scope of this book. The effects of the carbon content on the properties of plain carbon steels are shown in Figure 1.1.

Cast irons are also ferrous metals. They have substantially more carbon than the plain carbon steels. Grey cast irons usually have a carbon content between 3.2% and 3.5%. Not all this carbon can be taken up by the iron and some is left over as flakes of graphite between the crystals of metal. It is these flakes of graphite that gives cast iron its particular properties and makes it a 'dirty' metal to machine. The compositions and typical uses of plain carbon steels and a grey cast iron are summarized in Table 1.1.

Low carbon steels

These are also called mild steels. They are the cheapest and most widely used group of steels. Although they are the weakest of the steels, nevertheless they are stronger than most of the non-ferrous metals and alloys. They can be hot and cold worked and machined with ease.

Medium carbon steels

These are harder, tougher, stronger and more costly than the low carbon steels. They are less ductile than the low carbon steels and cannot be bent or formed to any great extent in the cold condition without risk of cracking. Greater force is required to bend and form them. Medium carbon steels hot forge well but close temperature control is essential. Two carbon ranges are shown. The lower carbon range can only be toughened by heating and quenching (cooling quickly by dipping in water). They cannot be hardened. The higher carbon range can be hardened and tempered by heating and quenching.

High carbon steels

These are harder, stronger and more costly than medium carbon steels. They are also less tough. High carbon steels are available as hot rolled bars and forgings. Cold drawn high carbon steel wire (piano wire) is available in a limited range of sizes. Centreless ground high carbon steel rods (silver steel) are available in a wide range of diameters (inch and metric sizes) in lengths of 333 mm, 1 m and 2 m. High carbon steels can only be bent cold to a limited extent before cracking. They are mostly used for making cutting tools such as files, knives and carpenters' tools.

Test your knowledge 1.2

Describe the effect of increasing the carbon content of steel on:

(a) the hardness
(b) the tensile strength
(c) the ductility of the material.

Test your knowledge 1.3

State the principal characteristics of:

(a) low carbon steel
(b) medium carbon steel
(c) high carbon steel.

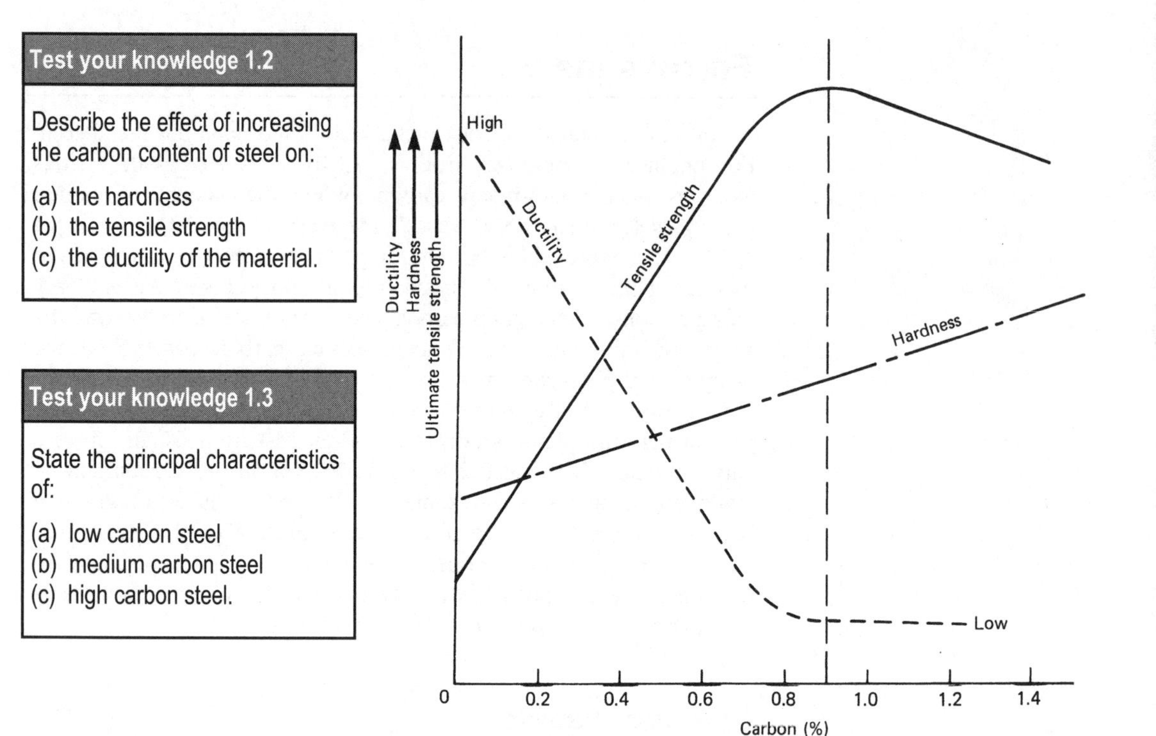

Figure 1.1 *Effect of carbon content on the properties of plain carbon steels*

Non-ferrous metals

Non-ferrous metals (i.e. metals that are *not* based on iron) include metals such as aluminium and zinc as well as alloys such as brass and bronze. We shall start by looking at copper—a material that is widely used in electrical engineering.

Copper

Pure copper is widely used for electrical conductors and switchgear components. It is second only to silver in conductivity but it is much more plentiful and very much less costly. Pure copper is too soft and ductile for most mechanical applications.

For general purpose applications such as roofing, chemical plant, decorative metal work and copper-smithing, tough-pitch copper is used. This contains some copper oxide which makes it stronger, more rigid and less likely to tear when being machined. Because it is not so highly refined, it is less expensive than high conductivity copper.

There are many other grades of copper for special applications. Copper is also the basis of many important alloys such as brass and bronze, and we will be considering these next. The general properties of copper are:

- relatively high strength

Table 1.1 *Ferrous metals*

Name	*Group*	*Carbon content (%)*	*Some uses*
Dead mild steel (low carbon steel)	Plain carbon steel	0.10–0.15	Sheet for pressing out components such as motor car body panels. General sheet-metal work. Thin wire, rod and drawn tubes.
Mild steel (low carbon steel)	Plain carbon steel	0.15–0.30	General purpose workshop rod, bars and sections. Boiler plate. Rolled steel beams, joists, angles, etc.
Medium carbon steel	Plain carbon steel	0.30–0.50 0.50–0.60	Crankshafts, forgings, axles, and other stressed components. Leaf springs, hammer heads, cold chisels, etc.
High carbon steel	Plain carbon steel	0.8–1.0 1.0–1.2 1.2–1.4	Coil springs, wood chisels. Files, drills, taps and dies. Fine-edge tools (knives, etc.)
Grey cast iron	Cast iron	3.2–3.5	Machine castings.

Test your knowledge 1.4

Select a suitable ferrous metal, and state its carbon content, for each of the following objects.

(a) A hexagon head bolt produced on a lathe
(b) a car engine cylinder block
(c) a cold chisel
(d) a wood-carving chisel
(e) a pressed steel car body panel.

- very ductile so that it is usually cold worked. An annealed (softened) copper wire can be stretched to nearly twice its length before it snaps
- corrosion resistant
- second only to silver as a conductor of heat and electricity
- easily joined by soldering and brazing. For welding, a phosphorous deoxidized grade of copper must be used.

Copper is available as cold-drawn rods, wires and tubes. It is also available as cold-rolled sheet, strip and plate. Hot worked copper is available as extruded sections and hot stampings. It can also be cast. Copper powders are used for making sintered components. It is one of the few pure metals of use to the engineer as a structural material.

Brass

Brass is an alloy of copper and zinc. The properties of a brass alloy and the applications for which you can use it depends upon the amount of zinc present. Most brasses are attacked by sea water. The salt water eats away the zinc (dezincification) and leaves a weak, porous, spongy mass of copper. To prevent this happening, a small amount of tin is added to the alloy. There are two types of brass that can be used at sea or on land near the sea. These are Naval brass and Admiralty brass.

Brass is a difficult metal to cast and brass castings tend to be coarse grained and porous. Brass depends upon hot rolling from cast

Table 1.2 *Properties and applications of brass alloys*

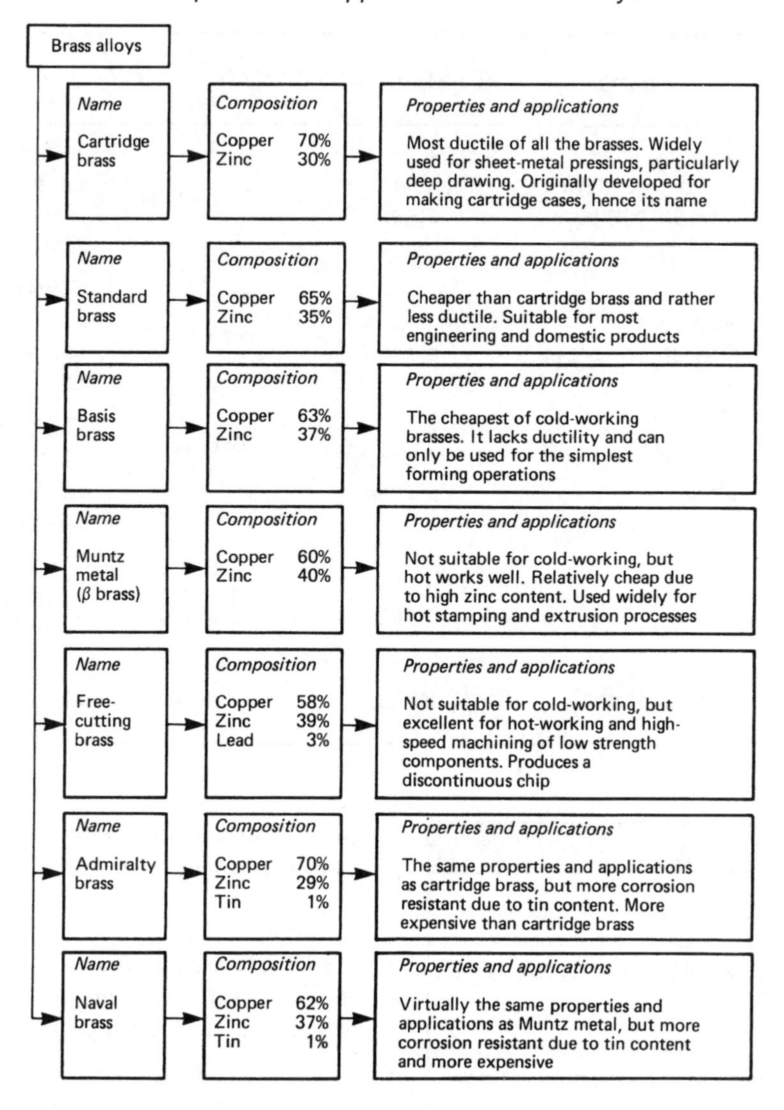

Brass alloys

Name	*Composition*	*Properties and applications*
Cartridge brass	Copper 70% Zinc 30%	Most ductile of all the brasses. Widely used for sheet-metal pressings, particularly deep drawing. Originally developed for making cartridge cases, hence its name
Standard brass	Copper 65% Zinc 35%	Cheaper than cartridge brass and rather less ductile. Suitable for most engineering and domestic products
Basis brass	Copper 63% Zinc 37%	The cheapest of cold-working brasses. It lacks ductility and can only be used for the simplest forming operations
Muntz metal (β brass)	Copper 60% Zinc 40%	Not suitable for cold-working, but hot works well. Relatively cheap due to high zinc content. Used widely for hot stamping and extrusion processes
Free-cutting brass	Copper 58% Zinc 39% Lead 3%	Not suitable for cold-working, but excellent for hot-working and high-speed machining of low strength components. Produces a discontinuous chip
Admiralty brass	Copper 70% Zinc 29% Tin 1%	The same properties and applications as cartridge brass, but more corrosion resistant due to tin content. More expensive than cartridge brass
Naval brass	Copper 62% Zinc 37% Tin 1%	Virtually the same properties and applications as Muntz metal, but more corrosion resistant due to tin content and more expensive

ingots, followed by cold rolling or drawing to give it its mechanical strength. It can also be hot extruded and plumbing fittings are made by hot stamping. Brass machines to a better finish than copper as it is more rigid and less ductile than that metal. Table 1.2 lists some typical brasses, together with their compositions, properties and applications.

Tin bronze

As the name implies, the tin bronzes are alloys of copper and tin. These alloys also have to have a deoxidizing element present to prevent the tin from oxidizing during casting and hot working. If the tin oxidizes the metal becomes hard and 'scratchy' and is weakened. The two deoxidizing elements commonly used are:

Table 1.3 *Properties and applications of bronze alloys*

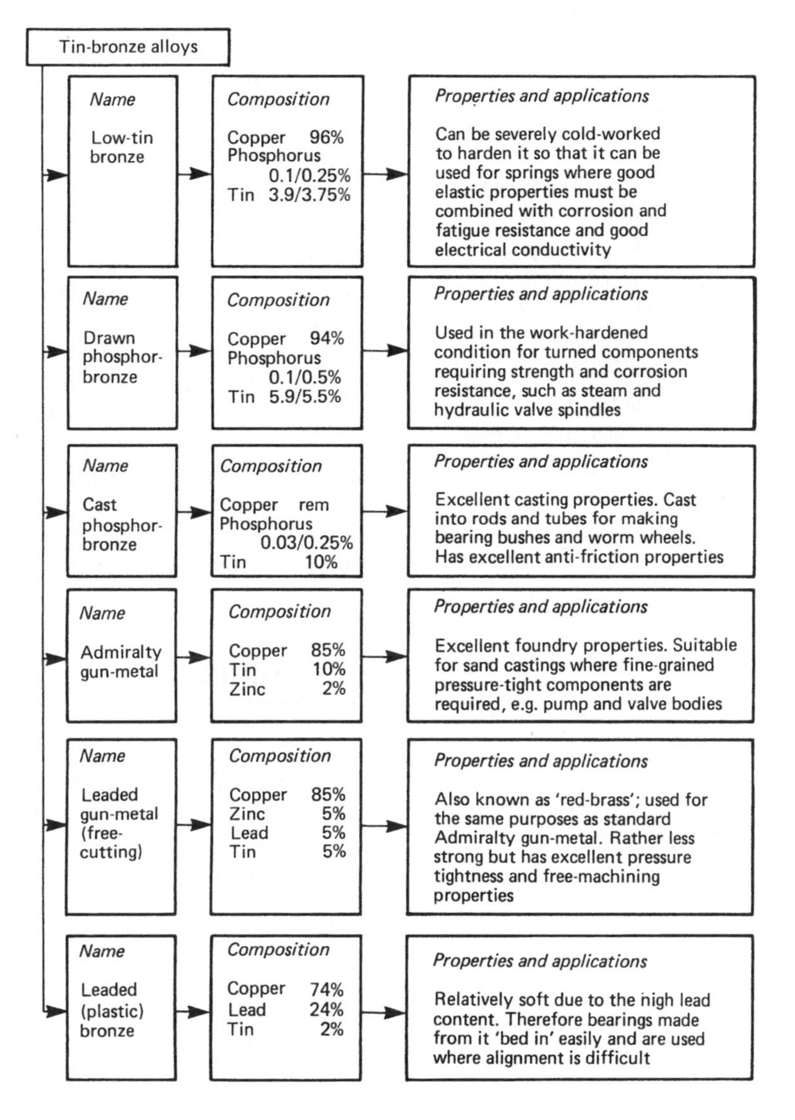

Tin-bronze alloys

Name	*Composition*	*Properties and applications*
Low-tin bronze	Copper 96% Phosphorus 0.1/0.25% Tin 3.9/3.75%	Can be severely cold-worked to harden it so that it can be used for springs where good elastic properties must be combined with corrosion and fatigue resistance and good electrical conductivity
Drawn phosphor-bronze	Copper 94% Phosphorus 0.1/0.5% Tin 5.9/5.5%	Used in the work-hardened condition for turned components requiring strength and corrosion resistance, such as steam and hydraulic valve spindles
Cast phosphor-bronze	Copper rem Phosphorus 0.03/0.25% Tin 10%	Excellent casting properties. Cast into rods and tubes for making bearing bushes and worm wheels. Has excellent anti-friction properties
Admiralty gun-metal	Copper 85% Tin 10% Zinc 2%	Excellent foundry properties. Suitable for sand castings where fine-grained pressure-tight components are required, e.g. pump and valve bodies
Leaded gun-metal (free-cutting)	Copper 85% Zinc 5% Lead 5% Tin 5%	Also known as 'red-brass'; used for the same purposes as standard Admiralty gun-metal. Rather less strong but has excellent pressure tightness and free-machining properties
Leaded (plastic) bronze	Copper 74% Lead 24% Tin 2%	Relatively soft due to the high lead content. Therefore bearings made from it 'bed in' easily and are used where alignment is difficult

- zinc in the gun-metal alloys.
- phosphorus in the phosphor–bronze alloys.

Unlike the brass alloys, the bronze alloys are usually used as castings. However low-tin content phosphor–bronze alloys can be extensively cold worked. Tin–bronze alloys are extremely resistant to corrosion and wear and are used for high pressure valve bodies and heavy duty bearings. Table 1.3 lists some typical bronze alloys together with their compositions, properties and applications.

Aluminium

Aluminium has a density approximately one third that of steel. However it is also very much weaker so its strength/weight ratio is inferior. For stressed components, such as those found in aircraft, aluminium alloys have to be used. These can be as strong as steel and

Test your knowledge 1.5

Select a suitable non-ferrous metal for each of the following objects. Give reasons for your choice.

(a) The body casting of a water pump
(b) screws for clamping the electric cables in the terminals of a domestic electric light switch
(c) a bearing bush
(d) a deep drawn, cup-shaped component for use on land
(e) a ship's fitting made by hot stamping.

nearly as light as pure aluminium.

High purity aluminium is second only to copper as a conductor of heat and electricity. It is very difficult to join by welding or soldering and aluminium conductors are often terminated by crimping. Despite these difficulties, it is increasingly used for electrical conductors where its light weight and low cost compared with copper is an advantage. Pure aluminium is resistant to normal atmospheric corrosion but it is unsuitable for marine environments. It is available as wire, rod, cold-rolled sheet and extruded sections for heat sinks.

Commercially pure aluminium is not as pure as high purity aluminium and it also contains up to 1% silicon to improve its strength and stiffness. As a result it is not such a good conductor of electricity nor is it so corrosion resistant. It is available as wire, rod, cold-rolled sheet and extruded sections. It is also available as castings and forgings. Being stiffer than high purity aluminium it machines better with less tendency to tear. It forms non-toxic oxides on its surface which makes it suitable for food processing plant and utensils. It is also used for forged and die-cast small machine parts. Because of their range and complexity, the light alloys based upon aluminium are beyond the scope of this unit.

Non-metals

Non-metallic materials can be grouped under the headings shown in Figure 1.2. In addition, wood is also used for making the patterns which, in turn, are used in producing moulds for castings. We are only going to consider some ceramics, thermosets and thermoplastics.

Ceramics

The word ceramic comes from a Greek word meaning potter's clay. Originally, ceramics referred to objects made from potter's clay. Nowadays, ceramic technology has developed a range of materials far beyond the traditional concepts of the potter's art. These include:

- glass products
- abrasive and cutting tool materials
- construction industry materials

Another view

Metals, both ferrous and non-ferrous, are conventional engineering materials and you will already be very familiar with their use. Other classes of material (such as ceramics, plastics and composites) are also widely used in engineering. You will find plenty of examples of their use in electrical and electronic engineering, chemical engineering, optical engineering, and so on.

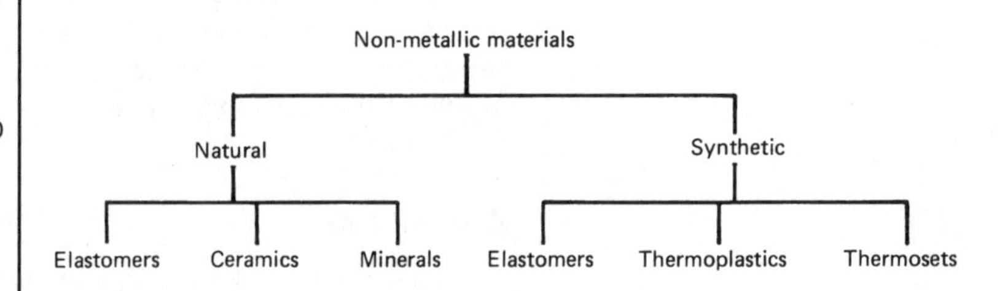

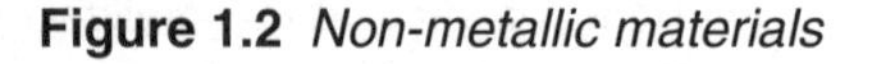
Figure 1.2 *Non-metallic materials*

Table 1.4 *Applications of ceramic materials*

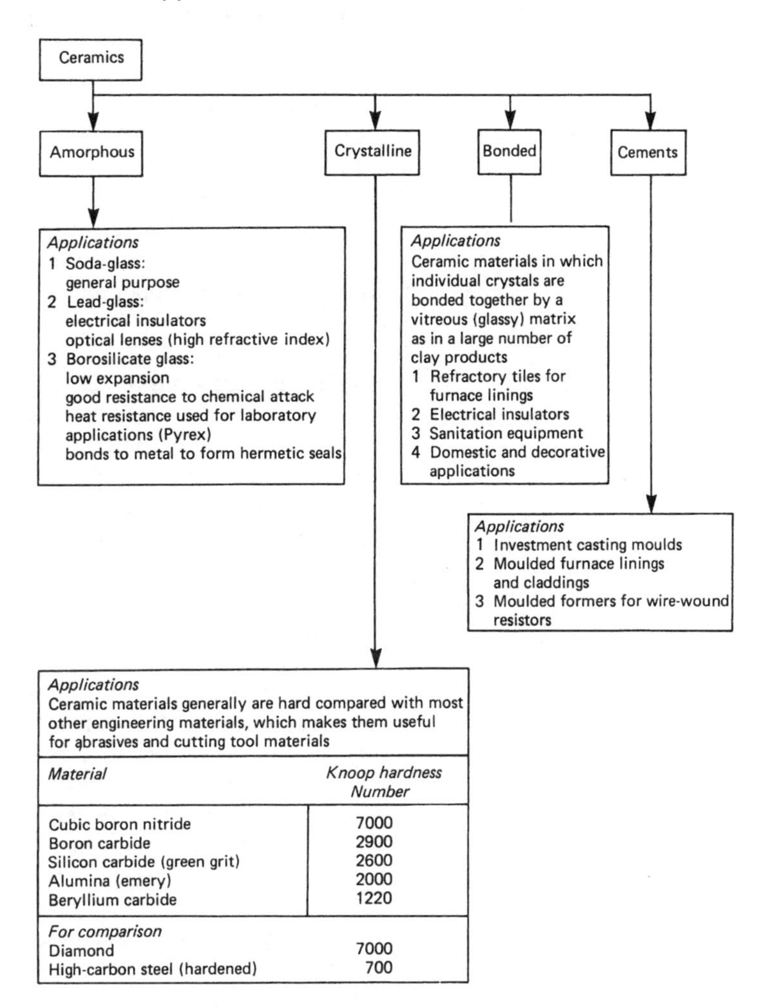

Material	*Knoop hardness Number*
Cubic boron nitride	7000
Boron carbide	2900
Silicon carbide (green grit)	2600
Alumina (emery)	2000
Beryllium carbide	1220
For comparison	
Diamond	7000
High-carbon steel (hardened)	700

- electrical insulators
- cements and plasters for investment moulding
- refractory (heat resistant) lining for furnaces
- refractory coatings for metals.

The four main groups of ceramics and some typical applications are summarized in Table 1.4. The common properties of ceramic materials can be summarized as follows:

Strength

Ceramic materials are reasonably strong in compression, but tend to be weak in tension and shear. They are brittle and lack ductility. They also suffer from micro-cracks which occur during the firing process. These lead to fatigue failure. Many ceramics retain their high compressive strength at very high temperatures.

Hardness

Most ceramic materials are harder than other engineering materials, as shown in Table 1.4. They are widely used for cutting tool tips and abrasives. They retain their hardness at very high temperatures that would destroy high carbon and high speed steels. However they have to be handled carefully because of their brittleness.

Refractoriness

This is the ability of a material to withstand high temperatures without softening and deforming under normal service conditions. Some refractories such as high-alumina brick and fireclays tend to soften gradually and may collapse at temperatures well below their fusion (melting) temperatures. Refractories made from clays containing a high proportion of silica to alumina are most widely used for furnace linings.

Electrical properties

As well as being used for weather resistant high-voltage insulators for overhead cables and sub-station equipment, ceramics are now being used for low-loss high-frequency insulators. For example they are being used for the dielectric in silvered ceramic capacitors for high frequency applications.

In all the previous examples the ceramic material is *polycrystalline*. That is, the material is made up of a lot of very tiny crystals.

For solid state electronic devices single crystals of silicon are grown under very carefully controlled conditions. The single crystal can range from 50 mm diameter to 150 mm diameter with a length ranging from 500 mm to 2,500 mm. These crystals are without impurities. They are then cut up into thin wafers and made into such devices as thermistors, diodes, transistors and integrated circuits. This is done by doping the pure silicon wafers with small, controlled amounts of carefully selected impurities.

Some types of impurity give the silicon n-type characteristics. That is they make the silicon electrically negative by increasing the number of electrons present. Other types of impurity give the silicon p-type characteristics. That is they make the silicon electrically positive by reducing the number of electrons present.

Thermosetting plastics

Themosetting plastics are also known as *thermosets*. These materials are available in powder or granular form and consist of a synthetic resin mixed with a 'filler'. The filler reduces the cost and modifies the properties of the material. A colouring agent and a lubricant are also added. The lubricant helps the plasticised moulding material to flow into the fine detail of the mould.

The moulding material is subjected to heat and pressure in the moulds during the moulding process. The hot moulds not only plasticize the moulding material so that it flows into all the detail of moulds, the heat also causes a chemical change in the material.

This chemical change is called *polymerisation* or, more simply, 'curing'. Once cured, the moulding is hard and rigid. It can never again be softened by heating. If made hot enough it will just burn. Some thermosets and typical applications are summarised in Table 1.5.

Another view

Plastic materials are not particularly new. In fact, the first semi-synthetic plastics appeared in the 1860s, and plastics made out of natural polymers have been used for centuries. Many other plastic materials were developed in the twentieth century, and some were in mass production well before the Second World War.

Table 1.5 *Thermosetting plastics*

Type	*Applications*
Phenolic resins and powders	The original 'Bakelite' type of plastic materials, hard, strong and rigid. Moulded easily and heat 'cured' in the mould. Unfortunately, they darken during processing and are only available in the darker and stronger colours. Phenolic resins are used as the 'adhesive' in making plywoods and laminated plastic materials (Tufnol)
Amino (containing nitrogen) resins and powders	The basic resin is colourless and can be coloured as required. Can be strengthened by paper-pulp fillers and are suitable for thin sections. Used widely in domestic electrical switchgear
Polyester resins	Polyester chains can be cross-linked by adding monomer such as styrene, when the polyester ceases to behave as a thermoplastic and becomes a thermoset. Curing takes place by internal heating due to chemical reaction and not by heating the mould. Used largely as the bond in the production of glass fibre mouldings.
Epoxy resins	The strongest of the plastic materials used widely as adhesives, can be 'cold cast' to form electrical insulators and used also for potting and encapsulating electrical components.

Thermoplastics

Unlike the thermosets we have just considered, thermoplastics soften every time they are heated. In fact, any material trimmed from the mouldings can be ground up and recycled. They tend to be less rigid but tougher and more 'rubbery' than the thermosetting materials. Some thermoplastics and typical applications are summarized in Table 1.5.

Reinforced plastics

The strength of plastics can be increased by reinforcing them with fibrous materials. Such materials include:

- *Laminated plastics (Tufnol).* Fibrous material such as paper, woven cloth, woven glass fibre, etc. is impregnated with a thermosetting resin. The sheets of impregnated material are laid up in powerful hydraulic presses and they are heated and squeezed until they become solid and rigid sheets, rods, tubes, etc. This material has a high strength and good electrical properties. It can be machined with ordinary metal working tools and machines. Tufnol is used for making insulators, gears and bearing bushes.
- *Glass reinforced plastics (GRP).* Woven glass fibre and chopped strand mat can be bonded together by polyester or by epoxy resins to form mouldings. These may range from simple objects such as

Test your knowledge 1.6

Name and state the properties of:

(a) TWO thermosetting plastics
(b) TWO thermoplastics.

Test your knowledge 1.7

Select a suitable plastic material for the following applications. Give reasons for your choice.

(a) The moulded cockpit cover for a light aircraft
(b) a moulded bearing for a computer printer
(c) a rope for rock climbing
(d) the body of a domestic light switch
(e) encapsulating (potting) a small transformer.

Another view

Plastics are composed of natural, modified natural, or completely synthetic polymers–long-chain molecules that determine the properties of the resulting materials. Various additives are added to the base polymer to produce other desirable qualities such as colour, bulk or improved longevity. There are two main types of plastics: the relatively soft and pliable thermoplastics that can soften and flow when reheated, and the normally harder and more brittle thermosetting plastics that do not.

crash helmets to complex hulls for ocean-going racing yachts. The thermosetting plastics used are cured by chemical action at room temperature and a press is not required. The glass fibre is laid up over plaster or wooden moulds and coated with the resin which is well worked into the reinforcing material. Several layers or 'plies' may be built up according to the strength required. When cured the moulding is removed from the mould. The mould can be used again. Note that the mould is coated with a release agent before moulding commences.

We shall return to this topic a little later when we describe composite materials as materials in their own right.

General properties of plastics

Although the properties of plastic materials can vary widely, they all have some general properties in common.

Strength/weight ratio

Plastic materials vary considerably in strength and some of the stronger (such as nylon) compare favourably with the metals. All plastics have a lower density than metals and, therefore, chosen with care and proportioned correctly their strength/weight ratio compares favourably with the light alloys.

Corrosion resistance

Plastic materials are inert to most inorganic chemicals and some are inert to all solvents. Thus they can be used in environments that are hostile to the most corrosion resistant metals and many naturally occurring non-metals.

Electrical resistance

All plastic materials are good electrical insulators, particularly at high frequencies. However their usefulness is limited by their softness and low heat resistance compared with ceramics. Flexible plastics such as PVC are useful for the insulation and sheathing of electric cables.

Composite materials

Composite materials are quite different from any of the materials that we have met so far. As their name suggests, composite materials are combinations of materials that, however, unlike their individual constituents, retain their separate identities and do not dissolve or merge together.

In structural applications composite materials have the following characteristics:

1. They generally consist of two or more physically distinct and mechanically separable materials.
2. They are made by mixing the separate materials in such a way as to achieve controlled and uniform dispersion of the constituents.
3. The mechanical properties of the composite material are

Table 1.6 *Thermoplastic materials*

Type	*Material*	*Characteristics*
Acrylics	Polymethyl-methacrylate	Materials of the 'Perspex' or 'Plexiglass' types. Excellent light transmission and optical properties, tough, non-splintering and can be easily heat-bent and shaped. Excellent high-frequency electrical insulators.
Cellulose plastics	Nitro-cellulose	Materials of the 'celluloid' type. Tough, waterproof, and available as preformed sections, sheets and films. Difficult to mould because of their high flammability. In powder form nitro-cellulose is explosive.
	Cellulose acetate	Far less flammable than nitro-cellulose and the basis of photographic 'safety' film. Frequently used for moulded handles for tools and electrical insulators.
Fluorine plastics (Teflon)	Polytetrafluoro-ethylene (PTFE)	A very expensive plastic material, more heat resistant than any other plastic. Also has the lowest coefficient of friction. PTFE is used for heat-resistant and anti-friction coatings. Can be moulded (with difficulty) to produce components with a waxy feel and appearance.
Nylon	Polyamide	Used as a fibre or as a wax-like moulding material. Tough, with a low coefficient of friction. Cheaper than PTFE but loses its strength rapidly when raised above ambient temperature. Absorbs moisture readily, making it dimensionally unstable and a poor electrical insulator.
Polyesters (Terylene)	Polyethylene-teraphthalate	Available as a film or in fibre form. Ropes made from polyesters are light and strong and have more 'give' than nylon ropes. The film makes an excellent electrical insulator.
Vinyl plastics	Polythene	A simple material, relatively weak but easy to mould, and a good electrical insulator. Used also as a waterproof membrane in the building industry.
	Polypropylene	A more complicated material than polythene. Can be moulded easily and is similar to nylon in many respects. Its strength lies between polythene and nylon. Cheaper than nylon and does not absorb water.
	Polystyrene	Cheap and can be easily moulded. Good strength but tends to be rigid and brittle. Good electrical insulation properties but tends to craze and yellow with age.
	Polyvinylchloride (PVC)	Tough, rubbery, practically non-flammable, cheap and easily manipulated. Good electrical properties and used widely as an insulator for flexible and semi-flexible cables.

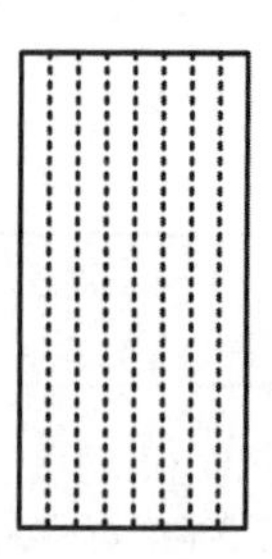

(a) Aligned continuous

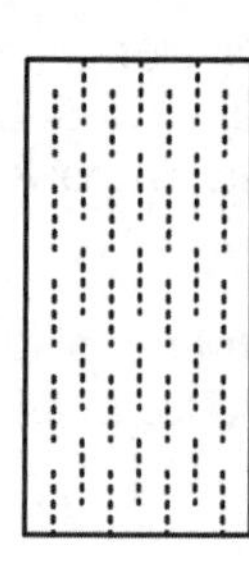

(b) Aligned discontinuous

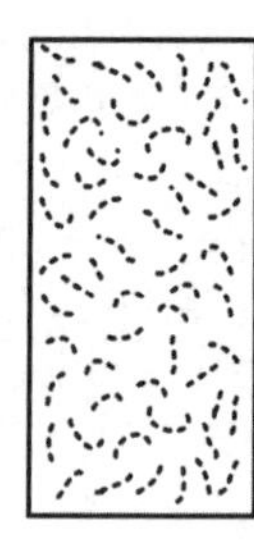

(c) Random discontinuous

Figure 1.3 *Fibre alignment in a composite material*

superior to (and in many cases quite different from) the properties of the constituent materials.

A good example of a composite material is glass-reinforced plastic (GRP). This material combines glass fibres with epoxy resin. The latter material is relatively weak and brittle and, although the glass fibres are strong and stiff, they can only be effectively loaded in tension as single fibres. However, when combined into a composite material, the resin and fibre provide us with a strong, stiff material with excellent toughness characteristics.

The following materials are commonly used as fibres:

- alumina
- boron
- carbon
- glass
- polyethylene
- polyamide.

The following materials are used as matrix materials:

- alumina
- aluminium
- epoxy
- polyester
- polypropylene.

The fibres may be arranged within the matrix material in various ways depending on the properties required and the intended application (see Figure 1.3). Note that, whenever the fibres are aligned in a specific direction, the properties of the material also become direction. Conversely, if the fibres are arranged in random orientation, the resulting material will have the same properties in all directions. This is an important point because we may sometimes require that a component has maximum strength in a particular

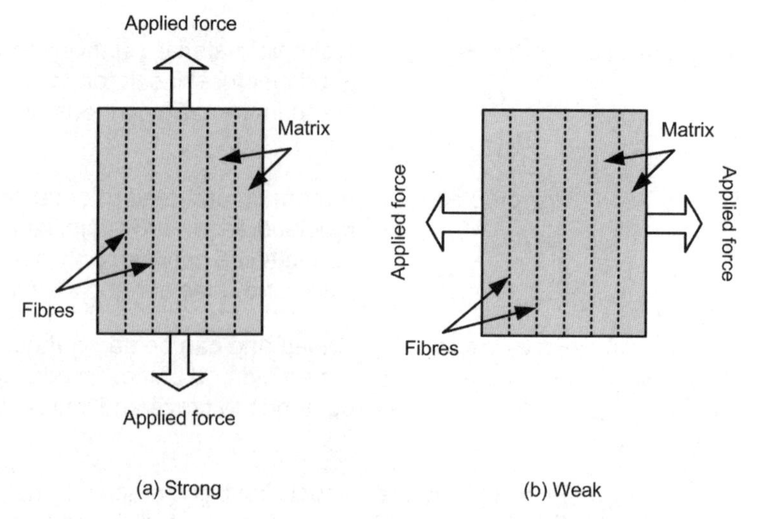

Figure 1.4 *Effect of applying a force to a composite material with aligned fibres*

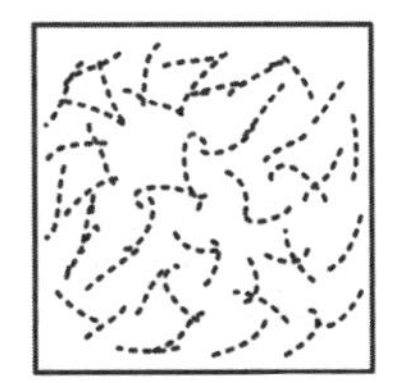

(a) Chopped strand mat

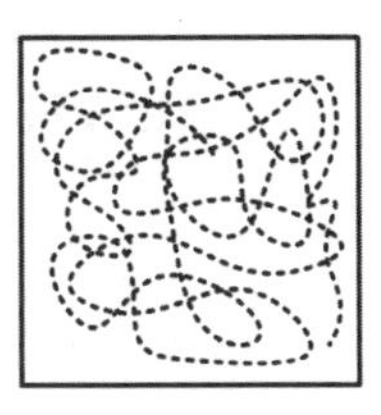

(b) Continuous filament mat

(c) Bi-directional woven mat

Figure 1.6 *Construction of fibre mats*

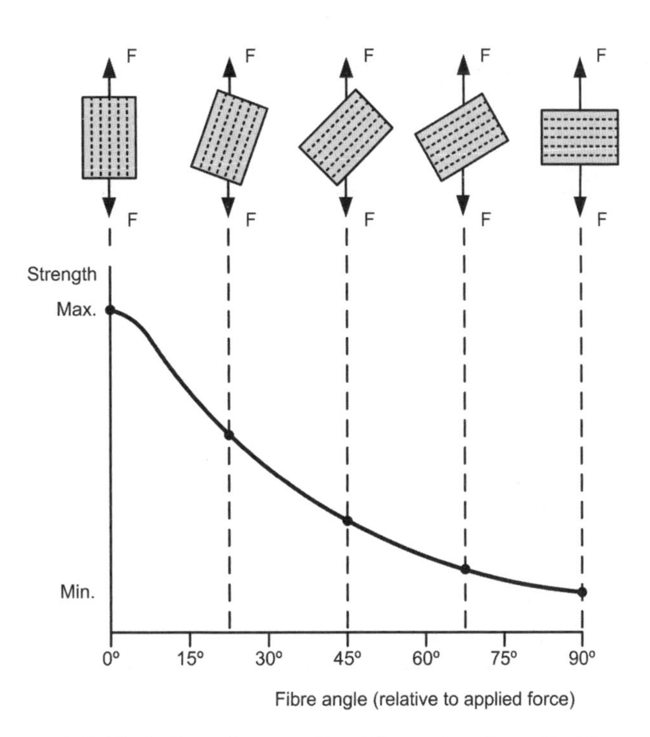

Figure 1.5 *Variation of strength with angle of applied force*

direction. In this case the correct alignment of the fibres becomes extremely important (see Figures 1.4 and 1.5). Where the material is required to have equal properties in all directions, fibre mats can be constructed as shown in Figure 1.6. We shall return to this important point later when we look at the processing of composite materials. Table 1.7 shows the properties of some reinforced polyester materials.

Figure 1.7 *Composite materials are widely used in the aerospace industry where strong but lightweight aerodynamic profiles are required*

Table 1.7 *Properties of reinforced polyester*

Type	*% of glass fibre by weight*	*Tensile modulus (GPa)*	*Tensile strength (MPa)*
Polyester (with no fibres added)	0	2 to 4	20 to 70
Polyester matrix with discontinuous, randomly aligned glass fibres	10 to 45	5 to 15	40 to 175
Polyester matrix with plain weave glass fibre cloth	45 to 65	10 to 20	250 to 350
Polyester matrix with long glass fibres	50 to 80	20 to 50	400 to 1250

Elastomers

Elastomers are a class of material that exhibit appreciable reversible strains when subject to stress. Perhaps the most obvious example of this sort of behaviour is that of a rubber band. This will often strength to four or five times its unstrained length without breaking and then return to its original length when the stress is remove.

The reason for this interesting (and useful) behaviour is that, unlike the linear, branched and cross-linked molecular chains of polymer materials, elastomers have long tangled polymer chains with very few cross-linked bonds. The properties of some common elastomers are shown in Table 1.8.

Table 1.8 *Properties of some common elastomers*

Material	*Percentage elongation*	*Resilience*	*Tensile strength (MPa)*
Natural rubber	800	Good	20
Butadiene-styrene	600	Good	24
Butyl	900	Fair	20
Polyurethane	650	Poor	40

Adhesives

The usual definition of an adhesive is a substance that, when placed between two surfaces, holds the surfaces together by means of a chemical reaction. When two surfaces are bonded together by means of an adhesive (rather than a mechanical or thermal method) the surfaces are held together by the intermolecular forces between the molecules present in the adhesive and in the surfaces. Additional strength may also result from the adhesive flowing into cracks and crevices in the surface. Because of this surfaces are often roughened prior to the application of adhesive.

Various types of adhesive are in common use. These include natural adhesives (vegetable glues), elastomer adhesives, thermoplastic adhesives (polyamides and cyanoacrylate 'super glues'), and thermoset adhesives (which include epoxy resins such as 'Araldite'). Adhesive bonding and jointing is described in detail on page 66.

Smart materials

Another class of materials that is becoming increasingly important is the new so-called 'smart' materials. These are materials that respond to changes in their environment by changing their properties. For this reason, they are probably more aptly referred to as *responsive materials*.

Depending on changes in some external conditions, smart materials change either their properties (mechanical, electrical, optical, appearance), their structure or composition, or their functions.

An example of responsive materials is a sub-class of materials known as *shape-memory alloys* (SMA). These are metals that, after being strained, will at a certain temperature revert back to their original shape. This change of shape results from a change in the material's crystal structure which occurs at a particular temperature known as the *transformation temperature*.

SMAs are also *superelastic* in that they are able to sustain a large deformation at a constant temperature and, when the deforming force is released, they return to their original undeformed shape.

Activity 1.3

An electrical equipment manufacturer has asked you to advise the company on the selection of materials to be used in the manufacture of an adjustable desk lamp which is to be fitted with a switch and is to accept a conventional 40 W light bulb. Sketch a suitable design and list each of the component parts required. Suggest, with reasons, a material to be used for each component used in your design. Present your work in the form of an illustrated report supported by appropriate sketches and diagrams.

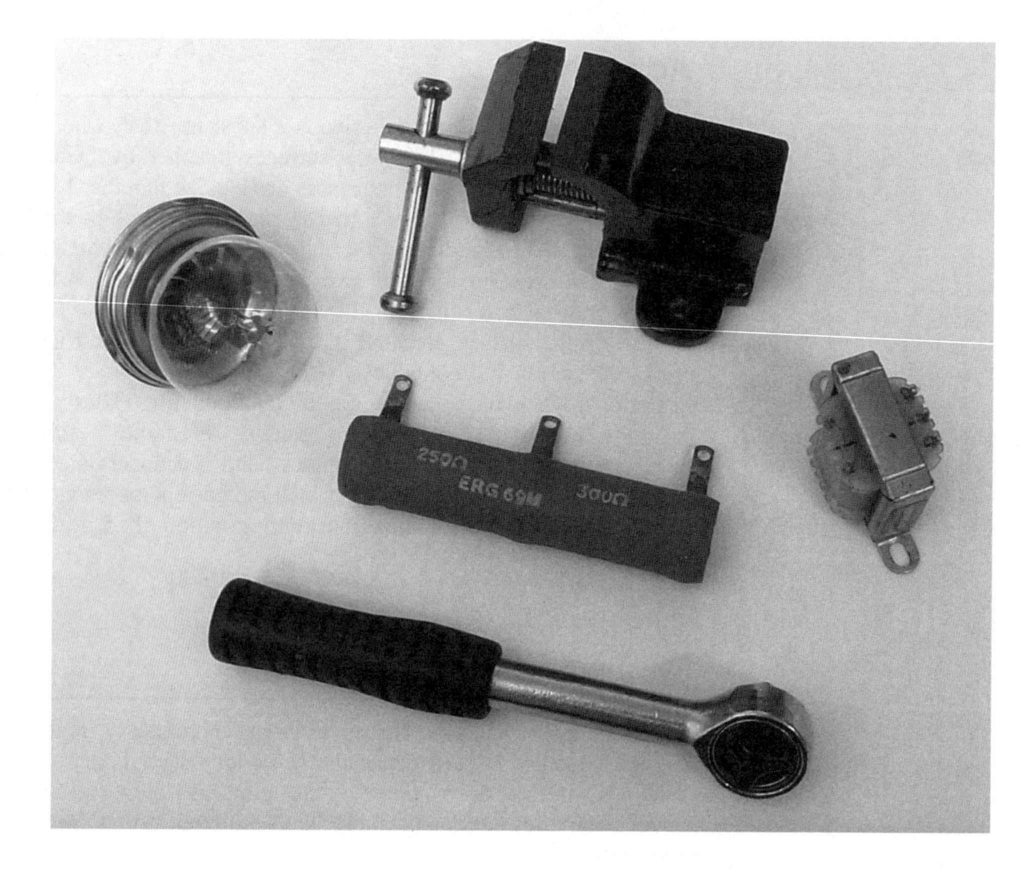

Figure 1.8 *See Activity 1.4*

Activity 1.4

Figure 1.8 shows the following engineering parts:

(a) a small bench vice
(b) a car headlamp bulb
(c) a high-power resistor
(d) a small transformer
(e) a ratchet handle.

Select THREE of these parts and for each part identify the materials used and give reasons for their use.

Activity 1.5

Use the Internet to carry out research into the following types of new material:

(a) optical fibres
(b) piezoelectric actuators
(c) heat-shrink material

Write a brief note about each and include relevant sketches.

Another view

MatSdata is a great way of building up your own database of information on a wide variety of engineering materials. However, don't forget that you can change the data or add your own records to the database! Also, it's important to be aware that the values given in the database are representative of those of the material in question and may not be exactly the same as those quoted in text books or other data sources.

Activity 1.6

There is a variety of materials databases accessible via the Internet, including matSdata (which was designed specifically for A-level Engineering students—see Figure 1.9 and Activity 1.2). Using either matSdata (or another on-line line database or reference book) find the name and describe the appearance of:

1. A material with a composition of Cu (70%) Zn (30%)
2. A material used for 'gilding'
3. A ceramic material
4. A naturally occurring and renewable composite material
5. A material used for 'galvanising' metals
6. A metal which has a melting point of 1668°C and which is widely used in the aerospace industry

Hint: Get into the habit of using the 'search' or 'find' facility as this can speed things up considerably!

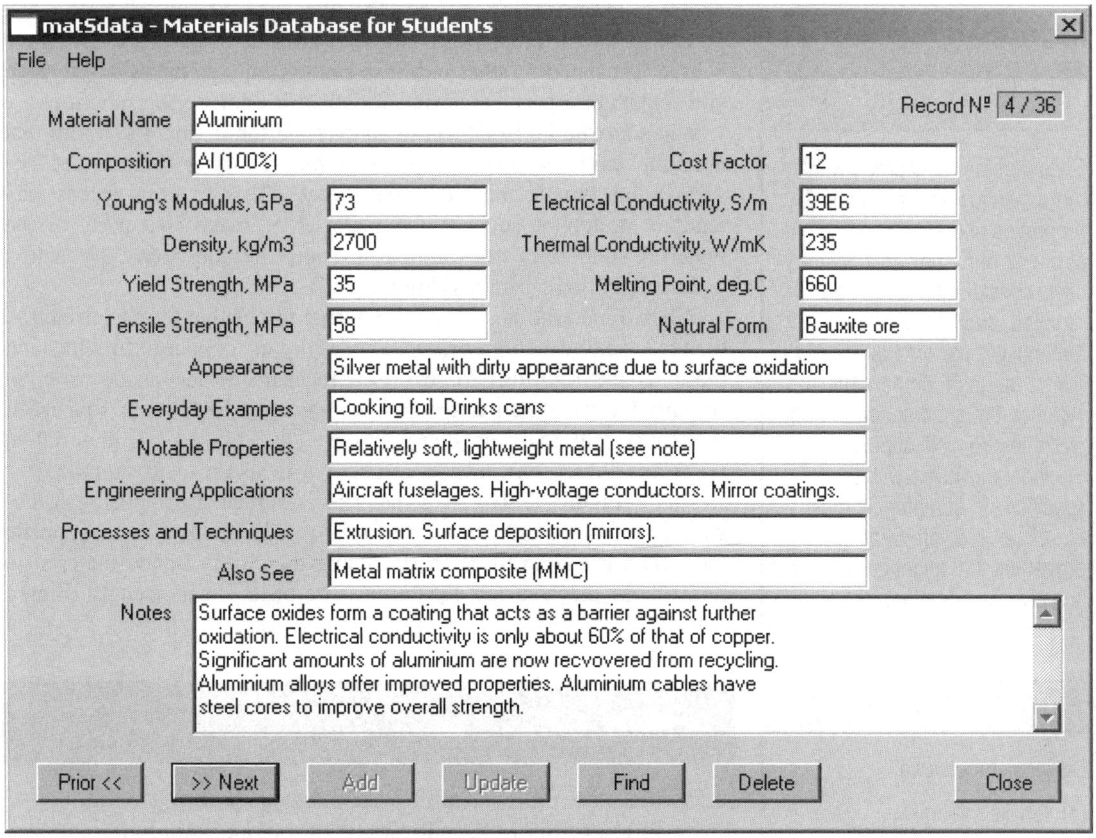

matSdata - Materials Database for Students

File Help

Record Nº 4 / 36

Material Name	Aluminium		
Composition	Al (100%)	Cost Factor	12
Young's Modulus, GPa	73	Electrical Conductivity, S/m	39E6
Density, kg/m3	2700	Thermal Conductivity, W/mK	235
Yield Strength, MPa	35	Melting Point, deg.C	660
Tensile Strength, MPa	58	Natural Form	Bauxite ore
Appearance	Silver metal with dirty appearance due to surface oxidation		
Everyday Examples	Cooking foil. Drinks cans		
Notable Properties	Relatively soft, lightweight metal (see note)		
Engineering Applications	Aircraft fuselages. High-voltage conductors. Mirror coatings.		
Processes and Techniques	Extrusion. Surface deposition (mirrors).		
Also See	Metal matrix composite (MMC)		
Notes	Surface oxides form a coating that acts as a barrier against further oxidation. Electrical conductivity is only about 60% of that of copper. Significant amounts of aluminium are now recvovered from recycling. Aluminium alloys offer improved properties. Aluminium cables have steel cores to improve overall strength.		

Prior << | >> Next | Add | Update | Find | Delete | Close

Figure 1.9 *The matSdata record for aluminium*

1.2 Properties of materials

Having examined the main classes of materials and described some of the more common materials in each class, we shall now move on to look at some of the more important properties of materials. These properties are often broken down into two major subdivisions; *mechanical properties* and *physical properties*. The latter includes; *electrical properties, magnetic properties, thermal properties* and *chemical properties*.

It can be argued that chemical properties are in themselves another major subdivision. To introduce the small amount of physical chemistry necessary to understand the environmental stability of materials, I will present the subject of corrosion and corrosion prevention separately.

We will be looking at all of the above properties but we will leave processing properties until we deal with material processing, where it fits more readily. Also, information on the cost of materials will be found later when we deal with the selection of materials for engineering products.

Mechanical properties

Mechanical properties are the behaviour of materials when subject to forces and include strength, stiffness, hardness, toughness, and ductility to name but a few. For the sake of completeness and precision most of the important mechanical properties are defined below. They make rather tedious reading, but are necessary to help you select appropriate materials for specific engineering functions.

When a material is subject to an external force, then the forces that hold the atoms of the material together (bonding forces), act like springs and oppose these external forces. These external forces may tend to stretch or squeeze the material or make two parts of the material slide over one another in opposite directions, by acting against the bonding forces (Figure 1.10).

When a material is subject to external forces that tend to stretch it, then it is said to be in *tension*. The ability of a material to withstand these tensile (pulling) forces is a measure of its *tensile strength* (Figure 1.11a). When a material is subject to forces that squeeze it then it is said to be in *compression*. The ability of a material to withstand these compressive forces is a measure of its *compressive strength* (Figure 1.11b). If a material is subject to opposing forces which are offset and cause one face to slide relative to an opposite face, then it is said to be in *shear* and the ability of the material to resist these shearing forces is a measure of its shear strength (Figure 1.11c).

Another view

Engineers are concerned with more than just the mechanical properties of materials (e.g. how strong or hard they are). Many other considerations may be required, such as whether a material will perform satisfactorily over a range of temperatures or whether it might deteriorate or be prone to chemical attack. Engineers involved with the specification of materials must therefore be aware of *all* of the properties that might be important in a given application.

Test your knowledge 1.8

Distinguish between:

(a) tensile strength
(b) compressive strength.

Illustrate your answer with a sketch.

Activity 1.7

Use the matSdata (or other) database to determine the tensile strength of the following materials: aluminium, phosphor bronze, carbon steel, copper, lead, titanium. Rank these materials (from the strongest to the weakest) in order of tensile strength.

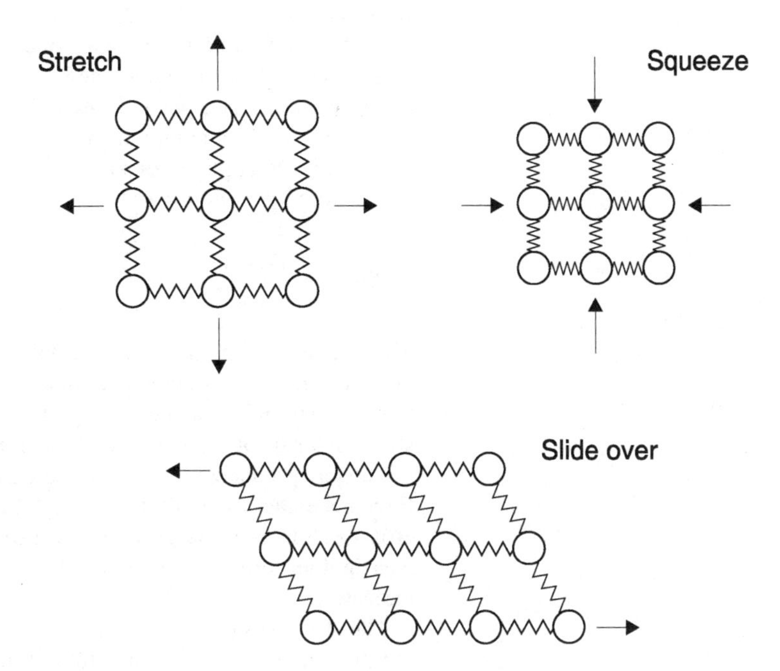

Figure 1.10 *Reaction of atomic bonds to external forces*

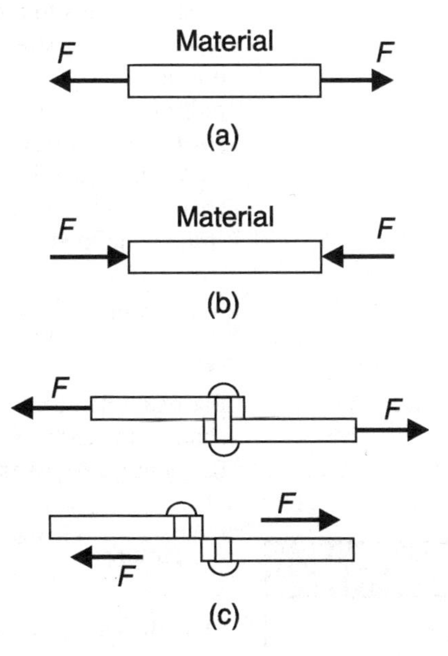

Figure 1.11 *(a) Tensile, (b) compressive and (c) shear*

In discussing the application of forces to materials, it is often desirable to be able to compare one material with another. For this reason we may not be concerned so much with the size of the force, as the force applied per unit area. Thus, for example, if we apply a tensile force F to a length of material over its cross-sectional area A, then the force applied per unit area is F/A. The term *stress* is used for the force per unit area.

$$\text{Stress} = \frac{\text{Force}}{\text{Area}}$$

The unit in which stress is measured is the Pascal (Pa), with 1 Pa being a force of 1 Newton per square metre, i.e. 1 Pa = 1 N/m^2. Note that, in materials science, it is perhaps more convenient to measure stress in terms of Newtons per square millimetre (N/mm^2). This unit, moreover, produces a value that is easier to appreciate, whereas the force necessary to break (for example) a steel bar one square metre in cross-section is so large as to be difficult to visualize in ordinary measurable terms. The Greek letter (sigma σ) is often used to indicate stress.

The stress is said to be *direct stress* when the area being stressed is at right angles to the line of action of the external forces, as when the material is in tension or compression. Shear stresses are not direct stresses since the forces being applied are in the same plane (parallel with) the area being stressed.

The area used for calculating the stress is generally the original area that existed before the application of forces. This stress is often referred to as the *engineering stress*. The term *true stress* being used for the force divided by the actual area that exists while the material is in the stressed state.

Strain refers to the proportional change produced in a material as a result of the stress applied. It is measured as the number of millimetres of deformation (change in dimension) suffered per millimetre of original dimension and is a numerical ratio, therefore *strain has no units*.

When a material is subject to tensile or compressive forces and a change in length results then the material has been strained, this strain is defined as:

$$\text{Strain} = \frac{\text{change in length}}{\text{original length}}$$

For example, if we have a strain of 0.02. This would indicate that the change in length is 0.02 × the original length. However, strain is frequently expressed as a percentage.

$$\text{Strain} = \frac{\text{change in length}}{\text{original length}} \times 100\%$$

Thus the strain of 0.02 as a percentage is 2%, i.e. here the change in length is 2% of the original length.

The Greek letter (epsilon ε) is the symbol normally used to represent strain. The symbol for length is (l) and the symbol for change in length is (δl), where δ is the Greek letter delta.

Test your knowledge 1.9

Define the terms:

(a) stress
(b) direct stress
(c) true stress.

Example 1.1

A copper bar has a cross-sectional area of 75 mm^2 and is subject to a compressive force of 150 N. What is the compressive stress?

The compressive stress is the force divided by the area:

$$\text{compressive stress } (\sigma) = \frac{150}{75} \text{ N/mm}^2 = 2 \text{ N/mm}^2$$

$$\text{or compressive stress } (\sigma) = \frac{150}{75 \times 10^6} = 2 \text{ MN/m}^2 = 2 \text{ MPa}$$

Example 1.2

A brass rod having a cross-sectional area of 50 mm^2 is to be subjected to a compressive stress (σ) of 750 kPa. What compressive force should be applied to the brass rod?

The compressive force is the compressive stress (σ) (in Pa) multiplied by the area (in m^2):

$$\text{compressive force} = (750 \times 10^3) \times (50 \times 10^{-6}) = 37.5 \text{ N}$$

We have already mentioned tensile, compressive and shear strength (Figure 1.7) but we did not give the general definition. The *strength* of a material is defined as *the ability of a material to resist the application of a force without fracturing*. The stress required to cause the material to fracture i.e. *fracture stress* is a *measure of its strength* and requires careful consideration.

Hooke's Law

Hooke's Law states that *within the elastic limit of a material the change in shape is directly proportional to the applied force producing it.* What this means is that a linear (straight line) relationship exists between the applied force and the corresponding strain while the material is being elastically deformed (in other words, while it is still able to return to its original size when the external load (i.e. force) is removed).

If the load (weight) is increased sufficiently, there will come a point when the internal forces holding the spring material together start to break or permanently stretch and so, even after the load has been removed, the spring will remain permanently deformed, i.e. *plastically strained.*

By considering Hooke's law, it follows that *stress* is also directly proportional to strain while the material remains elastic, because stress is no more than force (load) per unit area. The stress required to first cause plastic strain is known as the *yield stress* (Figure 1.13).

For *metals*, the *fracture stress* (i.e. the measure of *strength*) is considered to be identical to the 0.2% yield stress. This is because, for all engineering purposes, metals are only used within their elastic

Another view

A good example of the application of Hooke's Law is the *spring*. You might also be familiar with the *spring balance* where an increase in weight force causes a corresponding extension which can be measured on a calibrated scale, see Figure 1.12.

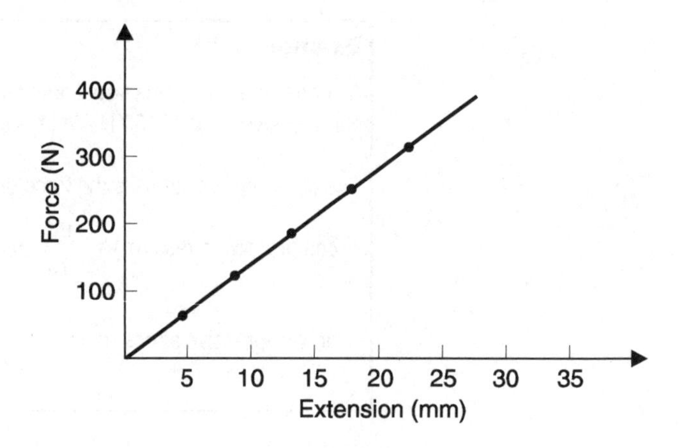

Figure 1.12 *Application of Hooke's Law to a spring*

range, so a strength measurement above the yield stress is of little value to engineers when deciding which metal to use. The *0.2% yield stress* is defined as the stress at which the stress–strain curve (Figure 1.13), for tensile loading, deviates by a strain of 0.2% from the linear elastic line. Note here that a permanent change in length (strain) has occurred, the material has been plastically deformed by 0.2% of its original length.
Where σ_f is the symbol for fracture stress (units MPa or MN/m^2) and σ_y is the yield stress (the stress needed to start to plastically deform or permanently strain the material), with the same units as fracture

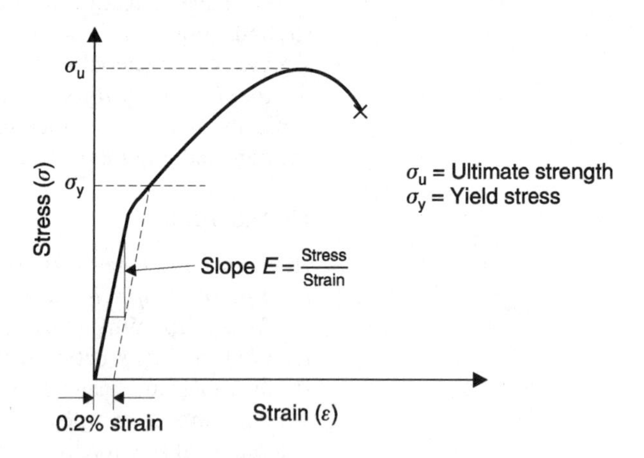

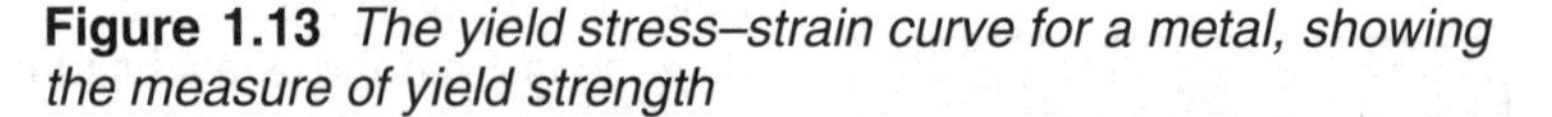

Figure 1.13 *The yield stress–strain curve for a metal, showing the measure of yield strength*

stress. For metals the fracture stress is the same in tension and compression.

For *polymers* the fracture stress is identified as the yield stress at which the stress–strain curve becomes markedly non-linear i.e., typically at strains of around 1%. Polymers are a little stronger (approximately 20%) in compression than in tension (Figure 1.14).

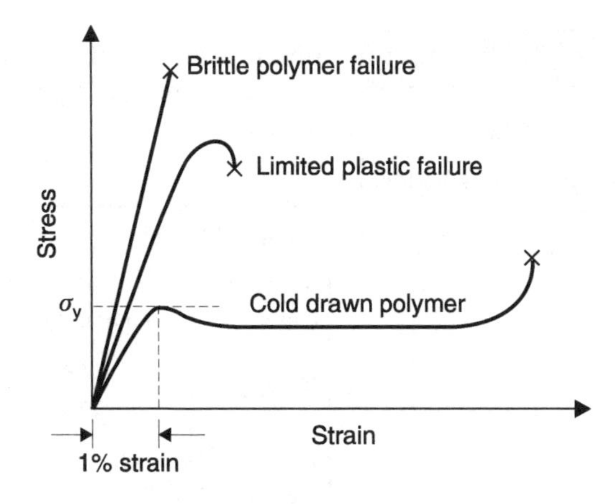

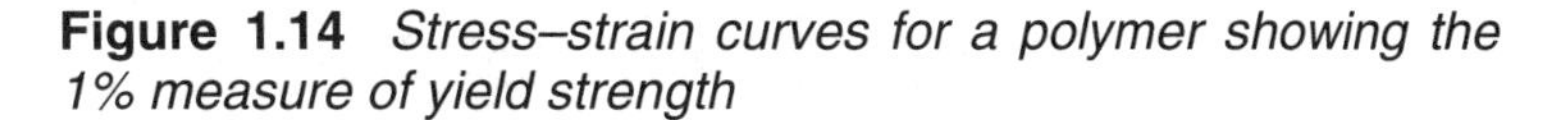

Figure 1.14 *Stress–strain curves for a polymer showing the 1% measure of yield strength*

For ceramics and glasses the strength depends strongly on the mode of loading. In tension, strength means the fracture strength given by the *tensile fracture stress* (symbol σ_f^t) . In compression it means the crushing strength, given by the *compressive fracture stress* (symbol σ_f^c) which is much larger than the tensile fracture stress, typically 15 times as large. The tensile and compressive fracture stresses for a typical ceramic are shown in Figure 1.15.

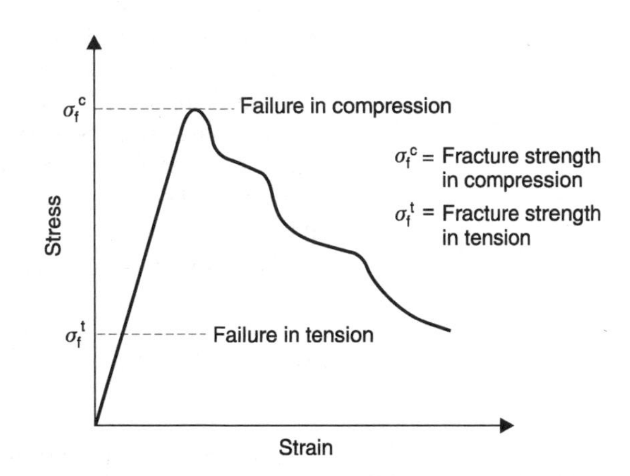

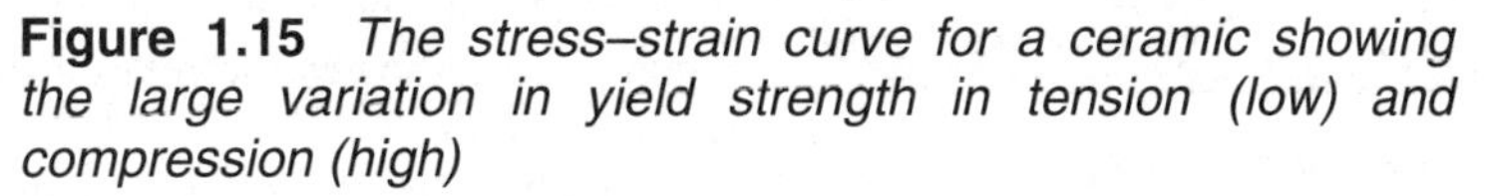

Figure 1.15 *The stress–strain curve for a ceramic showing the large variation in yield strength in tension (low) and compression (high)*

The symbol σ_u is used to indicate the *ultimate tensile strength*, measured by the nominal stress at which a bar of material loaded in tension separates (breaks). For brittle solids (ceramics, glasses and brittle polymers) it is the same as the fracture stress in tension (Figures 1.14 and 1.15). For metals, ductile polymers and most composites, it is larger than the fracture stress (σ_f) by a factor of between 1.1 and 3 because of work hardening (Figure 1.13), or in the case of composites, because of load transfer to the strong reinforcing

Test your knowledge 1.10

(a) State Hooke's Law

(b) Sketch a typical stress–strain curve for a metal.

fibres. More will be said about work hardening and fibre reinforcement of composites later when we deal with the structure of materials.

Example 1.3

A circular metal rod 10 mm in diameter has a yield stress of 210 MPa. What tensile force is required to cause yielding?

Since stress (s) $= \dfrac{\text{force}}{\text{area}}$

we must first calculate the cross-sectional area of the metal rod. Thus the area A is given by:

$A = \pi r^2 = \pi \times 5^2 = 78.54\ \text{mm}^2$

Therefore yield force required = yield stress × area
$= 210 \times 10^6 \times 78.54 \times 10^{-6} = 16{,}493$ N

Young's modulus

Stiffness can be defined as the ability of a material to resist deflection when loaded. For example, a material's resistance to bending is a measure of its stiffness. When a material is subject to external bending forces the less it gives, the stiffer it is. Thus stiffness is related to the stress imposed on the material and the amount of movement or strain caused by this stress. We mentioned earlier the linear relationship between stress and strain on the stress–strain graph. A measure of the stiffness of a material when strained in tension can be obtained from the graph by measuring the slope of the linear part of the graph (see Figure 1.13). This slope (stress)/(strain) is known as the *elastic modulus* or *Young's modulus of elasticity*.

$$\text{Modulus of elasticity} = \frac{\text{stress}}{\text{strain}}$$

The modulus of elasticity has units of stress (often quoted as GPa), since strain has no units. The symbol for the elastic modulus is *E*. Figure 1.13 indicates the relationship between, stress, strain and the elastic modulus.

In addition to the elastic modulus *E*, when dealing with three-dimensional solids, there are two other moduli that are worthy of mention. The *shear modulus*, *G*, describes the rigidity of a material subject to shear loading; and the *bulk modulus, K,* describes the effect of a material when subjected to external pressure.

So far I have given a fairly rigorous definition of the strength of a material, as well as defining stress, strain and material stiffness. You will meet this topic again in Unit 4 where you will have an opportunity to carry out some tests on materials and also to use your knowledge to solve some practical engineering problems.

Another view

Stress and strain are very important concepts in engineering. Later in Unit 4 (see page 271) you will be returning to this topic when you look at how elasticity modulus is measured. For now, it's sufficient to know the definitions of stress and strain and what they tell you about the properties of a material.

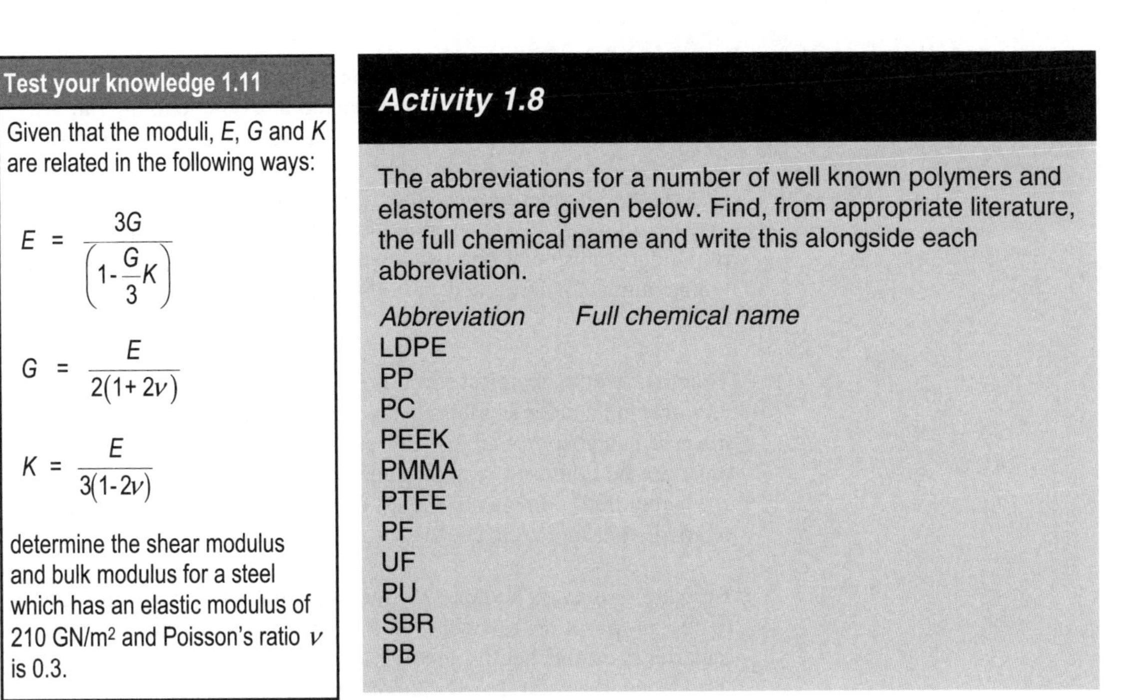

Test your knowledge 1.11

Given that the moduli, E, G and K are related in the following ways:

$$E = \frac{3G}{\left(1 - \frac{G}{3}K\right)}$$

$$G = \frac{E}{2(1 + 2\nu)}$$

$$K = \frac{E}{3(1 - 2\nu)}$$

determine the shear modulus and bulk modulus for a steel which has an elastic modulus of 210 GN/m^2 and Poisson's ratio ν is 0.3.

Activity 1.8

The abbreviations for a number of well known polymers and elastomers are given below. Find, from appropriate literature, the full chemical name and write this alongside each abbreviation.

Abbreviation	*Full chemical name*
LDPE	
PP	
PC	
PEEK	
PMMA	
PTFE	
PF	
UF	
PU	
SBR	
PB	

More mechanical properties

Let us now consider a few more mechanical properties, which are presented here with examples of materials that display these properties.

Ductility, the ability to be drawn into threads or wire. Examples include; wrought iron, low carbon steels, copper, brass and titanium.

Brittleness, the tendency to break easily or suddenly with no prior extension. Examples include; cast iron, high carbon steels, brittle polymers, concrete, and ceramic materials.

Malleability, the ability to plastically deform and shape a material by forging, rolling or by the application of pressure. Examples include; gold, copper and lead.

Elasticity, the ability of a material to deform under load and return to its original shape once the external loads have been removed. Internal atomic binding forces are only stretched, not broken and act like minute springs to return the material to normal, once the force has been removed. Examples include; rubber, mild steel and some plastics.

Plasticity is the readiness to deform to a stretched state when a load is applied. The *plastic deformation is permanent* even after the load has been removed. Plasticine exhibits plastic deformation.

Hardness, the ability to withstand scratching (abrasion) or indentation by another body. It is an indication of the wear resistance of a material. Examples of hard materials include; diamond high carbon steel and other materials that have undergone a hardening process.

Fatigue is a phenomenon by which a material can fail at much lower stress levels than normal when subjected to cyclic loading. Failure is generally initiated from micro-cracks on the surface of the material.

Creep may be defined as the time dependent deformation of a material under load, accelerated by increase in temperature. It is an important consideration where materials are subjected to high temperatures for sustained periods of time, e.g. gas turbine engine blades.

Toughness, in its simplest form, is defined as the ability of a material to withstand sudden loading. It is measured by the total energy that a material can absorb, that just causes fracture. Toughness (symbol, G) must not be confused with strength that is measured in terms of the stress required to break a standard test piece. Toughness has the units of energy per unit area, i.e. MN/m^2.

Fracture toughness (symbol K_c) measures the resistance of a material to the propagation (growth) of a crack. Fracture toughness for a material is established by loading a sample containing a deliberately introduced crack of length $2c$ (Figure 1.16), recording the tensile stress (σ_c) at which the crack propagates.

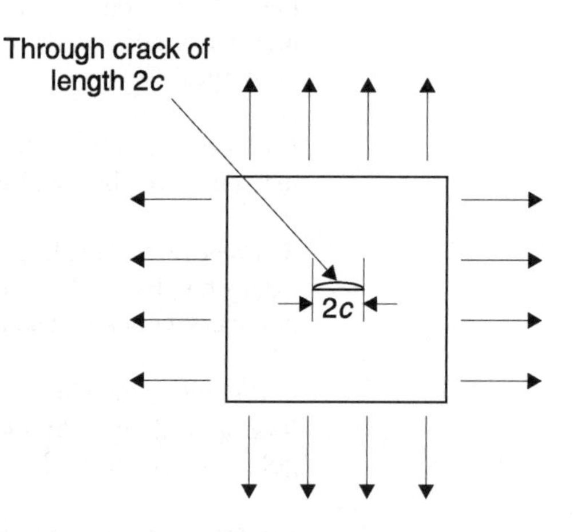

Figure 1.16 *Establishment of fracture toughness K_c by loading a sample with a deliberately introduced through-crack of length 2c and recording tensile stress*

The quantity K_c is then calculated from:

$$K_c = Y\sigma_c(\pi c)^{1/2}$$

normal units for K_c being $MPam^{1/2}$ or $MN/m^{3/2}$

The fracture toughness (K_c) is related to the toughness by the relationship:

$$G_c = \frac{K_c^2}{E\left(1 - \nu^2\right)}$$

Y is a geometric factor, near unity, which depends on the geometry of the sample under test, ν is known as Poisson's ratio where:

$$\text{Poisson's ratio } (\nu) = \frac{\text{lateral strain}}{\text{axial strain}} \text{ of a strained material under load}$$

Note: axial strain is strain along the longitudinal axis of the material.

Physical properties

In the previous section we dealt in some detail with the mechanical properties of materials, it is now time to concentrate on their physical properties, these are subdivided into *electrical* and *magnetic*, *thermal, density* and *optical*. With the exception of a brief statement on photoconduction, optical properties will not be considered in this course. We will start by briefly discussing some of the electrical and magnetic properties of materials including *electrical conductivity*, *superconductivity*, *semiconductors*, *magnetization*, and *dielectrics*.

In many applications, the electrical properties of a material are of primary importance. Copper wire is chosen for electrical wiring because of its extremely high electrical conductivity. You may also remember that copper is also a very ductile material and so is easily drawn or extruded into shape. These two properties make copper ideal for electrical wiring.

Conductors are materials having outer electrons that are loosely connected to the nucleus of their atoms and can easily move through the material from one atom to another. *Insulators* are materials whose electrons are held firmly to the nucleus.

For a conducting material the electromotive force (e.m.f.) measured in volts, V, the current measured in amperes, I, and the resistance measured in Ohms, R, are related by Ohm's law. This states that:

$$V = I\,R$$

The resistance to current flow in a circuit is proportional to the length, l, and inversely proportional to the cross-sectional area, A, of the component. The resistance can be defined as:

$$R = \frac{\rho l}{A} \quad \text{where } \rho \text{ is the electrical } \textit{resistivity}$$

or

$$R = \frac{l}{\sigma A} \quad \text{where } \sigma \text{ is the electrical } \textit{conductivity}$$

Test your knowledge 1.12

Explain the terms:

(a) ductility
(b) brittleness
(c) malleability
(d) elasticity
(e) plasticity.

Test your knowledge 1.13

Define Poisson's Ratio.

The unit of *resistivity* is the ohm metre (Ωm), and the unit of *conductivity* is the Siemen per metre (S/m). It can be seen that the electrical conductivity is the inverse of the electrical resistivity. Both of these properties, conductivity and resistivity, are inherent in each material. Typical tables of values normally show only the resistivity of a material. Very high values suggest very good resistive characteristics and, conversely very low values indicate good conductivity.

Metals which have many free electrons are very good electrical conductors, their *conductivity* is reduced with increase in temperature. The rate of increase of electrical resistivity is given by the temperature resistivity coefficient. The resistivity for some of the more common materials together with their temperature resistivity coefficients are given in Table 1.9.

Table 1.9 *Electrical resistivity and temperature resistivity coefficients for some of the more common materials*

Material	*Electrical resistivity at 20°C (Ωm)*	*Temperature resistivity coefficient (10^{-3}/K)*
Aluminium	27×10^{-9}	4.2
Brass	69×10^{-9}	1.6
Constantan	490×10^{-9}	0.02
Copper	17×10^{-9}	4.3
Duralumin	50×10^{-9}	2.3
Gold	23×10^{-9}	3.9
Lead	206×10^{-9}	4.3
Mild steel	120×10^{-9}	3.0
Polythene	100×10^{9}	–
Rubber	Approx. 10×10^{9}	–
Silver	16×10^{-9}	4.1
Tungsten	55×10^{-9}	4.6

Test your knowledge 1.14

Distinguish between the terms:

(a) resistivity
(b) conductivity.

Test your knowledge 1.15

Using the appropriate formula and Table 1.9, determine the resistance of a 50 m length of copper cable having a cross-sectional area of 1.5 mm².

Superconductivity – superconducting materials are those where the resistance falls to zero when they are cooled to a critical temperature and any magnetic fields are minimized. At these temperatures the superconducting material offers no resistance to the flow of current and so there are no wastages due to heat generation. All pure crystals would act as superconductors at zero degrees Kelvin, the problem of course is to maintain and use them at this temperature!

Serious research into raising the temperature at which selected materials will superconduct has taken place over the past two decades. By 1986 the critical temperature of a series of ceramic copper oxides had been raised to 77 K. This enabled liquid nitrogen to be used as the cooling medium, rather than liquid helium, the latter being expensive and providing only limited application of these materials. Since this date other compounds based on non rare-earth metal oxides have exhibited temperatures up to about 125 K. There is still a little way to go before we arrive at high temperature superconductors. As superconduction temperatures rise, then more and more industrial applications become feasible. Even today it seems possible that the new materials that are constantly being discovered, could be used to advantage in the electrical power generation industry. Other applications might include the rail transport industry, computing and electronics.

We have already discovered the fact that for metals and most other conductors, the conductivity decreases with increase in temperature. The resistance of insulators remains approximately constant with increase in temperature. *Semiconductors* behave in the opposite way to metals and as the temperature increases the conductivity of a semiconductor increases.

The most important semiconductor materials are *silicon* and *germanium,* both are used commonly in the electronics industry. As the temperature of these materials is raised above room temperature, the resistivity is reduced and ultimately a point is reached where effectively they become conductors. This increase in conductivity with temperature has made semiconductors an essential part of thermistors, where they can be used to sense the temperature and activate a signal when a predetermined temperature is reached. Other uses for semiconductors include, pressure transducers (energy converters), transistors and diodes. The effect of temperature on conductors, insulators and semiconductors is illustrated diagrammatically in Figure 1.17. Note that for a specimen of each of these materials it is assumed that they start with the same resistance at say room temperature (20°C). This, of course, implies that each specimen will have completely different physical dimensions (see definition of resistance).

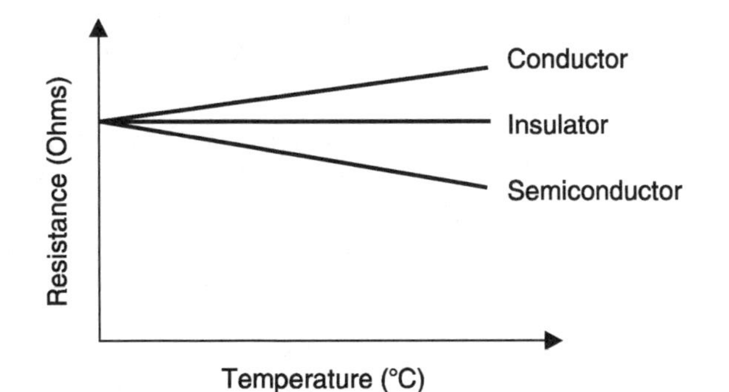

Figure 1.17 *The effect of temperature increase on conductors, insulators and semiconductors*

Another view

Because their electrical properties can be controlled during manufacture, semiconductors are a very interesting class of materials for engineers. When manufactured into a semiconductor device (such as a diode, thyristor, or transistor) these materials can act as either a conductor or an insulator by applying appropriate voltages and currents to them.

In order to control the conductivity of semiconductors, small amounts of impurity atoms are introduced and this is called *doping*. If antimony, arsenic or phosphorus are used as a doping agent then an n-type semiconductor is formed, since these doping agents add electrons to the parent semiconductor thus increasing its negative (n) charge. Conversely if gallium, indium or boron are added to the parent material then a p-type semiconductor is formed, with effectively an increased number of absences of electrons or *holes* which act like positive (p) charges.

In *photoconduction* a beam of light can be directed onto a semiconductor positioned in an electrical circuit. This can produce an electric current caused by the increased movement of electrons or holes in the atoms of the semiconductor material. This property is used in photoelectric components such as solar cells where the photoconductive property of the materials is used to produce power from the rays of light produced by the Sun. Other uses include electronic eyes that trigger the power to open garage doors, or automatic lighting where the photoelectric cell is activated by fading daylight.

Magnetization All substances are magnetized under the effect of a magnet field. A magnet dipole (an atom which has its own minute magnetic field) is formed by the rotation of each electron about its own axis (electron spin) and also its rotation about the nucleus of the atom (Figure 1.18). This induces a magnetic field in the material that is increased in strength when the dipoles are aligned.

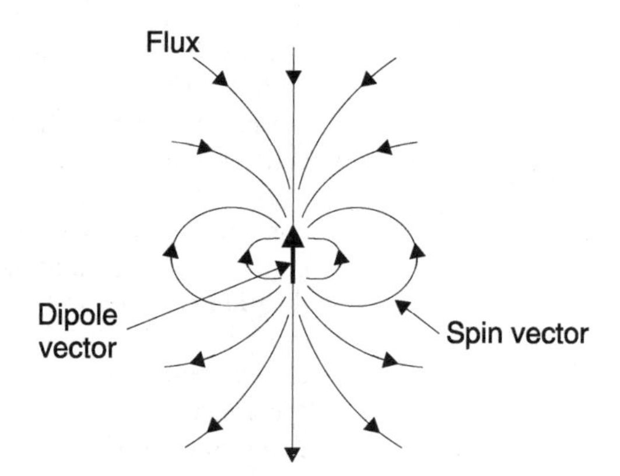

Figure 1.18 *Magnetic dipole moment of an electron (indicated by the direction of the arrow)*

Once they are aligned by an external magnetic field, the orientation of the dipoles will dictate what type of magnetism is created within the material being magnetized. Some materials called *paramagnetic*, magnetize with their dipole axis in the direction of the field and others, with their dipole axis perpendicular to the field these are *diamagnetic* materials. Iron, which shows a very pronounced magnetic effect and retains some residual magnetism when the magnetic field is removed, is *ferromagnetic*. Such materials can be used for permanent magnets.

If a bar of ferromagnetic material is wrapped with a conductor carrying a direct current, the bar will develop a north pole at one end and a south pole at the other (Figure 1.19).

The strength of the poles developed depends upon the material used and the magnetizing force. The magnetizing force is the result of the number of turns of conductor per unit length of the bar and the number of turns carried. The magnetic field produced as a result of the current and conductor length is denoted by the symbol, H, and is measured in amperes per metre (A/m). The magnetizing force is

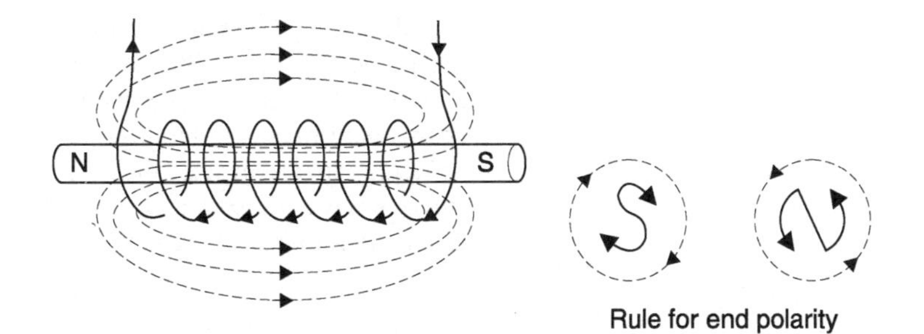

Figure 1.19 *Magnetic flux developed from the current-carrying conductor wound around the ferromagnetic material*

increased, decreased, or reversed by respectively increasing, decreasing, or reversing the current. The strength of the magnet so produced is called the *magnetic flux density,* which indicates the degree of magnetization that can be obtained. The magnetic flux density is denoted by the symbol, B, and is measured in Tesla (T); it is related to the applied field, H, by:

$$B = \mu H$$

where μ is the magnetic *permeability* and is measured in Henry/metre (H/m). The ratio of this magnetic permeability (m) to the permeability of free space (μ_0) measured in a vacuum, gives an indication of the degree of magnification of the magnetic field and is known as the relative permeability (μ_r).

Magnetization and demagnetization of magnetic materials results in producing a variable magnetic flux density, B, as a result of the magnetic field strength, H. A plot of B versus H results in a diagram known as a *hysteresis loop* (Figure 1.20). The area enclosed by the curve represents irreversibly lost electromagnetic energy which is converted to heat.

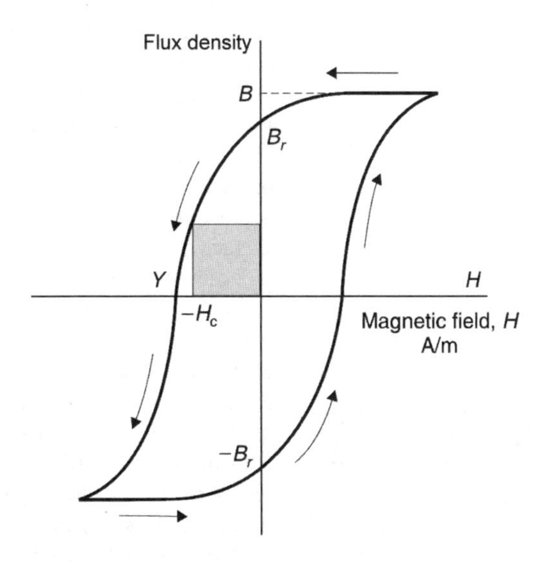

Figure 1.20 *Typical hysteresis curve for a ferromagnetic material. The area under the curve indicates the lost electromagnetic energy dissipated as heat*

Soft magnetic materials (temporary magnets) have a relatively small loop, because they are easily magnetized and demagnetized and so little heat is generated during the process. Hard magnetic (permanent magnets) have a large loop, due to the fact that they find difficulty in dissipating their residual magnetism. Thus *remanence* (or residual magnetism, B_r) and the *coercive force*, H_c, necessary to eliminate remanence are large in permanent magnets (Figure 1.21).

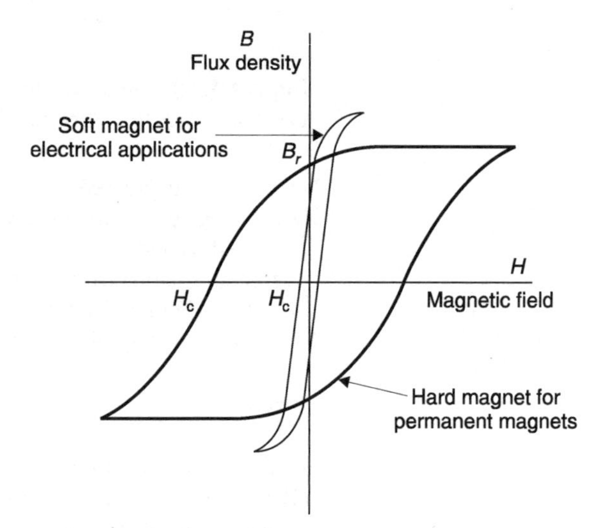

Figure 1.21 *Typical hysteresis curve for hard and soft magnetic materials*

Temporary magnets are soft iron or ferrous alloys; permanent magnets are hard steels, hard alloys, or metal oxides. Temporary magnets are used for alternating current applications such as a.c. motors, electromagnets and transformers.

A *dielectric* is an insulating medium separating charged surfaces. The most generally useful electric property of nonmetals is their high electrical resistance. Probably, the most common applications of ceramics, apart from their structural uses, depend upon their insulating properties. Breakdown of electrical insulators occurs either along the surface or through the body of the insulator. Surface breakdown is promoted by moisture and other surface contaminants. Water absorption can be minimized by applying a glaze to the surface of a ceramic.

Volume breakdown occurs at voltages high enough to accelerate individual free electrons to energies that will break the bonds holding the atoms of the material together, this causes a large number of electrons to break free at once and become charge carriers. The dielectric strength is the breakdown voltage for volume breakdown, it is measured in units of volts/m. The dielectric constant is the constant of proportionality between the charge stored between two plates of a capacitor separated by a dielectric, when compared with the charge stored when the plates are separated by a vacuum.

The ratio of the charge density, σ, to the electric field strength, E, is called *absolute permittivity*, ε, of a dielectric. It is measured in Farads/metre, F/m. The permittivity of free space measured in a

Test your knowledge 1.16

Distinguish between soft magnetic materials and hard magnetic materials. Use a B-H curve to illustrate your answer.

vacuum is a constant, given by:

$\varepsilon_0 = 8.55 \times 10^{-12}$ F/m

Relative permittivity is a ratio given by:

$$\varepsilon_r = \frac{\text{flux density of the field in the dielectric}}{\text{flux density of the field in a vacuum}}$$

Let us now continue with our discussion on the physical properties of materials by considering their *density*.

The *density*, ρ, of a substance is its mass per unit volume. The basic SI units of mass and volume are, respectively, the kilogramme and the metre cubed, so the basic units of density are kilogramme per cubic metre (kg/m^3).

The *specific weight*, *w*, of a substance is its weight per unit volume. Since the weight of a substance is mass multiplied by acceleration due to gravity, *g*, the relationship between specific weight and density is:

specific weight = density $\times g$

$$w = \rho g$$

Density is a property of the material that may be subject to small changes by mixing one material with another. Significant changes in density are only accomplished with composite materials where large percentages of each material can be altered. The density of a composite can be calculated from the proportions of the constituent materials. For example, consider a composite that consists by volume of 60% epoxy resin matrix material to which is added 40% by volume carbon fibre. If we assume that the carbon fibre has a density of 1500 kg/m^3 and the epoxy resin has a density of 1100 kg/m^3. Then the density of the composite is given by:

$$\rho_c = 0.6\rho_m + 0.4\rho_f = (0.6 \times 1800) + (0.4 \times 1400) = 1640 \text{ kg/m}^3$$

where ρ_c = density of the composite, ρ_m = density of the matrix material and ρ_f = the density of the reinforcing carbon fibre.

The above equation is often referred to as an *equation of mixtures*, it has many uses particularly in the study of composite materials.

Test your knowledge 1.17

What properties must dielectric materials have? Name a typical dielectric material.

Activity 1.9

A small engineering company, Thames Magnetic Components, has commissioned you to carry out some research into materials suitable for use in the cores of a new range of large industrial transformers that they intend to manufacture. They have asked you to summarize the essential electromagnetic properties of a suitable core material and explain how these properties differ from those used in the range of permanent magnets that make up their current product range. Write a briefing paper for the Technical Director, present your report in word processed form and include relevant diagrams and technical specifications.

Activity 1.10

An aerospace company, Archer Avionics, is about to manufacture an electronic control system that will be fitted to the propulsion unit of a deep space probe. This device is expected to be subject to extreme variations in temperature (e.g. from –50° to +150°C). Prepare a presentation (using appropriate visual aids) for the Board of Directors of Archer Avionics, explaining the effect of such a wide variation in temperature on the behaviour of the conductors, insulators and semiconductors that will make up the electronic system. Your presentation should last no more than 10 minutes and should include relevant graphs and material specifications. You should allow a further 5 minutes for questions.

Thermal properties

Let us now turn our attention to the thermal properties of materials. The way heat is transferred and absorbed by materials is of prime importance to the designer. For instance the walls of a house need to be constructed from materials that retain heat in winter and yet prevent the house from over heating in the summer. In a domestic refrigerator, the aim is to prevent the contents from absorbing heat. The thermal expansion of materials and their ability to withstand extremes of temperature are other characteristics that must be understood.

We therefore require knowledge of the most important thermal properties of materials, *specific heat capacity, thermal conductivity and thermal expansion*, as well as information on freezing, boiling and melting temperatures.

The specific heat capacity of a material is the quantity of heat energy required to raise the temperature of 1 kg of the material by one degree. The symbol used for specific heat capacity is c and the units are (J/kg °C) or (J/kg °K). If we just consider the units of specific heat capacity we can express it in words as:

$$\text{specific heat capacity} = \frac{\text{heat energy supplied}}{\text{mass} \times \text{temperature rise}}$$

Some typical values for specific heat capacity are given in Table 1.10.

Table 1.10 *Typical values for specific heat capacity for some common materials*

Material	*Specific heat capacity (c) at 0°C (J/kgK)*
Aluminium	880
Copper	380
Brass (65% Cu–35% Zn)	370
Iron	437
Lead	126
Mild steel	450
Silver	232
Tin	140
Brick	800
Concrete	1100
Polystyrene	1300
Porcelain	1100
Rubber	900

Thermal expansion takes place when heat is applied to a material and the energy of the atoms within the material increases, which causes them to vibrate more vigorously and so increase the volume of the material. Conversely, if heat energy is removed from a material contraction occurs in all directions.

The amount by which unit length of a material expands when the temperature is raised by one degree is called the *coefficient of linear expansion* of the material and is represented by the Greek letter (α) alpha. The units of the coefficient of linear expansion are usually quoted as just /K or K^{-1}.

Typical values for the coefficient of linear expansion of some materials are given in Table 1.11.

We can use the value of the coefficient of linear expansion of materials to determine changes in their length, as the temperature rises or falls. If a material has initial length l_1 at a temperature t_1 and has a coefficient of linear expansion α, then if the temperature is increased to t_2, the new length l_2 of the material is given by:

new length = original length + expansion, i.e.

$$l_2 = l_1 + l_1 \alpha (t_2 - t_1)$$

or since $t_2 - t_1$ is often expressed as Δt (change in temperature) we have new length $l_2 = l_1 + l_1 \alpha \Delta t$

Test your knowledge 1.18

Define specific heat capacity.

Table 1.11 *Typical values for coefficient of linear thermal expansion for some common materials*

Material	*Coefficient of linear thermal Expansion (α) (10^{-6}/K)*
Aluminium	23
Copper	16.7
Brass (65% Cu–35% Zn)	18.5
Invar	1
Iron	12
Lead	29
Magnesium	25
Mild steel	11
Silver	19
Tin	6
Brick	3–9
Concrete	11
Graphite	2
Polyethylene	300
Polyurethane foam	90
Polystyrene	60–80
Porcelain	2.2
PVC (plasticized)	50–250
Pyrex glass	3
Rubber	670

Example 1.4

In a domestic central heating system a 6 m length of copper pipe contains water at 8°C when the system is off, the temperature of the water in the pipe rises to 65°C when the system is in use. Calculate the linear expansion of the copper pipe.

From Table 1.11 we see that the coefficient of linear expansion for copper is 16.7 x 10^{-6}/°K, then using:

linear expansion of pipe = $l_1 \alpha t$ = 6 x 16.7 x 10^{-6} x (65 − 8)
= 0.0057 m

we see that the pipe expands by 5.7 mm over the 6 m length.

Thermal conductivity

Conduction is the transfer of energy from faster more energetic molecules to slower adjacent molecules, by direct contact. The ability of a material to conduct heat is measured by its *thermal conductivity, k*. The units of thermal conductivity are Watts per metre Kelvin (W/mK or $Wm^{-1}K^{-1}$). Typical values for some common materials are shown in Table 1.12.

Table 1.12 *Typical values for thermal conductivity for some common materials*

Material	*Thermal conductivity (k) at 0°C (W/m K)*
Aluminium	235
Copper	283
Brass (65% Cu–35% Zn)	120
Invar	11
Iron	76
Lead	35
Magnesium	150
Mild steel	55
Silver	418
Tin	60
Brick	0.4–0.8
Concrete	10.1
Graphite	150
Polyethylene	0.3
Polyurethane foam	0.05
Polystyrene	0.08–0.2
Porcelain	0.8–1.85
PVC (plasticized)	0.16–0.19
Pyrex glass	1.2
Rubber	0.15

A metal when left in a cold environment quickly feels cold to the touch and when brought into contact with heat quickly feels hot to the touch. Generally *metals* are *good conductors* of heat. Often if a material is a good thermal conductor, it is also a good electrical conductor; silver, copper and aluminium are all good conductors of both heat and electricity. Table 1.12 shows the low values of thermal conductivity for brick, porcelain, PVC and rubber, these are all good *insulators*. Air is also an excellent insulator, double glazing requires an airgap between the external and internal panes of glass, this provides good insulation from the cold and also helps prevent condensation forming between the panes. Expanded foams also use the properties of air to provide good insulation, foam cavity wall insulation for example.

Test your knowledge 1.19

Explain what is meant by thermal conductivity.

Modification of properties

In this section we are going to consider the processing of materials and the implications that these processing methods have on their property values. We will concentrate on a selection of metals, polymers and composite materials. Determining how, by modifying their grain structure, composition or by heat treatment, we are able to alter their properties.

Crystal structure, cold working and heat treatment

We have already learnt about how the atoms of metals bond together and form regular *lattice structures*. When a metal cools from the melt these lattice structures repeatedly join together three-dimensionally and grow *dendritically* to form *crystals* or *grains* (Figure 1.22). It is the size and nature of these grains that very much determines the properties of the parent material. Alloying metals also modifies their behaviour, during the manufacture of such alloys the metallurgist will try to maximize the good qualities that the individual elements bring to the alloy being produced.

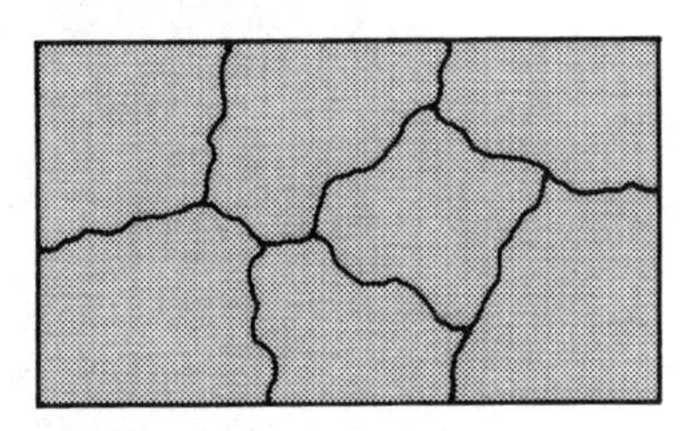

Figure 1.22 *Grains formed on solidification of the metal*

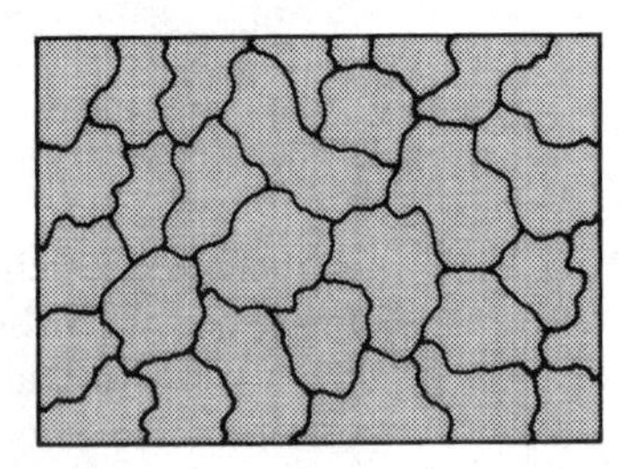

Figure 1.23 *Increase in grain boundary area as a result of reduction in grain*

The strength of metals and alloys can be improved by modifying their grain structures. For pure metals an increase in yield stress can be achieved by reduction in grain size or by cold working that not only increases the yield stress but also increases hardness.

A regular structure within the grains of a metal helps to maintain that metal in its lowest energy state (see bonding of materials). When this regular pattern is disturbed by either defects or an increase in grain boundary area, the internal energy of the metal is increased. Thus the energy needed to fracture the material is also increased. This increase in fracture strength is measured by a resulting increase in yield stress.

Therefore an increase in yield stress is achieved by reducing the grain size and so increasing the high energy grain boundary area (Figure 1.23).

If we cold work a metal say by cold rolling, we increase the dislocation density within the grains. That is we raise the energy of the metal by increasing the amount of line defects (dislocations) present. Further plastic deformation (resulting from lines of atoms slipping over one another) is impeded due to the increase in energy required to overcome the increased density of dislocations (line defects) that are already present due to the cold work. Figure 1.24 illustrates this idea.

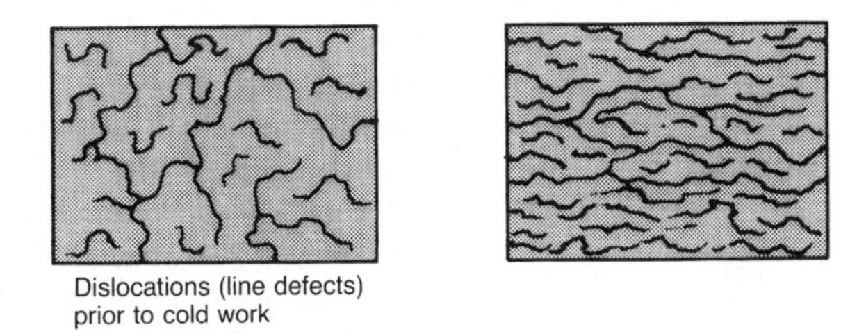

Figure 1.24 *The increase in dislocation density as a result of cold-working the material*

As we have already said an increase in cold work produces a subsequent increase in the number of dislocations which results in them becoming entangled with one another. It is the presence of these *dislocation tangles* that impedes the progress of further line defects and so plastic deformation only occurs at much higher stress levels. At the same time, for the same reasons, cold working also produces an increase in hardness.

When a cold-worked metal is heated to about a third of its melting temperature (measured on the Kelvin scale), then there is a marked reduction in tensile strength and hardness. What actually happens is that at this temperature recrystallization takes place (Figure 1.25). Further increase in temperature enables the crystals to grow until the original distorted crystals, resulting from the cold work, are all replaced. The term *annealing* is used for the heat treatment process whereby the material is heated to above the recrystallization temperature and more ductile properties replace those produced by the cold work. The greater the amount of cold work the smaller the grain size produced after heat treatment, with a subsequent increase in yield stress when compared to a metal which has not been cold worked.

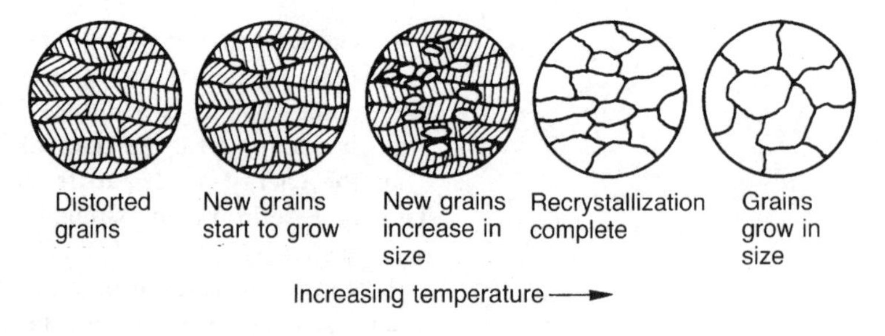

Figure 1.25 *Illustration of the recrystallization process with increasing temperature*

If copper is cold rolled then the yield stress and hardness will increase. At about 60% cold work the copper becomes so brittle that any further cold work results in fracture. So if further reduction in cross-section is required then the copper needs to be *process annealed* so that the original ductility and malleability is restored and the cold working process may continue.

Cold working involves plastically deforming the material at temperatures below the recrystallization temperature. Hot working as you might expect involves deforming the material at temperatures above the recrystallization temperature. So that as the grains deform they immediately recrystallize thus no hardening takes place and processing can continue without difficulty. Thus processing often involves hot working initially to produce the maximum amount of plastic deformation followed by cold rolling to improve the surface finish and increase the surface hardness as required. For example aluminium baking foil may be produced from blocks of aluminium in billet form, which are first hot rolled then cold rolled to produce thin even sheets.

Increase in yield stress for metal *alloys* is again achieved by modifying the grain structure of the host metal, but in a different way to that described above. Apart from line defects within the grain there may also be *point defects*. The various types of point defect are illustrated in Figure 1.26.

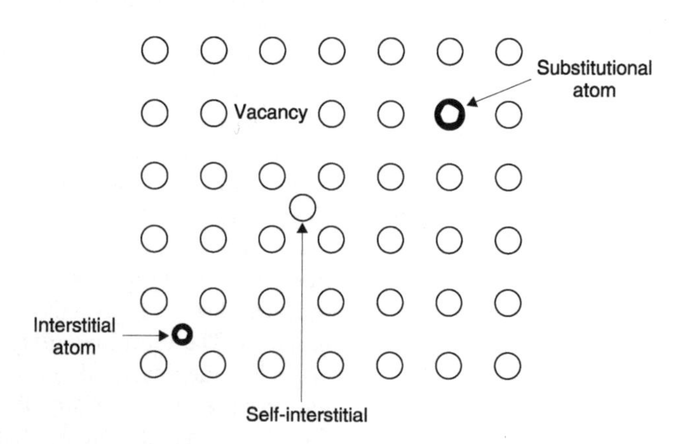

Figure 1.26 *Various types of point defect. A vacancy self-interstitial is known as a Frenkel defect. The simple vacancy on its own is known as a Schottky defect.*

These point defects may be caused by impurities in the melt or may be deliberately introduced by *alloying*. In either case the net result is to increase the energy within the grains or at the grain boundaries, resulting once again in increases in yield stress and hardness.

The above process is particularly useful in improving the properties of aluminium alloys, and other light alloys. The heat treatment process involved is known as *precipitation hardening*. The alloy is heated to above its recrystallization temperature, then quenched. The result is a distorted high energy lattice structure, where the alloying elements initially lodge within the grains. After a time the fine precipitate diffuses toward the grain boundaries where it stays. Once again the grain boundary energy is increased making slip very difficult, hence an increase in yield stress (a measure of strength) and an increase in hardness.

A classic example of this process is found in the Al–Cu alloy *duralumin* (named after the town Duran in Germany where it was first produced). In the annealed state this alloy has a tensile strength in the region of 180–190 MPa and a hardness around 40–50 HB. After precipitation hardening these values may climb to around 420 MPa and 100 HB, respectively. This alloy is often used in aircraft skin construction, where it is clad with pure aluminium to improve its corrosion resistance properties while at the same time increasing its strength and hardness over and above that of pure aluminium.

Steels may be made very hard by heating them to above their recrystallization temperature and then suddenly reducing their temperature by quenching in water or oil. The sudden reduction in temperature does not allow sufficient time for the carbon atom to diffuse to the grain boundaries and form cementite. Instead the excess carbon atoms are trapped within the lattice structure and form a separate very hard phase known as *martensite*. This very hard and brittle substance has the effect of increasing the hardness and brittleness of the steel. The above process is known as *quenching*. Quenched steels may be up to four times as hard as their annealed counterpart, dependent upon the original amount of carbon added to the alloy. For example, for a steel produced from 1% carbon, which in its annealed state has a hardness value of 200 HV, once quenched this figure may be as high as 800 HV.

After hardening a steel by quenching some of its ductility and resilience may be restored by *tempering*. Where the steel is reheated so that the carbon atoms can diffuse out and reduced the distortion within the lattice thus reducing the energy stored and so returning some ductility back to the steel. The higher the tempering temperature the more ductile the steel.

Activity 1.11

Use several reference sources to obtain information on duralumin alloy and add this as a new entry to the matSdata database (see Activities 1.2 and 1.6).

Test your knowledge 1.20

Distinguish between cold working and hot working and explain the effects of each type of process.

Deterioration of materials

Corrosion

Corrosion may be defined as a chemical process, where metals are converted back to the oxides, salts and other compounds from which they were first formed. The corrosion of metals is therefore a natural process and in trying to combat it, we are wrestling with the forces of nature! The chemical stability of materials (in particular metals) has long been the subject of much research, since the consequences of premature failure of materials by corrosive influences can be disastrous.

Corrosive attack frequently occurs in combination with other mechanisms of failure, such as corrosion-fatigue, erosion and stress corrosion. Many environmental factors help to promote corrosion including; moist air and industrial pollutants such as dirt, acids, dirty water and salts. Corrosion may occur at elevated temperatures in materials that at lower temperatures are inert.

Corrosion then is the chemical means by which metallic materials deteriorate and fail. Two basic mechanisms have been recognized: *direct chemical attack* and *electrochemical attack.*

Direct chemical attack results in a uniform reaction over the entire exposed surface. Usually, a scale or deposit of uniform thickness is produced on the metallic material. This deposit may adhere (stick) to the surface or remain as loose flakes, the rusting of an iron bar left in the open is an example of the later. An example of a material that is subject to direct chemical attack, where the products of corrosion (the corrosive oxides) adhere strongly to the surface of the metal, is aluminium. When this happens the oxide layer formed protects the metal underneath from further attack by adhering firmly to it. This process is known as *passivation* and we say that aluminium is a *passive* metal.

Electrochemical attack is characterized by the establishment of an electrochemical cell, this is formed when two metals in electrical contact are placed in a conducting liquid, the *electrolyte.* The cell permits electroplating or corrosion, depending on the source of electrical potential. In *electroplating* the electrochemical cell, consists of two electrodes (which may or may not be of the same material), the electrolyte and an *external* electric source such as a battery. It is found that an oxidation reaction takes place at electrode A (Figure 1.27).

Another view

Mechanical or thermal stress can cause physical damage to plastic materials, and latent stresses can be induced at manufacture and released at a later date. The result may be shrinkage, crazing or the appearance of surface deposits. More serious are chemical changes, usually caused by excess light and other unsuitable environmental conditions, poor manufacturing, or harmful materials or contaminants such as cleaning solvents. Symptoms of deterioration of plastics include colour change, brittleness, weakness, softness, surface 'blooms', and the release of gaseous breakdown products. Finally, combinations of materials (such as metals in close contact) and other circumstances can accelerate chemical degradation. For example, the harmful effects of ultraviolet radiation in light are substantially increased under high moisture or heat levels.

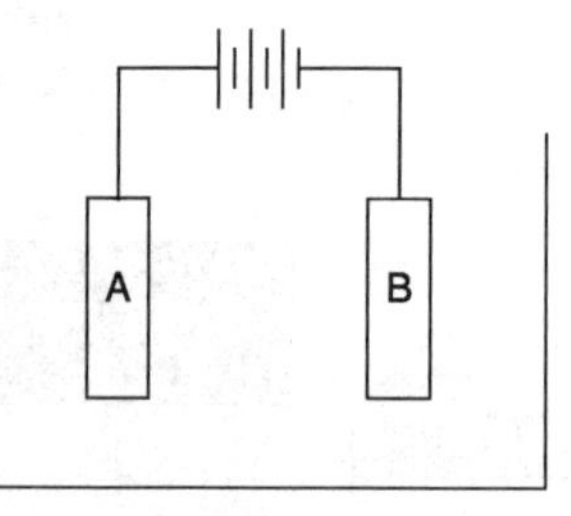

Figure 1.27 *In electroplating (external energy source) it is found that oxidation reaction takes place at electrode A*

An *oxidation reaction increases the energy of the atom.* In this case by forming an *ion* (an atom with an absence or excess of electrons) and a free electron. This may be represented by using a simple chemical equation:

$$M \rightarrow M^{+} + e^{-}$$

What we are saying is that the metal (M) has been subject to an oxidation reaction and form a metal ion (M^{+}), in this case an atom that has lost *one* electron (e^{-}). The electrode, at which an oxidation reaction takes place is defined as the *anode*. Note that, *corrosion always takes place at the anode*. Then at electrode (A), the ions go into the electrolyte and the electrons go into the external circuit. At electrode (B) a *reduction reaction* takes place. In this case, the ions in the electrolyte combine with electrons from the external circuit and form atoms which are deposited on the electrode. This can again be illustrated using a chemical formula, i.e.

$$M^{+} + e^{-} \rightarrow M$$

The electrode, at which the reduction reaction takes place is known as the *cathode*.

For electroplating, the battery provides the electrical potential to move the electrons, giving rise to an electric current in the external circuit. In practice, corrosion occurs in the absence of a battery. This can be demonstrated by a *galvanic cell*. A galvanic cell may be formed by two dissimilar metals in electrical contact immersed in an electrolyte, with no battery. Figure 1.28 shows a typical arrangement for a zinc–copper galvanic cell. The chemical symbol for zinc is (Zn) and for copper (Cu).

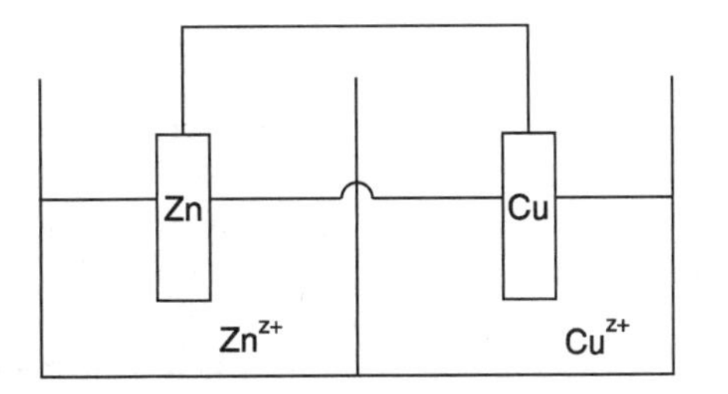

Figure 1.28 *In the copper–zinc galvanic cell an oxidation reaction takes place at the zinc electrode and it preferentially corrodes*

It is found that an oxidation reaction takes place at the Zn electrode. This electrode is thus the anode and it corrodes, i.e.

$$Zn \rightarrow Zn^{2+} + 2e^{-}$$

At the copper electrode, the cathode, a reduction reaction occurs, i.e.

$$Cu^{2+} + 2e^{-} \rightarrow Cu$$

The electrons in the external circuit flow between the two electrodes and so there must be an electrical potential difference between them.

Electroplating rarely occurs during electrochemical corrosion. The reduction reaction tends to form either a gas, liquid or solid by-product at the cathode.

A cathodic reaction is a reduction reaction, some possible cathodic reactions are given below.

1. Electroplating $M^+ + e^- \rightarrow M$

2. The hydrogen electrode – here hydrogen gas is liberated at the cathode:

$$2H^+ + 2e^- \rightarrow H_2$$

3. Water decomposition – here water is being broken down by the addition of electrons from the external circuit, to form hydrogen gas and hydroxyl (hydrogen–oxygen) ions

$$2H_2O + 2e^- \rightarrow H_2 + 2(OH)^-$$

Note that *chemical equations balance*. Taking the above equation, for example, two molecules of water plus two electrons form two atoms of hydrogen, diatomically bonded (H_2) and two hydroxyl ions.

4. The oxygen electrode – here oxygen combines with water and the external electrons to form hydroxyl ions.

$$O_2 + 2H_2O + 4e^- \rightarrow 4(OH)^-$$

5. The water electrode – in this case water is formed as a product.

$$O_2 + 4H^+ + 4e^- \rightarrow 2H_2O$$

When hydroxyl ions (OH^-) form as a result of a reaction, they can combine with other available ions to form a solid or sediment, e.g. Fe $(OH)_3$ which is rust.

Note that in all chemical equations the constituents that combine to make up the product, *the reactants*, are always placed at the tail of the arrow, the arrow indicates the direction of the reaction and *the products* (the result of the chemical reaction) are always placed after the head of the arrow.

If the zinc electrode, in the zinc–copper cell, is immersed in a 1 molar* electrolyte and the copper electrode (both at 25°C) is immersed in a 1 molar electrolyte, then a potential difference between the electrodes of 1.1 V occurs. The zinc electrode is at a lower potential than the copper electrode and so electrons flow, via the external circuit, from the zinc to the copper electrode.

The electrode potential of a metal is related to its ability to produce free electrons. As with all potential difference (PD) measurements, it is necessary to have a reference. In this case the PD of a metal is measured with respect to the *hydrogen half-cell*, i.e. zero potential is assigned to the oxidation reaction:

$H_2 \rightarrow 2H^+ + 2e^-$

The hydrogen half-cell consists of a platinum electrode immersed in a 1 molar solution of hydrogen ions through which hydrogen gas is bubbled. The other half of the galvanic cell consists of a metal electrode immersed in a 1 molar solution of its own ions. Both half cells are at 25°C. For example, if zinc is compared with the hydrogen half-cell under the conditions mentioned above, its potential difference is –0.76V. This implies that the electrons flow via the theoretical circuit, from the zinc electrode to the hydrogen electrode. Zinc is the anode and corrodes, the hydrogen electrode is the cathode.

In the case of silver (Ag), the potential difference, or electrode potential is found to be +0.8V. Thus silver becomes the cathode and corrosion does not occur. The electrode potential of metals have all been measured in this way and a table of values produced. This table is known as the *electrochemical* or *redox series*. Some of the more common metals are listed in Table 1.13.

Table 1.13 *The electrochemical (redox) series for some elements*

Element	*Potential E (V)*
Lithium	–3.05
Potassium	–2.93
Caesium	–2.92
Calcium	–2.87
Sodium	–2.71
Magnesium	–2.37
Aluminium	–1.66
Titanium	–1.63
Zinc	–0.76
Chromium	–0.74
Iron	–0.44
Cadmium	–0.40
Nickel	–0.25
Tin	–0.14
Lead	–0.13
Hydrogen	0.00
Copper	+0.34
Silver	+0.80
Palladium	+0.99
Platinum	+1.20
Gold	+1.50
Cobolt	+1.82

The standard conditions under which the electrode potential is measured are, 25°C and electrolytes of 1 molar solution.

Galvanic cells are classified into three groups; *composition cells, stress cells* and *concentration cells*.

Composition cells consist of two dissimilar metals in contact. The metal having the lower electrode potential becomes the anode and will corrode. Examples include galvanized steel (often used for manufacturing buckets) and tinplate. Galvanized steel consists of zinc coated mild steel sheet. Since zinc has a lower potential (–0.76 V) than the steel (Fe, –0.44 V), zinc becomes the anode that corrodes and so protects the mild steel. In the case of tinplate, which is a coating of tin (Sn, –0.14 V) on mild steel (Fe, –0.44 V), then iron becomes the anode which corrodes. The layer of tin provides a barrier to the corrosion, but if damaged, corrosion occurs. This is why damaged tin cans are removed from the shelves of shops, because corrosion of the underlying steel can contaminate the contents of the can.

Stress cells are formed where differences in stress within a material give rise to differences in electrode potential. For example regions within a component having different amounts of cold work during its forming process. The regions of strain hardening have a greater energy than annealed material and so become anodic and corrode. Thus stressed components are more likely to corrode than unstressed. More will be said about this when we look at the structure and cold working of materials.

An example where this may be a problem is with crimped metal joints, if the material is not given an appropriate heat-treatment after crimping, then high stress areas are created at the joints which will preferentially corrode.

Concentration cells arise due to differences of the ion concentrate in the electrolyte. The concentration cell accentuates corrosion, but it accentuates it where the concentration of the electrolyte is least. This relationship which occurs in practice can be proved theoretically.

In an oxidation type concentration cell, when the oxygen has access to a moist surface, a cathodic reaction can occur, such as:

$$2H_2O + O_2 + 4e^- \rightarrow 4(OH)^-$$

Thus regions where oxygen concentration is less become anodic and corrosion occurs (see Figure 1.29).

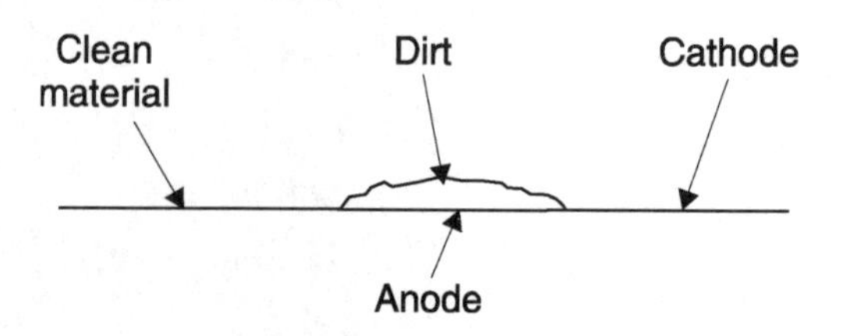

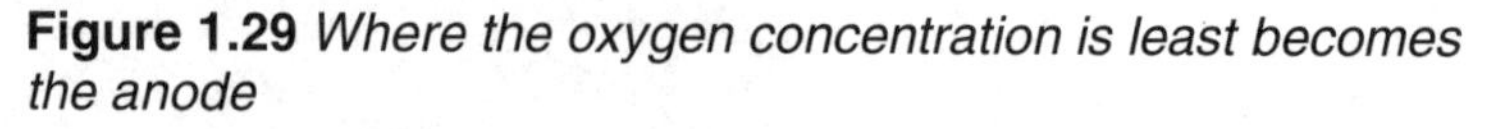

Figure 1.29 *Where the oxygen concentration is least becomes the anode*

An example of this behaviour may occur on a car's bodywork, if a small amount of dirt is left on the bodywork where the surface may have been damaged, the area under the dirt will corrode. The oxygen in this area has been excluded i.e., it is least when compared to the clean surface adjacent to the dirty area. This is one of the reasons why we try and keep motor vehicles clean!

We have discussed the mechanisms of electrochemical corrosion, so how as engineers may we prevent it? There are in fact four major ways in which to prevent, or at least help reduce, the effects of corrosion. By protective coatings, cathodic protection, design and materials selection.

Protective coatings, form a barrier layer between the metal and the electrolyte. They must be non-porous and non-conductive. There are many types of coating suitable for this purpose. *Organic coatings* such as polymeric paints, *ceramic coatings* like enamels (these are brittle so care must be taken to avoid damage), metal coatings like tinplate and *chemically deposited coatings* such as the formation of a phosphate layer. Phosphating tends to be porous, but forms a keying layer for subsequent metal deposition or paint.

There is also the naturally occurring *protective oxide layer* that is formed by *passive metals*, which provides an excellent barrier against further corrosion. Aluminium, stainless steel, titanium and nickel are all examples of passive metals.

Cathodic protection is the mechanism whereby the anode is made to act as the cathode. It is achieved by use of either a sacrificial anode or an impressed d.c. voltage.

The sacrificial anode is a metal, that when connected in the form of a galvanic cell to the component being protected, forms the anode (Figure 1.30). Examples of galvanized steel where zinc is the sacrificial anode, magnesium or zinc plates attached to ships hulls, and buried pipes.

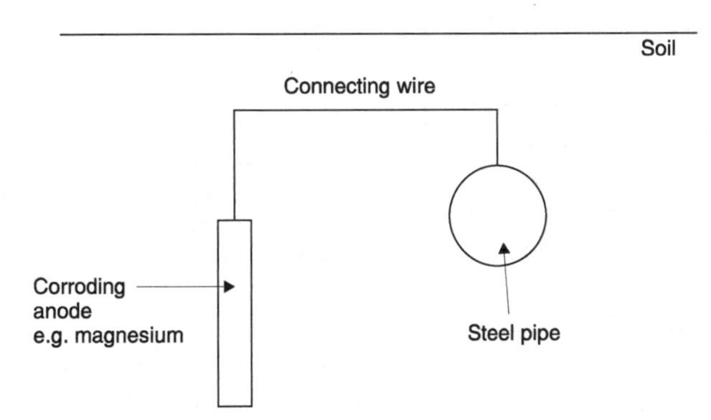

Figure 1.30 *Sacrificial anode*

For the impressed voltage (Figure 1.31), a d.c. source is connected to the metal to be protected and an auxiliary electrode, such that the electrons flow to the metal which then becomes the cathode. The auxiliary electrode has to be replaced from time to time.

There are numerous ways in which the designer can help prevent corrosion. If at all possible avoid the formation of galvanic cells, if dissimilar metals are to be brought into contact electrically insulate them. It is important to ensure that ingress of moisture is avoided.

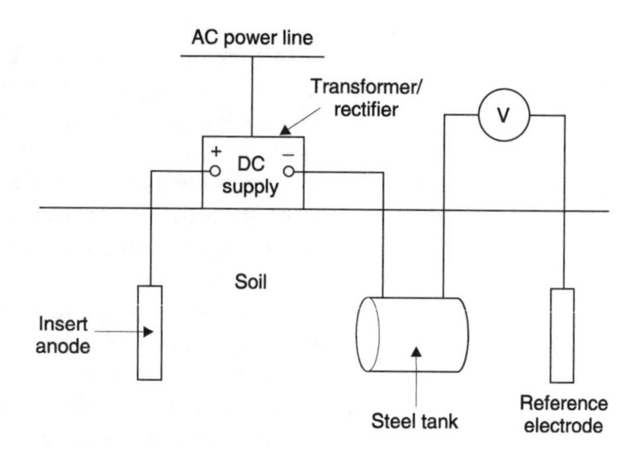

Figure 1.31 *Typical layout for impressed voltage corrosion protection*

Where possible, the anodic area should be made much larger than the cathodic area and copper rivets should be used to fasten the steel sheet, not the other way round. When installing fluid systems, try and ensure that system components and plumbing is enclosed to avoid oxygen pick-up. Avoid areas of stagnant liquid. Finally, avoid pipework or components being internally stressed, remember these areas of high stress are more energetic and therefore become anodic and subsequently corrode.

Corrosion can also be minimized by materials selection and appropriate heat treatment, particularly in steels. More will be said later on the subject in the element concerned with materials selection.

Finally in our discussion on corrosion, a number of the more commonly encountered forms of corrosion are detailed below:

Pitting is a localized form of corrosion resulting in small holes that may completely penetrate some members. It is encountered in aluminium and its alloys, copper and its alloys, stainless steel and high-nickel alloys. Pitting is an electrochemical form of attack involving either galvanic or concentration cells, and sometimes both types of cell.

Intergranular corrosion, a form of galvanic attack, occurs when grain boundaries (see structure of materials) of a metal are selectively corroded. It is the result of composition differences between the grain boundary and the grains themselves.

High-temperature corrosion is accelerated by alternate heating and cooling because brittle protective scales expand and contract at different rates compared to the base metals supporting them. This causes flaking to occur exposing fresh metal to attack.

Stress-corrosion is likely to occur when surface tensile stresses act in combination with a corrosive medium. Failure is believed to start at the high energy grain boundaries, which are anodic when compared to the grains themselves.

Test your knowledge 1.21

Explain the terms: concentration cell, composition cell and stress cell, with respect to galvanic corrosion. As part of your explanation give examples where each type of cell may cause corrosion.

Test your knowledge 1.22

Zinc and copper form the elctrodes of a galvanic cell.

(a) Which material is the anode and which is the cathode?

(b) Which material will corrode?

(c) Which way will the electrons flow around the external circuit?

Test your knowledge 1.23

Explain the difference between direct chemical attack and electrochemical attack, with respect to corrosion.

Test your knowledge 1.24

Under what conditions are the half-cell reactions of metals measured in order to produce the electrochemical series?

Corrosion-fatigue is caused by the action of a corrosive medium combined with variable cyclic stresses. In this type of failure, a corrosive agent attacks the metal surface, where imperfections produce stress raisers that start a fatigue failure.

This study on the mechanisms and prevention of corrosion brings to an end our discussion on material properties. We will now look at the structure of materials at the macroscopic (molecular) level and see how this affects their properties.

Activity 1.12

A material supplier, Beta Metals, has asked you to assist them in the production of some data sheets on corrosion protection that will be incorporated into the next edition of their catalogue. Identify four principal methods that can be used to protect against corrosion and, for each method, produce a summary that can be used to form the basis of the data sheets required by Beta Metals. Include relevant diagrams and sketches and make sure that your text is suitable for the non-technical reader.

1.3 Joining materials together

Thermal jointing methods

Metals and polymers are frequently jointed by means of the application of heat. The techniques are different according to the type of materials that are to be jointed and the particular application envisaged.

Fusion welding

Welding has largely taken over from riveting for many purposes such as ship and bridge building and for structural steelwork. Welded joints are continuous and, therefore, transmit the stresses across the joint uniformly. In riveted joints the stresses are concentrated at each rivet. Also the rivet holes reduce the cross-sectional areas of the members being joined and weaken them. However, welding is a more skilled assembly technique and the equipment required is more costly. The components being joined are melted at their edges and additional filler metal is melted into the joint. The filler metal is of similar composition to that of the components being joined. Figure 1.32(a) shows the principle of fusion welding.

High temperatures are involved to melt the metal of the components being joined. These can be achieved by using the flame of an oxy-acetylene blowpipe, as shown in Figure 1.32(b), or an electric arc, as shown in Figure 1.32(c). When oxy-acetylene welding (gas welding), a separate filler rod is used. When arc

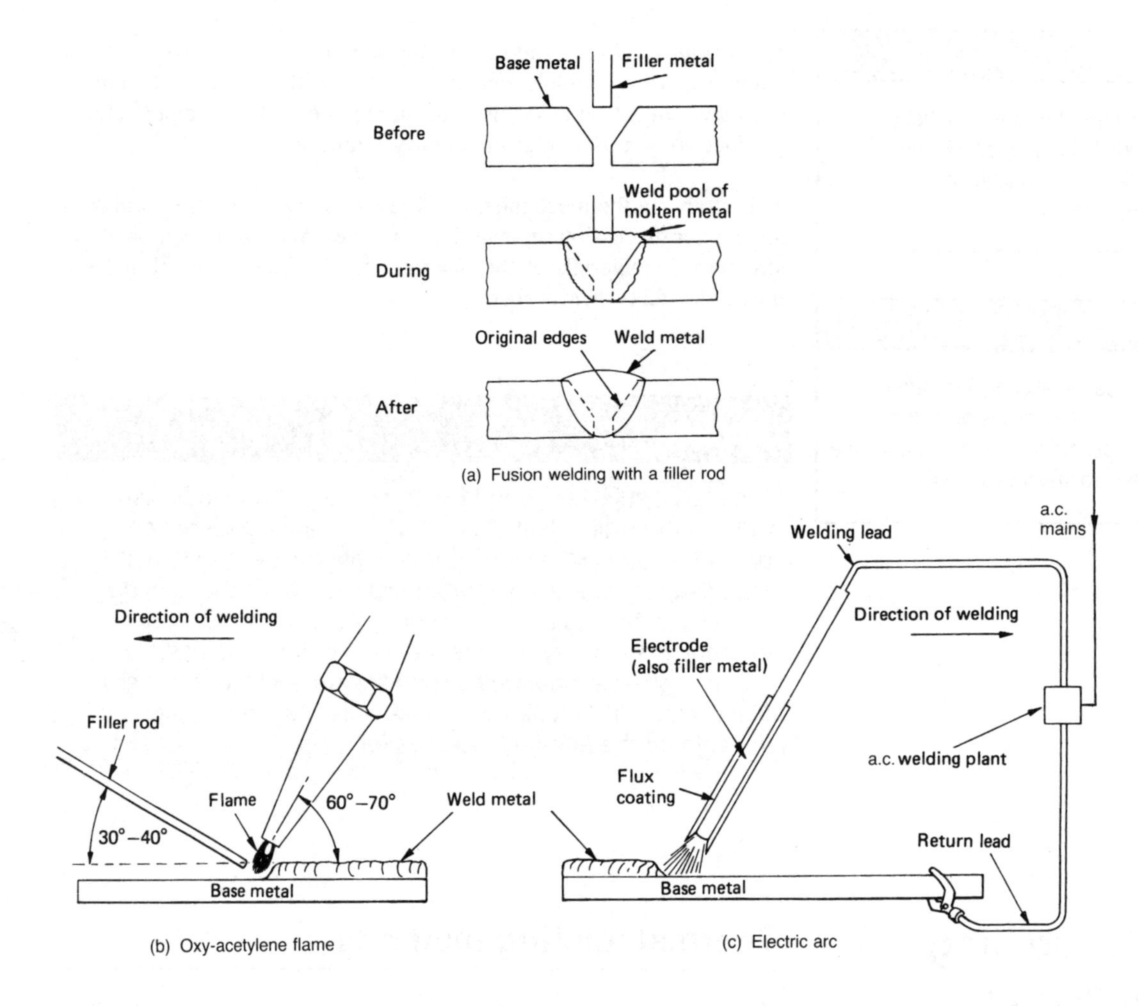

Figure 1.32 *Fusion welding*

welding, the electrode is also the filler rod and is melted as welding proceeds.

No flux is required when oxy-acetylene welding as the molten metal is protected from atmospheric oxygen by the burnt gases (products of combustion). When arc welding, a flux is required. This is in the form of a coating surrounding the electrode. This flux coating is not only deposited on the weld to protect it, it also stabilizes the arc and makes the process easier. The hot flux gives off fumes and adequate ventilation is required.

Protective clothing must be worn when welding and goggles or a face mask (visor) appropriate for the process must be used. These have optical filters that protect the user's eyes from the harmful radiations produced during welding. The optical filters must match the process.

The compressed gases used in welding are very dangerous and welding equipment must only be used by skilled persons or under close supervision. Acetylene gas bottles must only be stored and used in an upright position.

The heated area of the weld is called the weld zone. Because of the high temperatures involved, the heat affected area can spread back into the parent metal of the component for some distance from the actual weld zone. This can alter the structure and properties of the material so as to weaken it and make it more brittle. If the joint fails in service, failure usually occurs at the side of the weld in this heat affected zone. The joint itself rarely fails.

In *MIG welding* (metal inert gas welding) a wire (which also acts as a filler rod) is fed from a coil of wire. To prevent undesirable oxidation (which would otherwise occur due to a reaction between air and the molten metal) an inert gas (such as argon or carbon dioxide) is applied as a blanket to the weld area.

The MIG process requires no fluxing agent and a clean neat weld is produced. This method of welding is used extensively in the automotive and aerospace manufacturing industries.

TIG welding (tungsten inert gas welding) is similar to the MIG process but a separate filler rod is used. Once again, an inert gas (argon, helium or carbon dioxide) is fed to the weld area in order to prevent oxidation.

Soft soldering

Soft soldering is also a thermal jointing process. Unlike fusion welding, the parent metal is not melted and the filler metal is an alloy of tin and lead that melts at relatively low temperatures.

Soft soldering is mainly used for making mechanical joints in copper and brass components (plumbing). It is also used to make permanent electrical connections. Low carbon steels can also be soldered providing the metal is first cleaned and then *tinned* using a suitable flux. The tin in the solder reacts chemically with the surface of the component to form a bond.

Figure 1.33 shows how to make a soft soldered joint. The surfaces to be joined are first degreased and physically cleaned to remove any dust and dirt. Fine abrasive cloth or steel wool can be used. A flux is used to render the joint surfaces chemically clean and to make the solder spread evenly through the joint. Some soft soldering fluxes and their typical applications are listed in Table 1.14.

- The copper *bit* of the soldering iron is then heated. For small components and fine electrical work an electrically heated iron can be used. For joints requiring a soldering iron with a larger bit, a gas heated soldering stove can be used to heat the bit.
- The heated bit is then cleaned, fluxed and coated with solder. This is called *tinning* the bit.
- The heated and tinned bit is drawn slowly along the fluxed surfaces of the components to be joined. This transfers solder to the surfaces of the components. Additional solder can be added if required. The work should be supported on wood to prevent heat loss. The solder does not just 'stick' to the surface of the metal being tinned. The solder reacts chemically with the surface to form an amalgam that penetrates into the surface of the metal. This forms a permanent bond.
- Finally the surfaces are overlapped and 'sweated' together. That is, the soldering iron is re-heated and drawn along the joint as shown. Downward pressure is applied at the same time. The solder in the joint melts. When it solidifies it forms a bond between the two components.

Test your knowledge 1.25

(a) Describe the oxy-acetylene welding process. Illustrate your answer with a sketch.

(b) Explain the difference between oxy-acetylene and the MIG and TIG welding processes.

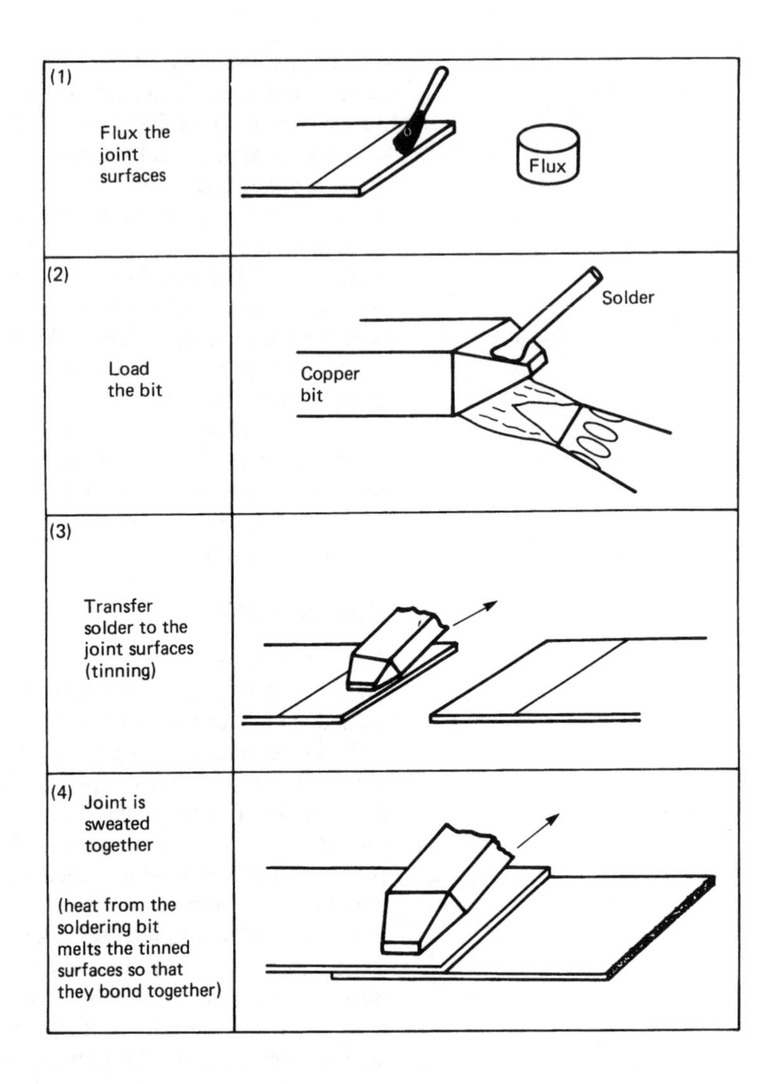

Figure 1.33 *Procedure for making a soft soldered joint*

Table 1.14 *Soldering fluxes*

Flux	*Metals*	*Characteristics*
Killed spirits (acidulated zinc chloride solution)	Steel, tin plate, brass and copper	Powerful cleansing action but leaves corrosive residue
Dilute hydrochloric acid	Zinc and galvanized iron	As above, wash after use
Resin paste or 'cored' solder	Electrical conductor and terminal materials	Only moderate cleansing action (passive flux) but non-corrosive
Tallow	Lead and pewter	As above
Olive oil	Tin plate	Non-toxic, passive flux for food containers, non-corrosive

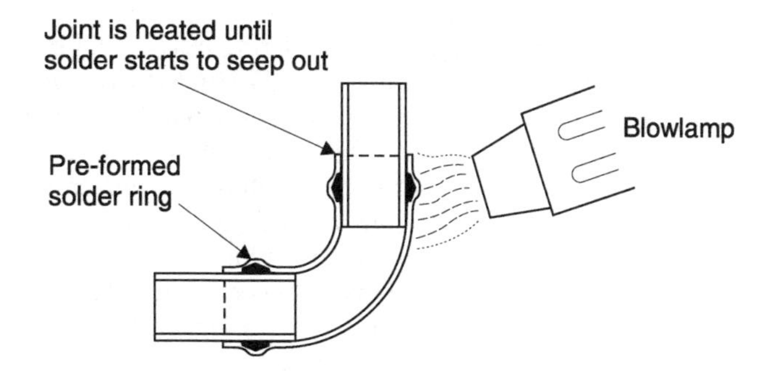

Figure 1.34 *'Sweating' a copper pipe*

Figure 1.34 shows how a copper pipe is sweated to a fitting. The pipe and the fitting are cleaned, fluxed and assembled. The joint is heated with a propane gas torch and solder is added. This is usually a resin-flux cored solder. The solder is drawn into the close fitting joint by capillary action.

Hard soldering

Hard soldering uses a solder whose main alloying elements are copper and silver. Hard soldering alloys have a much higher melting temperature range than soft solders. The melting range for a typical soft solder is 183–212°C. The melting range for a typical hard solder is 620–680°C. Hard soldering produces joints that are stronger and more ductile. The melting range for hard solders is very much lower than the melting point of copper and steel, but it is only just below the melting point of brass. Therefore great care is required when hard soldering brass to copper. Because the hard solder contains silver it is often referred to as 'silver solder'. A special flux is required based on borax.

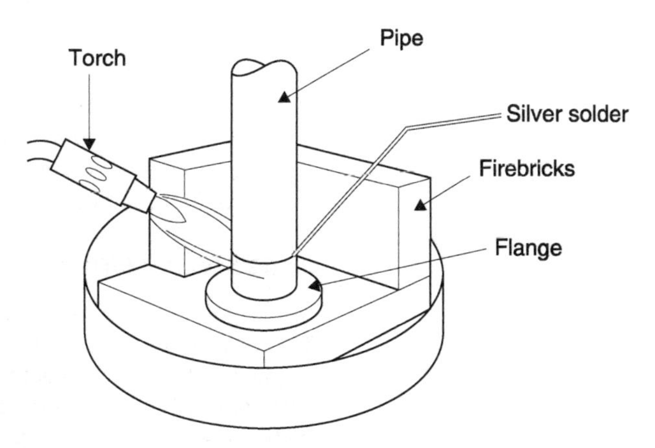

- The work is up to heat when the silver solder melts on contact with the work with the flame momentarily withdrawn.
- Add solder as required until joint is complete.

Figure 1.35 *Procedure for making a hard soldered joint*

Test your knowledge 1.26

Explain the difference between the soft soldering and hard soldering processes.

A soldering iron cannot be used because of the high temperatures involved. Heating is by a blow pipe. Figure 1.35 shows you how to make a typical hard soldered joint. Again cleanliness and careful surface preparation is essential for a successful joint. The joint must be close fitting and free from voids. The silver solder is drawn into the joint by capillary action.

Even stronger joints can be made using a brass alloy instead of a silver–copper alloy. This is called *brazing*. The temperatures involved are higher than those for silver soldering. Therefore, brass cannot be brazed. The process of brazing is widely used for joining the steel tubes and malleable cast iron fittings of bicycle frames.

Activity 1.13

Consult manufacturers' or suppliers' data and draw up a table showing the composition of several soft solders and their typical applications. Present your work in the form of an overhead projector transparency.

Joining thermoplastics

Engineering plastics can be joined using a variety of techniques including mechanical fastenings (see page 59) and adhesives (see page 66) and welding processes. Mechanical fastenings are quick and simple but they do not provide a leak tight joint and the localized stresses may cause them to pull free of the polymer material. Adhesives, on the other hand, can provide good jointing properties but they can be difficult to apply and may take considerable time to cure. Welding is generally superior to both adhesive bonding methods and mechanical fastening and the mechanical properties of the joint can approach those of the parent material.

Plastic welding processes can be divided into two main classes:

- processes involving mechanical movement such as ultrasonic welding, friction welding, and vibration welding
- processes involving external heating such as hot plate welding and hot gas welding.

Ultrasonic welding

Ultrasonic welding uses mechanical vibrations to form a joint between the two plastic materials. The parts to be assembled are aligned and then held together under pressure and then subjected to ultrasonic vibrations (in the frequency range 20 kHz to 40 kHz) at right angle to the contact area. These vibrations cause friction and localised heating and the result is a good quality weld. Due to the relatively high cost of the tooling this method tends to be reserved for high volume production. Furthermore, the method is really only appropriate to small components where the weld lengths are relatively small.

Friction welding

Friction welding (also referred to as *spin welding*) involves rotating one of the component parts whilst the other remains fixed. The resulting friction generated between the parts causes the polymer to melt (as with ultrasonic welding) and the result is an efficient weld on cooling. This process is relatively simple and reproducible and is ideal when at least one of the components has a circular cross-section.

Vibration welding

Vibration welding (also referred to as *linear friction welding*) is a similar process to pin welding but it involves vibrating two parts together until enough heat is generated in order to melt the polymer in the localised region of the joint. Vibration welding can be applied to almost all thermoplastic materials.

Hot plate welding

Hot plate welding is the simplest of the mass production methods used to joing plastics. In this process, a hot plate is fixed between the two surface to be joined until they are softened. The plate is then removed and the surfaces are brought together under pressure. The fused surfaces are then allowed to cool in order to form a strong joint between the two materials. This technique is commonly used to weld together the butt ends of plastic pipes and also to add components and connectors to blow moulded tanks and containers.

Hot gas welding

Hot gas welding is very similar to oxy-acetylene welding of metals. The main difference is that the open flame of oxy-acetylene welding is replaced by a stream of hot gas (typically compressed air, nitrogen, or carbon dioxide) which is pre-heated to a temperature of between 200°C to 400°C by passing it through an electrical heating element contained in the body of the welding gun (see Figure 1.36). This process is only applicable to those plastic materials that can be heated and melted repeatedly, namely thermoplastics. The hot gas is directed towards the components that are to be joined and this softens or melts the polymer. A *filler rod*, heated in the same stream of hot gas, is then fed in its molten state into the space between the two parts. Like oxy-acetylene welding this process requires manual skill and dexterity for it to be successful.

Test your knowledge 1.27

Describe THREE methods of joining thermoplastic materials.

Mechanical joining methods

Screwed fastenings

Screwed fastenings refer to nuts, bolts, screws and studs (see Figure 1.37). These come in a wide variety of sizes and types of screw thread. When selecting a screwed fastening for any particular purpose, you should ask yourself the following questions.

- Is the fastening strong enough for the application?

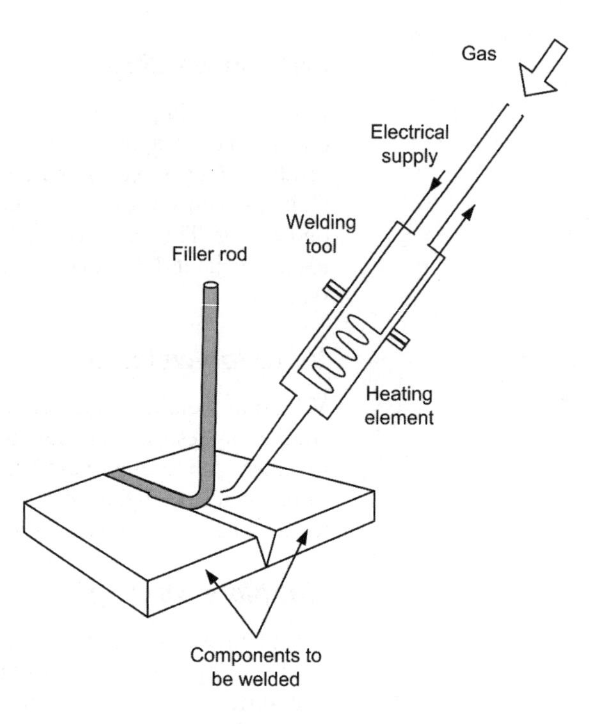

Figure 1.36 *Hot gas welding of thermoplastic components*

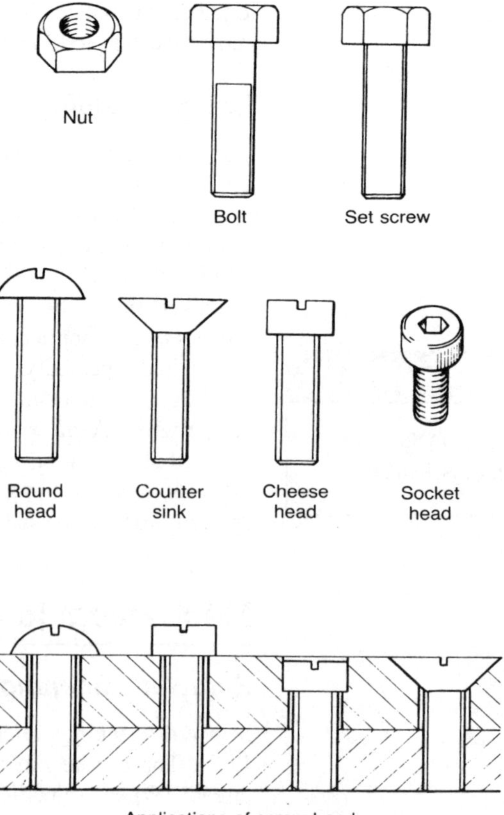

Figure 1.37 *Various types of nut, bolt and screw*

Test your knowledge 1.28

List THREE factors you would need to consider when selecting a screwed fastening for a particular application.

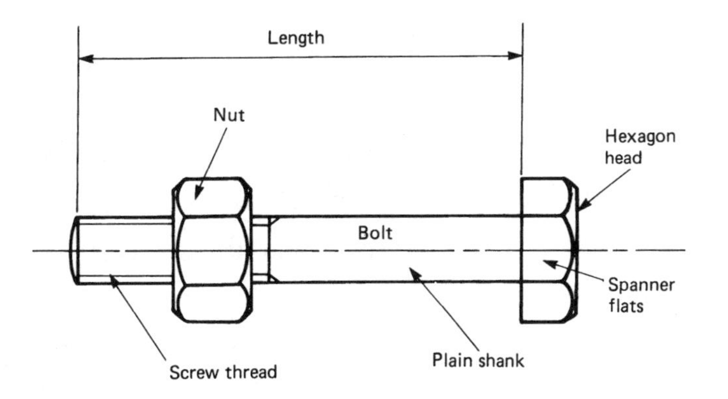

(a) HEXAGON HEAD BOLT AND NUT

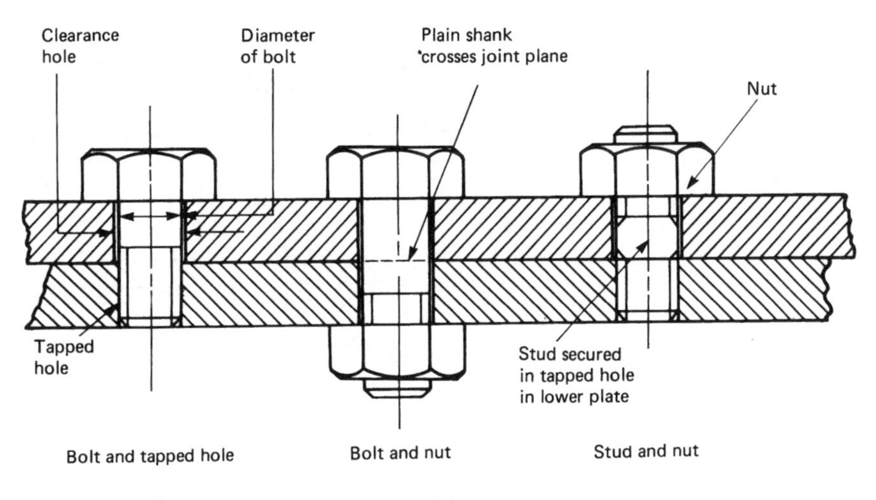

(b) TYPES OF SCREWED JOINT

Figure 1.38 *Screwed fastenings*

- Is the material from which the fastening is made corrosion resistant under service conditions and is it compatible with the metals being joined?
- Is the screw thread chosen, suitable for the job? Coarse threads are stronger than fine threads, particularly in soft metals such as aluminium. Fine threads are less likely to work loose.

Figure 1.38 shows some typical screwed fastenings and it also shows how they are used.

There are a large variety of heads for screwed fastenings, and the selection is usually a compromise between strength, appearance and ease of tightening. The hexagon head is usually selected for general engineering applications. The more expensive cap-head screw is widely used in the manufacture of machine tools, jigs and fixtures, and other highly stressed applications. These fastenings are forged from high-tensile alloy steels, thread rolled and heat treated. By

Test your knowledge 1.29

The cylinder head of a motor cycle engine is held on by studs. The studs are screwed into the cylinder casting at one end and the cylinder head is secured by nuts at the other end. The nuts must not vibrate loose. Which end of the stud should have a coarse thread? Which end of the stud should have a fine thread? Explain your answer.

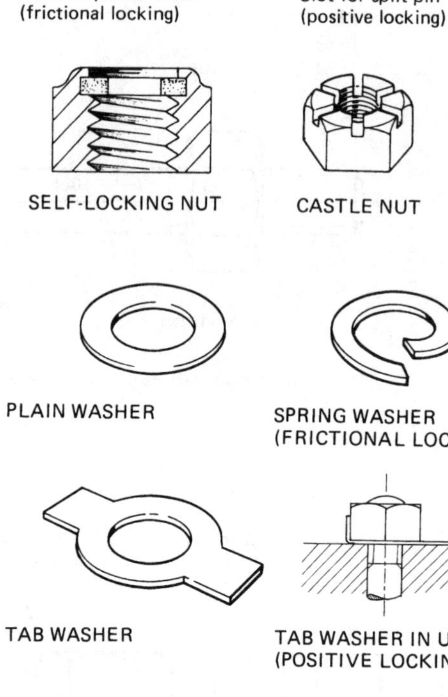

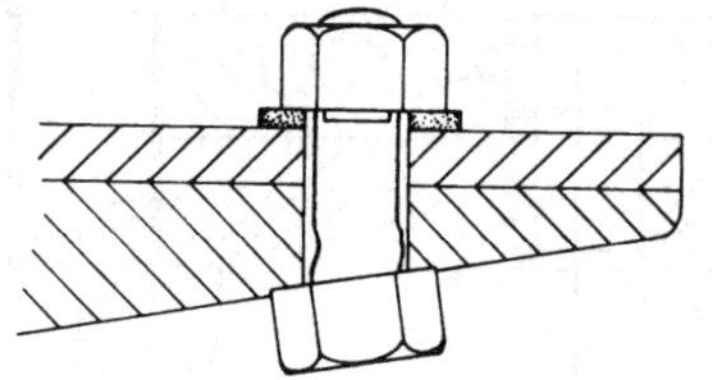

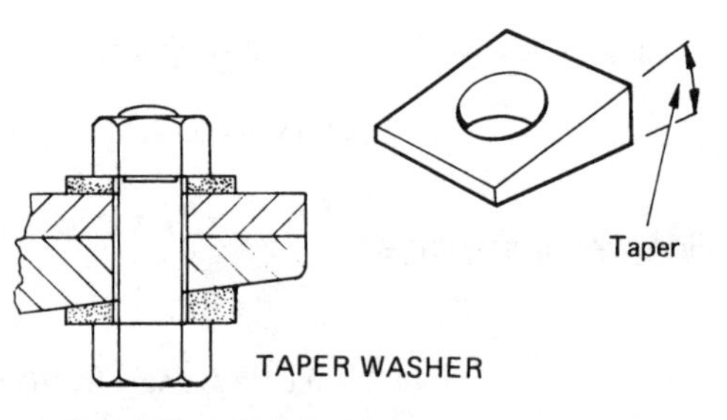

Figure 1.39 *Various nuts and washers*

recessing the cap-head, a flush surface is provided for safety and easy cleaning. Figure 1.39 shows various types of nuts and washers.

Riveted joints

Figure 1.40 shows some typical riveted joints. Riveted joints are very strong providing they are correctly designed and assembled. The joint must be designed so that the rivet is in shear and not in tension. Consider the head of the rivet as only being strong enough to keep

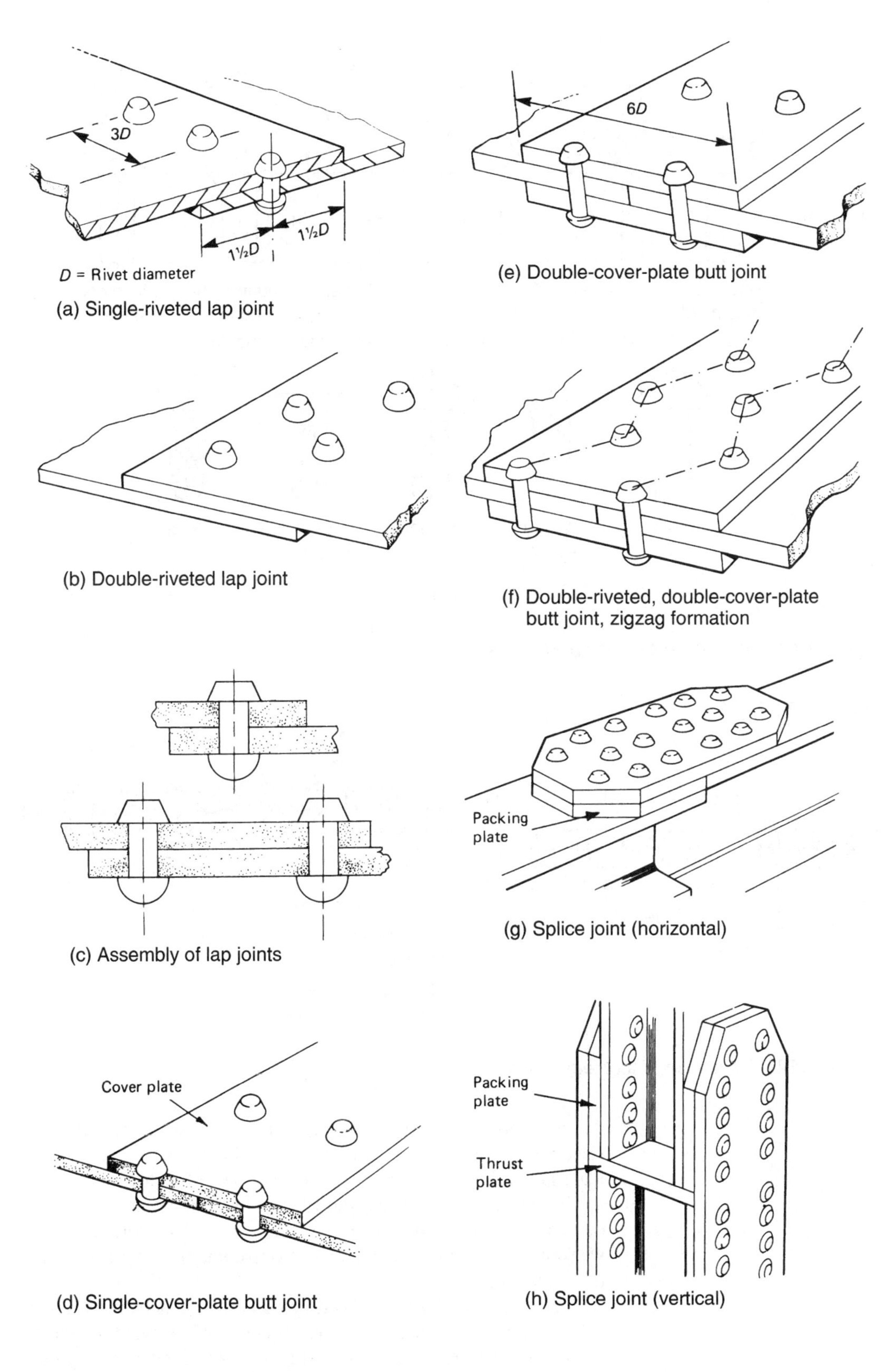

Figure 1.40 *Typical riveted joints*

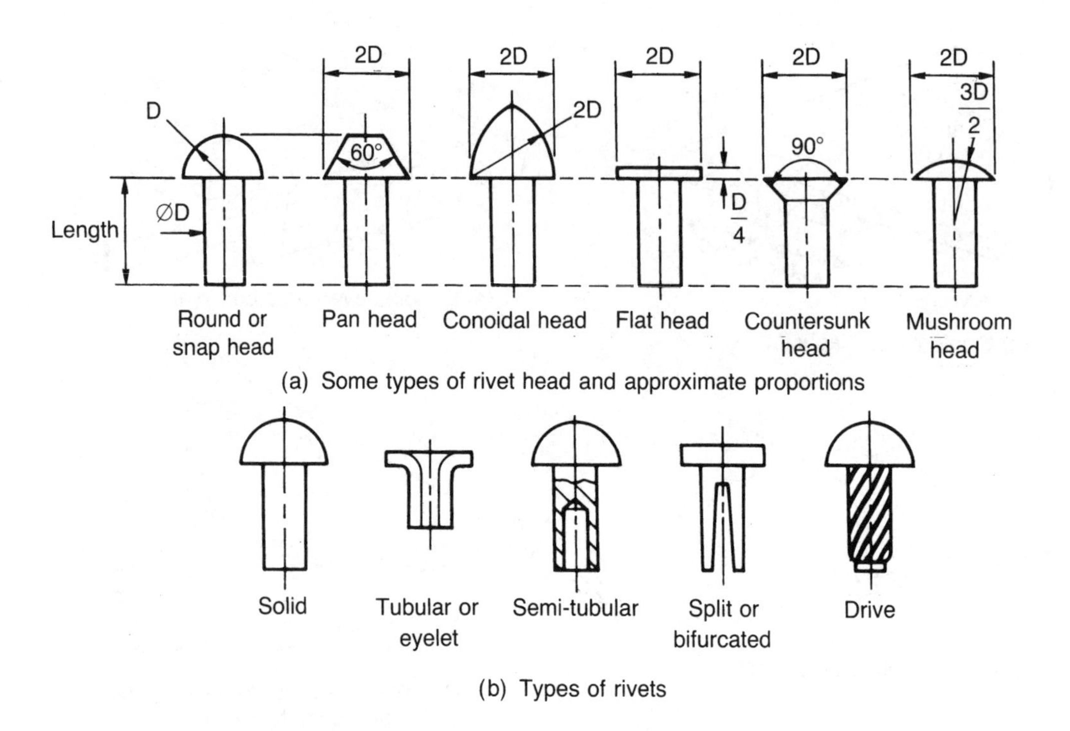

Figure 1.41 *Typical rivet heads and types*

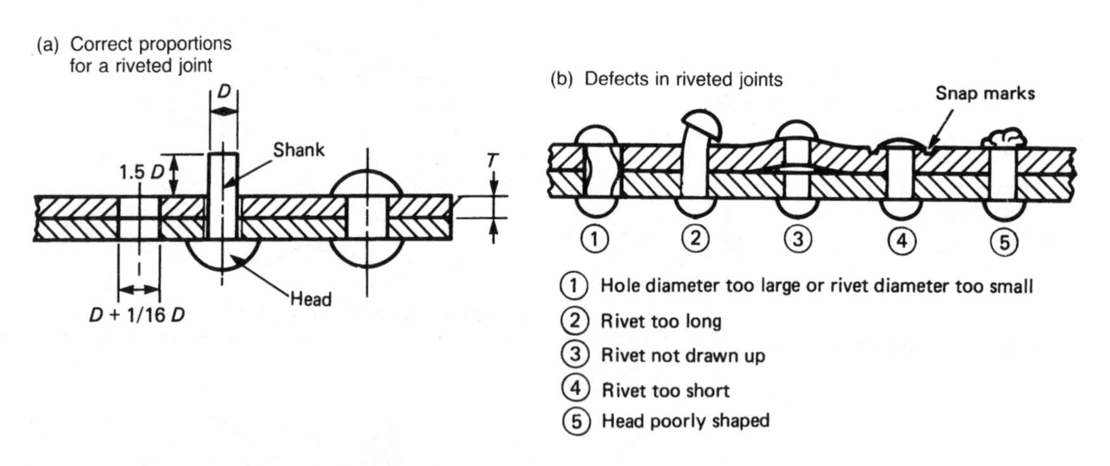

Figure 1.42 *Correct and incorrect riveted joints*

the rivet in place. You must consider a number of factors when selecting a rivet and making a riveted joint. These include the material used for the rivet as well as the shape of its head.

The material from which the rivet is made must not react with the components being joined as this will cause corrosion and weakening. Also the rivet must be strong enough to resist the loads imposed upon it.

The rivet head chosen is always a compromise between strength and appearance. In the case of aircraft components, wind resistance must also be taken into account. Figure 1.41 shows some typical rivet heads and rivet types whilst Figures 1.42 and 1.43 provide further information on how riveted joints are made.

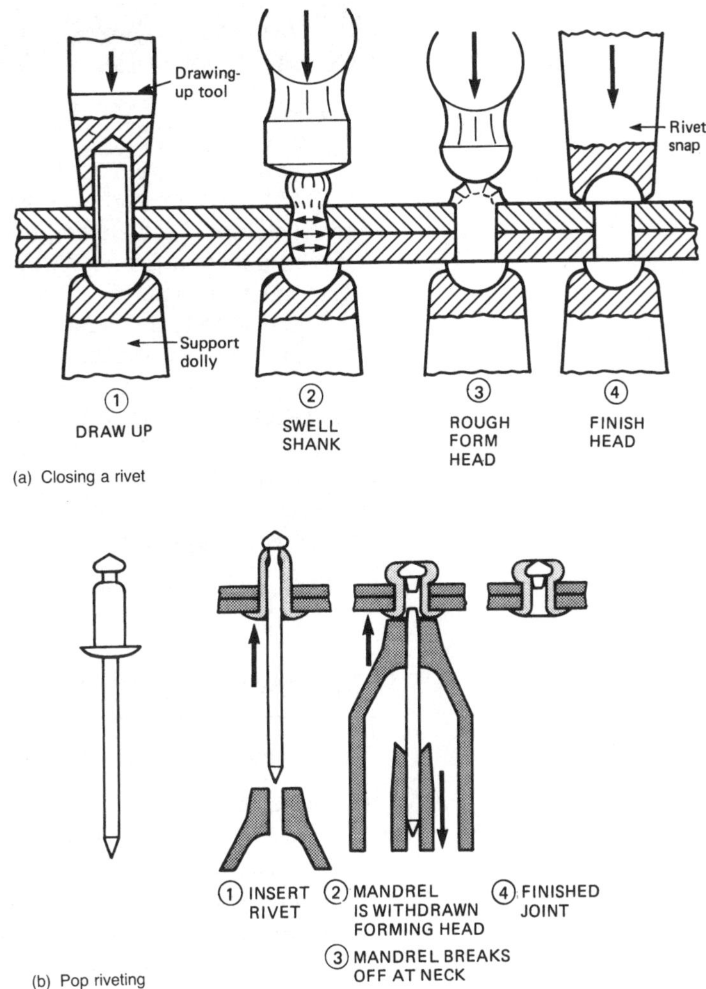

Figure 1.43 *Correct procedure for riveting*

Test your knowledge 1.30

State whether a rivet should be loaded in shear or in tension. Explain your answer.

Test your knowledge 1.31

State the main advantage of a screwed joint over a riveted joint.

Test your knowledge 1.32

List THREE factors you would need to consider when selecting a rivet for a particular application.

Activity 1.14

A fume extraction system is to be fitted in a small engineering workshop where welding, soldering and brazing processes take place. The extraction system is to be based on aluminium ducting which is to run the full length of the workshop at ceiling height. The ducting will require construction from sheet metal and, when assembled, will require joining at regular intervals.

Describe TWO methods of assembling the ducting and summarise the advantages and disadvantages of each method. Present your answer in the form of a brief written report together with illustrations and a final recommendation for the workshop manager.

Activity 1.15

A small engineering company, Ace Pumps, produces a range of submersible pumps. The pumps are housed in enclosures that are manufactured from a thermoplastic material with inlet and outlet hoses attached by galvanised metal clamps. These clamps have been found to be prone to corrosion and the company is considering replacing them with plastic fittings.

The company has asked you to investigate the use of a suitable welding process for attaching the new plastic hose fittings to the pump enclosures.

Prepare a brief report for the technical director of Ace Pumps describing the various processes available and summarising the advantages and disadvantages of each process. Include a final recommendation and illustrate your report with appropriate sketches.

Adhesive bonding

The advantages of adhesive bonding can be summarized as follows.

- The temperature rise from the curing of the adhesive is negligible compared with that of welding. Therefore the properties of the materials being joined are unaffected.
- Similar and dissimilar materials can be joined.
- Adhesives are electrical insulators. Therefore they reduce or prevent electrolytic corrosion when dissimilar metals are joined together.
- Joints are sealed and fluid tight.
- Stresses are transmitted across the joint uniformly.
- Depending upon the type of adhesive used, some bonded joints tend to damp out vibrations.

Bonded joints have to be specially designed to exploit the properties of the adhesive being used. You cannot just substitute an adhesive in a joint designed for welding, brazing or soldering. Figure 1.44(a) shows some typical bonded joint designs that provide a large contact area. A correctly designed bonded joint is very strong. Major structural members in modern high performance airliners and military aircraft are adhesive bonded. Figure 1.44(b) defines some of the jargon used when talking about bonded joints.

The strength of a bonded joint depends upon two factors.

Adhesion

This is the ability of the adhesive to 'stick' to the materials being joined (the *adherends*). This can result from physical keying or interlocking, as shown in Figure 1.45(a). Alternatively specific bonding can take place. Here, the adhesive reacts chemically with the surface of the adherends, as shown in Figure 1.45(b). Bonding occurs through intermolecular attraction.

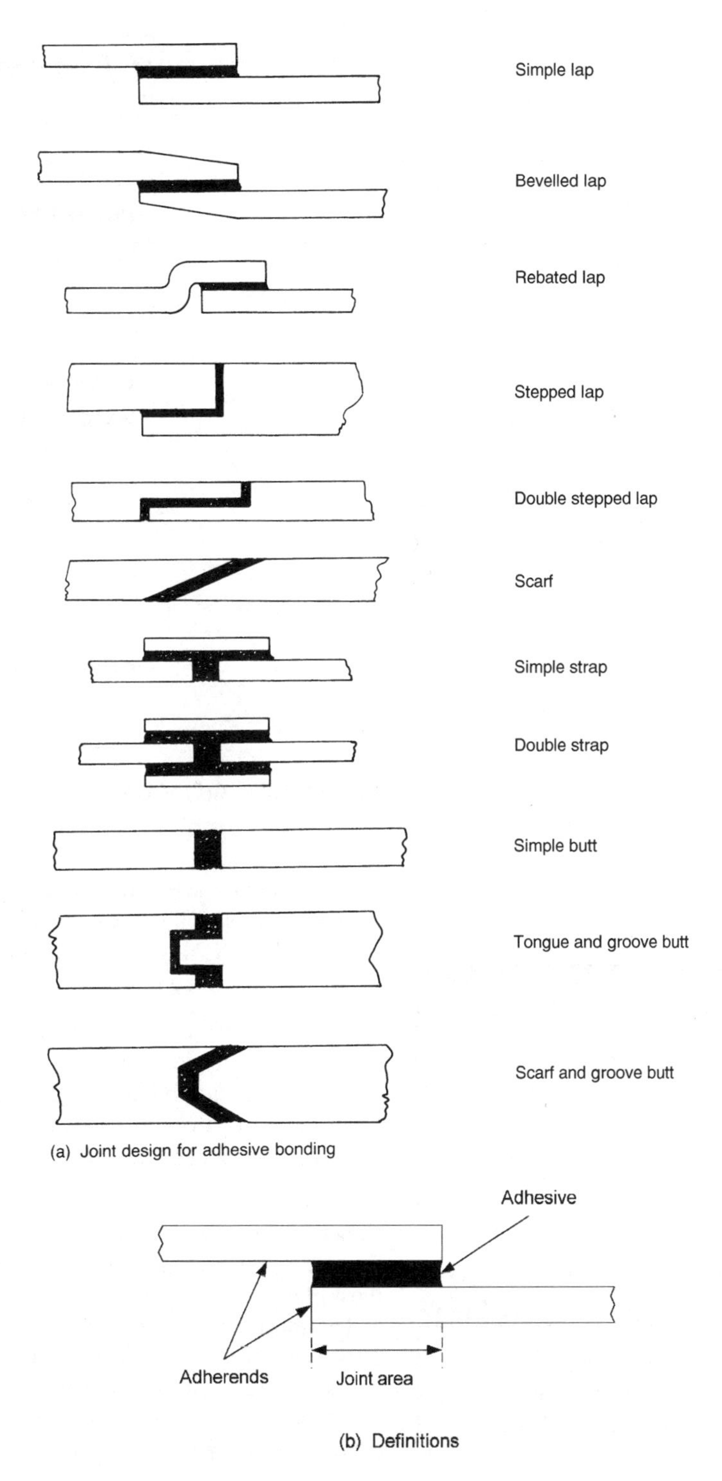

Figure 1.44 *Typical bonded joints*

Cohesion

This is the internal strength of the adhesive. It is the ability of the adhesive to withstand forces within itself. Figure 1.45(c) shows the failure of a joint made from an adhesive that is strong in adhesion but

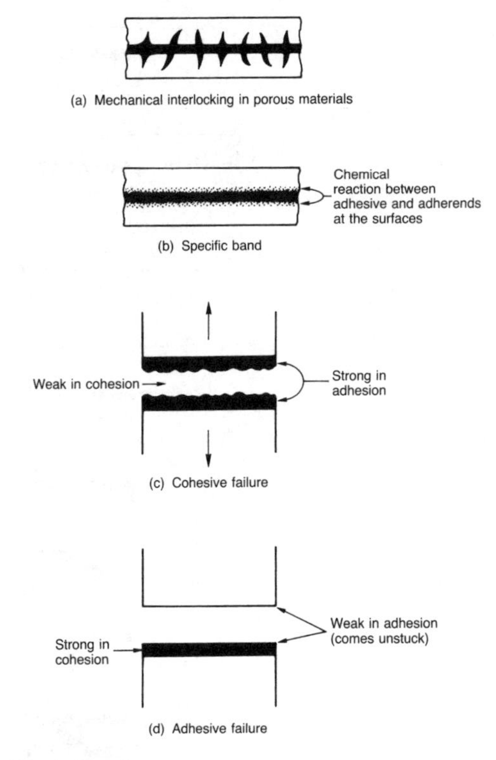

Figure 1.45 *Adhesion*

weak in cohesion. Figure 1.45(d) shows the failure of a joint that is strong in cohesion but weak in adhesion.

As well as the design of the joint, the following factors affect the strength of a bonded joint:

- The joint must be physically clean and free from dust, dirt, moisture, oil and grease.
- The joint must be chemically clean. The materials being joined must be free from scale or oxide films.
- The environment in which bonding takes place must have the correct humidity and be at the correct temperature.

Bonded joints may fail in four ways. These are shown in Figure 1.46. Bonded joints are least likely to fail in tension and shear. They are most likely to fail in cleavage and peel.

The most efficient way to apply adhesives is by an adhesive gun. This enables the correct amount of adhesive to be applied to the correct place without wastage or mess. It also prevents the evaporation of highly flammable and toxic solvents whilst the adhesive is waiting to be used.

Test your knowledge 1.33

Explain the terms:

(a) adhesion
(b) cohesion

in relation to adhesive bonding.

Test your knowledge 1.34

State THREE advantages of adhesive bonding over other method of joining materials.

Joining (electrical and mechanical)

Again, joints may be permanent or temporary. Permanent joints are soldered or crimped. Temporary joints are bolted, clamped or plugged in.

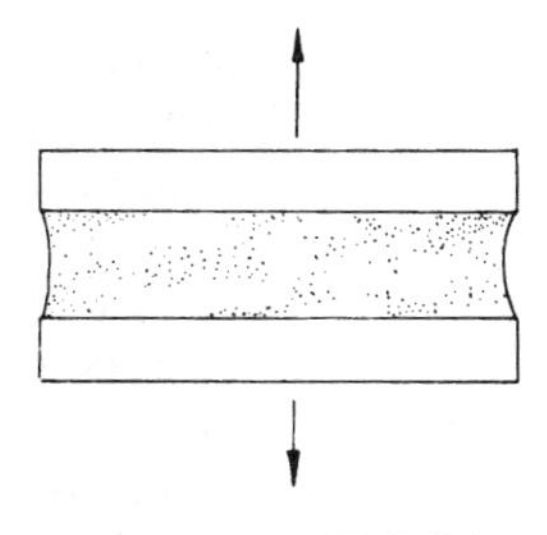

(a) JOINT IN TENSION

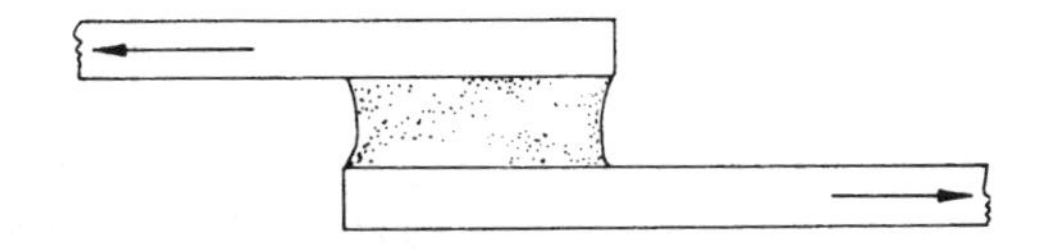

(b) JOINT IN SHEAR

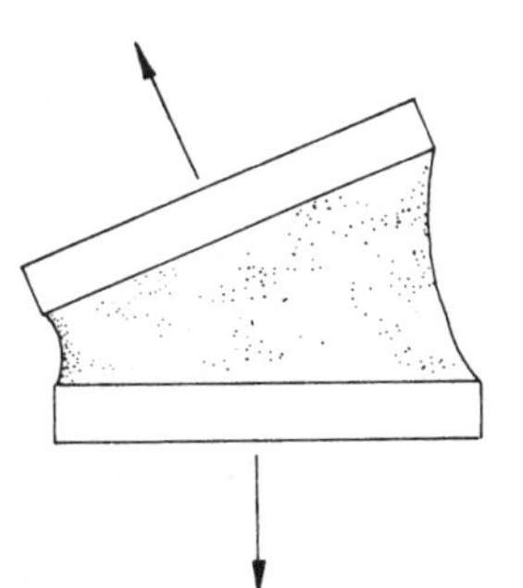

(c) JOINT IN CLEAVAGE

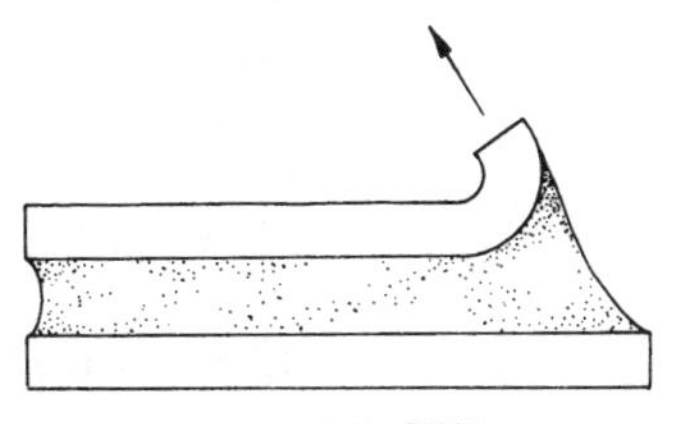

(d) JOINT IN PEEL

Figure 1.46 *Ways in which bonded joints can fail*

Soldered joints

When soldering electrical and electronic components it is important not to overheat them. Overheating can soften thermoplastic insulation and completely destroy solid state devices such as diodes and transistors. Very often some form of heat sink is required when soldering solid state devices.

A high tin content low melting temperature solder with a resin flux core should be used. This is a passive flux. It only protects the joint. It contains no active, corrosive chemicals to clean the joint. Therefore the joint must be kept clean whilst soldering. Even the natural grease from your fingers is sufficient to cause a high resistance 'dry' joint.

Figure 1.47(a) shows how a soldered connection is made to a solder tag. Note how the lead from the resistor is secured around the tag before soldering. This gives mechanical strength to the connection. Soldering provides the electrical continuity.

Figure 1.47(b) shows a prototype electronic circuit assembled on a matrix board. The board is made from laminated plastic and is pierced with a matrix of equally spaced holes. Pin tags are fastened into these holes in convenient places and the components are soldered to these pin tags.

Figure 1.47(c) shows the same circuit built up on a strip board. This is a laminated plastic board with copper tracks on the underside. The wire tails from the components pass through the holes in the board and are soldered to the tracks on the underside. The copper tracks are cut wherever a break in the circuit is required.

Figure 1.47(d) shows the underside of a printed circuit board (PCB). This is built up as shown in Figure 1.47(c), except that the tracks do not need to be cut since they are customized for the circuit.

Large volume assembly of printed circuit boards involves the use of pick and place robots to install the components. The assembled boards are then carried over a flow soldering tank on a conveyor. A roller rotates in the molten solder creating a 'hump' in the surface of the solder. As the assembled and fluxed board passes over this 'hump' of molten solder the components tags are soldered into place.

Wire wrapping

Wire wrapping is widely used in telecommunications where large numbers of fine conductors have to be terminated quickly and in close proximity to each other. Soldering would be inconvenient and the heat could damage the insulation of adjoining conductors. Also soldered joints would be difficult to disconnect. A special tool is used that automatically strips the insulation from the wire and binds the wire tightly around the terminal pins. The terminal pins are square in section with sharp corners. The corners cut into the conductor and prevents it from unwinding. The number of turns round the terminal is specified by the supervising engineer.

Crimped joints

For power circuits, particularly in the automotive industry, cable lugs and plugs are crimped onto the cables. The sleeve of the lug or the plug is slipped over the cable and then indented by a small pneumatic or hydraulic press. This is quicker than soldering and, as no heat is involved, there is no danger of damaging the insulation. Portable equipment is also available for making crimped joints on site. Hand operated equipment can be used to fasten lugs to small cables by crimping, as shown in Figure 1.48.

Clamped connections

Finally, we come to clamped connections using screwed fastenings. You will have seen many of these in domestic plugs, switches and lamp-holders. For heavier power installations, cable lugs are bolted to solid copper bus-bars using brass or bronze bolts, as shown in Figure 1.49.

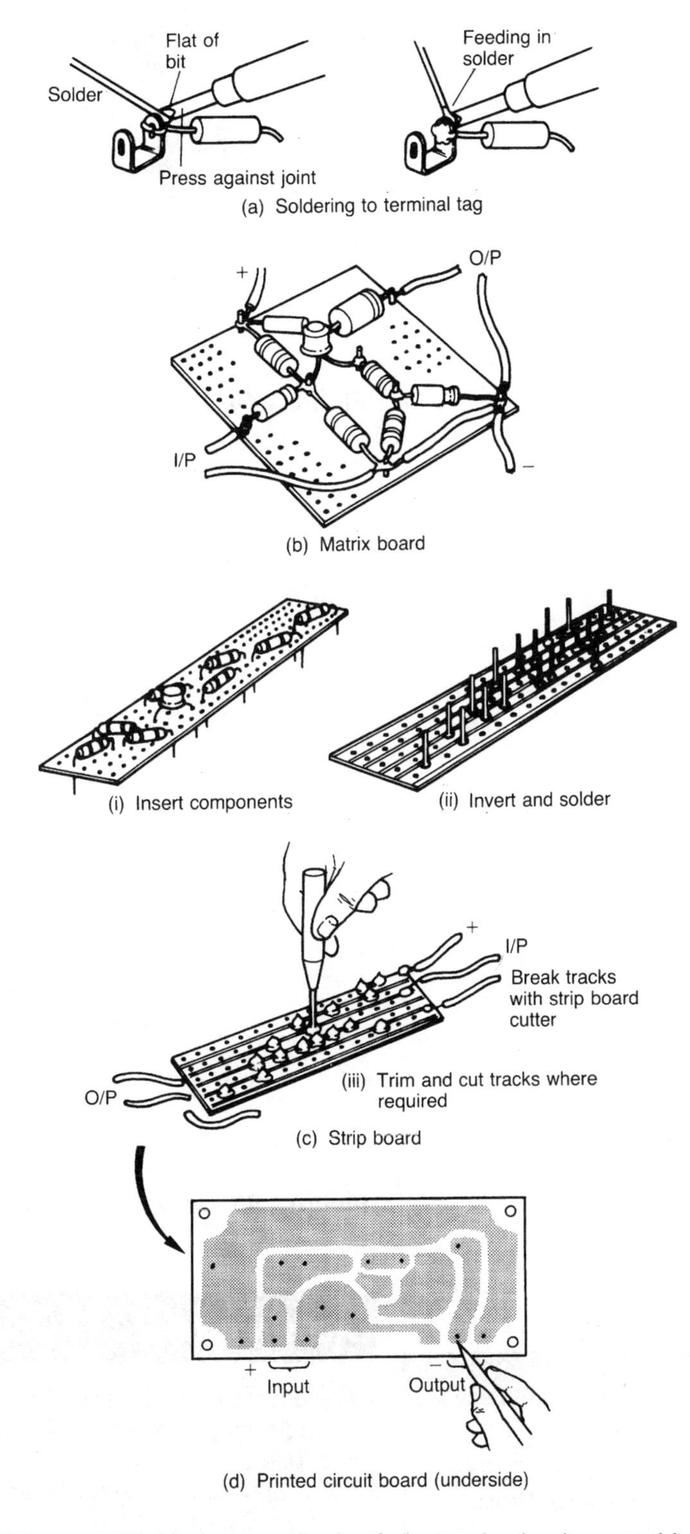

Figure 1.47 *Various methods of electronic circuit assembly*

Another view

Whilst the methods of electronic circuit construction shown in Figure 1.47 are suitable for prototyping and 'one-off' production, they are not used for modern electronic manufacturing where many small components must be mounted on a small space. Instead, Surface Mounting Technology (SMT) is now widely used. In this technology, components do not have conventional wire leads but instead they have contact pads which are attached to the printed circuit board (PCB) using automated soldering techniques. (see page 116 for more details).

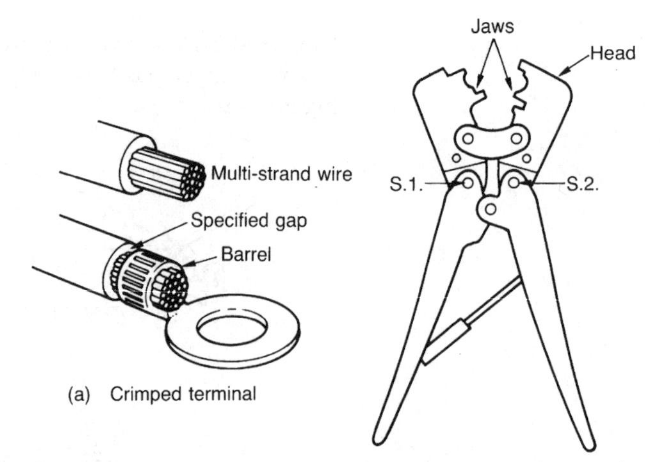

Figure 1.48 *Crimping*

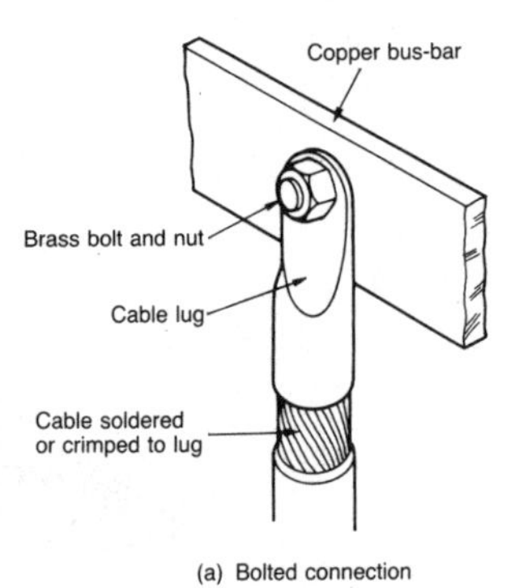

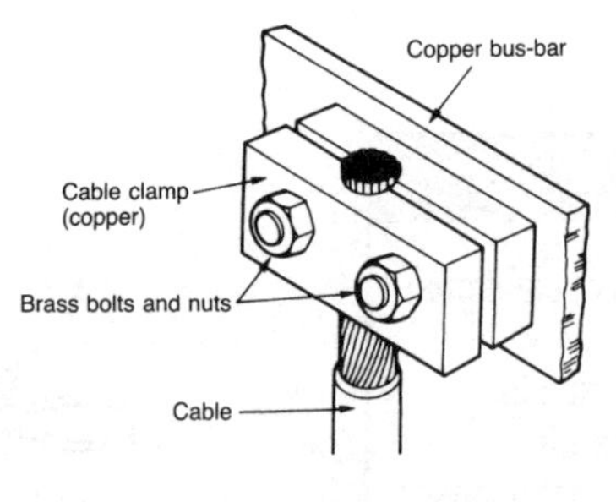

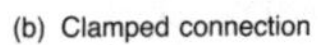
(b) Clamped connection

Figure 1.49 *Bolted and clamped connections*

Activity 1.16

A manufacturer of electronic kits has asked you to produce a single page instruction sheet on 'How to solder your kit'. Produce a word processed instruction sheet and illustrate it using appropriate sketches and drawings. Include a section headed 'Safety'.

Test your knowledge 1.35

Describe, with the aid of appropriate sketches, THREE methods of making electrical connections.

Health risks and safety precautions

It is essential to be aware of the health risks and safety precautions associated with handling and processing materials. Even apparently benign materials can be hazardous if they exist in certain forms (for example, as fine airborne particles that can be easily inhaled). Some of the most common hazards include:

- Exposure to fibres (either glass or carbon) may cause skin rashes (occupational dermatitis) as well as irritation to the eyes, nose and throat
- Contact with adhesives and uncured resins may cause skin sensitisation
- Resin fumes and solvent vapours may cause irritation to the eye and nose
- Sanding dust can be an irritant if it is inhaled
- Protective coatings, paints and solvents may produce vapours that cause irritation to the eyes, nose, throat and lungs
- Etching solutions and acids can cause skin irritation and burns.
- Powdered material (and dust) may cause explosions and an increased risk of static discharge
- Use of welding, brazing and soldering equipment may result in the production of toxic fumes
- Use of welding, brazing, hot gas and soldering equipment may cause burns
- Incorrect use of tools (particularly machine tools, drills, lathes, etc.) may cause injury to operators and other personnel.

Effective safeguards must be in place in order to minimise hazards. These safeguards will depend on the type of activity that is undertaken and the type of materials and processes that are in use.

Risk Assessments

Risk Assessments should be performed in order to identify hazardous activities and to inform action plans designed to improve safety and eliminate hazards. Correct storage and handling of materials is essential and protective clothing (gloves, goggles, overalls, etc), correct lighting and workspace organisation, adequate ventilation and efficient fume and dust extraction all have a part to play in making the workplace safe.

Material Safety Data Sheets

All hazardous materials should have a Material Safety Data Sheet (MSDS) supplied by, or at least available from, the material manufacturer or supplier. The sheet should be identified by the description and part number of the product and should contain comprehensive information about the product including details relating to handling, use and transportation. Details of the protective clothing or equipment required to work safely with the product should also be included. Personnel working with hazardous materials should have access to the MSDS for the materials concerned. An example of an MSDS is shown in Figure 1.50.

Howard Associates

SAFETY DATA SHEET

July 2003 **FDS 0703**

1. Identification of the chemical product

FERRIC CHLORIDE Ref : **AR37 - AR38 - AR371 – AR381**

Howard Associates
Brooklands Road
Weybridge
KT13 8TU

Emergency telephone number :
60080076767 (Europe) 603214989463 (Europe)

Identification of the substance or the preparation

Product name :	FERRIC CHLORIDE (Solution 37-46 %)
Chemical name :	Iran trichloride (solution 37-46 %)
Synonym(s) :	Iron chloride III (solution 37-46 %), Iron perchloride (solution 37-46 %)
Commercial Name :	SOLFLOC (R)
Formula :	FeCl3
Molecular Weight :	161.5

2. Composition information on ingredients

Ferric chloride :

- CAS Number :	7705-08-0
- EC Number (EINECS)	231-729-4
- Symbols	C
- Phrases R	34, 22, 52/53
- Concentration	37.00 – 46.00 %

3. Hazards identification

Toxicity effects principally related to its corrosive properties.
Hazardous product for the aquatic environment.
in case of decomposition, releases dangerous products.

4. First aid measures

Genera recommendations

- Personal protective equipment required for rescuers (ses section 8).
- In case of product splashing into the eyes and face, treat eyes first.
- Submerge soiled clothing in a basin of water.

Effects
Main affects

Irritating to skin; corrosive to mucous membrane and eyes.
The seriousness of the lesions and the prognosis of intoxication depend directly on the concentration and duration of exposure.
Risk of liver effects.
Fatalities have been observed after a single dose of 30 grams and more taken by an adult weighing 70 kg.
Chronic exposure to the product can induce iron accumulation in tissues characterized by redbrown deposits.

Inhalation

Severe irritation of the nose and the throat.
Cough and difficulty in breathing.
At high concentrations, risk of chemical pneumonitis, pulmonary (o)edema.
In case of repeated or prolonged exposure: risk of sore throat, nose bleeds, chronic bronchitis.
In case of repeated or prolonged exposure: risk of brown colouration of teeth.

Figure 1.50 *Extract from a Material Safety Data Sheet for Ferric Chloride solution*

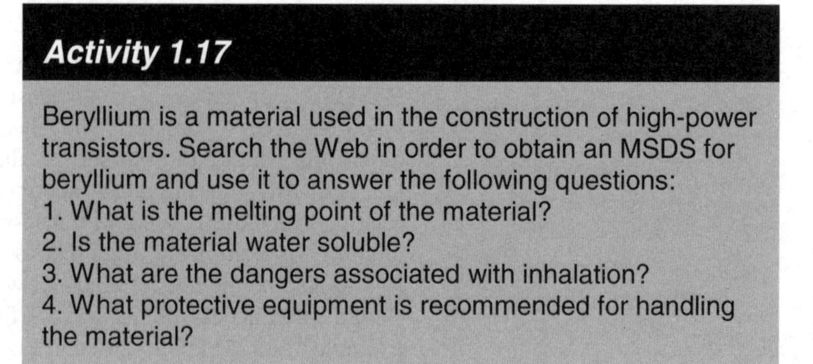

Activity 1.17

Beryllium is a material used in the construction of high-power transistors. Search the Web in order to obtain an MSDS for beryllium and use it to answer the following questions:

1. What is the melting point of the material?
2. Is the material water soluble?
3. What are the dangers associated with inhalation?
4. What protective equipment is recommended for handling the material?

1.4 Materials processing

Metals

Processing metals

Metals may be processed or fabricated by two major methods, either *shaped* or *joined*, to form a component or structure. Metals may be shaped into something approaching the final form by one of several operations including, casting, rolling, drawing, forging, extruding, cutting, grinding and sintering. Or fabricated by joining with adhesives, fasteners, soldering, welding or brazing. Time does not permit all of the processes to be covered comprehensively in this text. We will, however, concentrate on a few processing techniques leaving you to research further processes as an exercise.

Casting of metals requires the metal in liquid form to be poured into a mould and then allowed to solidify before breaking the mould open to reveal the cast product. In order to ensure that the liquid metal reaches all parts of the mould, we either choose alloying elements that provide a low viscosity alloy (that is an alloy which easily flows) or, we use some external pressure to force the molten metal into the mould, as in pressure die casting.

The grain structure of the metal, within the mould, is dependent upon the rate of cooling. If the metal is rapidly cooled only small crystals have time to form. As the cooling rate is reduced so the crystal size grows. Metals that cool within a mould may cool quickly near the surface of the mould and due to heat flow may form long crystals which grow towards the centre of the mould (columnar crystals). At the centre of the mould due to thermal convection the molten metal is continually on the move, this results in an even heat distribution, which results in the production of equal sized round crystals (equiaxed) being formed. The resultant structure of cast product moulded in an ingot is illustrated in Figure 1.51.

The method used for casting metals will depend on the size, number and complexity of the final product. Sand casting is suitable for small batch production of large components. If large numbers of castings are required a far superior product, with better dimensional accuracy, is produced by using a metal mould. In permanent mould casting known as *gravity die casting*, the molten alloy is allowed to run into the mould under gravity, while in *pressure* die casting the charge is forced into the mould under considerable pressure.

The use of die casting is confined mainly to aluminium or zinc based alloys. Metal dies are expensive to produce, so that die casting in all its forms is only economically viable for large scale production. Pressure die casting produces components with good dimensional accuracy, uniform grain structure and good surface finish. So the metallic components produced by this method require little or no further processing. Sand casting does have the advantage of producing intricate shapes, because of the possibility of using destructible cores.

Since most metals become considerably softer and more malleable as temperature rises less energy is needed to produce a given amount of deformation, this is why *hot-working* is extensively used to shape metals. Drop-forging involves the use of a shaped die, one half being attached to the hammer and other to the anvil (Figure 1.52). When producing complex shapes by this method a series of dies may be used.

Test your knowledge 1.36

Explain why hot working is extensively used to shape metals.

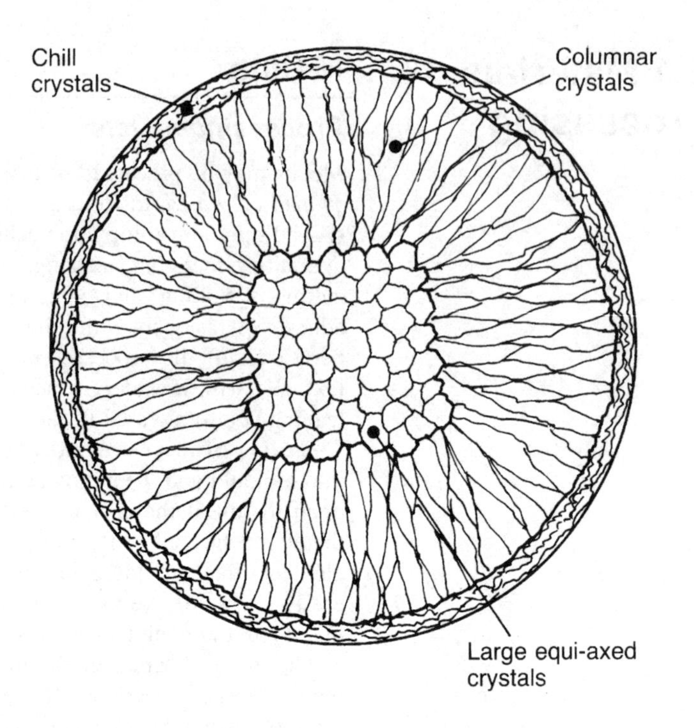

Figure 1.51 *Resultant crystal structure of a casting moulded in an ingot*

The hammer may be operated mechanically, pneumatically or hydraulically, dependent on the nature of the task.

The *extrusion* process is used for shaping a variety of ferrous and non-ferrous metals and alloys. The metal billet is heated to the required temperature, the ram is then driven hydraulically with sufficient pressure to force the metal through a hardened steel die. The solid metal section exudes from the die in a similar manner to that of toothpaste being squeezed from a tube. Figure 1.53 illustrates this process.

The extrusion process is able to produce a wide variety of sections, including round and hexagonal rod, curtain rails, tubes, bearing sections and ordinary wire.

Cold pressing and deep-drawing are closely related to each other and it is difficult to differentiate between them. The operations range from making a suitable pressing in one stage to cupping followed by a number of drawing operations as shown in Figure 1.54. In each case the components are produced from sheet metal. Car bodies, bullet cases and general metal containers are examples of components that are easily produced using this process.

Sintering from a powder has become an important way of producing metallic structures. The metals to be sintered, in the form of a fine powder, are mixed together and then placed in a hardened steel die and compressed. The pressures used depend upon the metals to be sintered but are usually between 70 and 700 MN/m^2. At these high pressures a degree of cold welding takes place between the metals. The brittle compressed mass is then heated in a furnace to a temperature at which sintering (grain growth across the cold welds) takes place.

Test your knowledge 1.37

Distinguish between gravity die casting and pressure die casting.

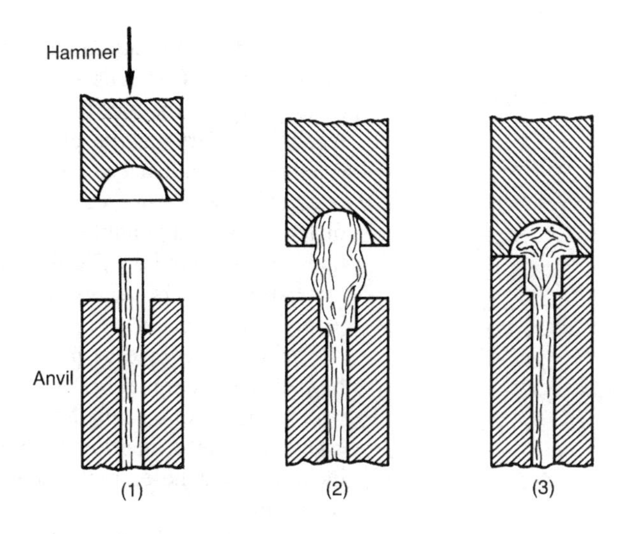

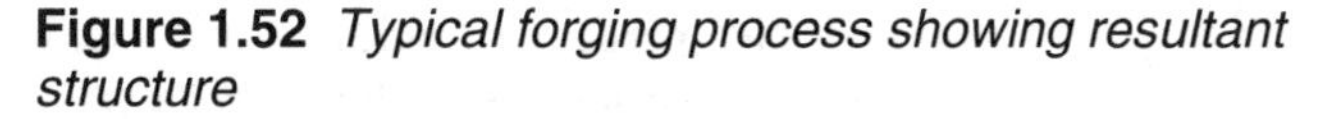
Figure 1.52 *Typical forging process showing resultant structure*

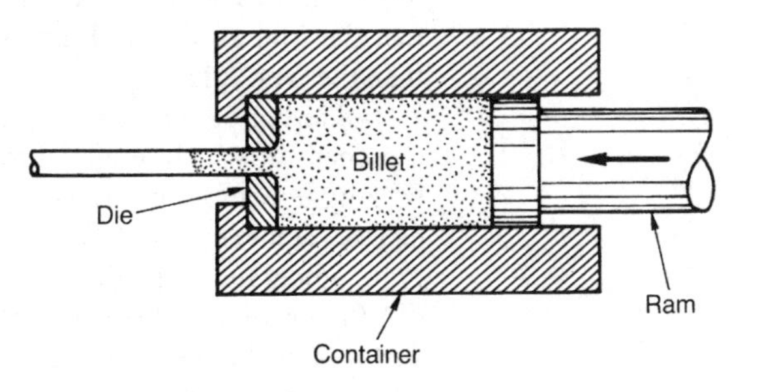

Figure 1.53 *An extrusion process*

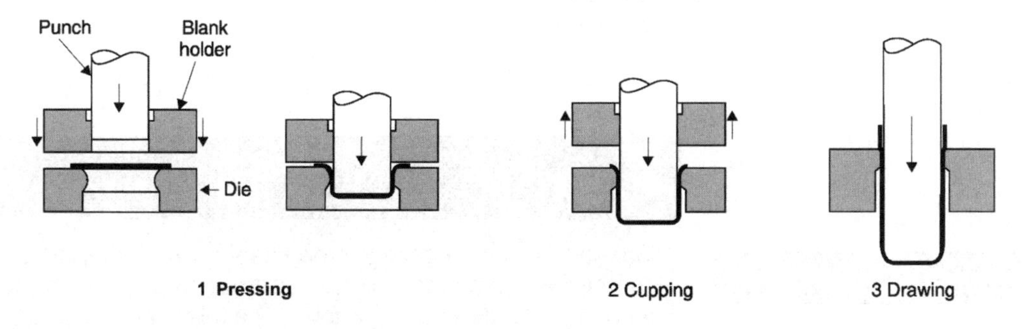

Figure 1.54 *A deep-drawing process*

Tungsten is compacted and sintered in this way and the resulting sintered rod may be drawn into a fine wire to produce tungsten filaments for light bulbs. Tungsten carbide products for machine tools are also produced in this way. Cobalt being used with the tungsten carbide to produce a tough shock resistant bonding agent between the particles.

Sintering may also be used to produce bronze bearings. Here copper, tin and graphite are used in the sintering process to produce self-lubricating bearings.

Let us now turn our attention to just one metal cutting process, that of *machining*. This is essentially a cold working process in which the cutting edge of the tool forms chips or shavings of the material being machined. Very ductile alloys do not machine well, because local fracture does not occur ahead of the cutting tool edge. Thus brittle materials are considered to have good machining properties. Ideal materials have a suitable concentration of small isolated particles in their microstructure. These particles have the effect of setting up local stress raisers, as the cutting edge approaches them and minute local fractures occur.

The graphite in cast iron and particles of a hard compound in bronze are examples of the presence of secondary stress raisers, which improve machinability. Elements may be added to alloys, other than those mentioned above, to improve machinability. These include manganese, molybdenum, zirconium, sulphur, carbon and selenium.

Let us now look briefly at one or two methods of joining materials, in particular welding and soldering. In welding, brazing and soldering fusion takes place at the surfaces of the metals being joined. Soldering and brazing are fundamentally similar processes in that the joining material always melts at a temperature that is lower than the work piece.

Soldering or *soft* soldering as it is often called can be described as a process in which temperatures below 450°C are involved, whereas brazing temperatures are generally between 600 and 900°C. Welding can be achieved by hammering the surfaces in contact together at high temperatures, so that crystal growth takes place.

When *soldering* the solder must be capable of spreading across the surfaces of the metals to be joined, the solder is often assisted in this process by use of a *flux* or *wetting agent*. Solders produced from tin–lead alloys have good flow properties and melt at temperatures (183 to 250°C) that are comfortably below the temperatures of the metals to be joined.

Activity 1.18

Visit your school or college's engineering workshop and identify ONE example of a tool, part or component that has been manufactured using each of the following processes or techniques:

(a) casting
(b) drop-forging
(c) extrusion
(d) drawing
(d) pressing
(e) machining
(f) soldering.

For each item explain why the process or technique was chosen.

Test your knowledge 1.38

State processes that are applicable to the manufacture of each of the following:

(a) a continuous length of wire
(b) a light-alloy metal box
(c) the body of a lorry cab
(d) an engineer's spanner
(e) a self-lubricating bearing
(f) the base of a pillar drill
(g) an H-section channel made from a light alloy.

In *fusion welding* processes (see page 53) either thermo-chemical sources, electric arc or some form of radiant energy can be used to melt the weld metal. In gas welding the surfaces to be joined are melted by a flame from a gas torch, the gases most commonly used being suitable mixtures of oxygen and acetylene.

In gas welding and other fusion welding processes a welding rod is used to supply the necessary metal for the weld (the weld joint being suitably prepared prior to the commencement of welding). Other fusion welding processes apart from gas welding include, electric arc where the arc is struck between a carbon electrode and the work itself.

Solid phase welding processes include smith welding (described earlier), ultrasonic welding, friction welding and diffusion welding. I will just mention the diffusion welding process. In diffusion welding the sheets to be joined are held together under light pressure in a vacuum chamber. The temperature is then raised sufficiently for diffusion to occur across the interface so that the surfaces become joined by a region of solid solution. Steel can be clad with brass in this way.

Some of the other joining processes that we met in section 1.3 involved riveting, and joining by adhesives. These methods all have their own particular advantages and disadvantages and will therefore be more appropriate for some applications than others.

Possibly the principal advantage of an adhesive bond is that the adhesive fastens to the entire bonded surface, thereby distributing the load more evenly and thus avoiding high localized stresses. Adhesives may also be used to advantage instead of riveting. Rivets add weight to structures, increase the likelihood of corrosion and can look unsightly on many domestic products. Materials with different coefficients of expansion can be joined by elastomer type adhesives that take the strain at the adhesive joint rather than the materials being joined. Adhesives can also be cured at relatively low temperatures, so preventing any unnecessary damage to the materials being joined (adherends).

Adhesive joints also have one or two disadvantages; primarily they are very much restricted in use at high temperatures. Also components cannot easily be dismantled for maintenance. Finally surface cleanliness and process control are very important, this usually requires a considerable amount of equipment.

Polymer adhesives are by far the most commonly used for engineering and industrial applications.

That concludes our study of the processing of metals and metal alloys and we shall now focus our attention on the processing of non-metals.

Non-metals

Polymer processing

An important attraction of polymer compounds is that they can be readily converted into a variety of useful shapes. Polymer processing is concerned with the technology needed to make articles from polymer compounds. Three common themes underly most of the methods. The first involves making the appropriate compound, usually in liquid form, from the raw ingredients. The second is concerned with transforming the compound into useful shapes. Finally the third theme is to ensure that once the product has been formed it retains its shape and dimensions.

So the first theme is concerned with mixing the raw ingredients to produce the correct compound. There are two basic stages in the mixing process, dispersing the additives in the polymer and achieving a uniform shapeable state. The two main processes used for good dispersion of the polymers ingredients, particularly in the rubber industry, are the two-roll mill and the intensive mixer. In the intensive mixer two rotors counter-rotate within a robust casing. There is only a small clearance between the tips of the rotors and the casing. Ingredients are fed in through an opening in the top of the machine, which may be closed during use. The fully mixed compound being removed through a port in the underside of the casing.

The two-roll mill consists of two heavy horizontal cylindrical polished rollers fitted with water-cooling channels (Figure 1.55). The rollers counter-rotate and are separated by a narrow gap. The feed is dropped between the two rollers. The sticky rubber substance is dragged down between the two rollers and adheres to them in the form of a band. This band is cut on a helix at an angle to the roll-axis and it is simply folded over.

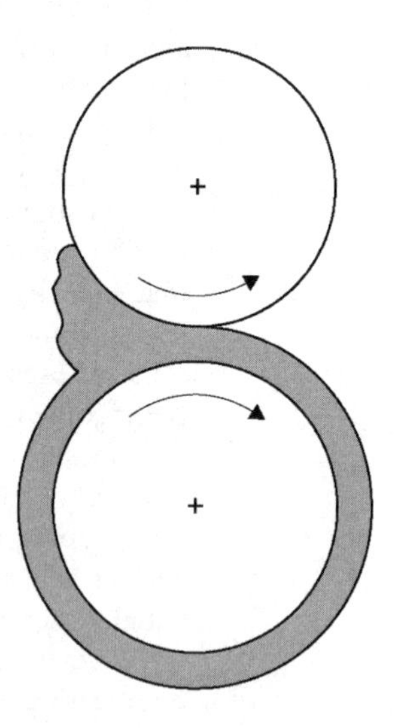

Figure 1.55 *Polymer being processed in a two-roll mill*

Because of the high viscosity of the rubber compound, the power to drive the rollers is high, but because the water cooled rollers have a large surface area in contact with the thin band of polymer, this prevents excessive temperature rise. It is therefore safe to add the vulcanizing (cross-linking) agents to the mix.

Extrusion processes are used to make continuous products of constant cross-section, usually from thermoplastics or rubber compounds. Two basic types of equipment are commonly used, these are the *calender* and the *screw extruder*.

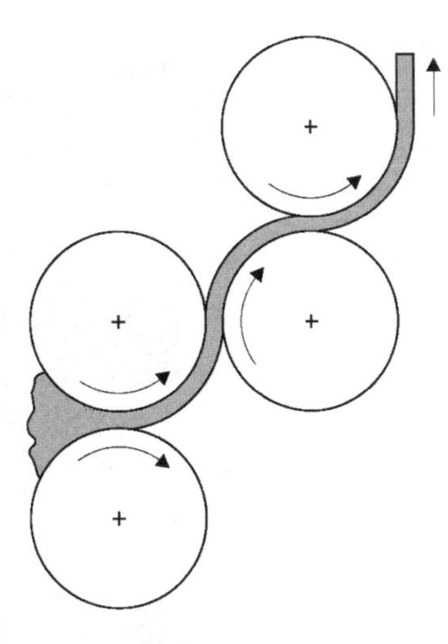

Figure 1.56 *A typical calendering process for a polymer*

The *calender* consists of four heavy rollers that are sometimes known as *bowls* (Figure 1.56). The top two bowls act just like the two-roll mill previously described. The compound is drawn through the first two rollers and adheres to them. It is thus transferred to the next pair where its cross-section is further reduced. Finally the required surface finish, either plain or embossed is supplied as the polymer mix passes through the final stage rollers. Sheet up to a few millimetres thick and a metre or so wide can be produced in this way.

The *screw extruder* consists of one or sometimes two screws that rotate inside a close-fitting barrel of constant diameter. The screw is driven by a large electric motor, being connected to the screw by a reduction gearbox. The screw has to deliver a steady stream of molten polymer to the die. The screw acts as a pump, which develops drag flow and thus heat which helps to melt or plasticize the feed. A hopper is used to supply the feed to the screw. A typical screw extruder is shown in Figure 1.57.

Compression moulding of polymers is similar to cold pressing metals. In that a thermosetting polymer compound known as the charge is fed into the jaws of a moulding die. Pressure is applied and

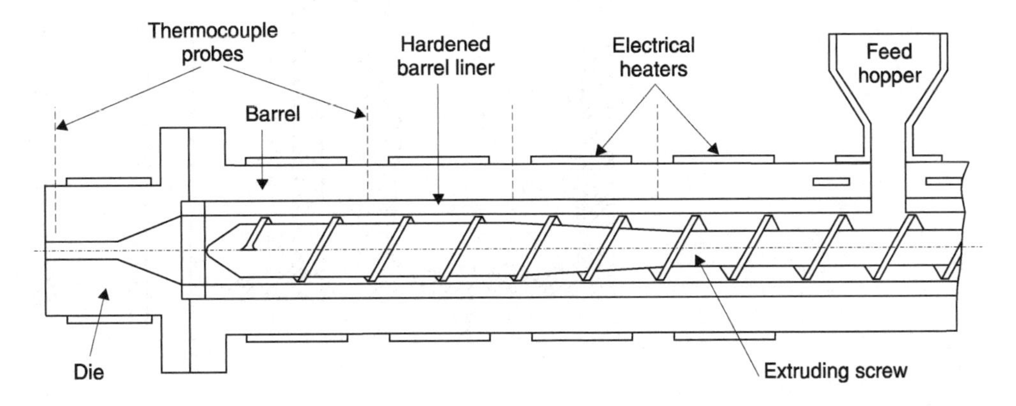

Figure 1.57 *A typical polymer screw extruder*

the polymer takes up exactly the shape of the die to form a three-dimensional product. The stages of this process are shown in Figure 1.58. The final product is ejected from the mould after completion of the process.

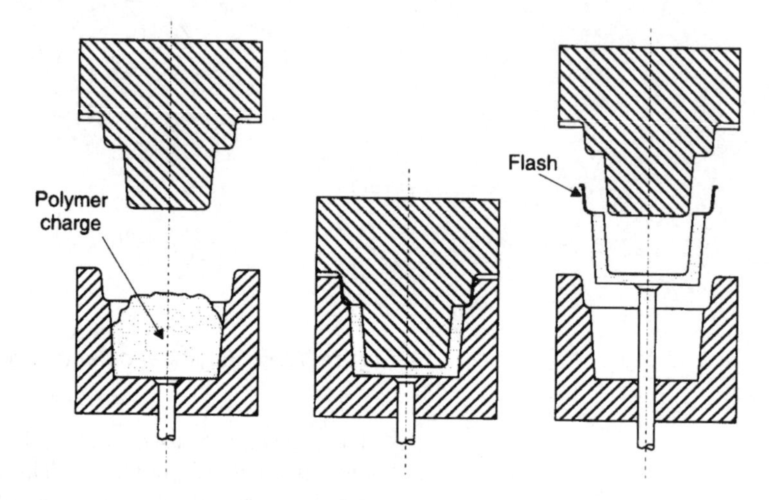

Figure 1.58 *Polymer compression moulding process*

In *extrusion blow moulding* a thick walled thermoplastic called a *parison* is extruded vertically downwards between the open faces of a cold split mould, which produces a hollow cavity. The mould is then closed and sealed and the still warm parison is inflated with compressed air so that the outside of the tube takes up the shape of the mould (Figure 1.59). Plastic bottles and other useful containers are often made in this way.

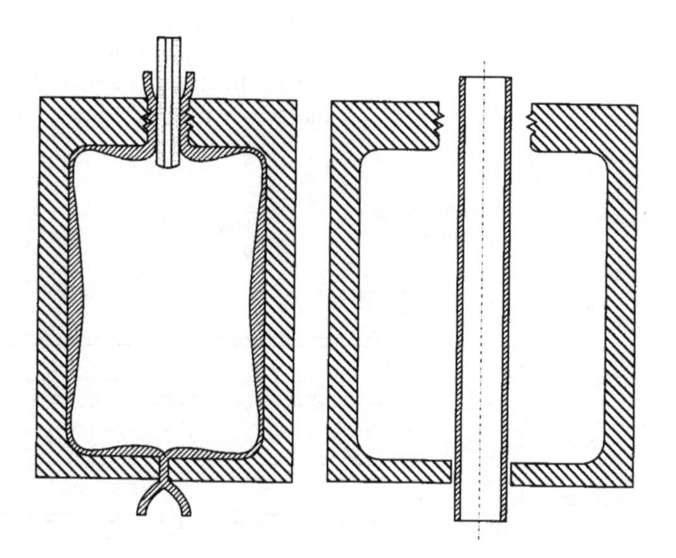

Figure 1.59 *Extrusion blow moulding*

Let us consider one final moulding process, that of *injection moulding*. In this process polymer melt is injected into an impression within a closed split mould which has the dimensions required for the finished product, see Figure 1.60.

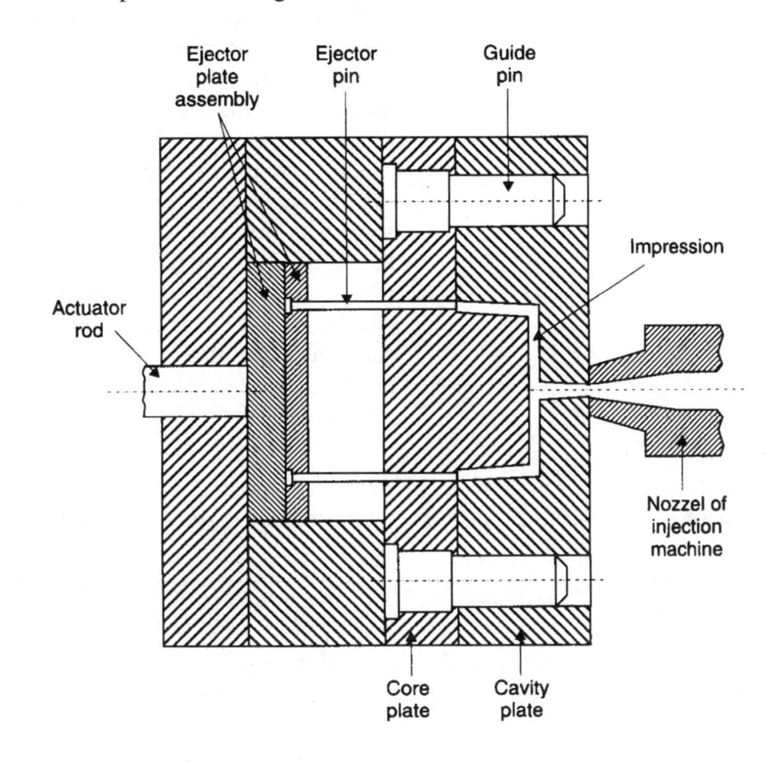

Figure 1.60 *Typical injection moulder*

Test your knowledge 1.39

Describe THREE methods or processing polymer materials.

Test your knowledge 1.40

State processes and techniques that are appropriate to the manufacture of each of the following:

(a) a flexible plastic liner designed to be used in the construction of garden ponds
(b) a food storage box
(c) a plastic plant container for use on a patio.

Thermoplastics, thermosets and elastomers can all be injection moulded. Many aspects of the construction of compression and injection moulds are similar in principle.

We could consider many other pieces of polymer processing equipment, such as hand lay-up techniques, but these are perhaps better covered in the next section, when I deal with composite materials.

Composite materials and their processing

This range of materials has been left to last because *composites* are a mixture of two or more constituents or phases, rather than a distinct class. I have already introduced you to composites when we talked initially about classes of material. However that definition was not really sufficient and three other criteria have to be satisfied in order to classify a material as a composite. First both constituents have to be present in reasonable proportions, rather than fractions of a per cent. Second, the constituent phases will have different properties, so that the properties of the resulting composite are different from its component parts. Finally, a composite is usually produced by intimately mixing and combining the constituents by various means.

So, for example, a metal alloy which has a two-phase microstructure which is produced during solidification from the melt, or by subsequent heat treatment, has not involved intimate mixing and cannot therefore be classified as a true composite material.

The continuous constituent of a composite is known as the *matrix*, the matrix is often but not always present in the greater quantity. A composite may have a ceramic, metallic or polymeric matrix. The mechanical properties of composites produced from these matrices differ completely, as Table 1.15 shows.

The second constituent within the composite is referred to as the *reinforcing phase*, because this phase strengthens or enhances in some way the properties of the matrix. The reinforcement may take the form of small particles, chopped strand, continuous strand or woven matting. Dependent on how it is mixed, will generally dictate whether or not the composite has uni-directional or multi-directional strength and stiffness characteristics. Figure 1.61 illustrates this point.

Table 1.15 *Some properties of ceramics, metals and polymers*

Material	*Density* (Mg/m^3)	*Young's Modulus* (GPa)	*Strength* (MPa)	*Ductility* (%)	*Toughness* K_{IC} ($MPa\ m^{1/2}$)	*Specific modulus* (GPa)/(Mg/m^3)	*Specific strength* (MPa)/(Mg/m^3)
Ceramics							
Alumina	3.87	382	332	0	4.9	99	86
Magnesia	3.60	207	230	0	1.2	58	64
Silicon nitride		166	210	0	4.0		
Zirconia	5.92	170	900	0	8.6	29	152
β-Sialon	3.25	300	945	0	7.7	92	291
Metals							
Aluminium	2.70	69	77	47	~30	26	29
Aluminium alloy	2.83	72	325	18	~25–30	25	115
Brass	8.50	100	550	70	–	12	65
Nickel alloy	8.18	204	1200	26	~50–80	25	147
Steel mild	7.86	210	460	35	~50	27	59
Titanium alloy	4.56	112	792	20	~55–90	24	174
Polymers							
Epoxy	1.12	4	50	4	1.5	4	36
Nylon 6.6	1.14	2	70	60	3–4	18	61
Polyetheretherketone	1.30	4	70		1.0	3	54
Polymethylmethacrylate	1.19	3	50	3	1.5	3	42
Polystyrene	1.05	3	50	2	1.0	3	48
Polyvinylchloride rigid	1.70	3	60	15	4.0	2	35

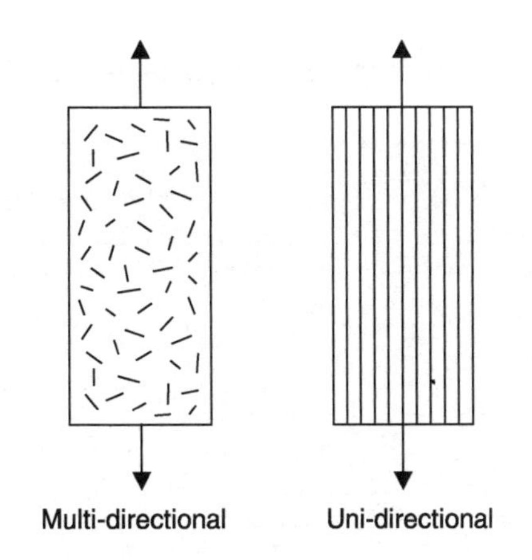
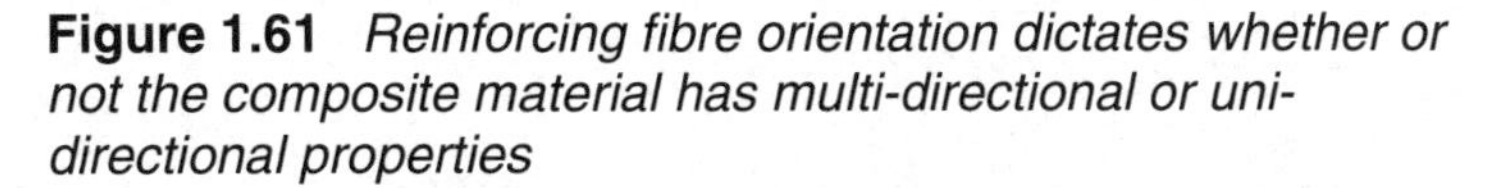

Figure 1.61 *Reinforcing fibre orientation dictates whether or not the composite material has multi-directional or uni-directional properties*

In this text I will be concentrating on just one major group of composites *polymer matrix composites* (PMCs). This group has a wide range of engineering applications and by studying them, the general principles underlying their properties and processing can be extended to other groups of composite such as those with a ceramic or metal matrix. PMCs use all three classes of polymer; thermoplastics, thermosets and elastomers, although thermosetting polymers dominate the market for this type of composite.

Carbon fibre reinforced plastic (CFRP) is a very well known polymer matrix composite, where the reinforcement is carbon fibre. Other reinforcing materials used with PMCs include glass, polyethylene, boron, and Kevlar (a type of aramatic polyamide). A range of reinforcing fibres and matrices, with their mechanical properties is given in Table 1.16.

Let us now concentrate on one or two processing methods for PMCs. In *hand lay-up* the reinforcement is put down to line a mould previously treated with some form of release agent to prevent sticking. The reinforcement can be in many forms, such as chopped strand mat, woven mat, etc. The liquid thermosetting resin is mixed with a curing (setting) agent and applied with a brush or roller, ensuring that the resin is thoroughly worked into the reinforcement. The most commonly used resins are polyesters and curing normally takes place at room temperature. Hand lay-up is labour intensive but requires little capital equipment. For these reasons it is often used to produce one-off specialist articles or large components such as swimming pools and boat hulls.

Another manual method of production is known as *spray-up*. In this method a spray gun charged with the matrix resin, chopped fibres, and curing agent, is used. This method is quick, cheap and efficient, although at suitable intervals the sprayed composite has to be rolled to release trapped air.

Let us consider just two moulding methods for the production of composite components, *die-moulding* and *bag-moulding*. Die-moulding is widely used for long production runs for components ranging in size from small domestic items to large commercial vehicle panels. The material to be shaped is pressed between heated

Table 1.16 *Some properties of typical reinforcing fibres and matrix materials*

Material	*Relative density*	*Young's modulus* (GPa)	*Tensile strength* (GPa)
Reinforcing fibres			
E Glass	2.55	72	1.5–3.0
S Glass	2.5	87	3.5
Carbon-pitch	2.0	380	3.0-3.6
Carbon-pan	1.8	220–240	2.3–3.6
SiC whisker	3.2	480	7.0
Kevlar	1.47	130–180	2.6–3.5
Nomex	1.4	17.5	0.7
Polyamide (typical)	1.4	5.0	0.9
Matrix materials			
Steel	7.8	210	0.34–2.1
Aluminium alloys	2.7	70	0.14–0.62
Epoxy resin	1.2	2–3.5	0.05–0.09
Polyester resin	1.4	2–3.0	0.04–0.08

matched dies. The pressure required may be as high as 50 MPa. The feed material flows into the contours of the mould and when the temperature is high enough it rapidly cures. Good dimensional accuracy and detail are possible with this method, depending on the quality of the die used. The feed material, which already contains all necessary ingredients and the curing agent, may be in the form of a sheet or simply fed in as a dough.

In the *bag-moulding* process only one half of the mould is used to shape the component in *vacuum-bagging* the laid-up material, which consists of heated pre-impregnated reinforcement, is sealed by a bag over the component. When vacuum pressure is applied to the bag, sufficient pressure is applied to the work piece to cure it (Figure 1.62).

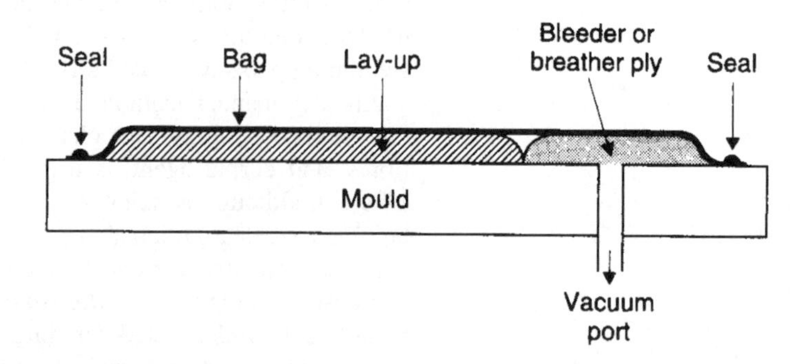

Figure 1.62 *Typical lay-up procedure using the vacuum-bagging process*

Autoclave moulding is a modification of vacuum forming where pressure in excess of atmospheric is used to produce high density products for critical applications, such as racing car and aircraft components. An *autoclave* is a re-circulating oven, that is pressurized by an inert gas, often nitrogen. The sealed bag is still used to stabilize the component against the mould (Figure 1.63).

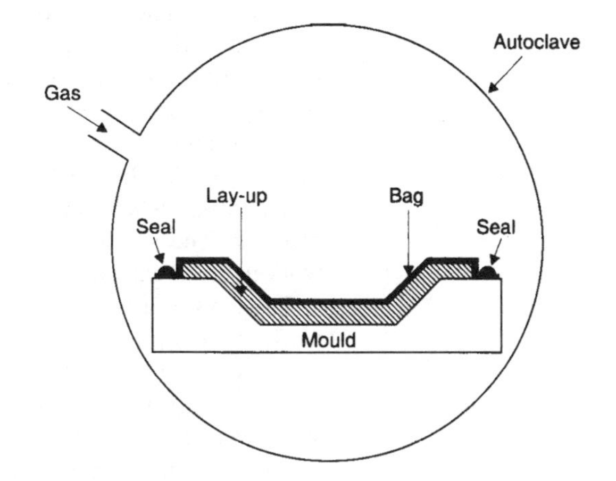

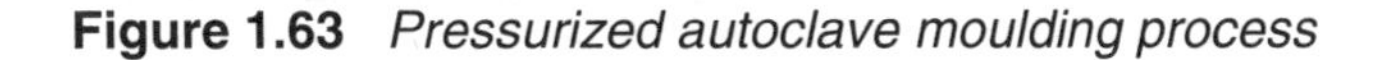

Figure 1.63 *Pressurized autoclave moulding process*

Activity 1.19

Use several reference sources to obtain information on Nylon 6.6 and add this as a new entry to the matSdata database (see Activities 1.2 and 1.6).

Activity 1.20

Investigate the use of carbon fibre composite materials in the manufacture of the Mercedes-Benz SLR McLaren high performance car. Use the Internet to search for information and present your findings in the form of a written report which describes the materials and techniques used in the manufacture of the SLR. List any reference sources used.

Hint: Start by visiting the following websites:

- www.mercedes-benz.com
- www.mclarencars.com
- http://www.concept-car.co.uk/news/technology/cardesignnews11.php

Test your knowledge 1.41

Describe a typical lay-up procedure for manufacturing a component from carbon fibre composite material.

Material removal

Before we look at methods used for removing material it is important to briefly introduce a method for work holding. In mist cases this mundane but nevertheless essential task can be performed using a fitter's bench and vice.

A fitter's bench should be substantial and rigid. This is essential if accurate work is to be performed on it. It should be positioned so that it is well lit by both natural and artificial light without glare or shadows. It should be equipped with a fitter's vice. A plain screw vice is shown in Figure 1.64(a) and a quick action vice is shown in Figure 1.64(b). In the latter type of vice, the jaws can be quickly pulled apart when the lever at the side of the screw handle is released. The screw is used for closing the jaws and clamping the work in the usual way.

The jaws of a fitter's vice are serrated and hardened to prevent the work from slipping. This also marks the surfaces of the work. For fine work with finished surfaces, the serrated jaws should be replaced with hardened and ground smooth jaws. Alternatively vice shoes can be used. These are faced with a fibre compound and can be slipped over the serrated jaws when required. A pair of typical vice shoes are shown in Figure 1.64(c).

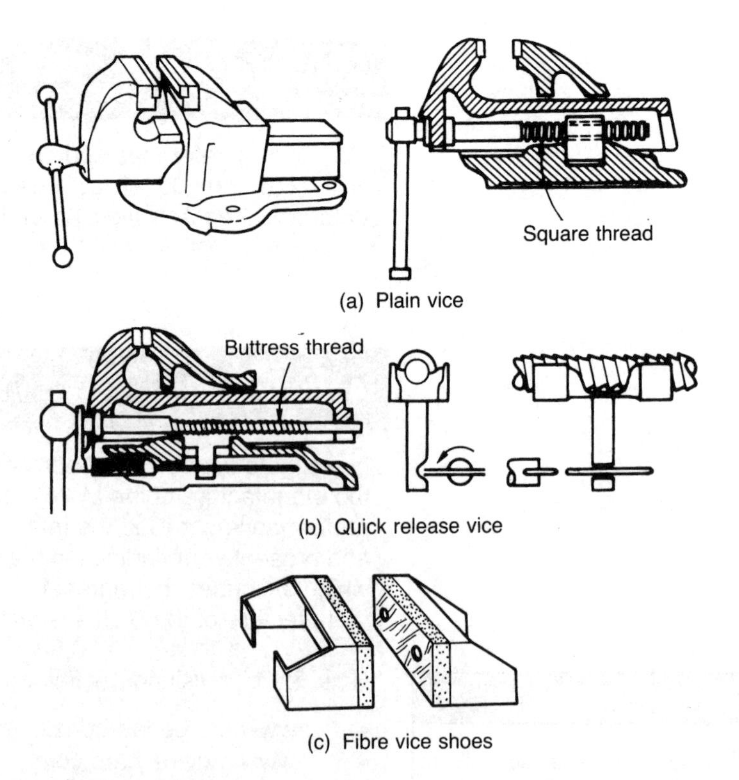

Figure 1.64 *A fitter's vice*

Cutting tools

Before we can discuss the cutting tools we use for bench fitting we need to look at the way metal is cut. Here are the basic facts.

Wedge angle

If you look at a hacksaw blade, as shown in Figure 1.65(a), you can see that the teeth are wedge shaped. Figure 1.65(b) shows how the wedge angle increases as the material gets harder. This strengthens the cutting edge and increases the life of the tool. At the same time it reduces its ability to cut. Try cutting a slice of bread with a cold chisel!

Clearance angle

If you look at the hacksaw blade in Figure 1.65(a), you can see that there is a clearance angle behind the cutting edge of the tooth. This is to enable the tooth to cut into the work.

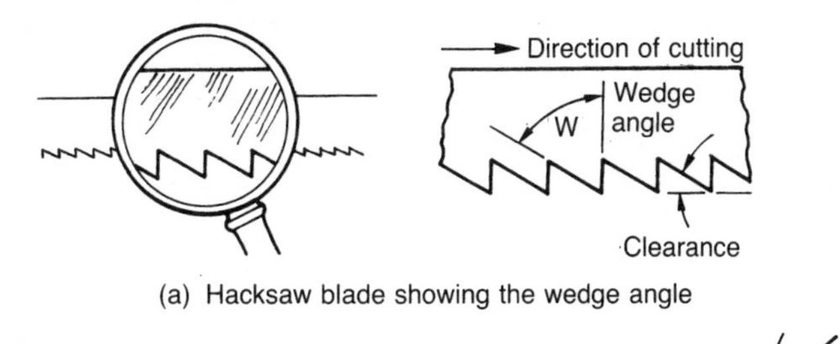

(a) Hacksaw blade showing the wedge angle

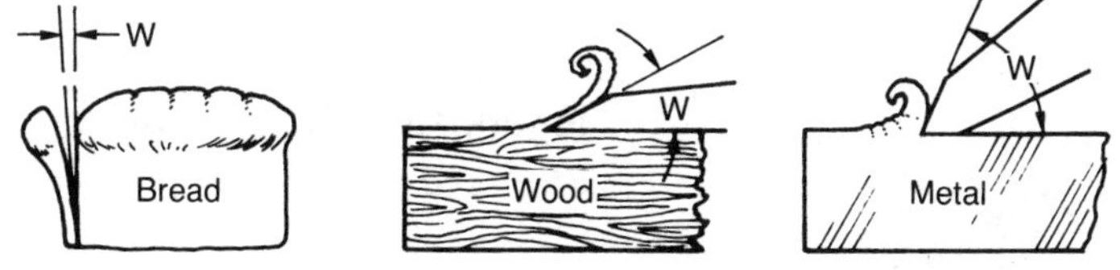

(b) Wedge angles for various materials

Figure 1.65 *Hacksaw blade and wedge angles*

Rake angle

This angle controls the cutting action of the tool. It is shown in Figure 1.66. I hope you can see that the wedge angle, clearance angle and rake angle always add up to 90°. This is true even when the rake angle is zero or negative as shown in Figure 1.66(b).

The greater the rake angle the more easily the tool will cut. Unfortunately, the greater the rake angle, the smaller the wedge angle will be and the weaker the tool will be. Therefore the wedge and rake angles have to be a compromise between ease of cutting and tool strength and life. The clearance angle remains constant at between 5° and 7°. Typical rake angles for high speed cutting tools are shown in Table 1.17.

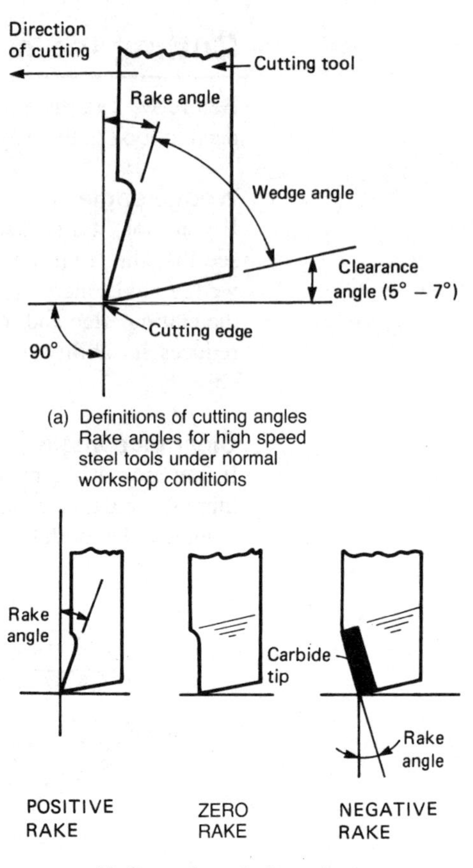

(a) Definitions of cutting angles
Rake angles for high speed steel tools under normal workshop conditions

(b) Comparison of rake angles

Figure 1.66 *Rake angle*

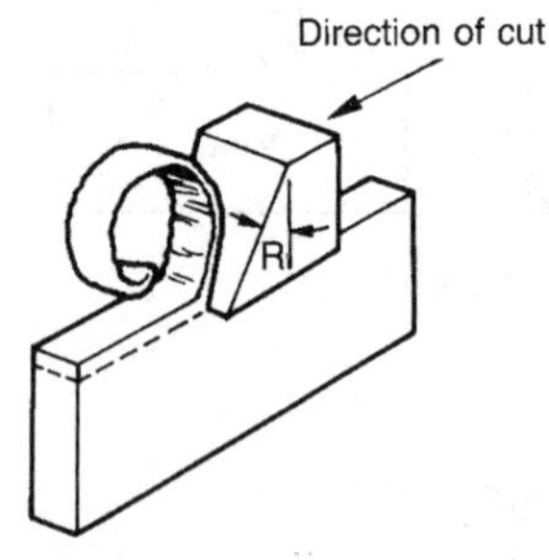

(a) Orthogonal cutting

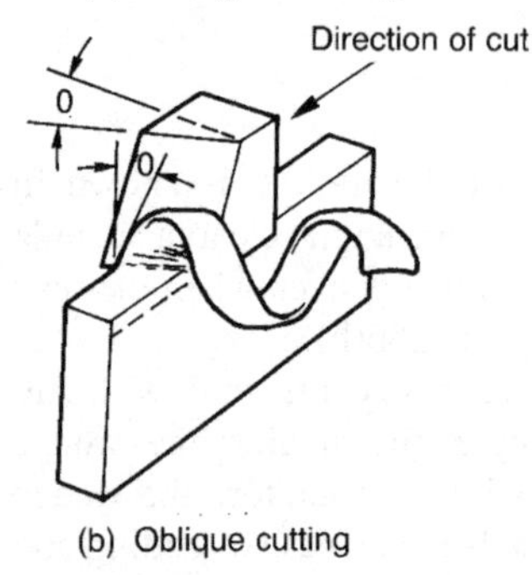

(b) Oblique cutting

Figure 1.67 *Orthogonal and oblique cutting*

Orthogonal and oblique cutting

Figure 1.67(a) is a pictorial representation of the single point cutting tool shown in Figure 1.65. Notice how the cutting edge is at right-angles to the direction in which the tool is travelling along the work. This is called orthogonal cutting. Now look at Figure 1.67(b). Notice how the cutting edge is inclined at an angle to the direction of cut. This is called oblique cutting. Oblique cutting results in a better finish than orthogonal cutting, mainly because the chip is thinner for a given rate of metal removal. This reduced thickness and the geometry of the tool allows the chip to coil up easily in a spiral.

Apart from threading operations, it is very rare to use a coolant or lubricant when using hand tools. However the conditions are very different when using machine tools. Large amounts of metal are removed quickly, considerable energy is used to do this, and this energy is largely converted into heat at the cutting zone. The rapid temperature rise of the work and the cutting tool can lead to inaccuracy and short tool life. A coolant is required to prevent this. The chip flowing over the rake face of the tool results in wear. A lubricant is required to prevent this. Usually coolants are poor lubricants, and lubricants are poor coolants.

Test your knowledge 1.42

Explain why metal cutting tools need a clearance angle.

Test your knowledge 1.43

Explain why metal cutting tools have a larger wedge angle than wood cutting tools.

Test your knowledge 1.44

Calculate the wedge angle for a tool if the clearance angle is 5° and the rake angle is 17°.

For general machining an emulsion of cutting oil (which also contains an emulsifier) and water is used. This has a milky-white appearance and is commonly known as 'suds'. On no account try to use a mineral lubricating oil. This cannot stand up to the temperatures and pressures found in the cutting zone. It is completely useless as a cutting lubricant or as a coolant. It gives off clouds of noxious fumes and it is a fire risk.

Table 1.17 *Rake angle for high speed tools under normal conditions*

Material	*Rake angle*
Aluminium alloy	30°
Brass (ductile)	14°
Brass (free-cutting)	0°
Cast iron	0°
Copper	20°
Phosphor bronze	8°
Mild steel	25°
Medium carbon steel	15°

Files and filing

Files are the most widely used and important tools for the fitter. The main parts of a file are named in Figure 1.68(a). Files are forged to shape from 1.2% plain carbon steel. After forging, the teeth are machine cut by a chisel shaped tool, as shown in Figure 1.68(b). The teeth of a single-cut are wedge shaped with the rake and clearance angles essential for metal cutting. Most files used in general engineering are double-cut. That is they have two rows of cuts at an angle to each other, as shown in Figure 1.68(c).
Files are classified by the following features:

- length
- kind of cut
- grade of cut (roughness)
- profile
- cross-sectional shape or most common use.

The grades of cut are: rough, bastard, second, smooth and dead smooth. These cuts vary with the length of a file. For example a short, second cut file will be smoother than a longer, smooth file. For further information on file cuts, see Table 1.18. The profiles and cross-sectional shapes of some typical files are shown in Figure 1.69. Figure 1.70 shows how a file should be held and used. To file flat is

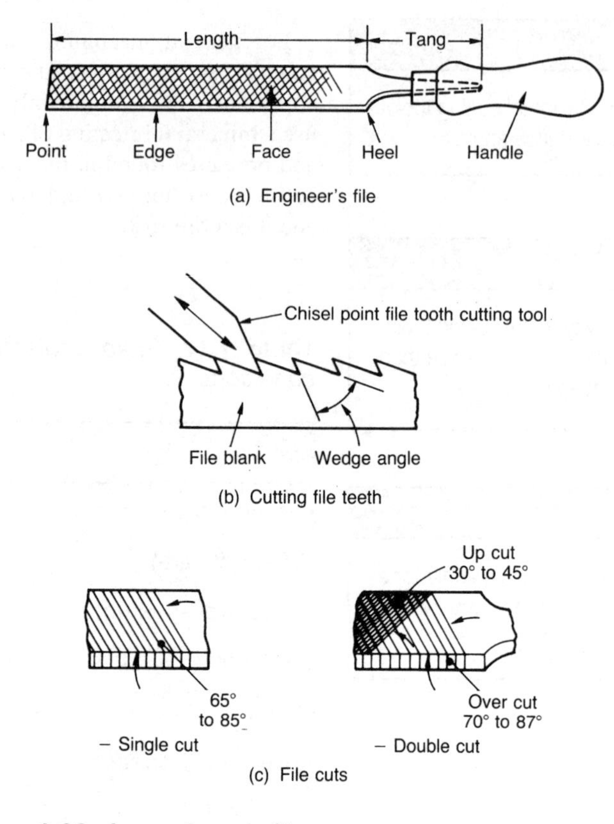

Figure 1.68 *An engineer's file*

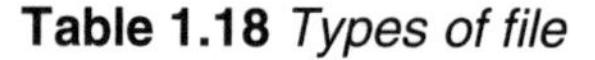
Table 1.18 *Types of file*

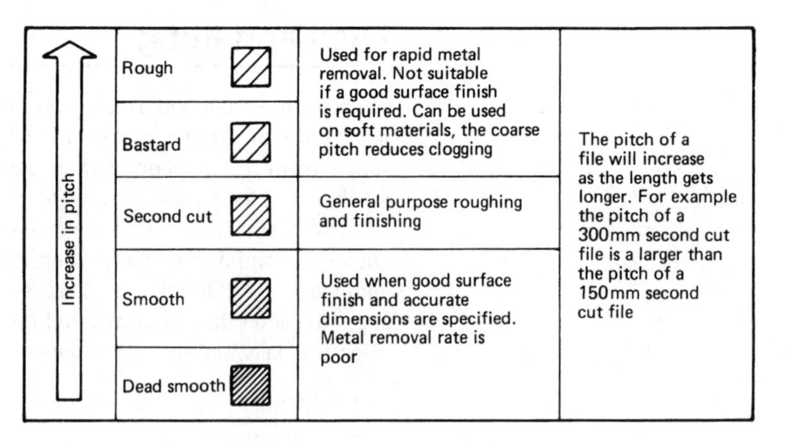

Increase in pitch	Type	Use	
↑	Rough	Used for rapid metal removal. Not suitable if a good surface finish is required. Can be used on soft materials, the coarse pitch reduces clogging	The pitch of a file will increase as the length gets longer. For example the pitch of a 300 mm second cut file is a larger than the pitch of a 150 mm second cut file
	Bastard		
	Second cut	General purpose roughing and finishing	
	Smooth	Used when good surface finish and accurate dimensions are specified. Metal removal rate is poor	
	Dead smooth		

very difficult and the skill only comes with years of continual practice. Cross-filing is used for rapid material removal. Draw-filing is only a finishing operation to improve the surface finish. It removes less metal per stroke than cross filing and can produce a hollow surface, unless care is taken.

The spaces between the teeth of a file tend to become clogged with bits of metal. This happens mostly when filing soft metals. It is called 'pinning'. The clogged teeth tend to leave heavy score marks

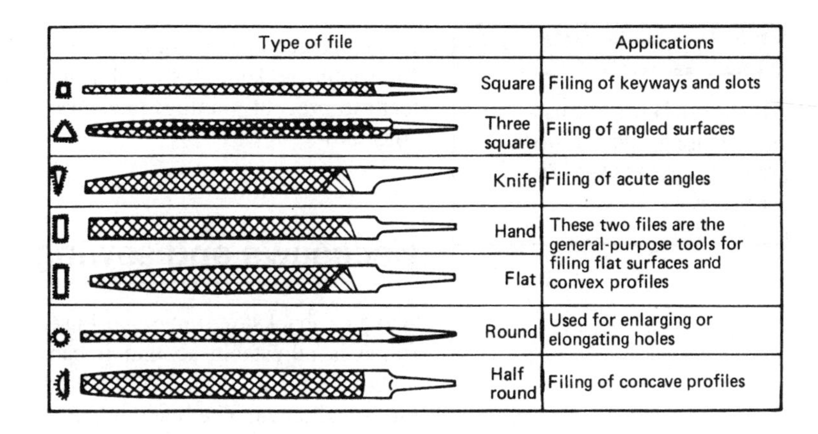

Type of file	Applications
Square	Filing of keyways and slots
Three square	Filing of angled surfaces
Knife	Filing of acute angles
Hand	These two files are the general-purpose tools for filing flat surfaces and convex profiles
Flat	
Round	Used for enlarging or elongating holes
Half round	Filing of concave profiles

Figure 1.69 *Types of file and applications*

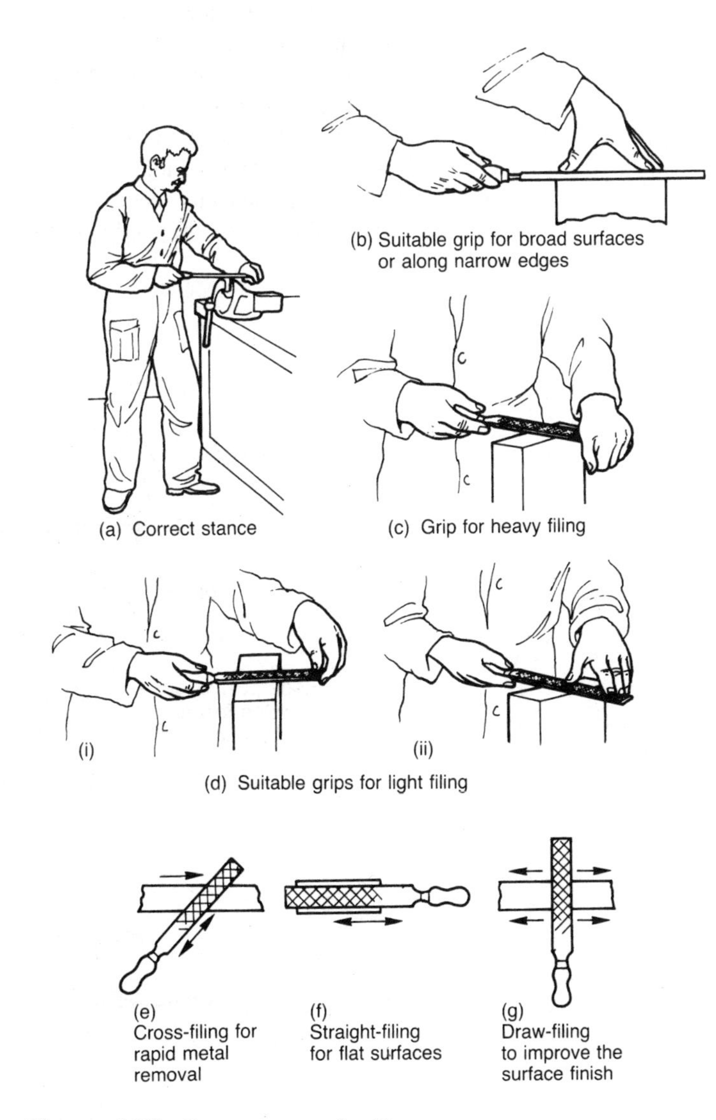

Figure 1.70 *Correct use of a file*

Test your knowledge 1.45

Distinguish between cross-filing, straight-filing, and draw-filing. Illustrate your answer with sketches.

in the surface of the work. These marks are difficult to remove. The file should be kept clean and a little chalk should be rubbed into the teeth to prevent pinning. Files are cleaned with a file brush called a 'file card'.

Hacksaws and sawing

A typical hacksaw frame and blade is shown in Figure 1.71(a). The frame is adjustable so that it can be used with blades of various lengths. It is also designed to hold the blade in tension when the wing nut is tightened. The blade is put into the frame so that the teeth cut on the forward stroke. Figure 1.71(b) shows how a hacksaw should be held when being used.

There are a variety of blade types available:

- High speed steel 'all hard' blades are the most rigid and give the most accurate cut. However they are brittle and easily broken when used by an inexperienced person.
- High speed steel 'soft back' blades have a good life and, being more flexible, are less easily broken.
- Carbon steel flexible blades are satisfactory for occasional use on soft non-ferrous metals. They are cheap and not easily broken. Unfortunately they only have a limited life when cutting steels.

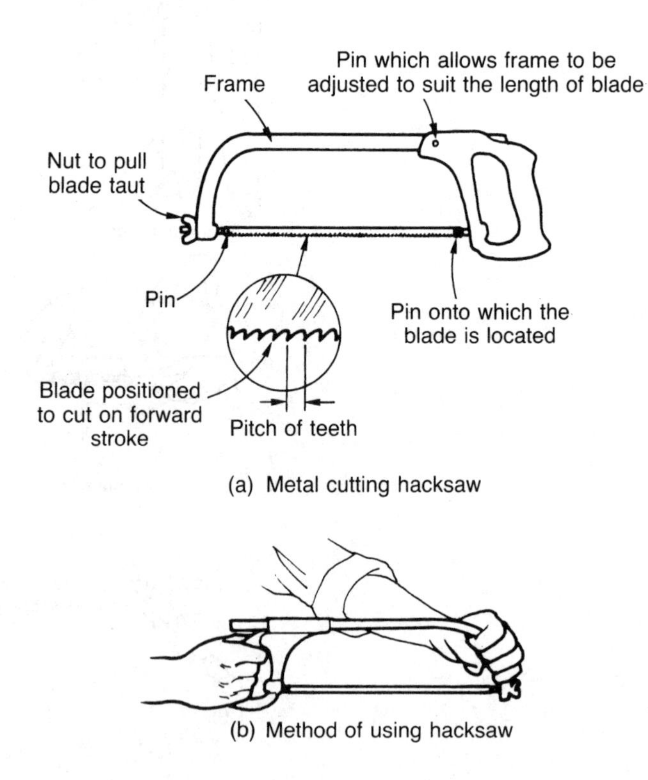

Figure 1.71 *A typical hacksaw frame and blade*

To prevent the blade from jamming in the slot it makes as it cuts, all saw blades are given a 'set'. This is shown in Figure 1.72. Coarse pitch hacksaw blades and power-saw blades have the individual teeth set to the left and to the right with either the intermediate teeth or every third tooth left straight to clear the slot. This is shown in Figure 1.72(a). This is not possible with fine pitch blades, and the blade as a whole is given a 'wave' set as shown in Figure 1.72(b). The effect of set on the cut being made is shown in Figure 1.72(c). If you have to change a blade part way through a cut, never continue in the old slot. Because the set of the old blade will have worn, the new blade will jam in the old cut and break. Always start a new cut to the side of the failed cut.

The sizes of hacksaw blades are now given in metric sizes. The length (between the fixing hole centres), the width and the thickness. However, the pitch of the teeth is still given as so many teeth per inch.

The fewer teeth per inch the coarser will be the cut, the more quickly will the metal be removed, and the greater will be the set so that there is less chance of the blade jamming. However, there should always be at least three teeth in contact with the work at any one time. Therefore, the thinner the metal being cut the finer the pitch of the blade that should be used. Some typical examples are given in Table 1.19.

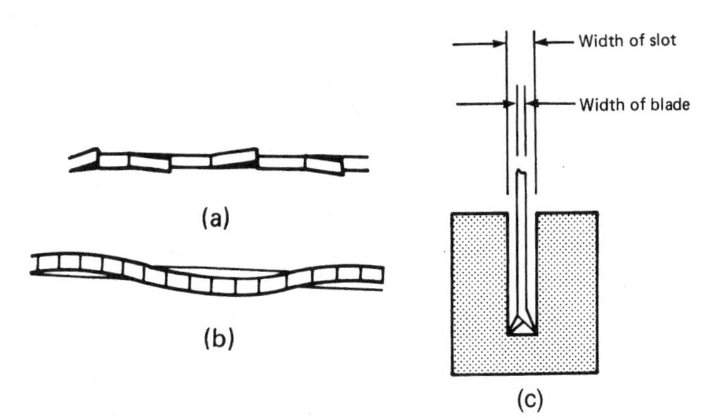

Figure 1.72 *The 'set' of a hacksaw blade*

Screw thread cutting

Internal screw threads are cut with taps. A set of straight fluted hand taps are shown in Figure 1.73. The difference between them is the length of the lead. The taper tap should be used first to start the thread. Great care must be taken to ensure that the tap is upright in the hole and it should be checked with a try square. The second tap is used to increase the length of thread and can be used for finishing if the tap passes through the work. The third tap is used for 'bottoming' in blind holes.

The hole to be threaded is called a 'tapping size' hole and it is the same size or only very slightly larger than the core diameter of the thread. Drill diameters for drilling tapping size holes for different screw threads can be found in sets of workshop tables. For example, the tapping size drill for an M10 × 1.5 thread is 8.5 mm diameter.

Table 1.19 *Hacksaw blade sizes and applications*

Teeth per inch	Material to be cut	Blade applications
32	Up to 3 mm	Thin sheets and tubes Hard and soft materials (thin sections)
24	3 mm to 6 mm	Thicker sheets and tubes Hard and soft materials (thicker sections)
18	6 mm to 12 mm	Heavier sections such as mild steel, cast iron, aluminium, brass, copper, bronze
14	Greater than 12 mm	Soft materials (such as aluminium, brass, copper, bronze) with thicker sections

A tap wrench is used to rotate the taps. There are a variety of different styles available depending upon the size of the taps. An example of a suitable wrench for small taps is shown in Figure 1.74. Taps are very fragile and are easily broken, particularly in the small sizes. Once a tap has been broken into a hole, it is virtually impossible to get it out without damaging or destroying the workpiece.

Taps are relatively expensive and should be looked after carefully. High speed steel ground thread taps are the most expensive. However, they cut very accurate threads and, with careful use, last a long time. Carbon steel cut thread taps are less accurate and less expensive and have a reasonable life when cutting the softer non-ferrous metals. Whichever sort of taps are used, they should always be well lubricated. Traditionally tallow was used, but nowadays proprietary screw-cutting lubricants are available that are more effective.

External threads are cut using split button dies in a die holder, as shown in Figure 1.75. One face of the die is always marked up with details of the thread and the maker's logo. This should be visible when the die is in the die holder. Then the lead is on the correct side for starting the cut. Screw A is used to spread the die for the first cut. The screws marked B are used to close the die until it gives the correct finishing cut. This is judged by using a standard nut or a screw thread gauge. The nut or gauge should run up the thread without binding or without undue looseness.

Again, the die must be started square with the workpiece or a 'drunken' thread will result. Also, a thread cutting lubricant should be used. Like thread cutting taps, dies are available in carbon steel cut thread and high speed steel ground thread types. For both taps and dies, each set only cuts one size and pitch of thread and one thread form.

Test your knowledge 1.46

Give TWO reasons why a tap must be started square in its hole.

Test your knowledge 1.47

Explain how a button die is adjusted to cut the required diameter in a thread.

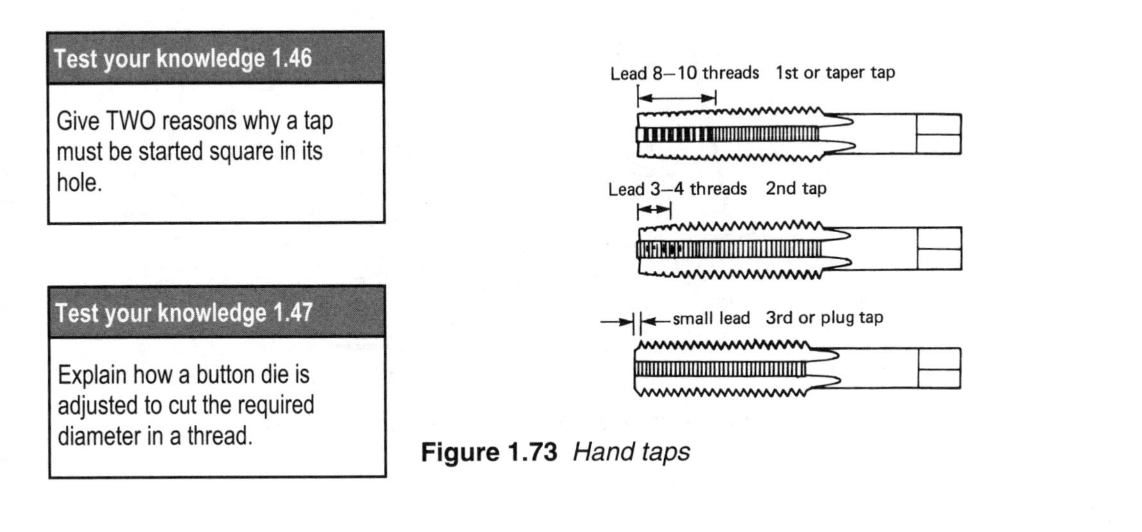

Figure 1.73 *Hand taps*

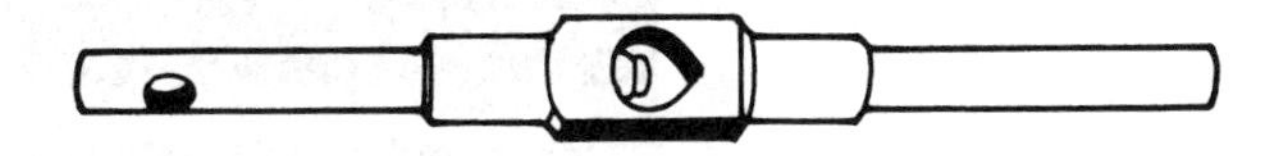

Figure 1.74 *A tap wrench*

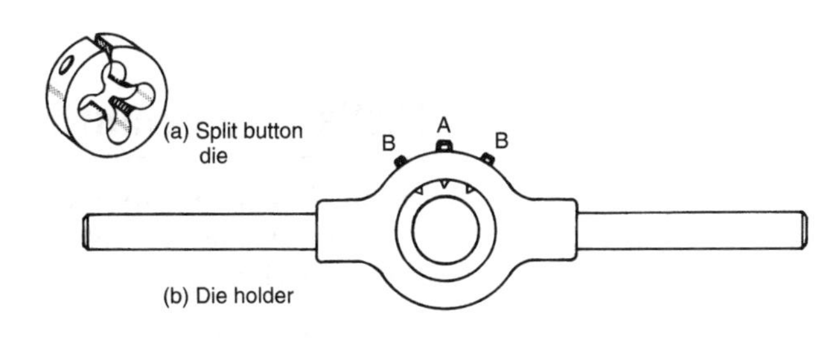

Figure 1.75 *A die holder*

Spanners and keys

In addition to cutting tools a fitter should also have a selection of spanners and keys available for dismantling and assembly purposes. Figure 1.76 shows a selection of spanners and keys. These are carefully proportioned so that a person of average strength will be able to tighten a screwed fastening correctly.

Use of a piece of tubing to extend a spanner or key is very bad practice. It strains the jaws of the spanner so that it becomes loose and may slip. It may even crack the jaws of the spanner so that they break. In both cases this can lead to nasty injuries to your hands and even a serious fall if you are working on a ladder. Also it over stresses the fastening which will be weakened or even broken. Always check a spanner for damage and correct fit before using it. A torque spanner should be used to tighten important fastenings.

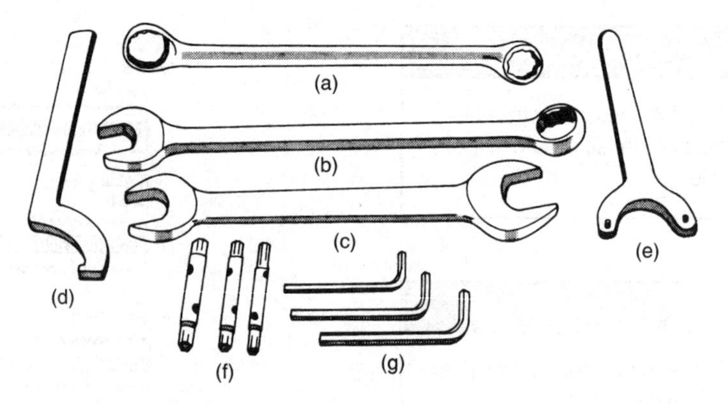

Figure 1.76 *A selection of spanners and keys*

Activity 1.21

Use a tool supplier's catalogue to identify each of the spanners and keys shown in Figure 1.76. State the typical application for each of these tools.

Drills and drilling

Drilling is a process for producing holes. The holes may be cut from the solid or existing holes may be enlarged. The purpose of the drilling machine is to:

- Rotate the drill at a suitable speed for the material being cut and the diameter of the drill.
- Feed the drill into the workpiece.
- Support the workpiece being drilled; usually at right-angles to the axis of the drill. On some machines the table may be tilted to allow holes to be drilled at a pre-set angle.

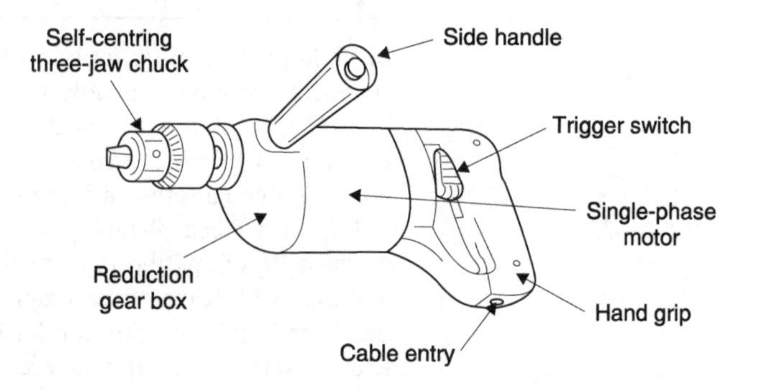

Figure 1.77 *An electric power drill*

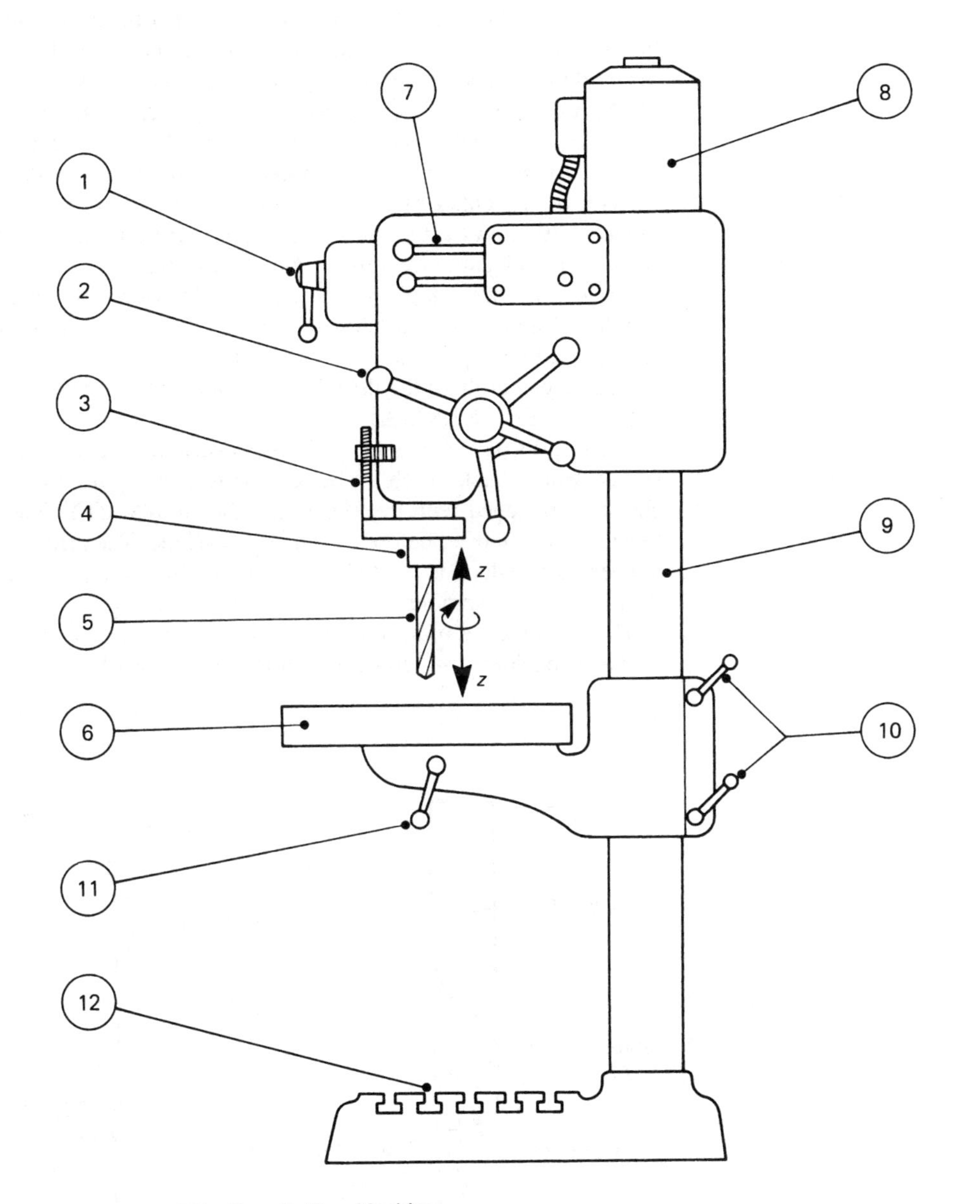

Parts of the Pillar Type Drilling Machine

1 Stop/start switch (electrics).
2 Hand or automatic feed lever.
3 Drill depth stop.
4 Spindle.
5 Drill.
6 Table.
7 Speed change levers.
8 Motor.
9 Pillar.
10 Vertical table lock.
11 Table lock.
12 Base.

Figure 1.78 *A pillar drill*

Drilling machines come in a variety of types and sizes. Figure 1.77 shows a hand-held, electrically driven, power drill. It depends upon the skill of the operator to ensure that the drill cuts at right-angles to the workpiece. The feed force is also limited to the muscular strength of the user. Figure 1.78 shows a more powerful, floor mounted machine. The spindle rotates the drill. It can also move up and down in order to feed the drill into the workpiece and withdraw the drill at the end of the cut. Holes are generally produced with twist drills. Figure 1.79 shows a typical straight shank drill and a typical taper shank drill and names their more important features.

Large drills have taper shanks and are inserted directly into the spindle of the machine, as shown in Figure 1.80(a). They are located and driven by a taper. The tang of the drill is for extraction purposes only. It does not drive the drill. The use of a drift to remove the drill is shown in Figure 1.80(b).

Small drills have straight (parallel) shanks and are usually held in a self-centring chuck. Such a chuck is shown in Figure 1.80(c). The chuck is tightened with the chuck key shown. SAFETY: The chuck key must be removed before starting the machine. The drill chuck has a taper shank which is located in, and driven by, the taper bore of the drilling machine spindle.

The cutting edge of a twist drill is wedge-shaped, like all the tools we have considered so far. This is shown in Figure 1.81.

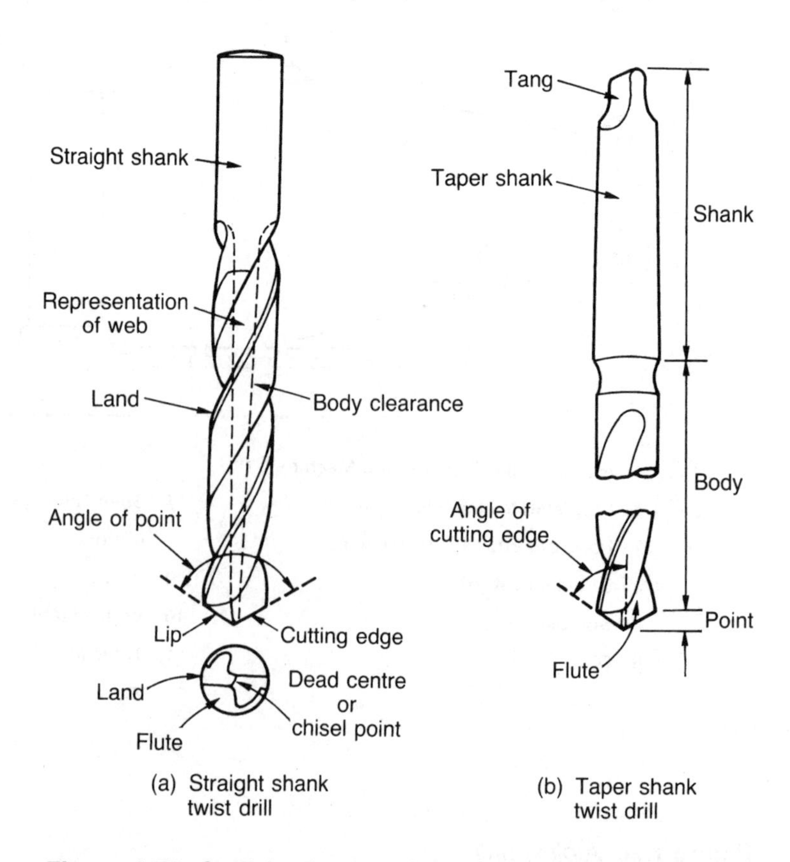

Figure 1.79 *Straight shank and taper shank twist drills*

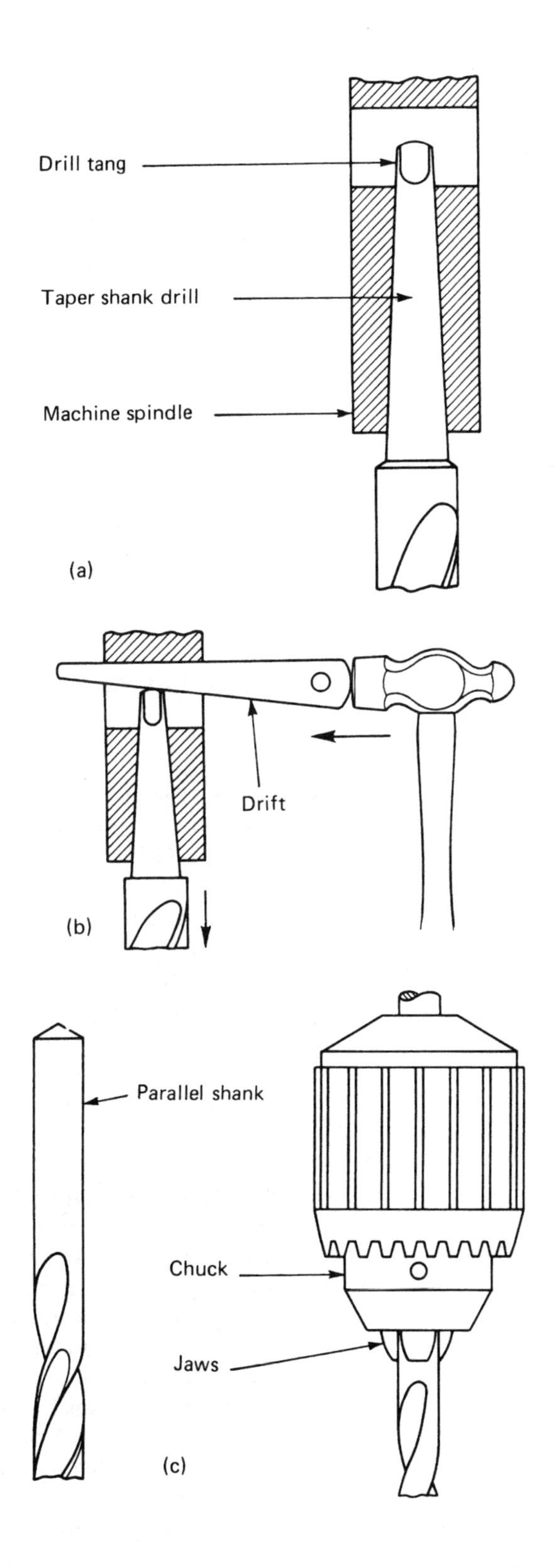

Figure 1.80 *Methods of holding a twist drill*

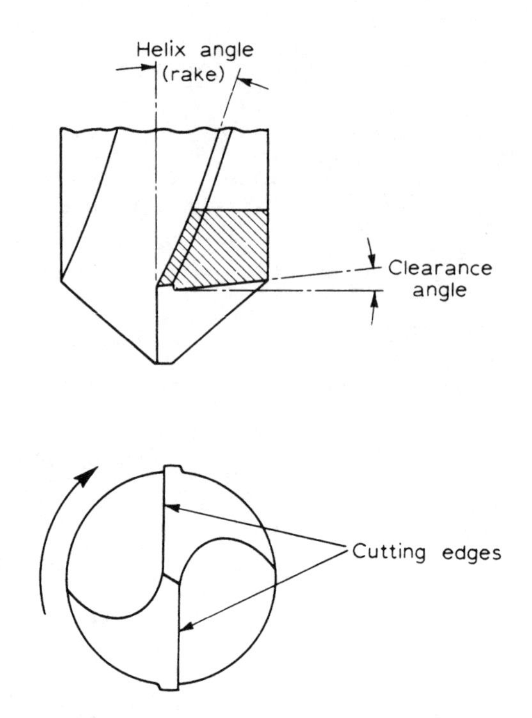

Figure 1.81 *Cutting edges of a twist drill*

When regrinding a drill it is essential that the point angles are correct. The angles for general purpose drilling are shown in Figure 1.82(a). After grinding, the angles and lip lengths must be checked as shown in Figure 1.82(b). The point must be symmetrical. The effects of incorrect grinding are shown in Figure 1.82(c).

If the lip lengths are unequal, an oversize hole will be drilled when cutting from the solid. If the angles are unequal, then only one lip will cut and undue wear will result. The unbalanced forces will cause the drill to flex and 'wander'. The axis of the hole will become displaced as drilling proceeds. If both these faults are present at the same time, both sets of faults will be present and an inaccurate and ragged hole will result.

Work-holding when drilling

It is dangerous to hold work being drilled by hand. There is always a tendency for the drill to grab the work and spin it round. Also the rapidly spinning *swarf* can produce some nasty cuts to the back of your hand. Therefore the work should always be securely fastened to the machine table.

Nevertheless, small holes in relatively large components are sometime drilled with the work handheld. In this case a stop bolted to the machine table should be used to prevent rotation.

Small work is usually held in a machine vice which, in turn, is securely bolted to the machine table. This is shown in Figure 1.83(a).

Larger work can be clamped directly to the machine table, as shown in Figure 1.83(b). In both these latter two examples the work is supported on parallel blocks. You mount the work in this way so

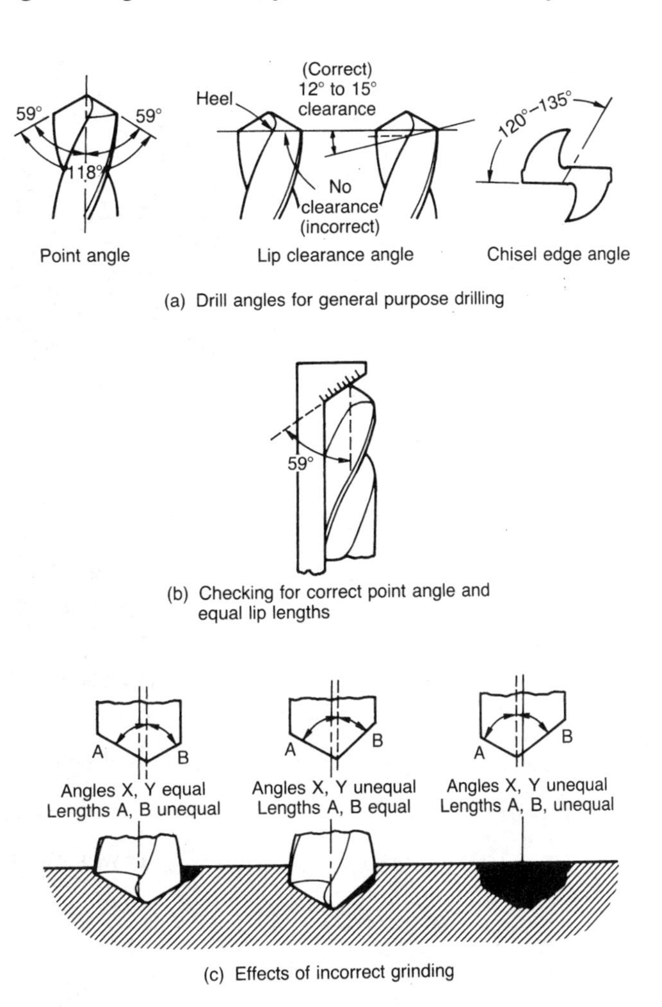

Figure 1.82 *Point angles for a twist drill*

Test your knowledge 1.48

Identify, with the aid of a sketch, the point and clearance angles for a twist drill.

Test your knowledge 1.49

Explain why some drills have tapered shanks.

that when the drill 'breaks through' the workpiece it does not damage the vice or the machine table.

Figure 1.83(c) shows how an angle plate can be used when the hole axis has to be parallel to the datum surface of the work. Finally, Figures 1.84(a) and 1.84(b) show how cylindrical work is located and supported using vee blocks.

Miscellaneous drilling operations

Figure 1.85 shows some miscellaneous operations that are frequently carried out on drilling machines.

Countersinking

Figure 1.85(a) shows a countersink bit being used to countersink a hole to receive the heads of rivets or screws. For this reason the included angle is 90°. Lathe centre drills are unsuitable for this operation as their angle is 60°.

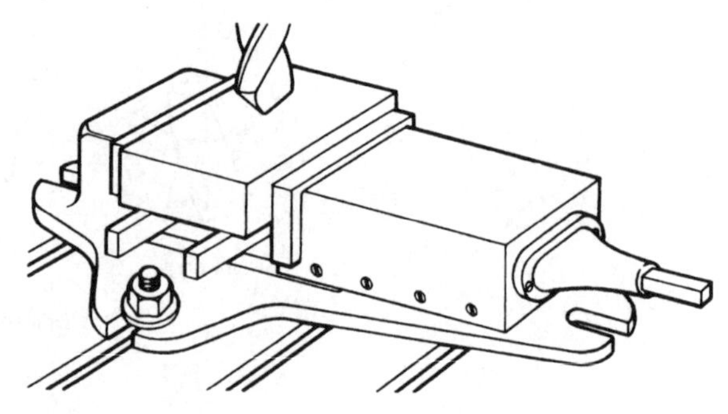

(a) Machine vice

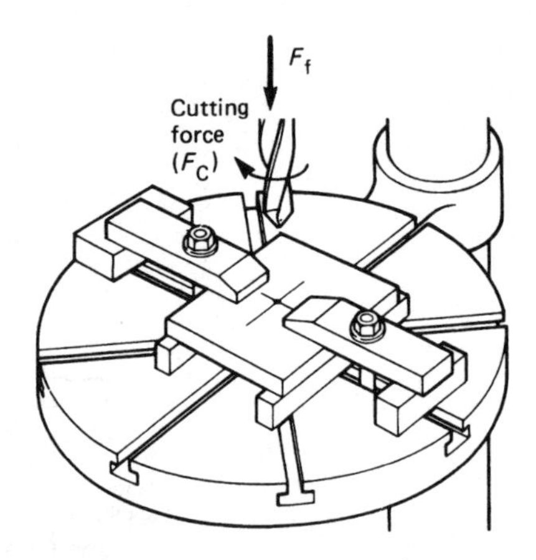

(b) Work supported on parallels and clamped to table

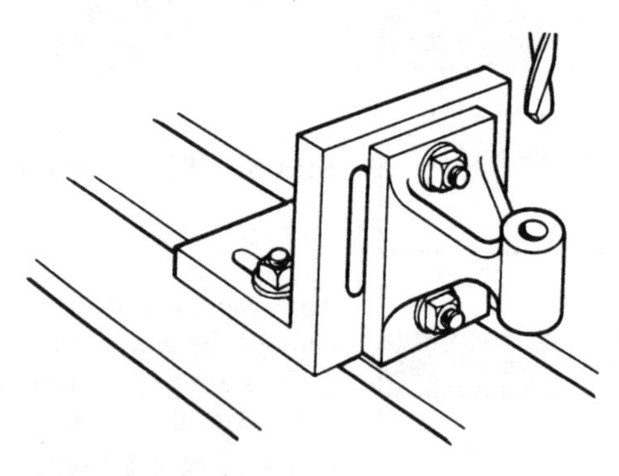

(c) Use of angle plate

Figure 1.83 *Work-holding*

Activity 1.22

Make a sketch of a machine vice and identify all of the materials and processes used in its construction. Present your findings in the form of a brief word processed report with supporting sketches.

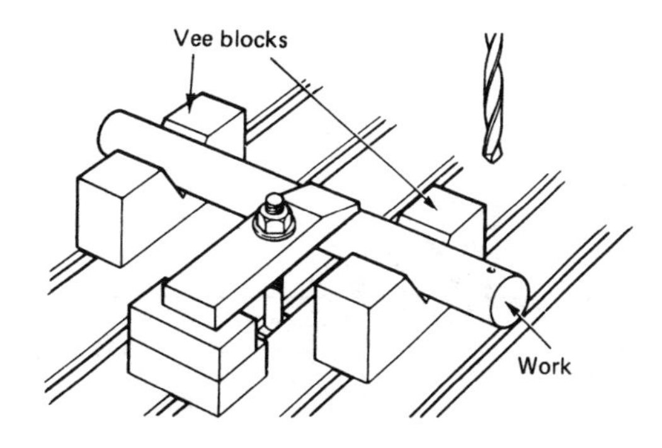

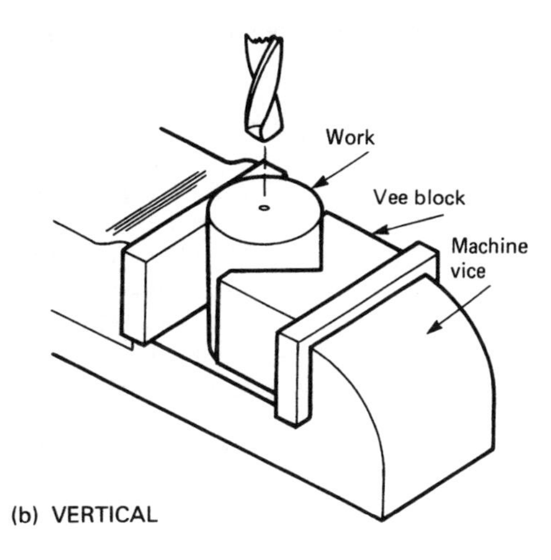

Figure 1.84 *Work-holding cylindrical components*

Counterboring

Figure 1.85(b) shows a piloted counterbore being used to counterbore a hole so that the head of a capscrew or a cheese-head screw can lie below the surface of the work. Unlike a countersink cutter, a counterbore is not self-centring. It has to have a pilot which runs in the previously drilled bolt or screw hole. This keeps the counterbore cutting concentrically with the original hole.

Spot-facing

This is similar to counterboring but the cut is not as deep. It is used to provide a flat surface on a casting or a forging for a nut and washer to 'seat' on. Sometimes, as shown in Figure 1.85(c), it is used to machine a boss (raised seating) to provide a flat surface for a nut and washer to 'seat' on.

The main purpose of a centre lathe is to produce external and internal cylindrical and conical (tapered) surfaces. It can also produce plain surfaces and screw threads.

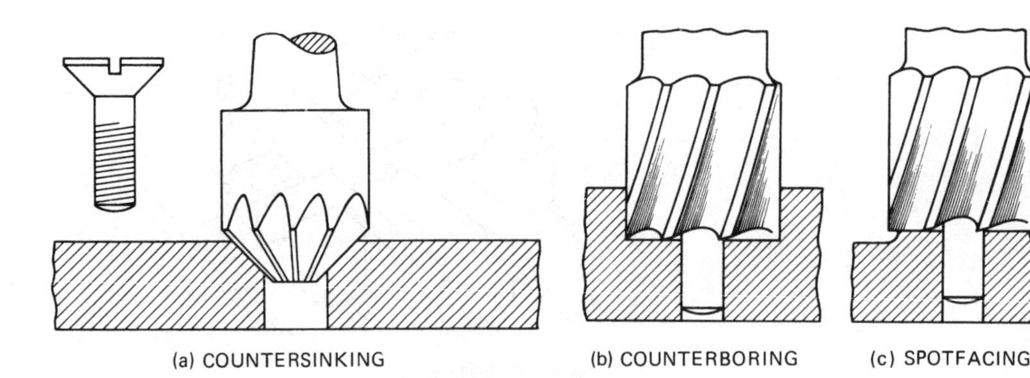

Figure 1.85 *Countersinking, counterboring and spot-facing*

The centre lathe

The main purpose of a centre lathe is to produce external and internal cylindrical and conical (tapered) surfaces. It can also produce plain surfaces and screw threads.

Figure 1.86(a) shows a typical centre lathe and identifies its more important parts.

- The bed is the base of the machine to which all the other sub-assemblies are attached. *Slideways* accurately machined on its top surface provide guidance for the saddle and the tailstock. These slideways also locate the headstock so that the axis of the spindle is parallel with the movement of the saddle and the tailstock. The saddle or carriage of the lathe moves parallel to the spindle axis as shown in Figure 1.86(b).
- The cross-slide is mounted on the saddle of the lathe. It moves at 90° to the axis of the spindle, as shown in Figure 1.86(c). It provides in-feed for the cutting tool when cylindrically turning. It is also used to produce a plain surface when facing across the end of a bar or component.

The top-slide (*compound-slide*) is used to provide in-feed for the tool when facing. It can also be set at an angle to the spindle axis for turning tapers, as shown in Figure 1.86(b). Table 1.20 summarises the basic cutting movements provided by a centre lathe.

Work-holding in the lathe

The work to be turned can be held in various ways. We will now consider the more important of these.

Between centres

The centre lathe derives its name from this basic method of work-holding. The general layout is shown in Figure 1.87(a). Centre holes are drilled in the ends of the bar and these locate on centres in the headstock spindle and the tailstock barrel. A section through a

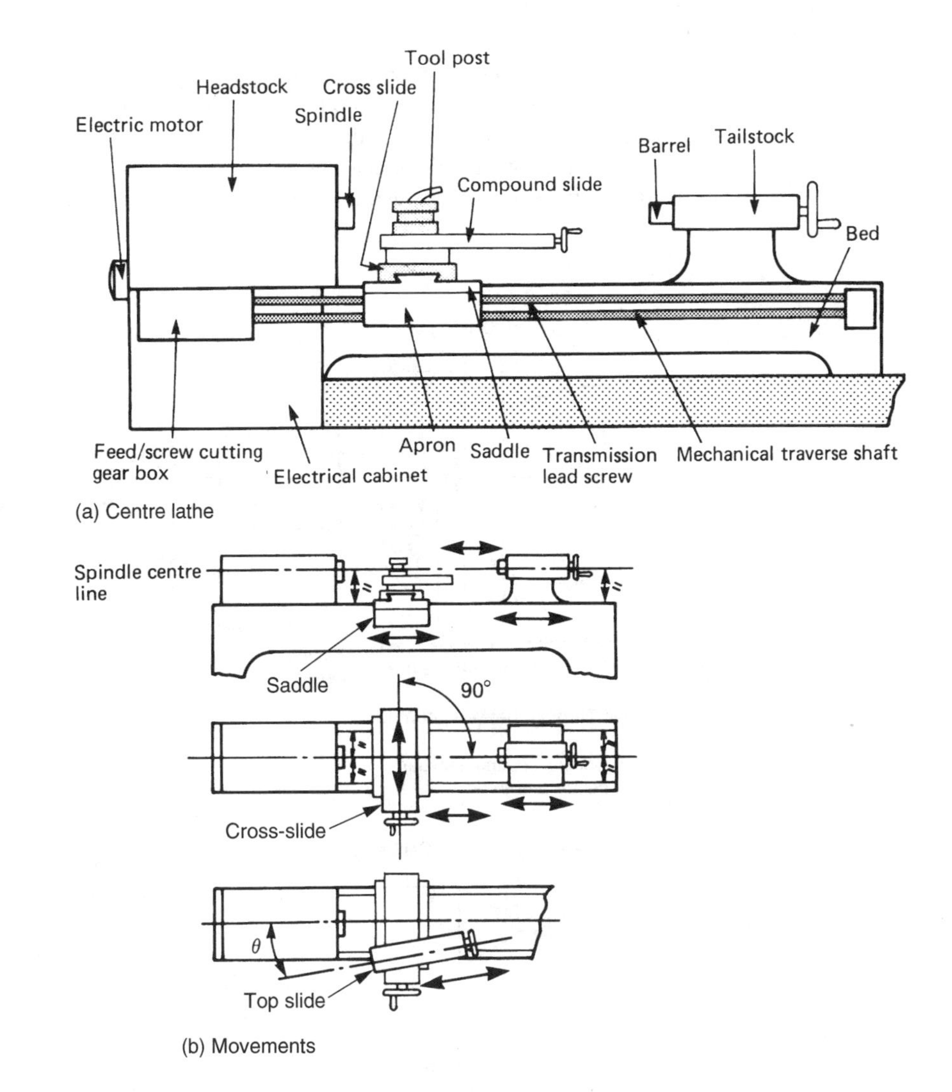

Figure 1.86 *A centre lathe*

Table 1.20 *Centre lathe movements*

Cutting movement	*Hand or power traverse*	*Means by which movement is achieved*	*Turned feature*
Tool parallel to the spindle centre line	Both	The saddle moves along the bed slideways	A parallel cylinder
Tool at 90° to the spindle centre line	Both	The cross-slide moves along a slideway machined on the top of the saddle	A flat face square to the spindle centre line
Tool at an angle relative to the spindle centre line	Hand	The compound slide is rotated and set at the desired angle relative to the centre line	A tapered cone

correctly centred component is shown in Figure 1.87(b). The centre-hole is cut with a standard centre-drill. The main disadvantage of this method of work-holding is that no work can be performed on the end of the component. Work that has been previously bored can be finish turned between centres using a taper mandrel as shown in Figure 1.87 (c).

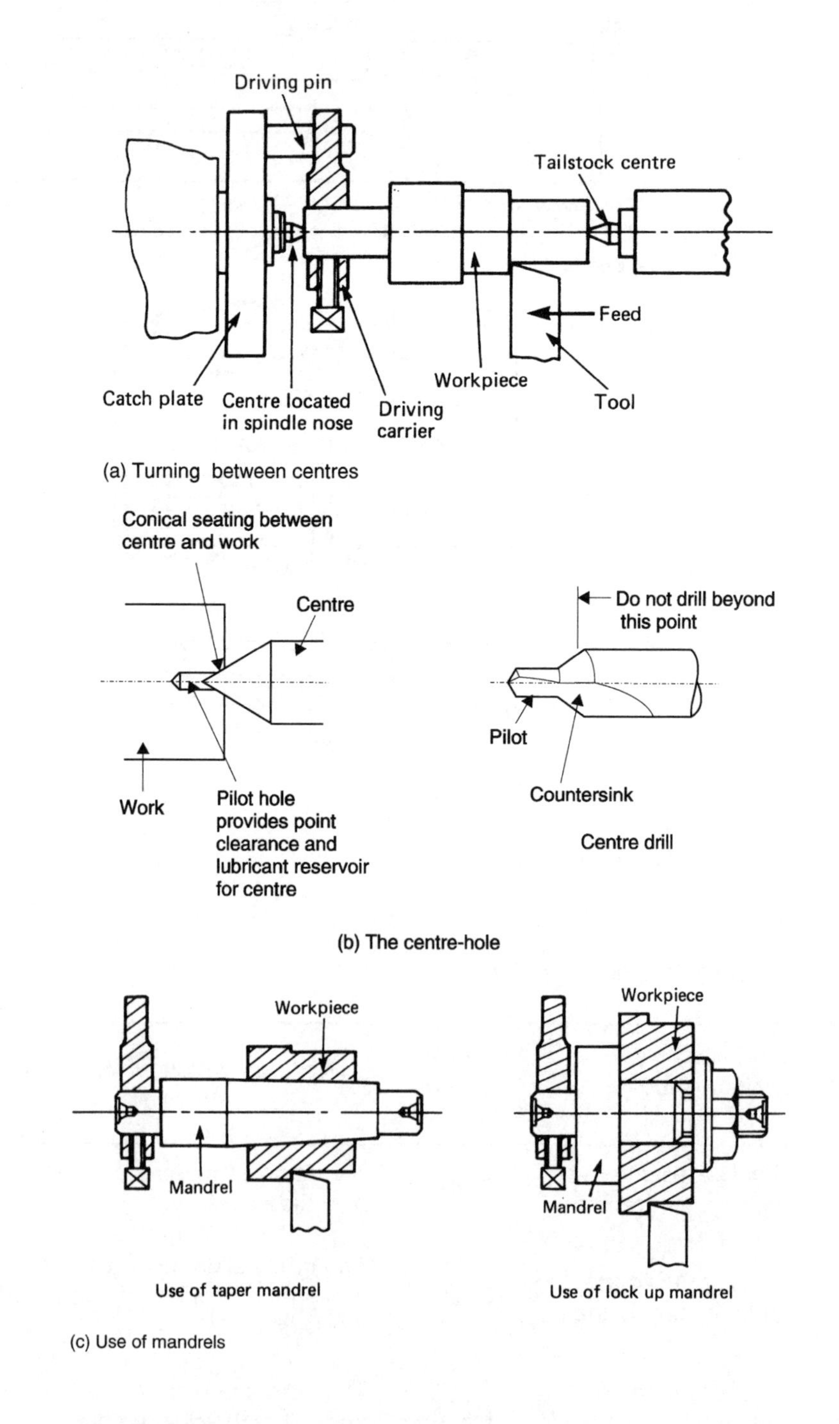

Figure 1.87 *Work-holding in a centre lathe*

Four-jaw chuck

Chucks are mounted directly onto the spindle nose and hold the work securely without the need for a back centre. This allows the end of the work to be faced flat. It also allows for the work to have holes bored into it or through it.

In the four-jaw chuck, the jaws can be moved independently by means of jack-screws. As shown in Figure 1.88(a), the jaws can also be reversed and the work held in various ways, as shown in Figure 1.88(b). As well as cylindrical work, rectangular work can also be held, as shown in Figure 1.88(c).

Because the jaws can be moved independently, the work can be set to run concentrically with the spindle axis to a high degree of accuracy. Alternatively the work can be deliberately set off-centre to produce eccentric components as shown in Figure 1.88(d).

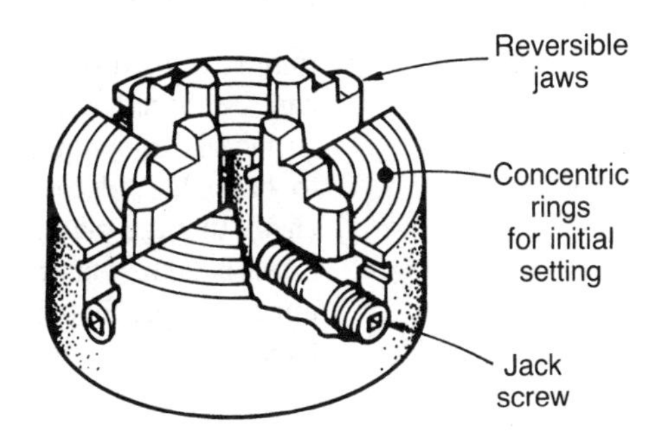

(a) Independent four-jaw chuck

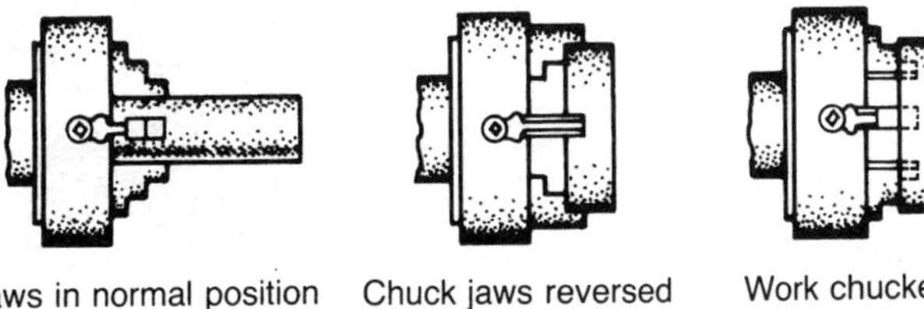

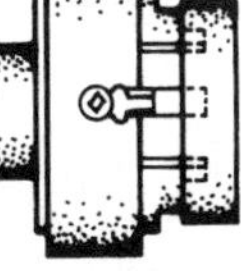

(b) Methods of holding work in a chuck

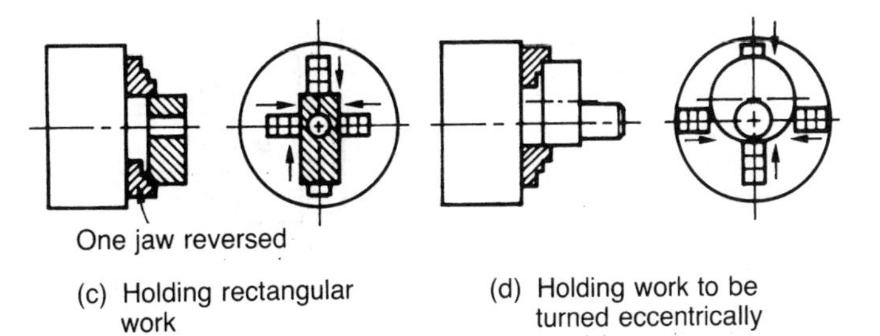

(c) Holding rectangular work

(d) Holding work to be turned eccentrically

Figure 1.88 *Four-jaw chuck*

Three-jaw chuck

The self-centring, three-jaw chuck is shown in Figure 1.89(a). The jaws are set at 120° and are moved in or out simultaneously (at the same time) by a scroll when the key is turned. SAFETY: This key must be removed before starting the lathe or a serious accident can occur. When new and in good condition this type of chuck can hold cylindrical and hexagonal work concentric with the spindle axis to a high degree of accuracy. In this case the jaws are not reversible, so it is provided with separate internal and external jaws. In Figure 1.89(a) the internal jaws are shown in the chuck, and the external jaws are shown at the side of the chuck. Again the chuck is mounted directly on the spindle nose of the lathe.

Work to be turned between centres is usually held in a three-jaw chuck whilst the ends of the bar are faced flat and then centre drilled, as shown in Figure 1.89(b).

Face-plate

Figure 1.90 shows a component held on a face-plate so that the hole can be bored perpendicularly to the datum surface. This datum surface is in contact with the face-plate. Note that the face-plate has to be balanced to ensure smooth running. Care must be taken to check that the clamps will hold the work securely and do not foul the machine. The clamps must not only resist the cutting forces, but they must also prevent the rapidly rotating work from spinning out of the lathe.

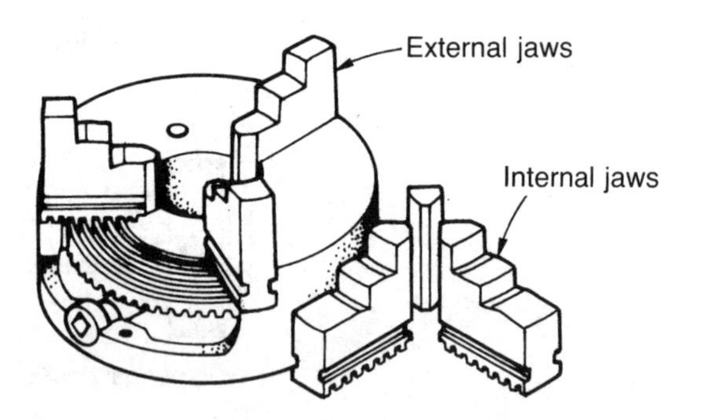

(a) Three jaw self-centring chuck

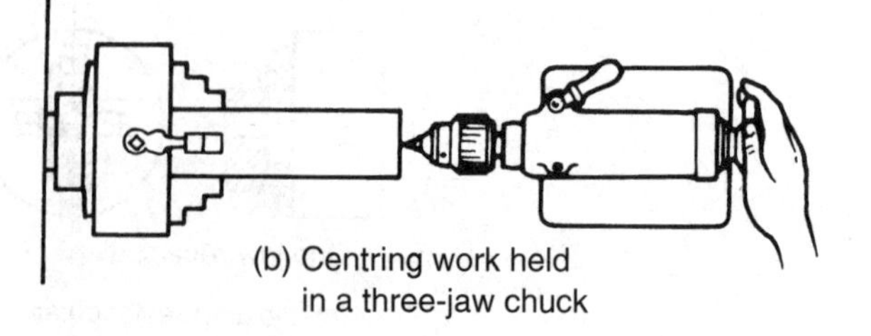

(b) Centring work held in a three-jaw chuck

Figure 1.89 *Three-jaw chuck*

Test your knowledge 1.50

Sketch the construction of:

(a) a three-jaw chuck
(b) a four-jaw chuck

as fitted to a lathe. Label your sketch.

Test your knowledge 1.51

With the aid of a sketch, identify the rake and clearance angles for a turning tool.

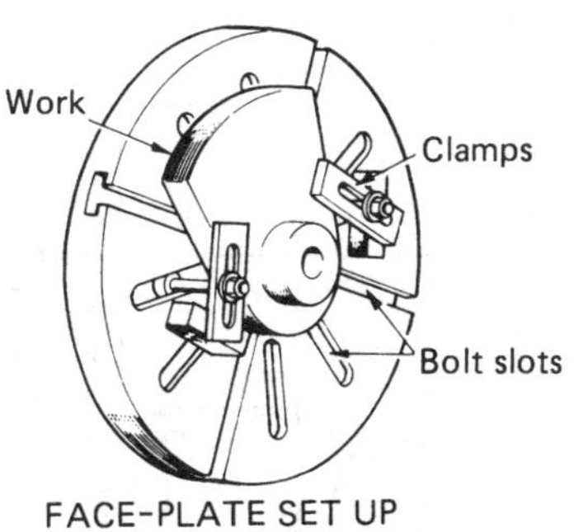

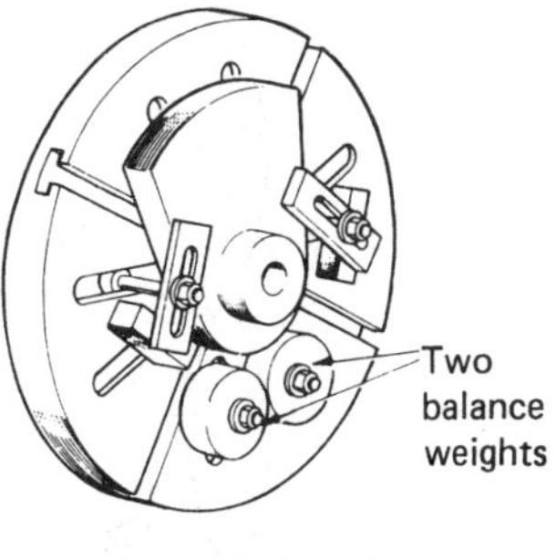

Figure 1.90 *Face-plate set up*

Surfacing and boring

Figure 1.91(a) shows a range of turning tools and some typical applications. Figure 1.91(b) shows how the metal-cutting wedge also applies to turning tools. Turning tools are fastened into a tool-post which is mounted on the top slide of the lathe. There are many different types of tool-post. The four-way turret tool-post shown in Figure 1.91(c) allows four tools to be mounted at any one time.

Surfacing

A *surfacing* (facing or perpendicular-turning) operation on a workpiece held in a chuck is shown in Figure 1.92. The saddle is clamped to the bed of the lathe and the tool motion is controlled by the cross-slide. This ensures that the tool moves in a path at right-angles to the workpiece axis and produces a plain surface. In-feed of the cutting tool is controlled by micrometer adjustment of the top-slide.

Boring

Figure 1.93 shows how a drilled hole can be opened up by using a *boring* tool. The workpiece is held in a chuck and the tool movement is controlled by the saddle of the lathe. The in-feed of the tool is controlled by micrometer adjustment of the cross-slide. The pilot hole is produced either by a taper shank drill mounted directly into the tailstock barrel (poppet), or by a parallel shank drill held in a drill chuck. The taper mandrel of the drill chuck is inserted into the tailstock barrel.

Activity 1.23

It is sometimes necessary to turn long cylindrical components and to ensure that the component remains truly cylindrical with no taper. Investigate a means of carrying out such a *parallel turning* operation and illustrate the process with sketches.

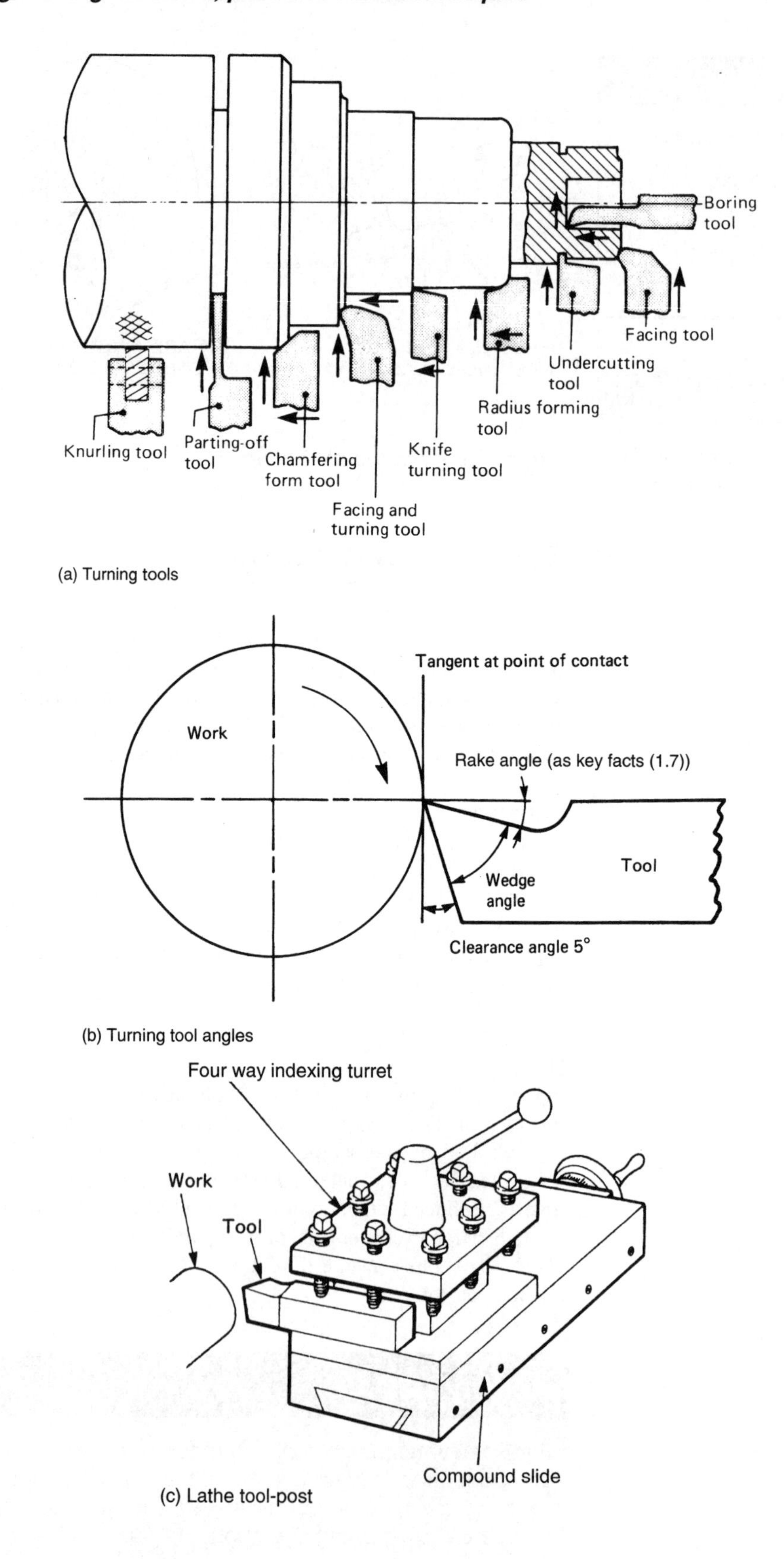

Figure 1.91 *Turning tools*

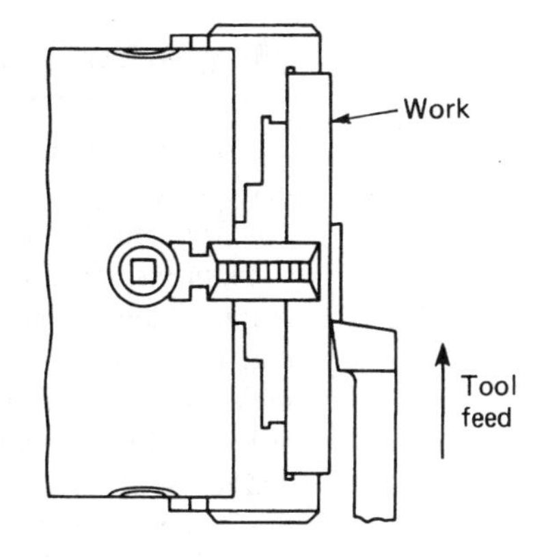

Figure 1.92 *Surfacing*

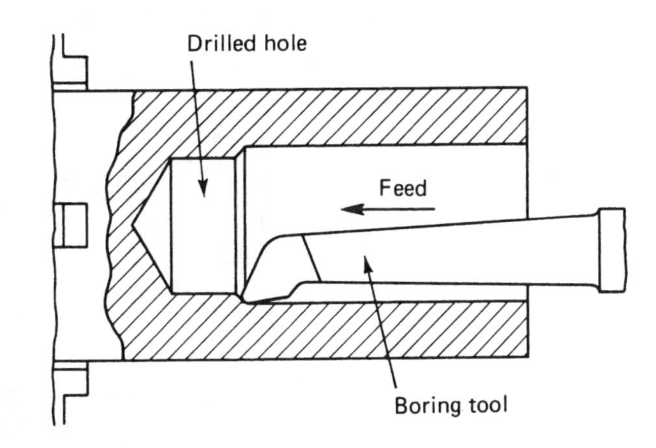

Figure 1.93 *Boring*

Conical surfaces

Chamfers on the corners of a turned component are short conical surfaces. These are usually produced by using a chamfering tool, as shown in Figure 1.91(a). Longer tapers can be produced by use of the top-slide. Use of the top (compound) slide is shown in Figure 1.94. The slide is mounted on a swivel base and it is fitted with a protractor scale. It can be swung round to the required angle and clamped in position. The taper is then cut as shown.

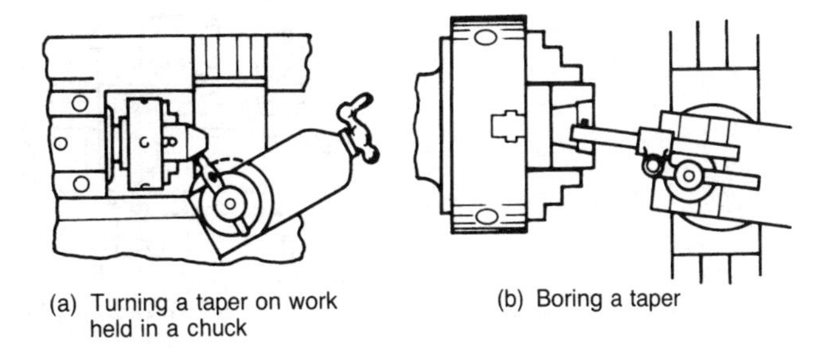

(a) Turning a taper on work held in a chuck

(b) Boring a taper

Figure 1.94 *Producing a chamfer*

Miscellaneous turning operations

Apart from basic metal turning, several other operations can be performed using a centre lathe. These operations include *reaming*, *tapping*, and *threading*.

Figure 1.95(a) shows a hole being reamed in a lathe. A machine reamer is being used and it is held in the barrel of the tailstock. Note that reamers are sizing tools and they actually remove very little metal. Because a drilled hole invariably runs out slightly, the pilot hole should be single-point bored in order to correct the position and geometry of the hole. It is finally sized using the reamer. Only the minimum amount of metal for the surface of the hole to clean up should be left in for the reamer to remove.

Standard, non-standard and large diameter screw threads can be cut in a centre lathe by use of the lead-screw to control the saddle movement. This is a highly skilled operation.

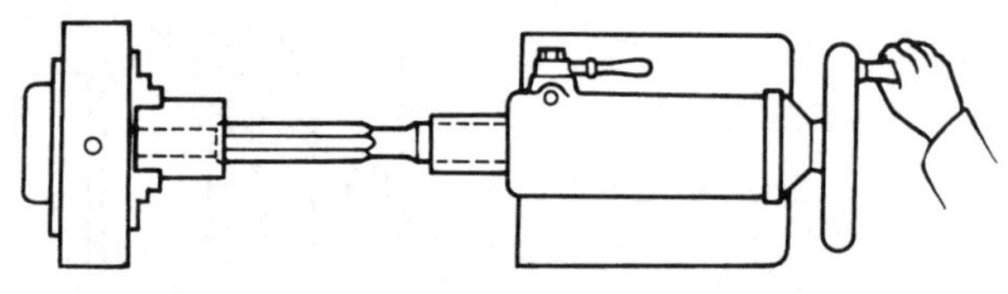

(a) Machine reamer supported in the tailstock

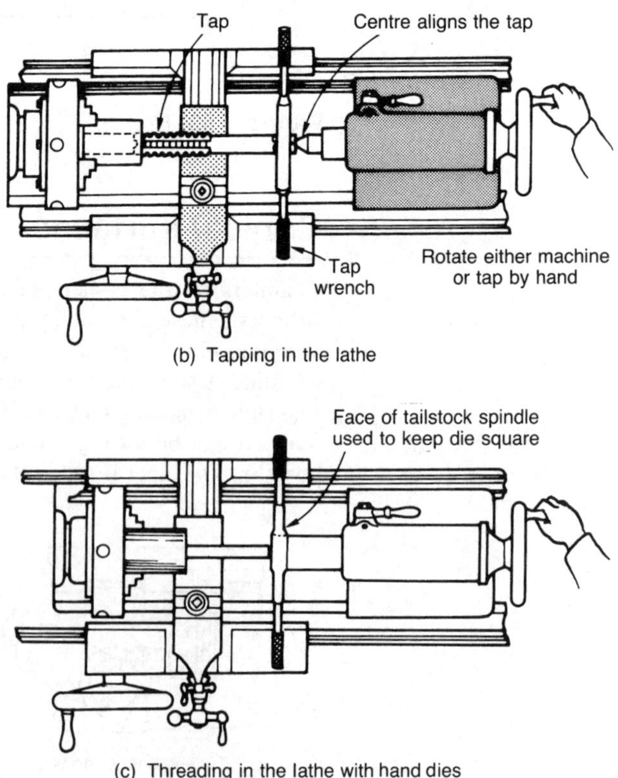

(b) Tapping in the lathe

(c) Threading in the lathe with hand dies

Figure 1.95 *Miscellaneous turning operations*

Test your knowledge 1.52

List FIVE main operations that can be performed on a centre lathe.

Test your knowledge 1.53

Describe, with the aid of sketches, THREE different types of turning tool for use with a centre lathe.

Test your knowledge 1.54

With the aid of a sketch explain how you would hold the component shown in Figure 1.86 (b) in order to machine the 50 mm diameter hole.

However, standard screw threads of limited diameter can be cut using hand threading tools as shown in Figures 1.95(b) and 1.95(c). Taps are very fragile and the workpiece should be rotated by hand with the lathe switched off and the gears fully disengaged.

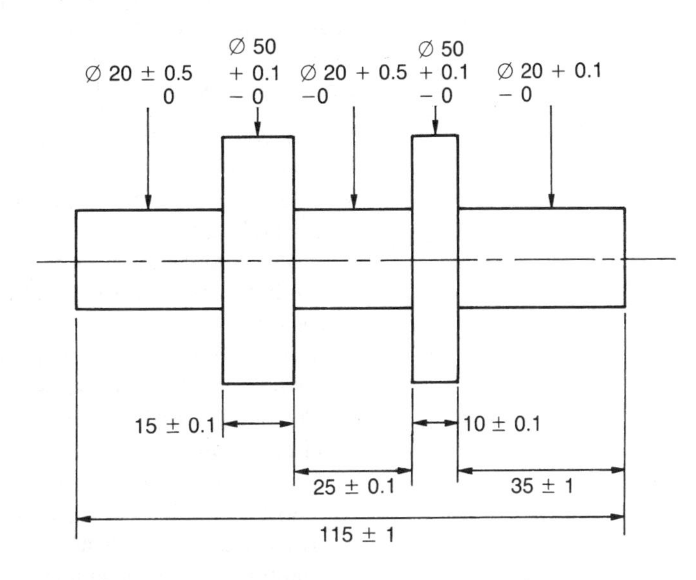

Dimensions in millimetres
Material: Phosphor bronze: Blank size ∅ 60 × 125
Hint: Turn between centres, after facing and centring

(a)

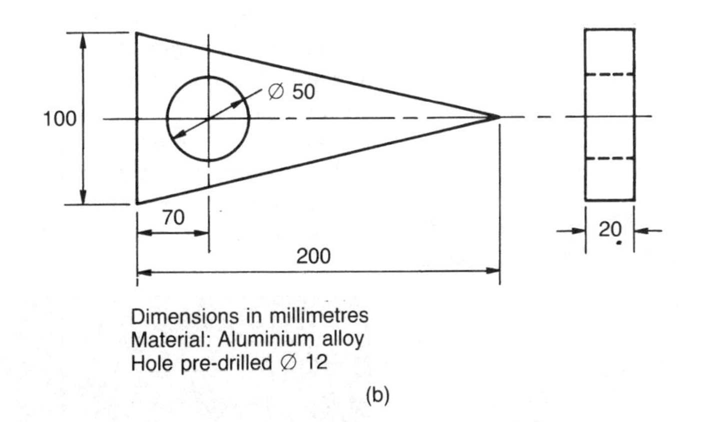

Dimensions in millimetres
Material: Aluminium alloy
Hole pre-drilled ∅ 12

(b)

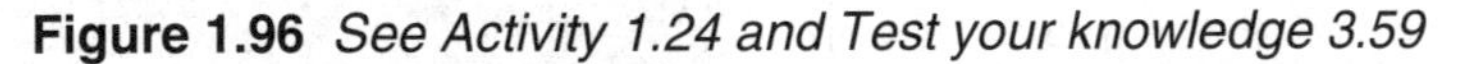

Figure 1.96 *See Activity 1.24 and Test your knowledge 3.59*

Activity 1.24

Select suitable tools and measuring equipment (giving reasons for your choice) and draw up a plan for making the component shown in Figure 1.96. Present your work in the form of a brief word processed report with supporting hand-drawn sketches.

Etching

Printed circuit boards have already been introduced in the section on assembly. First the circuit is drawn out by hand or designed using a computer aided design (CAD) package. A typical printed circuit board is shown in Figure 1.97. The master drawing of the circuit is then photographed to produce a transparent negative. The printed circuit board is made as follows:

- The copper coated laminated plastic (Tufnol) or fibre glass board is coated with a photoresist by dipping or spraying
- The negative of the circuit is placed in contact with the prepared circuit board. They are then exposed to ultraviolet light. The light passes through the transparent parts of the negative. The areas of the board exposed to the ultraviolet light will eventually become the circuit
- The exposed circuit board is then developed in a chemical solution that hardens the exposed areas
- The photoresist is stripped away from the unexposed areas of the circuit board
- The circuit board is then placed in a suitable etchant. Ferric chloride solution can be used as an etchant for copper. This eats away the copper where it is not protected by the hardened photoresist. The remaining copper is the required circuit
- The circuit boards are washed to stop the reaction. The remaining photoresist is removed so as not to interfere with the tinning of the circuit with soft solder and the soldering of the components into position.

SAFETY: This process is potentially dangerous. Ultraviolet light is very harmful to your skin and to your eyes. The ferric chloride solution is highly corrosive to your skin. The various processes also give off harmful fumes. Therefore you should only carry out this process under properly supervised, controlled and ventilated conditions. Appropriate protective clothing should be worn.

A similar process can be used for the chemical engraving of components with their identification numbers and other data.

Another view

Surface mounted components are now extensively used in electronic equipment manufacture. Such devices (resistors, capacitors, diodes, transistors and integrated circuits) are located on copper pads on the surface of the printed circuit board and held in place by means of a solder paste. The entire printed circuit board is then dipped in a bath of molten solder. The solder paste quickly reflows in order to simultaneously produce a large number of good quality soldered joints.

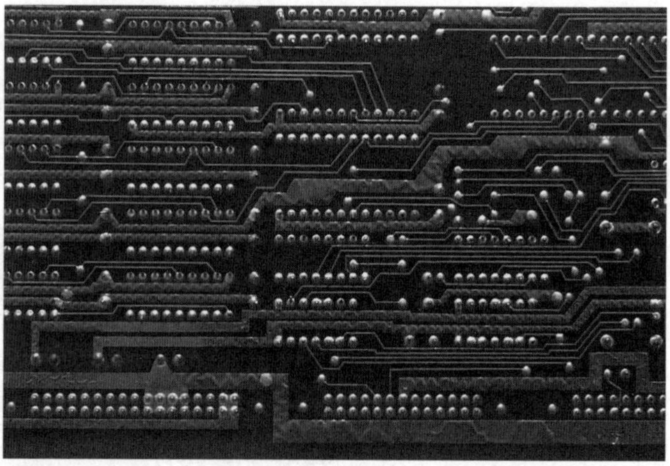

Figure 1.97 *A printed circuit board*

Electroplating

The component to be plated is placed into a plating bath as shown in Figure 1.98. The component is connected to the negative terminal of a direct current supply. This operates at a low voltage but relatively heavy current. The anode is connected to the positive terminal of the supply. The anode is usually made from the same metal as that which is to be plated onto the component. The electrolyte is a solution of chemical salts. The composition depends upon the process being carried out.

When the current passes through the bath the component is coated with a thin layer of the protective and/or decorative metal. This metal comes from the chemicals in the electrolyte. The process is self-balancing. The anode dissolves into the electrolyte at the same rate as the metal taken from the electrolyte is being deposited on the component. This applies to most plating processes such as zinc, copper, tin and nickel plating.

An exception is chromium plating. A neutral anode is used that does not dissolve into the electrolyte. Additional salts have to be added to the electrolyte from time to time to maintain the solution strength. Chromium is not usually deposited directly onto the component. The component is usually nickel plated and polished and then a light film of chromium is plated over the nickel as a decorative and sealing coat.

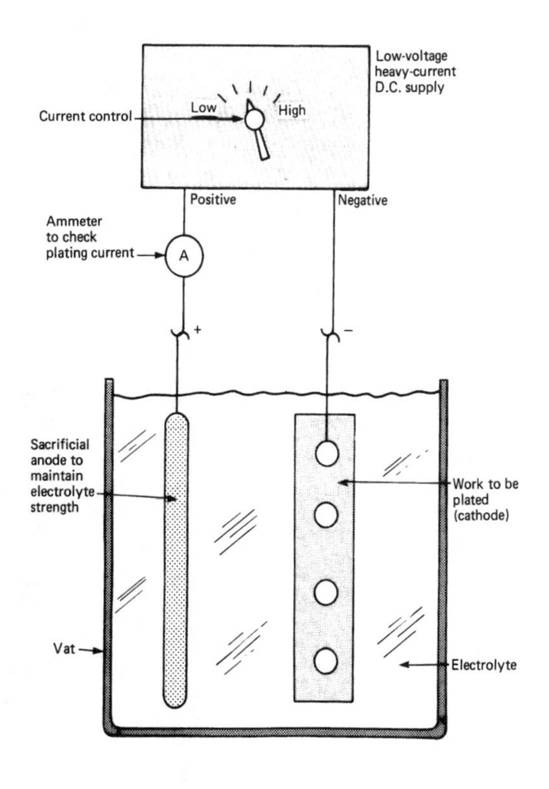

Figure 1.98 *Electroplating*

Test your knowledge 1.55

Give TWO examples of where electroplating is used. Name the type of plating used in each case.

A surface finish can be applied to many engineered products in order to improve its appearance, durability or corrosion resistance. We shall look at a variety of different finishes and how they are applied.

Grinding

A grinding wheel consists of abrasive particles bonded together. It does not 'rub' the metal away, it cuts the metal like any other cutting tool. Each abrasive particle is a cutting tooth. Imagine an abrasive wheel to be a milling cutter with thousands of teeth. Wheels are made in a variety of shapes and sizes. They are also available with a variety of abrasive particle materials and a variety of bonds. It is essential to choose the correct wheel for any given job.

Figure 1.99(a) shows two types of electrically powered portable grinding machines used for dressing welds and for fettling castings. Care must be taken in its use and the operator should wear a suitable grade of protective goggles and a filter type respirator.

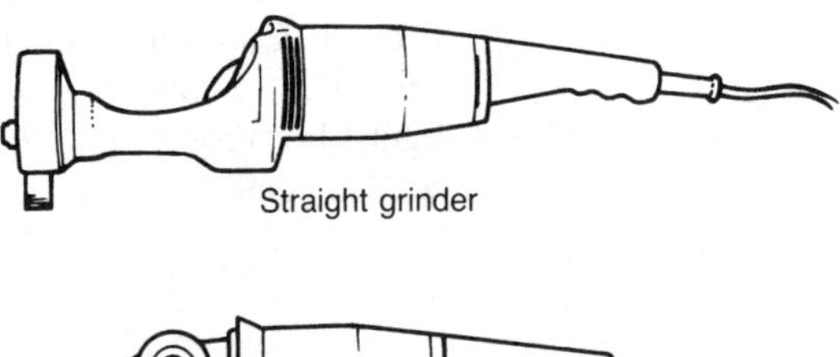

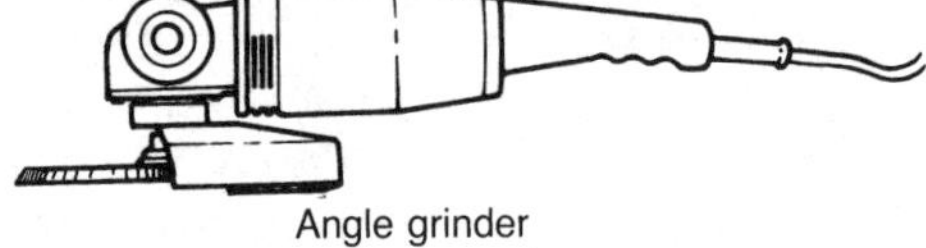

(a) Portable grinding machines

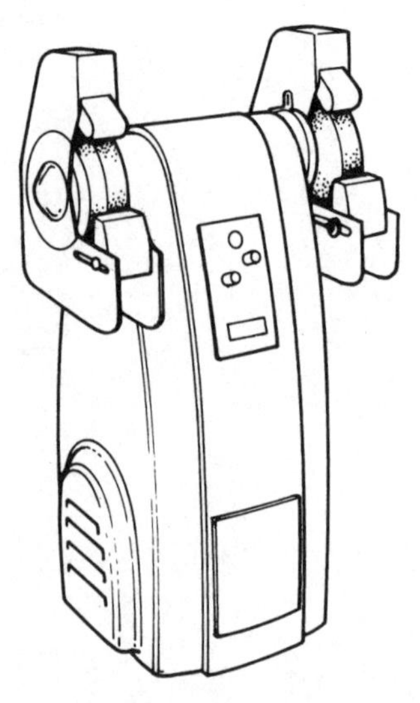

(b) Off-hand grinding machine

Figure 1.99 *Grinding machines*

Figure 1.99(b) shows a double-ended, off-hand tool grinder. This is used for sharpening drills and lathe tools and other small tools and marking out instruments. It is essential to check that the guard is in place and that the visor and tool rest are correctly adjusted before commencing to use the machine. Grinding wheels can only be changed by a trained and certificated person.

Polishing

Polishing produces an even better finish than grinding but only removes the minutest amounts of metal. It only produces a smooth and shiny surface finish, the geometry of the surface is uncontrolled. Polishing is used to produce decorative finishes, to improve fluid flow through the manifolds of racing engines, and to remove machining marks from surfaces that cannot be precision ground. This is done to reduce the risk of fatigue failure in highly stressed components.

Figure 1.100 shows a typical polishing lathe. It consists of an electric motor with a double-ended extended spindle. At each end of the spindle is a tapered thread onto which you screw the polishing mops. The mops may be made up from discs of leather (basils) or discs of cloth (calico mops). Polishing compound in the form of 'sticks' is pressed against the mops to impregnate them with the abrasive.

The components to be polished are held against the rapidly spinning mops by hand. Because the mops are soft and flexible they will follow the contours of complex shaped components. It is essential that dust extraction equipment is fitted to the polishing lathe and that the operator wears eye protection and a filter type dust mask. The process should only be carried out by a skilled polisher or under close supervision.

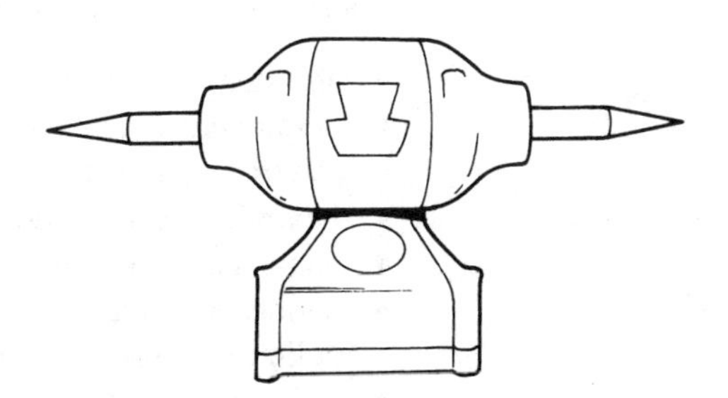

Figure 1.100 *A typical polishing lathe*

Coating

Electroplating has already been discussed and is the coating of metal components with another metal that is more decorative and/or corrosion resistant. Hot-dip galvanizing coats low carbon steels with zinc without using an electroplating process.

Hot-dip galvanizing

Hot-dip galvanizing is the original process used for zinc coating buckets, animal feeding troughs and other farming accessories. It is also used for galvanized sheeting. The work to be coated is chemically cleaned, fluxed and dipped into the molten zinc. This forms a coating on the work. A small percentage of aluminium is added to the zinc to give the traditional bright finish. The molten zinc also seals any cut edges and joints in the work and renders them fluid tight. Metal components may also be coated with non-metallic surfaces.

Oxidizing

Oil blueing

Steel components have a natural oxide film due to their reaction with atmospheric oxygen. This film can be thickened and enhanced by heating the steel component until it takes on a dark blue colour. Then immediately dip the component into oil to seal the oxide film. This process does not work if there is any residual mill scale on the metal surfaces.

Chemical blacking

Alternatively, an even more corrosion resistant oxide film can be applied to steel components by chemical blacking. The components are cleaned and degreased. They are then immersed in the oxidizing chemical solution until the required film thickness has been achieved. Finally the treated components are rinsed, dewatered and oiled. Again, the process only works on bright surfaces.

Plastic coating

Plastic coatings can be both functional, corrosion resistant and decorative. The wide range of plastic materials available in a wide variety of colours and finishes provides a designer with the means of achieving one or more of the following:

- abrasion resistance
- cushioning effects with coatings up to 6 mm thick
- electrical and thermal insulation
- flexibility over a wide range of temperatures
- non-stick properties (Teflon PTFE coatings)
- permanent protection against weathering and atmospheric pollution, resulting in reduced maintenance costs
- resistance to corrosion by a wide range of chemicals
- the covering of welds and the sealing of porous castings.

To ensure success, the surfaces of the work to be plasticized must be physically and chemically clean and free from oils and greases. The surfaces to be plasticized must not have been plated, galvanized or oxide treated.

Fluidized bed dipping

Finely powdered plastic particles are suspended in a current of air in a fluidizing bath as shown in Figure 1.101. The powder continually bubbles up and falls back and looks as though it is boiling. It offers

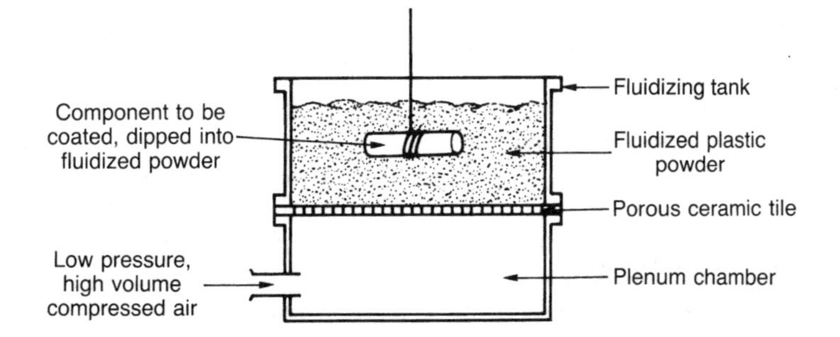

Figure 1.101 *Fluidized bed dipping*

no resistance to the work to be immersed in it. The work is pre-heated and immersed in the powder. A layer of powder melts onto the surface of the metal to form a homogeneous layer.

Liquid plastisol dipping

This process is limited to PVC coating. A plastisol is a resin powder suspended in a plastisol and no dangerous solvents are present. The pre-heated work is suspended in the PVC plastisol until the required thickness of coating has adhered to the metal surface.

Painting

Painting is used to provide a decorative and corrosion resistant coating for metal surfaces. It is the easiest and cheapest means of coating that can be applied with any degree of permanence. A paint consists of three components:

- *Pigment* The finely powered pigment provides the paint with its opacity and colour.
- *Vehicle* This is a film-forming liquid or binder in a volatile solvent. This binder is a natural or synthetic resinous material. When dry (set) it must be flexible, adhere strongly to the surface being painted, corrosion resistant and durable.
- *Solvent (thinner)* This controls the consistency of the paint and its application. It forms no part of the final paint film as it totally evaporates. As it evaporates it increases the concentration of catalyst in the 'vehicle' causing it to change chemically and set.

A paint system consists of the following components:

- *Primer* This is designed to adhere strongly to the surface being painted and to provide a key for the subsequent coats. It also contains anti-corrosion compounds.
- *Putty or filler* This is used mainly on castings to fill up and repair blemishes. It provides a smooth surface for subsequent paint coats.
- *Undercoat* This builds up the thickness of the paint film. To produce a smooth finish, more than one undercoat is used with careful rubbing down between each coat. It also gives richness and opacity to the colour.
- *Top coat* This coat is decorative and abrasion resistant. It also

Another view

The term 'paint system' refers to the method of application, and the number of layers of various types of paint required for a particular application. In the case of engineered parts and components, a number of layers are required, starting with a base coat known as primer. It is the final top coat that determines the appearance (colour and surface finish) of the part.

Test your knowledge 1.56

State the essential difference between grinding and polishing.

Test your knowledge 1.57

State a suitable coating medium and describe its application process for each of the following engineered components:

(a) painting refrigerator body panels
(b) painting pressed steel angle brackets
(c) plastic cladding metal tubing for a bathroom towel rail.

seals the paint film with a waterproof membrane. Modern top coats are usually based on acrylic resins or polyurethane rubbers.

There are four main groups of paint:

- *Group 1* The vehicle is polymerized (see thermosetting plastics) into a solid film by reaction with atmospheric oxygen. Paints that set naturally in this way include traditional linseed oil based paints, oleo-resinous paints, and modern general purpose air drying paints based on modified alkyd resins.
- *Group 2* This group of paints is based on amino-alkyd resins that do not set at room temperatures but they have to be *stoved* at 110–150°C. When set these paints are much tougher than group 1 air-drying paints.
- *Group 3* These are the 'two-pack' paints. Polymerization starts to occur as soon as the catalyst is mixed with the paint immediately before use. Modern 'one-pack' versions of these paints have the catalyst diluted with a volatile solvent as mentioned earlier. The solvent evaporates after the paint has been spread and, when the concentration of the catalyst reaches a critical level, polymerization takes place and the paint sets.
- *Group 4* These paints dry by evaporation of the solvent and no polymerization occurs. An example is the cellulose paint used widely at one time for spray painting motor car body panels. Lacquers also belong to this group but differ from all other paints in that dyes are used as the colorant instead of pigments.

Paints may be applied by brushing, spraying or dipping. Whatever method is used, great care must be taken to ensure adequate ventilation. Not only can the solvents produce narcotic effects, but inhaling dried particles and liquid droplets of paint can cause serious respiratory diseases.

The appropriate protective clothing, goggles and face masks must be used. Paints are also highly flammable and the local fire-prevention officer must be consulted over their storage and use. On no account can smoking be tolerated anywhere near the storage or use of paints.

Activity 1.25

Carry out a detailed risk analysis of your school or college's engineering workshop or engineering science laboratory. Start by thinking about the activities that are undertaken, the materials and processes used, and the facilities (such as tools, benches, lighting, fume extraction, that are available). You should also take into consideration the skills, training and experience of those who use the facilities and any safety warnings or advisory notices that may be present. Present your findings in the form of a detailed written report. Hint: You may also find it useful to consult your school or college Safety Officer before you begin this exercise.

Review questions

1 Explain, in simple terms, each of the following properties of materials:
(a) strength
(b) toughness
(c) elasticity
(d) hardness
(e) rigidity.

2 Distinguish between the terms *ductility* and *malleability* when applied to engineering materials.

3 Sketch graphs to show how tensile strength and ductility varies with carbon content for a plain carbon steel. Label your graph.

4 Explain what is meant by the term *composite material*. Give THREE examples of common composite materials.

5 Name, and briefly describe the properties of, THREE main types of *polymer material*.

6 Name, and briefly describe the properties of, THREE *ceramic materials*.

7 Classify each of the materials listed below as either metals, polymers, ceramics or composite materials:
(a) aluminium
(b) clay
(c) polyvinyl chloride (PVC)
(d) rubber
(e) glass
(f) tungsten
(g) brick
(h) wood.

8 Give examples of suitable non-ferrous metals (with reasons for your choice) for each of the following applications:
(a) an instrument case used to house a portable electronic test set
(b) a water pump to be used with a marine engine
(c) a bus-bar to carry electric current in a steelworks
(d) a screw terminal used in an electric light fitting.

9 Distinguish between the terms *thermosetting plastic* and *thermoplastic* when applied to engineering materials.

10 Give examples of suitable plastic materials (with reasons for your choice) for each of the following applications:
(a) a car tow rope
(b) an engine cover for use on a light aircraft
(c) an electrical junction box
(d) the body of a 'ride-on' toy.

11 State *Hooke's Law*.

12 Define *Young's modulus*.

13 Sketch typical stress–strain graphs for the following types of material:
(a) a brittle polymer
(b) a cold-drawn polymer
(c) a ceramic material.

14 A steel reinforcing rod has a cross-sectional area of 240 mm^2. Determine the compressive stress on the rod if it is subject to a compressive force of 720 N.

15 In a tensile test a specimen is subjected to a strain of 0.05%. Determine the change in length of the specimen if it has an unstrained length of 250 mm.

16 A metal bar has a cross-sectional area of 350 mm^2. If the material has a yield stress of 225 MPa, determine the tensile force that must be applied in order to cause yielding.

17 Define *Poisson's ratio.*

18 Briefly explain each of the following terms in relation to the electrical properties of materials:
(a) resistivity
(b) conductivity.

19 State TWO examples of *semiconductor materials.*

20 Sketch graphs showing how the electrical resistance of each of the following types of material varies with temperature:
(a) conductors
(b) insulators
(c) semiconductors.

21 Sketch a typical hysteresis curve for a ferromagnetic material. Label your drawing clearly.

22 Describe the essential properties of materials that will be used in each of the following electrical/electronic applications:
(a) the magnetic core of a transformer
(b) the field coil windings in a generator
(c) the dielectric between the plates of a capacitor
(d) the material used for fabricating an integrated circuit.

23 State TWO materials that have a low value of thermal conductivity and TWO materials that have a high value of thermal conductivity. Suggest a typical engineering application for each of these classes of material.

24 Sketch the construction of a riveted lap joint.

25 Explain, with the aid of a diagram, what is meant by the term *rake angle* when applied to a cutting tool.

26 Sketch a pillar drill. Label your drawing showing each of the main features.

27 Name the two engineering operations shown in Figure 1.102.

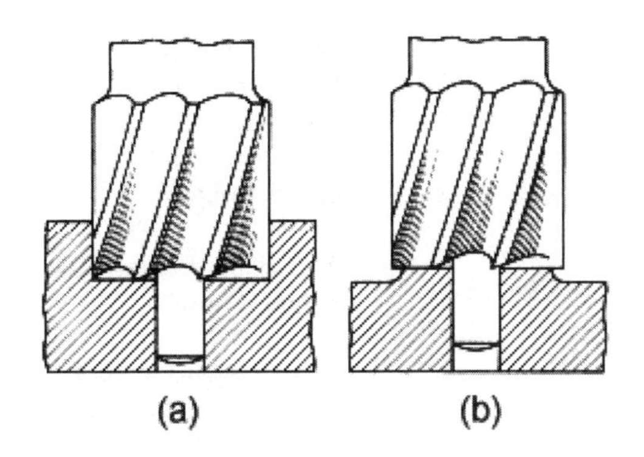

Figure 1.102 *See Question 27*

28 Describe THREE operations that can be performed using a centre lathe.

29 Name the process that is performed with the engineering workstation shown in Figure 1.103.

Figure 1.103 *See Question 29*

30 Explain what is meant by the term *soft soldering*. Why is this process different from *fusion welding*?

31 Describe THREE different methods of making electrical joints.

32 Explain what is meant by the terms:
(a) hardening
(b) tempering.

33 Describe, briefly, the main stages in the production of a printed circuit board.

34 List and briefly explain the function of the components used in a paint system.

35 Describe THREE types of corrosion that can affect metals. Explain how each type of corrosion is caused.

36 What material is best suited to the manufacture of the aircraft component shown in Figure 1.104. Give reasons for your answer.

Figure 1.104 *See Question 36*

Unit 2 The role of the engineer

Summary

The chapter provides a broad introduction to the role of the engineer. It has been designed for use as preparatory reading before you carry out your own investigation of an engineered product or service. Therefore, each of the unit topics has been treated with a broad range of engineering applications in mind and it is left for you to make the appropriate links to the products or engineering services that you have chosen for your own investigation. Every engineer needs to have an understanding of the full range of engineering activities and how these contribute to the economy. We shall begin with an introduction to the various engineering sectors and why they are important to the economy of the UK and Europe. We continue by looking at some of the diverse job roles of engineers within engineering companies and how the companies themselves are organised. We also look at how technology is used in engineering and the constraints imposed on engineers including the need to meet British and European standards.

2.1 Engineering activities

Engineering affects all industries that make or use engineered products. Engineering activities therefore span a huge number of industries and sectors, including the following key engineering industries:

- electrical and electronic engineering
- civil engineering
- mechanical engineering
- aerospace and aeronautical engineering
- communications and telecommunications engineering
- automotive engineering

- bioengineering
- chemical and petroleum engineering
- marine engineering.

You must be able to recognize which industry a particular engineering company belongs to by considering either what the company produces or what its main functions are. It can also be useful to have an awareness of the economic importance of engineering activities and how engineering has changed due to changing market needs and the introduction of new technology. We shall begin by introducing the various sectors in which engineers and engineering companies operate.

Engineering sectors

The engineering sectors and typical engineered products with which you need to be familiar are as follows:

Chemical and petroleum engineering
Fertilizers, pharmaceuticals, plastics, petrol, etc. Companies in this field include Fisons, Glaxo, ICI and BP.

Mechanical engineering
Bearings, agricultural machinery, gas turbines, machine tools and the like from companies such as RHP, GKN and Rolls-Royce.

Electrical and electronic engineering
Electric generators and motors, consumer electronic equipment (radio, TV, audio and video), power cables, computers, etc. produced by companies such as GEC, BICC and ICL.

Civil engineering
Concrete bridges and flyovers, motorways, tunnels, docks, factories, power stations, dams, etc. from companies like Bovis, Wimpey and Balfour-Beatty.

Aerospace and aeronautical engineering
Civil and military aircraft, satellites, space vehicles, missiles, etc. from companies such as British Aerospace, Westland and Rolls-Royce.

Communications and telecommunications engineering
Telephone and radio communication, data communications equipment, etc. from companies like Nokia, NTL, GEC, Plessey and BT.

Automotive and motor vehicle engineering
Cars, commercial vehicles (lorries and vans), motorcycles, tractors and specialized vehicles from companies such as Vauxhall and McLaren.

Activity 2.1

Marine engineering is an important engineering sector. Prepare a brief word processed report suitable for a foreign investor who is considering investing in the UK marine engineering sector. You need not include detailed financial information but you should include a description of the sector that identifies *at least four* marine engineering activities in the UK. You should also include contact and other information for *at least three* UK engineering companies that are active in this field.

Activity 2.2

A group of Estate Agents has asked you to carry out some research for them based on your home town. They are particularly interested in the size and scope of local engineering firms with a view to future developments.

(a) Identify the three largest engineering firms in the area of your home town.
(b) Find out who owns these firms and whether they are part of a larger group of companies.
(c) Allocate each firm to the appropriate engineering sector.
(d) Ascertain the annual sales turnover (its output in money terms) of each firm.
(e) Draw a histogram showing the individual engineering sector turnover in your area.

Present your findings in the form of a brief word processed report and include relevant diagrams and tables.

Activity 2.3

Visit your local museum and obtain information concerning the engineering companies that were active in your area (city, town or county) in 1900 and in 1950. Allocate each of these companies to a particular engineering sector and compare your findings with the present day situation. Identify any new industries that have appeared in your area since 1950 and explain why they have developed. Present your findings in the form of a brief class presentation supported by appropriate visual aids.

Engineering and the economy

To put your investigation of a product or engineering service into context, it is important to be aware of the contribution made by engineering to the economy at four different economic levels: local, regional, national and European. In each of these economies the engineering activities appear to be spread unevenly. This is the case whether we examine a local town or rural area, a larger region of possibly two or three counties, the whole of the UK on a national basis or the European Union.

Local economy

For the first half of the twentieth century, engineering was generally located within cities. Since then there has been a tendency for any new engineering enterprise to be located in an industrial estate on the periphery of a town rather than in the town itself. This is because:

- the town centre is already too congested to allow for additional new industry;
- of the advantages of being located in a ready, purpose-built industrial accommodation on a site having good road links with the national motorway network;
- engineering activities which may involve noise and other pollutants are best kept away from the commercial and domestic centres of towns.

In general, the engineering industries that remained in the city centres have slowly become outdated and, in many cases, have closed down. The impact of this migration from the city centres to the suburbs has been to leave derelict buildings, unemployment and social deprivation for the city residents. For the outer suburbs receiving the new engineering industries, the impact has not always been positive. The decentralization of engineering from the city centres has contributed to urban sprawl, and this has led to conflict for land on the city's boundary between engineering, farming and recreation. Also, it has tended merely to move the problem of engineering pollution from the city centre to its suburbs.

Regional economy

The regional economy comprises many local economies but the change in the engineering pattern is much the same. While there is still a great deal of engineering activity to be found in and around many large cities and built-up areas, there is a definite migratory move towards the small town and rural areas. This trend is to be found in most economically developed countries and has been a consistent feature of the last 25 years.

National economy

At the national level the uneven spread of engineering is between the different regions. The processes that caused this variation are historic. Very often they are directly connected to the availability of natural resources. For example, in the nineteenth century, regions rich in coal were favoured with engineering expansion because of the local availability of coal to fire boilers to drive the steam engines that powered the factories. The technical skills acquired by the workers in the coal bearing regions were the same skills required for other industries and enterprises and cumulative expansion took place. This expansion, and the highly paid work it created, attracted labour from the less industrialized regions so exacerbating the regional disparities.

However, over the last 50 years there has been a shift of engineering away from the old industrial regions such as the North-East and Midlands of England and parts of Scotland to more convenient locations such as the Thames Valley along the M4 motorway and along the M11 motorway north of London. The reasons for the regional shift are many and varied and include such factors as:

- because of its cost and pollution causing record, coal is no longer a popular fuel;
- with natural gas and electrical power being available almost anywhere in the country, new engineering activities can be located in regions having pleasant natural and social environments;
- because of the ubiquitous motor car, good roads and frequent air services, commuter and business communications to most regions are no longer a major problem;
- the availability of a pool of technologically skilled labour in places where high-technology companies are clustered together.

European Union

Within the EU, engineering activities have the usual varied pattern. The favoured countries are those which were the first to industrialize in the eighteenth and nineteenth centuries. Britain, Germany, France and Italy are predominant in Europe with the main concentration lying within a rough triangle formed by London, Hamburg and Milan. Ireland, Spain, Southern Italy and Greece lie outside this triangle and tend to be less industrialized.

The past 30 years has seen a shift in some of the major engineering activities which used to be concentrated in Europe, North America and Japan. In particular, much of the electronics and printing industries have migrated to the *Pacific rim* countries such as Hong Kong, Singapore, Taiwan, Thailand and more recently into Indonesia. The main reason for this shift is the low labour costs to be found in the Far East.

Another prime example of the shift of engineering activities out of Europe is that of shipbuilding. Britain's contribution in particular has fallen and is now virtually nonexistent except for the manufacture of oil platforms and ships for the Royal Navy.

However, the traffic in engineering activities has not been all negative. The Japanese, wanting to sell their motor cars in Europe, have established several engineering production plants in Britain. The firms of Toyota (Deeside and Burnaston), Nissan (Sunderland) and Honda (Swindon) are three good examples. All occupy rural sites and have access to skilled and well-educated workforces. Road communications are good and, in the case of Nissan, the site is in an assisted area where substantial Government grants are available.

Activity 2.4

A company based in the Far East has asked you to carry out some research in order to help them investigate some investment opportunities in Europe. The Managing Director has asked you to produce a broad comparison of the performance of the various countries that comprise the European Union (EU).

(a) Draw a map of Europe (Hint: Photocopy or scan this from a suitable atlas).
(b) Mark in the countries that constitute the EU.
(c) Mark in the capital cities of each EU country.
(d) Ascertain the GDP for each EU country then list the countries in descending order of GDP. What is the position of the UK?

Produce your findings in the form of a set of overhead projector transparencies and handouts to be used at a Board meeting.

Test your knowledge 2.1

Explain why it can be advantageous for Japanese firms to manufacture goods in the UK.

Engineering roles

The Engineering Council divides engineers into three specific categories: Chartered Engineer, Incorporated Engineer and Engineering Technician. These categories are not used in all branches of engineering but the roles are generally well understood and serve as useful benchmarks with which to compare the roles of engineers working in a wide variety of engineering sectors. The roles can be summarised as follows:

- *Chartered Engineers:* Chartered engineers are characterised by their ability to develop appropriate solutions to engineering problems, using new or existing technologies. Engineers are variously engaged in technical and commercial leadership and possess effective interpersonal skills.
- *Incorporated Engineers:* Incorporated engineers maintain and manage applications of current and developing technology, and may undertake engineering design, development, manufacture, construction and operation. Incorporated Engineers are variously engaged in technical and commercial management and possess effective interpersonal skills.
- *Engineering Technicians:* Engineering technicians are involved in applying proven techniques and procedures to the solution of practical engineering problems. They carry supervisory or technical responsibility, and are competent to exercise creative aptitudes and skills within defined fields of technology. Engineering Technicians contribute to the design, development, manufacture, commissioning, operation or maintenance of products, equipment, processes or services.

Other skilled staff may operate equipment such as lathes and many 'crafts' or 'trades people' use a range of skills to require manual dexterity or hand tool skills. As part of the assessment of this unit you should ask a practising engineer about his or her role and how this relates to the company and engineering sector that he or she works in. A visit to the workplace can be extremely useful in helping you to understand the work role and the way in which engineering companies are organised.

Finally, because engineers may be associated with several different departments in an engineering company we shall now look at the way these different departments work and the work of engineers within them.

Test your knowledge 2.2

Describe the THREE types of engineer defined by the Engineering Council.

Design and development

New product design and development is often a crucial factor in the survival of a company. In an industry that is fast changing, firms must continually revise their design and range of products. This is necessary because of the relentless progress of technology as well as the actions of competitors and the changing preferences of customers. This area is the province of the *design engineer*.

A good example of this situation is the motor industry. The British motor industry has gone through turbulent times, caused by its relative inefficiency compared with Japan and Germany and also because the quality of its products was below that of its competitors. This difficult position was then made worse by having products that lacked some of the innovative features of the competition.

Strategies for product development

There are three basic ways of approaching design and development:

- driven by marketing
- driven by technology
- coordinated approach.

A system driven by marketing is one that puts the customer needs first, and only produces goods which are known to sell. Market research is carried out which establishes what is needed. If the development is technology driven then it is a matter of selling what it is possible to make. The product range is developed so that production processes are as efficient as possible and the products are technically superior, hence possessing a natural advantage in the market place. Marketing's job is therefore to create the market and sell the product.

Both approaches have their merits, but each of them omit important aspects, hence the idea that a coordinated approach would be better. With this approach the needs of the market are considered at the same time as the needs of the production operation and of design and development. In many businesses this inter-functional system works best, since the functions of R&D, production, marketing, purchasing, quality control and material control are all taken into account. However, its success depends on how well the interface between these functions is managed and integrated. Sometimes committees are used, as are matrix structures or task forces (the latter being set up especially to see in new product

Another view

This unit is all about the work that engineers do. However, it's very important to realise that this work can be extremely diverse. The Cambridge International Dictionary of English defines an engineer as "a person whose job is to design or build machines, engines or electrical equipment, or things such as roads, railways or bridges ....". The definition ends with " ... an engineer is also an engine driver." Perhaps it shouldn't be much of a surprise to find that the public at large has some difficulty in understanding the role of the engineer!

developments). In some parts of the motor industry a function called *programme timing* coordinates the activities of the major functions by agreeing and setting target dates and events using network planning techniques.

The development process

The basic process is outlined as follows:

- idea generation
- selection of suitable products
- preliminary design
- prototype construction
- testing
- final design.

This is a complex process and involves co-operative work between the design and development engineers, marketing specialists, production engineers and skilled engineering technicians to name some of the major players.

Ideas can come from the identification of new customer needs, the invention of new materials or the successful modification of existing products. Selection from new ideas will be based on factors like:

- market potential
- financial feasibility
- operations compatibility.

This means screening out ideas that have little marketability, are too expensive to make at a profit and which do not fit easily along-side current production processes.

After this, preliminary designs will be made within which trade-offs between cost, quality and functionality will be made. This can involve the processes of *Value Analysis* and *Value Engineering*. These processes look at both the product and the methods of production with a view to maintaining good product performance and durability whilst achieving low cost.

Test your knowledge 2.3

List each of the main stages in the development of a new product.

Prototypes are then produced, possibly by hand and certainly not by using mass production methods. This is followed by rigorous testing to verify the marketing and technical performance characteristics required. Sometimes this process will involve test marketing to check customer acceptance of the new product.

Final design will include the modifications made to the design as a result of prototype testing. The full specification and drawings will be prepared so that production can be planned and started.

Production and manufacture

Once the final design has been completed the focus passes to the next stage, i.e. that of actually manufacturing the product. This is the role of the *production engineer*. New products will have different characteristics, and will perhaps be made from different materials from previous products with similar functionality. They may also require different processes in their manufacture. All of this will require liaison between *design engineers* and *production engineers* on methods for production and in deciding what manufacturing equipment, machine tools and processes are required. Detailed process sheets will be required that show how products are to be assembled or made.

Whilst the particular methods of production are the province of production management, the designer has to be aware of the implications for his design of different methods of manufacture, whether this be batch production, assembly lines or one-off projects. Detailed specifications of the new and changed product will be communicated and there may be liaison on temporary and permanent deviations to original specifications in order to facilitate production.

When quality problems appear and are related to faulty design there will be liaison on ways in which design modifications can be phased into production as soon as possible.

There will be proposals for the replacement of machines and equipment used for manufacturing and production. This function may require quite sophisticated techniques for what is called *investment appraisal* so that the company can choose the best methods of manufacture from several alternatives.

Also important is the control of raw materials and component stocks, especially the levels of *work-in-process*. Finance will want to restrict stock levels to reduce the amount of capital tied up in stocks, whilst the production manager will be concerned with having sufficient stock to maintain production, but avoiding congestion of factory floor space.

Budgetary control of production cost centres will involve regular contact and advice from the Finance function. Matters of interest will be costs of production, wastage rates, labour costs, obsolescent stock, etc. Typical reporting lines within the production and manufacturing functions of a large engineering company are shown in Figure 2.1.

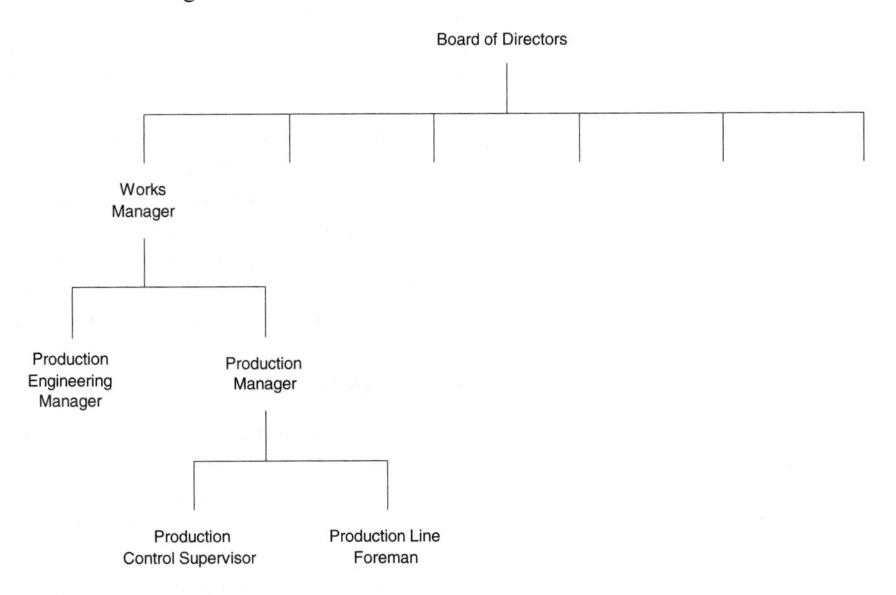

Figure 2.1 *Typical reporting lines within the production and manufacturing function of a large engineering company*

The production or manufacturing operation is at the heart of the business. It translates the designs for products, which are based on market analysis, into the goods wanted by customers.

Process and facilities management

Decisions have to be made by *production engineers* in relation to location of the factory and the design and layout of production facilities. The design of production processes also requires interaction with design engineers and marketing functions.

Selecting the process of production is important and is strategic in nature. This means that it has a wide impact on the operation of the entire business. Decisions in this area bind the company to particular kinds of equipment and labour force because the large capital investments that have to be made limit future options. For example a motor manufacturer has to commit very large expenditures to lay down plant for production lines to mass produce cars.

Once in production, the company is committed to the technology and the capacity created for a long time into the future. There are three basic methods for production processes:

- line flow
- intermittent flow
- project.

Line flow is the type of system used in the motor industry for assembly lines for cars. It also includes continuous type production of the kind that exists in the chemicals and food industries. Both kinds of line flow are characterized by linear sequences of operations and continuous flows and tend to be highly automated and highly standardized.

Intermittent flow is the typical batch production or job shop, which uses general purpose equipment and highly skilled labour. This system is more flexible than line flow, but is much less efficient than line flow. It is most appropriate when a company is producing small numbers of non-standard products, perhaps to a customer's specification.

Finally *project-based production* is used for unique products which may be produced one at a time. Strictly speaking there is not a flow of products, but instead there is sequence of operations on the product which have to be planned and controlled. This system of production is used for prototype production in R&D and is used in some engineering companies that produce major machine tool equipment for other companies to use in their factories.

Capacity planning

Once facilities for production have been put in place the next step is to decide how to flex the capacity to meet predicted demand. Production managers will use a variety of ways to achieve this from maintaining excess capacity to making customers queue or wait for goods to having stocks to deal with excess demand. The process is complex and may require the use of forecasting techniques, together with careful planning. The primary goal of scheduling is that of meeting customer delivery dates whilst at the same time making efficient use of resources.

Sales and marketing

Marketing

Marketing is arguably the most important function in any business, since if customers cannot be found for an engineering company's products the company will go out of business, regardless of how financially well run or efficient it is or how good its products are!

Although sales is considered separately later, it is really part of the marketing function. Marketing is all about matching company products with customer *needs*. If customer needs are correctly identified and understood, then products can be made which will give the customer as much as possible of what he wants. Companies which view the customer as *sovereign* are those companies that tend to stay in business, because customers will continue to buy products that meet their requirements.

Hence marketing activities are centred around the process of filling customers' known needs, discovering needs the customer does not yet know he or she has and exploiting this by finding out how to improve products so that customers will buy this company's products in preference to other goods. Some of the most important activities are:

- market research
- monitoring trends in technology and customer tastes
- tracking competitors' activities
- promotion activities
- preparing sales forecasts.

Remember that in some businesses the marketing activity is directed at end consumers, members of the public. This has to be done by national forms of advertising, such as TV commercials, direct mail, newspapers or through major retailers selling to the consumer. The methods used may be somewhat different if the customers are other companies. Although the principle of meeting customer needs is the same, the approach taken may be much more technical and may include the services of *sales engineers* to provide technical back-up and advice. The publicity methods are more likely to be centred around trade fairs, exhibitions, advertisements in the trade press or technical journals, for example. You should note these two distinct marketing approaches are respectively called consumer marketing and industrial marketing.

Test your knowledge 2.4

Understanding the needs of customers invariably underpins all other marketing activities. Explain why this is.

Sales

The sales department is concerned with advertising and selling goods. It will have procedures for controlling sales and the documentation required. The documents used are the same as for purchasing, described later, except from the supplier's viewpoint rather than from the customer's. It may employ commercial travelers or have a resident sales force. It is involved with many possible ways of publicizing the company's products such as trade fairs, wholesalers' displays, press and TV advertising, special campaigns, promotional videos, etc. It will also be concerned about the quality of goods and services as well as administering warranty and guarantee services, returns and repairs, etc. Sales will maintain contacts with customers that will entail the following customer services:

- technical support
- after-sales service, service engineering
- product information, prices and delivery
- maintaining customer records.

A typical organizational structure for the marketing and sales function in a large engineering company is shown in Figure 2.2.

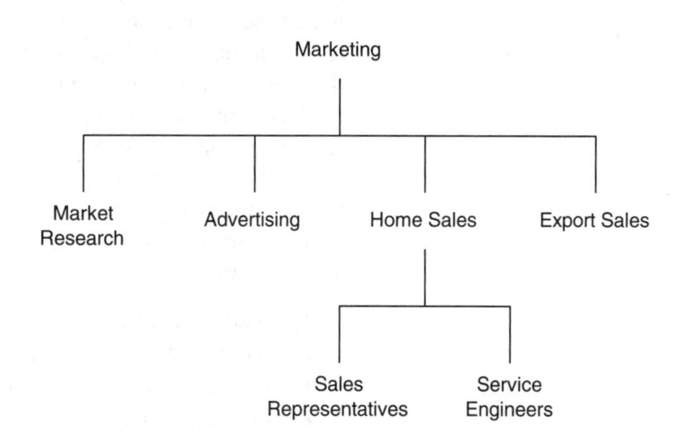

Figure 2.2 *Organization of the marketing and sales functions in a large engineering company*

Activity 2.5

A local engineering company has asked you to assist with the development of a new product. This product is to be aimed at the DIY market and is to consist of a combination steel rule, drill gauge and spirit level.

(a) Suggest ways of determining the market potential and financial feasibility of this product.

(b) Identify the materials and processes that could be used to manufacture the product – are these materials and processes available in your school or college?

(c) Explain why it would be necessary to test market this product.

Present your findings in a brief word processed report.

Constraints

Engineers have to operate under a variety of different constraints. However, these will be different for different engineering roles. For example, the constraints that might face a *design engineer* could involve the need to comply with European legislation whilst keeping costs as low as possible. In relation to the same product, the constraints facing a *production engineer* might centre around the high costs associated with a particular manufacturing process or the need to use materials that are hazardous. A *sales engineer* supporting the same product might face the need to become fluent in a foreign language or to provide support for a range of products that will soon become obsolete because of the introduction of new technology.

When you carry out your own investigation of an engineer (and the products or engineering services with which he or she is concerned) it is important to understand that some constraints result from the external environment (for example, European legislation) whilst others may result from internal factors (such as the availability of a particular process). External constraints are something that most companies have little control over. Internal constraints, on the other hand, can often be resolved by mutual agreement. When questioning your chosen subject, some of the questions that you might wish to ask include:

- What are the particular constraints associated with the product or service and how do you overcome them?
- What materials and processes are used to manufacture the product or deliver the service and why are they used?
- What technologies are used in the design and manufacture of the product or service and why are they used?
- What standards and legislation (e.g. Health and Safety legislation) relates to the product or service and what impact has this had on design, manufacture and working practices within the company?

Activity 2.6

A small engineering company manufactures power supplies that are used in conjunction with a rack-mounted computer system. Each power supply is fitted with a large backup battery. The Marketing Manager has indicated that the battery should be a rechargeable nickel-cadmium (NiCd) type but the Chief Design Engineer has specified a sealed lead-acid battery because this type of battery is currently used in several other products manufactured by the company. The company's Production Manager has asked you to help resolve this problem and help her to arrive at a final decision as to which type of battery is to be used. List FIVE factors that should be taken into consideration and, for each factor specify any internal or external constraints that might apply. Present your work as a briefing note that will be used at a meeting.

Test your knowledge 2.5

List THREE constraints under which engineers may have to work.

2.2 The application of technology in engineering

Technology has had a huge impact on engineering and the engineering industry. Not only has it revolutionised the way that we design and manufacture products and services but it has also given us an exciting array of new materials, processes and techniques.

We have already looked at new materials and processes in Unit 1, in this section we shall explore the ways in which engineers make extensive use of new technology and, in particular, how it has been instrumental in changing the way that engineers design and manufacture products. A prime mover in all of this has been the widespread availability of the microcomputer.

Computer aided engineering

You will probably already know a little about computer aided design (CAD) however this is just one aspect of computer aided engineering (CAE). Computer aided engineering is about automating *all* of the stages that go into providing an engineered product or service. When applied effectively, CAE ties all of the functions within an engineering company together. Within a true CAE environment, information (i.e. data) is passed from one computer aided process to another. This may involve computer simulation, computer aided drawing (drafting), and computer aided manufacture (CAM).

The term, CAD/CAM, is used to describe the integration of computer aided design, drafting and manufacture. Another term, CIM (computer integration manufacturing), is often applied to an

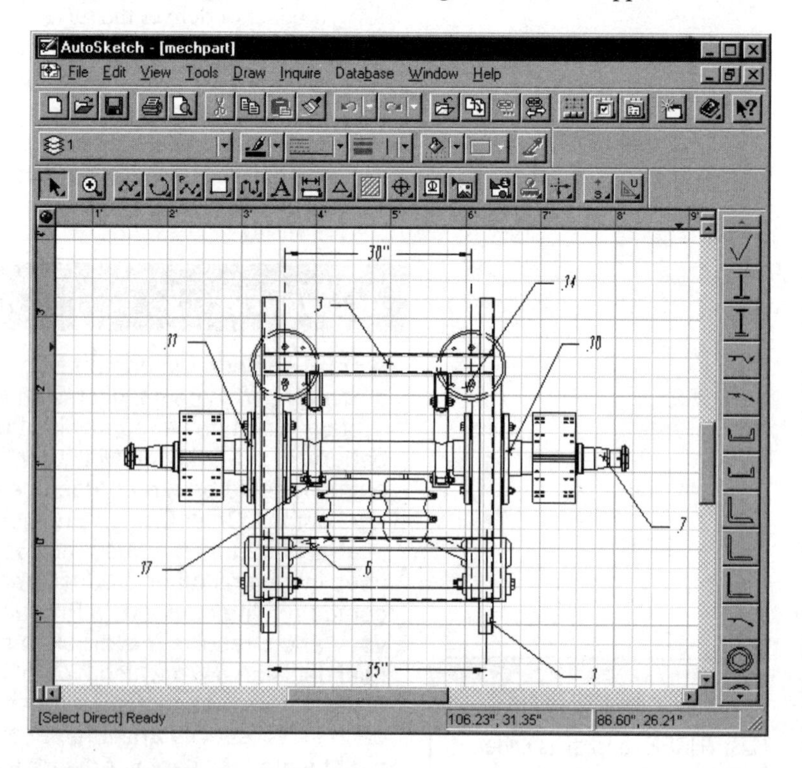

Figure 2.3 *A drawing of a mechanical part using a popular 2D CAD package*

environment in which computers are used as a common link that binds together the various different stages of manufacturing a product, from initial design and drawing to final product testing.

Whilst all of these abbreviations can be confusing (particularly as some of them are often used interchangeably) it is worth remembering that 'computer' appears in all of them. What we are really talking about is the application of computers within engineering. Nowadays, the boundaries between the strict disciplines of CAD and CAM are becoming increasingly blurred and fully integrated CAE systems are becoming commonplace in engineering companies.

We have already said that CAD is often used to produce engineering drawings. Several different types of drawing are used in engineering. Some examples of different types of CAD drawing are shown in Figures 2.3 to 2.8.

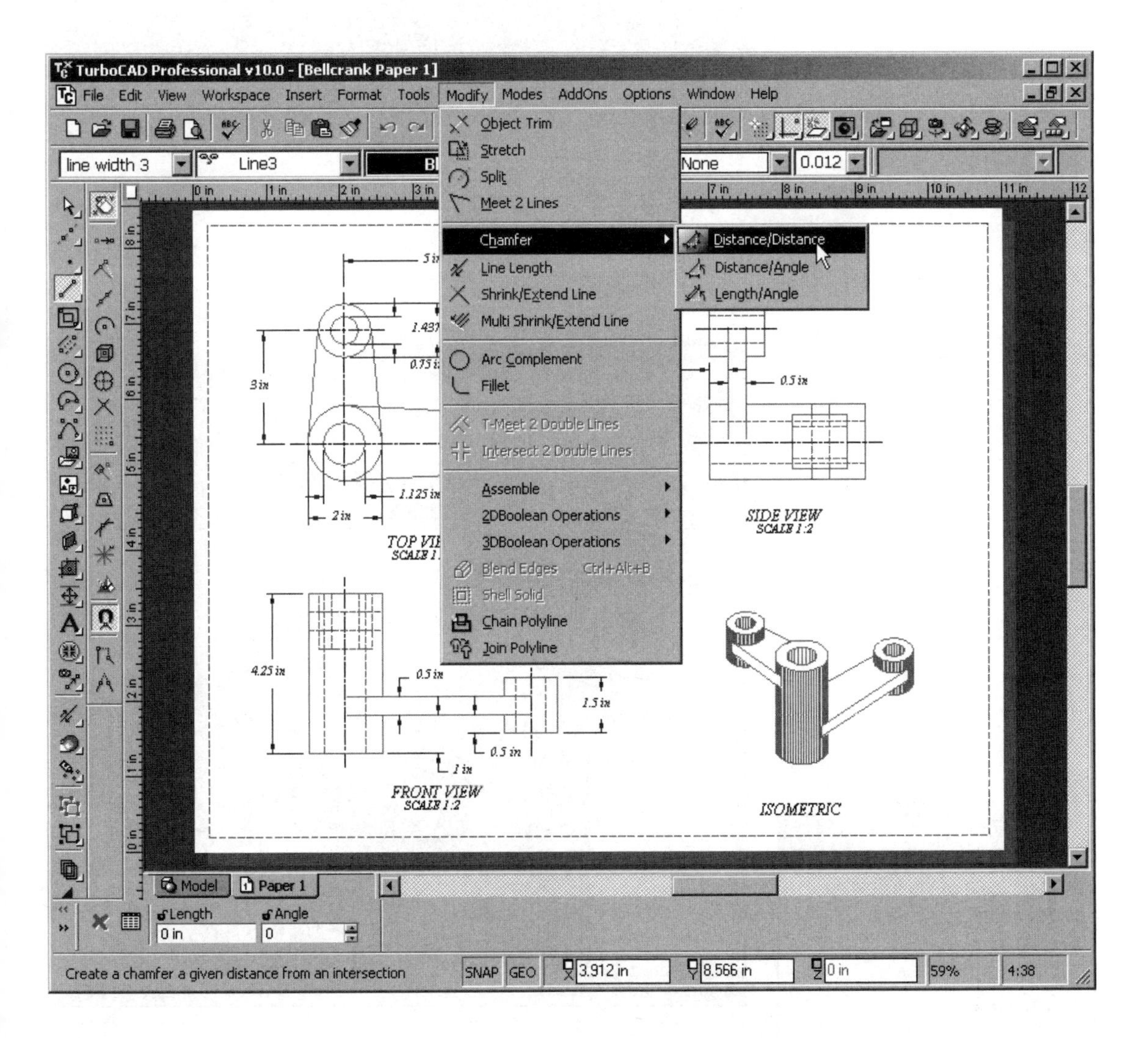

Figure 2.4 *Drop-down menus and extensive toolbars are available in this powerful CAD package which incorporates both 2D and 3D features*

Figure 2.5 *A rendered 3D representation of a car using a popular 3D CAD package*

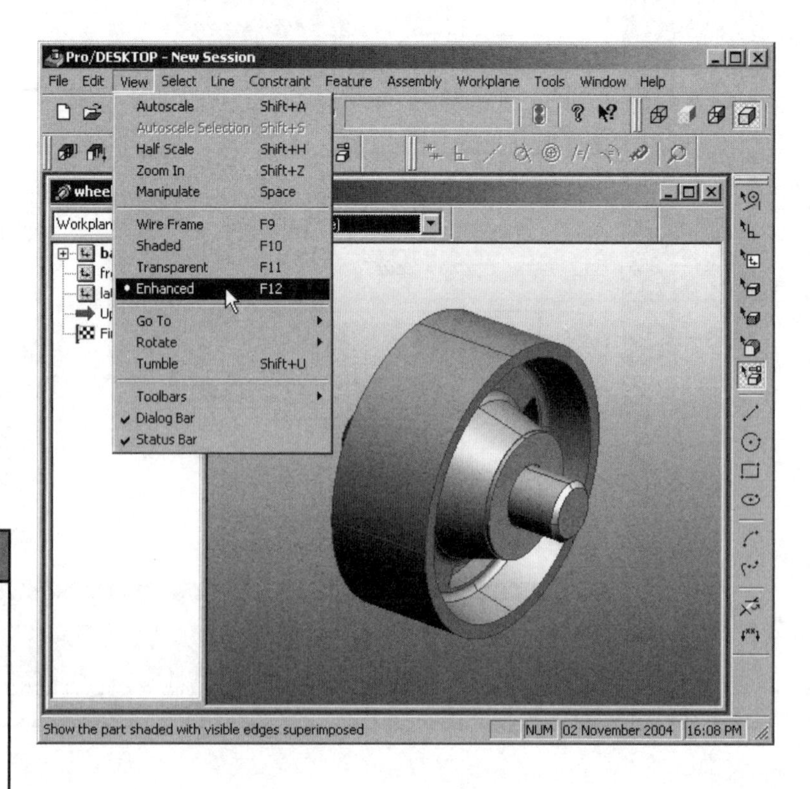

Figure 2.6 *A rendered 3D representation of a flywheel*

Test your knowledge 2.6

What do each of the following abbreviations stand for?

(a) CAD
(b) CAM
(c) CAD/CAM
(d) CIM
(e) CNC
(f) CAE.

2D CAD

Two-dimensional (2D) CAD packages are widely used by engineers for creating conventional 'flat' drawings. Typical of these 2D packages are 'industry standard' packages like Autosketch, AutoCAD, TurboCAD, DesignCAD and FastCAD.

Older 2D CAD programs accept text commands or a combination of buttons following by coordinates or other parameters entered as text. More modern packages make more use of the graphical user in interface (GUI) available in modern computers but they may still require precise dimensions to be entered from the keyboard during a drawing session. For example, in the A9CAD 2D CAD package the following commands draw a circle inscribed into a square with a side domension of five units (see Figure 2.7):

Command: RECTANGLE (by clicking on the rectangle tool button)
First corner: 0,0
Other corner: 10,10
Draws a square with a side of length 10 units starting at the origin, 0,0.
Command: CIRCLE (by clicking on the circle tool button)
Circle center point: 5,5
Circle radius: 5
Draws a circle with a radius of 5 units centred on 5,5.

Most modern CAD programs are reasonably intuitive and will allow you to quickly create drawings using standard templates and symbol libraries. A variety of drawing tools will be provided which will allow you to assign different properties to drawing entities (such as a dashed line or a hatched rectangle) or to snap, glue or group entities together. Intelligent connectors (which rebuild automatically when

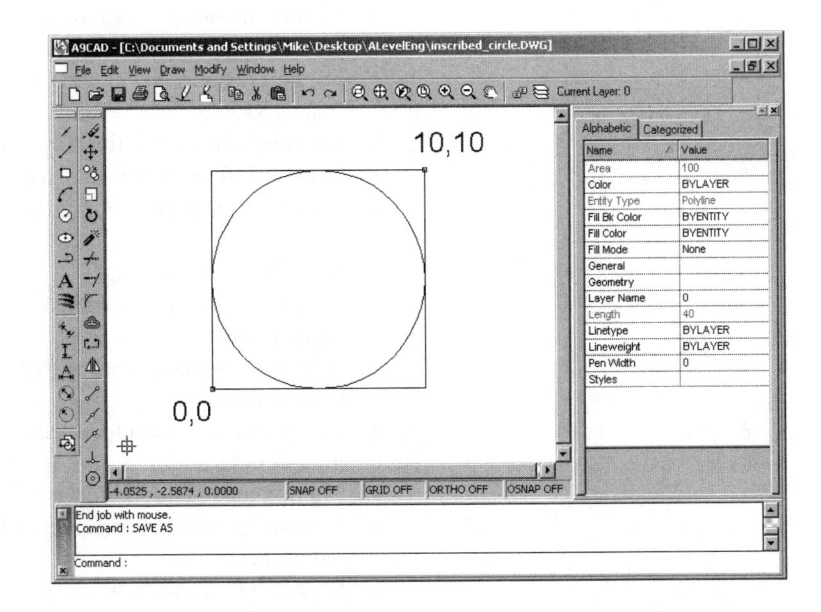

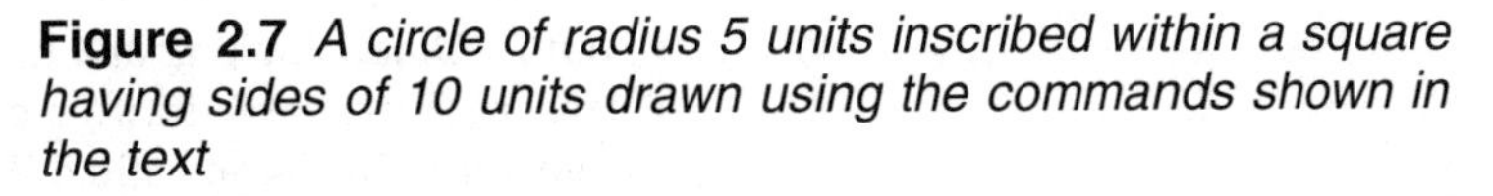

Figure 2.7 *A circle of radius 5 units inscribed within a square having sides of 10 units drawn using the commands shown in the text*

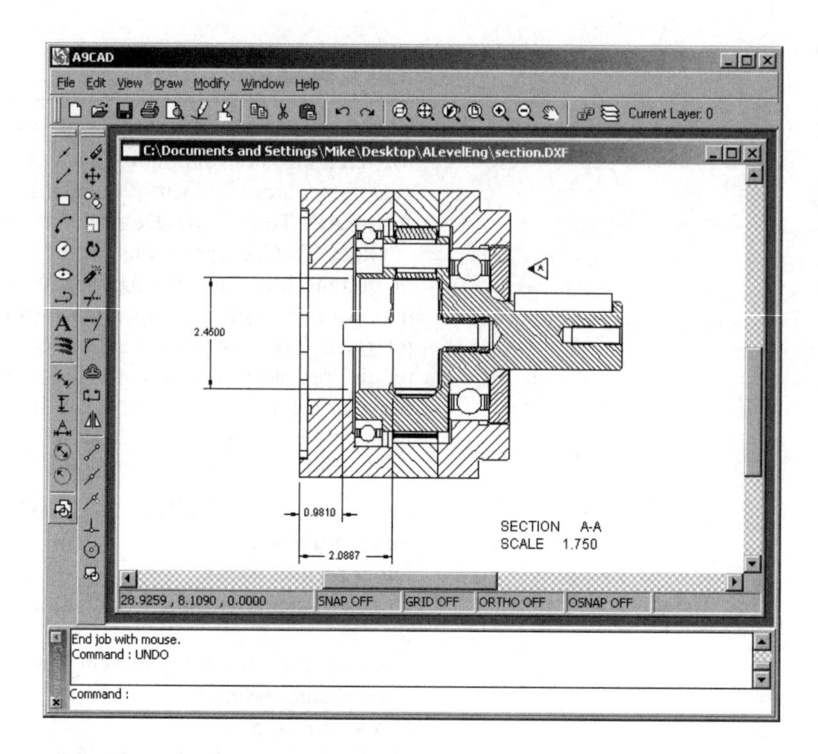

Figure 2.8 *A more complex 2D drawing showing the use of hatching and dimensioning*

you reposition objects and avoid crossing other objects) allow you to easily create charts and schematics.

The features available from a modern 2D CAD package include the following:

- Various modes for creating lines, arcs, circles, ellipses, parallel line, angle bisectors, etc.
- Support (import and export) for standard CAD file formats (e.g. DXF)
- Standard and user-defined symbol libraries
- Text in different fonts and sizes
- Automatic dimensioning of distances, angles, diameters, and tolerances
- Solid fills and hatching
- Support for layers and blocks (which can be manipulated as a separate entity)
- Selection and modification tools (move, rotate, mirror, trim, stretch, etc.)
- Snapping to grid and to objects (endpoints, centers, intersections, etc.)
- Multiple undo and redo levels
- Support for various units including metric, imperial, degrees, radians, etc.
- Import and export of bitmaps and other images in various format (e.g. BMP, JPEG, PNG, etc.)
- OLE (Object Linking and Embedding) compatibility (this means that a CAD drawing can be inserted and edited within any other OLE-compatible Windows application).

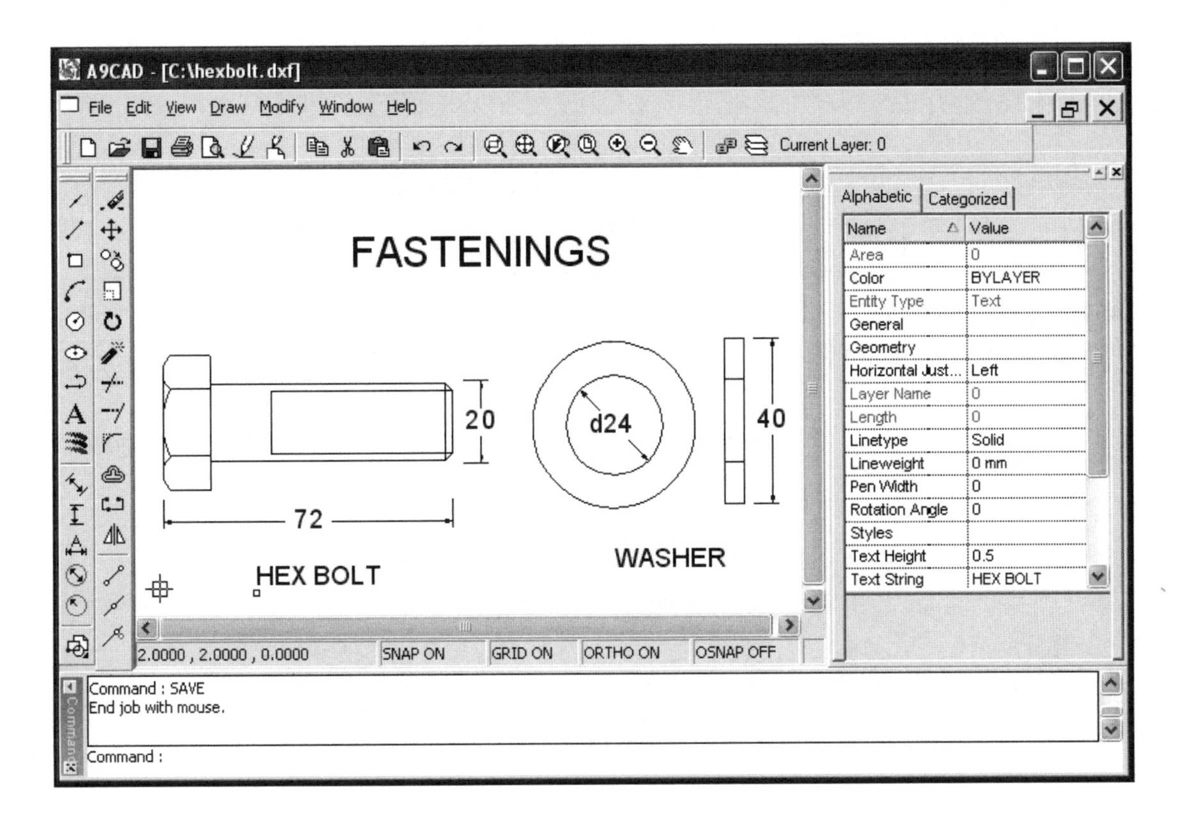

Figure 2.9 *See Activity 2.7*

Test your knowledge 2.7

List SIX features of a modern 2D CAD package.

Activity 2.7

A9CAD is a general purpose two-dimensional CAD program which is available from our companion website (see Preface for details). Use A9CAD (or a similar 2D CAD package) to create the drawing of the fastenings shown in Figure 2.9. Save and print your completed drawing.

3D CAD

A 3D CAD package produces a model of a component part or product which can be viewed from any desired angle. 3D CAD packages may operate in *wireframe mode* (in which you will see a skeletal outline of the component or product) or in *render mode* (in which you will see a more realistic shaded image). Packages will usually provide you with both *orthographic views* (see Figures 2.11 and 2.12) and *isometric views* (as shown in Figure 2.13). These views will often be displayed in multiple windows so that it is possible to view a model from several different points at the same time. In addition, more advanced packages provide a *camera view* that can be

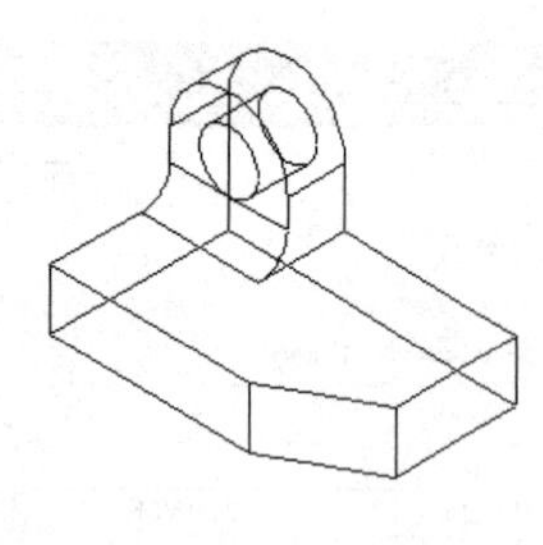

Figure 2.10 *A 3D wire-frame view of a component*

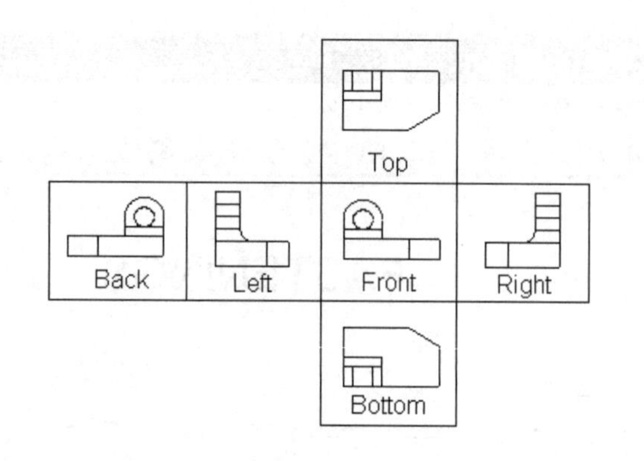

Figure 2.11 *Orthographic views of the component shown in Figure 2.10*

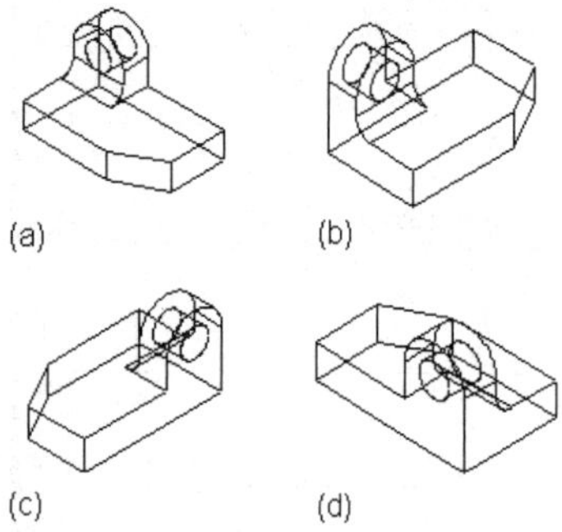

(a) (b) (c) (d)

Figure 2.12 *Isometric views of the object shown in Figure 2.10*

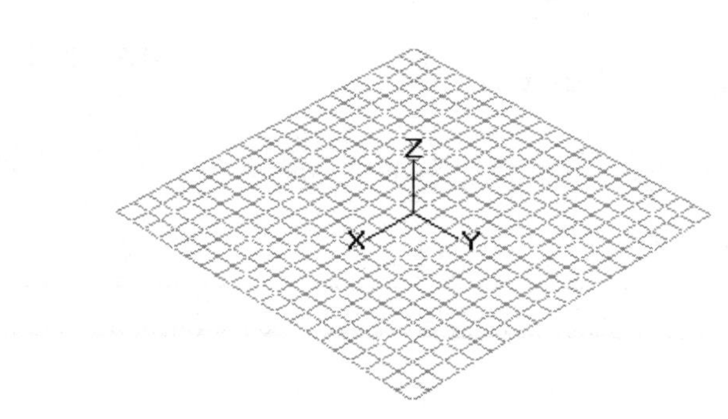

Figure 2.13 *A 3D workplane showing reference axes*

used to provide additional perspective and which will allow you to examine a component part or a product from any desired angle.

3D CAD packages use a three-axis coordinate system and a plane of reference (the *workplane*) as shown in Figure 2.13. In addition, some packages also allow you to define your own user coordinate system which has a different workplane that travels with an object and which is usually defined first in 2D mode.

The workplane is the plane in which a 2D object is initially created (i.e. as a 2D drawing). In 2D mode, you will normally do all your work in the same workplane but in 3D mode it is frequently necessary to change the workplane in order to perform all the required commands.

One of the most important tools for controlling the view of a 3D model is that which allows you to render an object so that it looks both solid and realistic. In normal *rendering* modes, all 3D objects are displayed as shaded (with or without hidden lines). Higher level rendering will enable you to view materials and textures, thus providing a more realistic image of what a component or your model will actually look like. In order to create a realistic rendered view lighting effects must be added and the positions of the light sources are usually made adjustable. In high-end packages, it may also be possible to further enhance the rendering of the image by assigning materials and luminance qualities to objects. This adds further realism to the finally generated image.

Test your knowledge 2.8

Sketch a 3D workplane and label the axes.

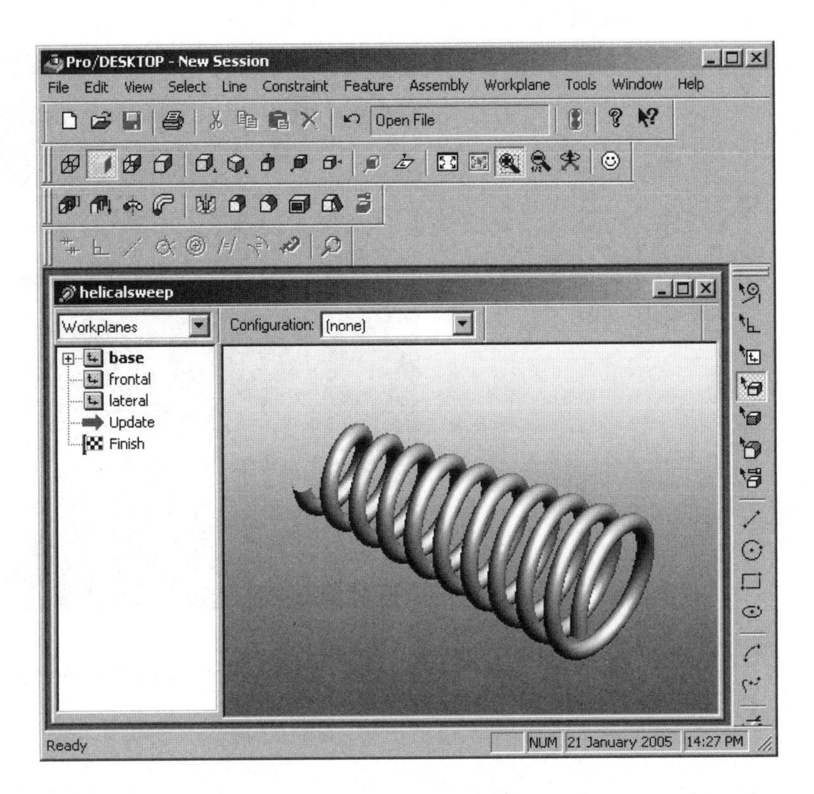

Figure 2.14 *A 3D rendered model of a helical spring*

Computer aided manufacture

Computer aided manufacture (CAM) encompasses a number of more specialized applications of computers in engineering including computer integrated manufacturing (CIM), manufacturing system modeling and simulation, systems integration, artificial intelligence applications in manufacturing control, CAD/CAM, robotics, and metrology.

Computer aided engineering analysis can be conducted to investigate and predict mechanical, thermal and fatigue stress, fluid flow and heat transfer, and vibration/noise characteristics of design concepts to optimize final product performance. In addition, metal and plastic flow, solid modeling, and variation simulation analysis are performed to examine the feasibility of manufacturing a particular part.

In addition, all of the machine tools within a particular manufacturing company may be directly linked to the CAE network through the use of centrally located floor managers which monitor machining operations and provides sufficient memory for complete machining runs.

Manufacturing industries rely heavily on computer controlled manufacturing systems. Some of the most advanced automated systems are employed by those industries that process petrol, gas, iron and steel. The manufacture of cars and trucks frequently involves computer-controlled robot devices. Industrial robots are

Figure 2.15 *A large CNC milling machine*

Figure 2.16 *Some typical components manufactured using CNC processes and steel bar*

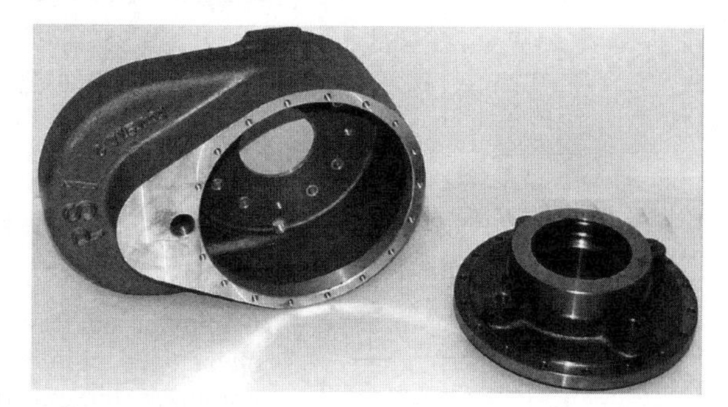

Figure 2.17 *Some typical parts manufactured using CNC processes on cast alloy components*

used in a huge range of applications that involve assembly or manipulation of components.

The introduction of CAD/CAM has significantly increased productivity and reduced the time required to develop new products. When using a CAD/CAM system, an engineer develops the design of a component directly on the display screen of a computer. Information about the component and how it is to be manufactured is then passed from computer to computer within the CAD/CAM system. After the design has been tested and approved, the CAD/CAM system prepares sequences of instructions for computer numerically controlled (CNC) machine tools and places orders for the required materials and any additional parts (such as nuts, bolts or

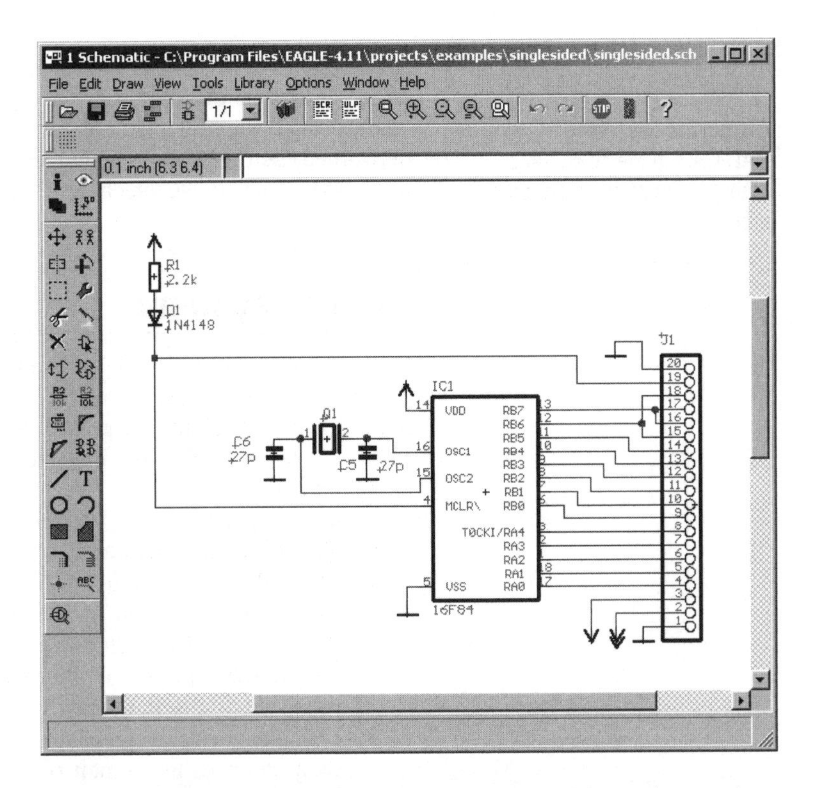

Figure 2.18 *Printed circuit board design and manufacture is a typical application for CAD/CAM. Here the electronic circuit schematic has been drawn prior to automatically generating the circuit layout.*

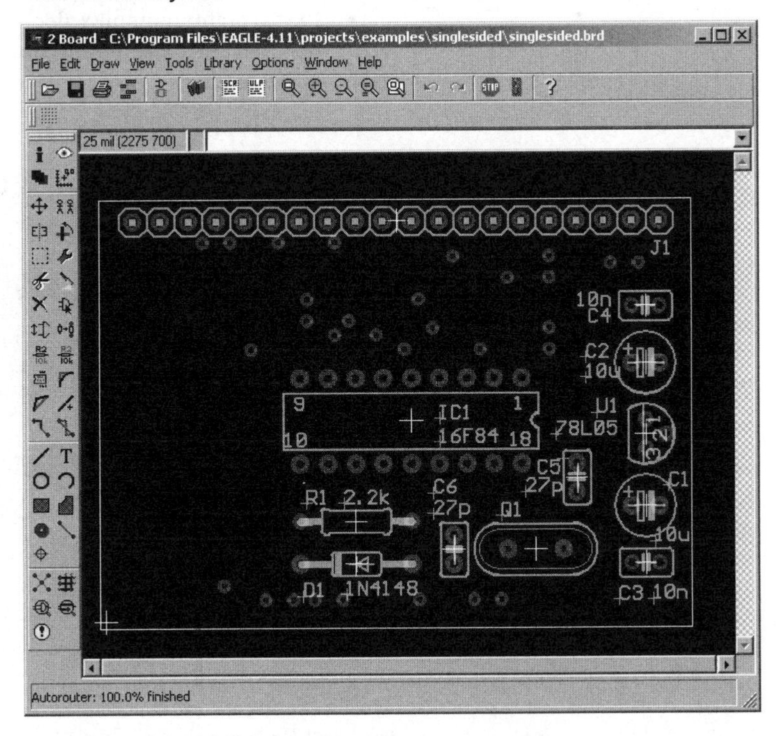

Figure 2.19 *The PCB track layout and component overlay automatically generated from the circuit schematic shown in Figure 2.18.*

Test your knowledge 2.9

Explain the advantage of using CAD/CAM in the design and manufacture of an engineered product.

adhesives). The CAD/CAM system allows an engineer (or, more likely, a team of engineers) to perform all the activities of engineering design by interacting with a computer system (invariably networked) before actually manufacturing the component in question using one or more CNC machines linked to the CAE system.

Control systems

Control systems are widely used in manufacturing and in process control acting as an intelligent controller. All that is required is sufficient peripheral hardware and the necessary software to provide an interface with the production/test environment.

Computer-based control systems offer several advantages over manual systems including:

- Flexibility and adaptability (systems can be easily extended or re-configured for different applications), Accuracy and repeatability (there is no need to rely on the judgement of a human operator which can be impaired due to fatigue or lack of experience)
- Low-cost (operating costs are usually significantly less than when using human operators and the technology is well known)
- Rugged and reliable (systems can be easily ruggedised for operation in harsh environments and systems can be designed so that they are 'failsafe')
- Fast (highly complex decisions can be made in fractions of a second, and production data can be stored for later analysis).

Figure 2.20 *A versatile single-board PC-compatible microcontroller designed for use in a wide range of industrial control system (Photo courtesy of Arcom)*

Test your knowledge 2.10

List THREE advantages of computer-based control systems.

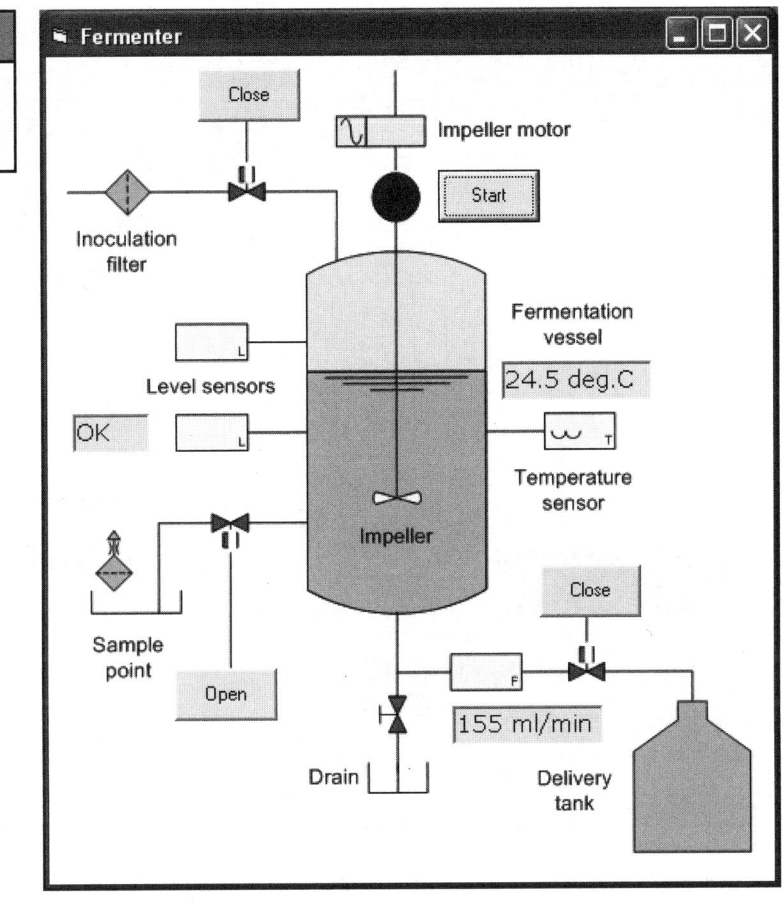

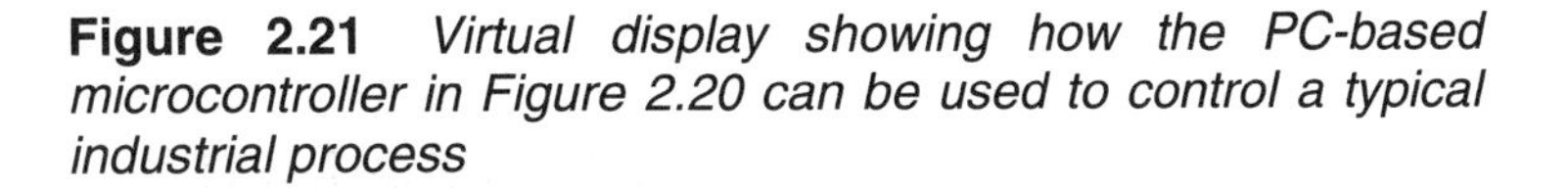

Figure 2.21 *Virtual display showing how the PC-based microcontroller in Figure 2.20 can be used to control a typical industrial process*

Simulation

We have already seen how modern technology is invaluable in the design and manufacture of products. Another important application of this technology is in the simulation and analysis of complex systems. Put simply, this means that you don't actually have to build or run the system in order to be able to carry out tests and measurements on it. This means that, for example, it is possible to determine what will happen when a malfunction occurs in a critical process (such as nuclear power plant). At the other extreme, it can be highly cost-effective to 'build' a virtual prototype of a simple electronic circuit (using virtual models of the components) in order to check that it meets the design specification before actually having to assemble a real circuit. Using this technique it becomes possible to determine the effect of 'what if' situations, such as wide variations in the supply voltage or a short-circuit imposed on the output.

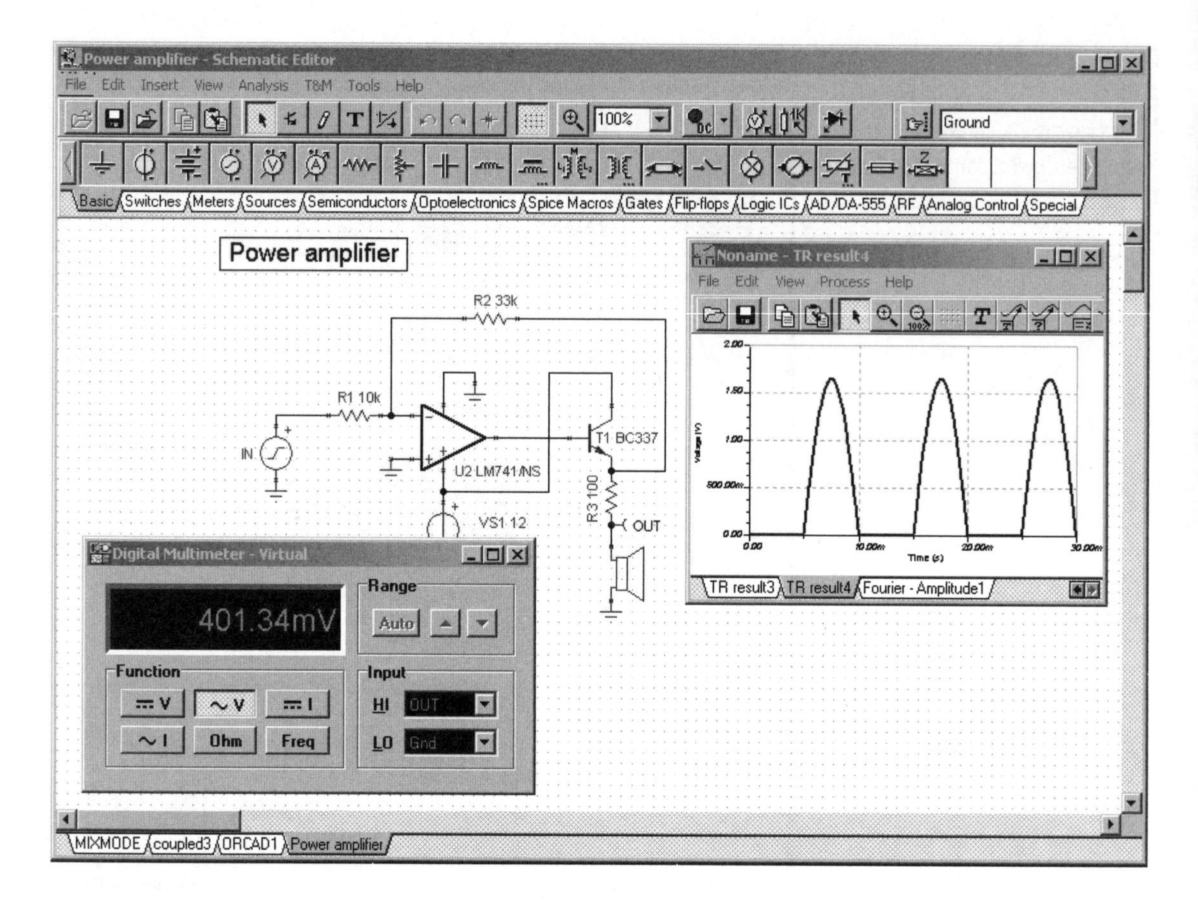

Figure 2.22 *Simulation of a simple electronic circuit prior to building the real circuit*

Virtual instruments

Computer-based instruments (i.e. *virtual instruments*) are rapidly replacing items of conventional test equipment in many of today's test and measurement applications. Virtual instruments can be used to build complex automated test systems but they can also be used in stand-alone laboratory, bench and field service applications.

Currently available virtual instruments include *digital storage oscilloscopes* (many of which incorporate additional features such as spectrum analysis, digital voltmeters and digital frequency meters), event counters and timers, waveform generators, and digital logic analysers.

It is important to remember that virtual instruments comprise both hardware and software. The hardware provides a means of interfacing to the computer and of connecting external inputs. Additional controls may also be present (for example, input selection, trigger selection, attenuation, etc). The software provides a means of controlling the instrument, collecting data from it, processing the data, and then displaying, analysing and recording it. When specifying virtual instruments for a particular application it is important to ensure that *both* the hardware *and* the software fully meet the specification.

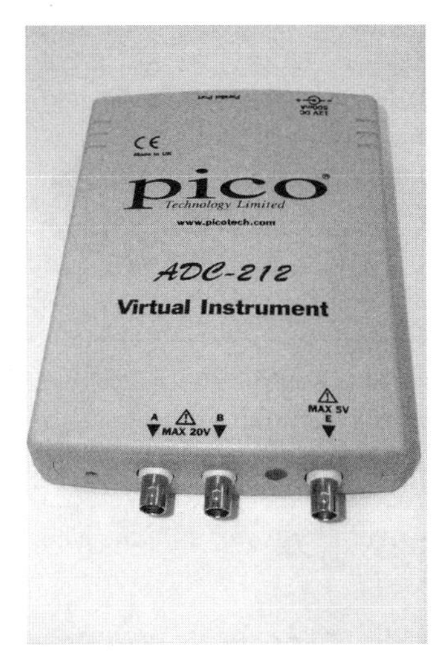

Figure 2.23 *A virtual oscilloscope test instrument*

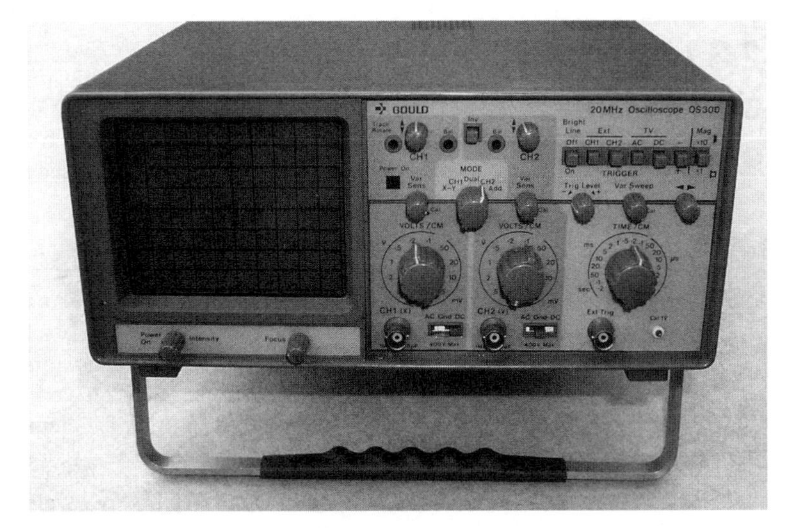

Figure 2.24 *A conventional oscilloscope test instrument*

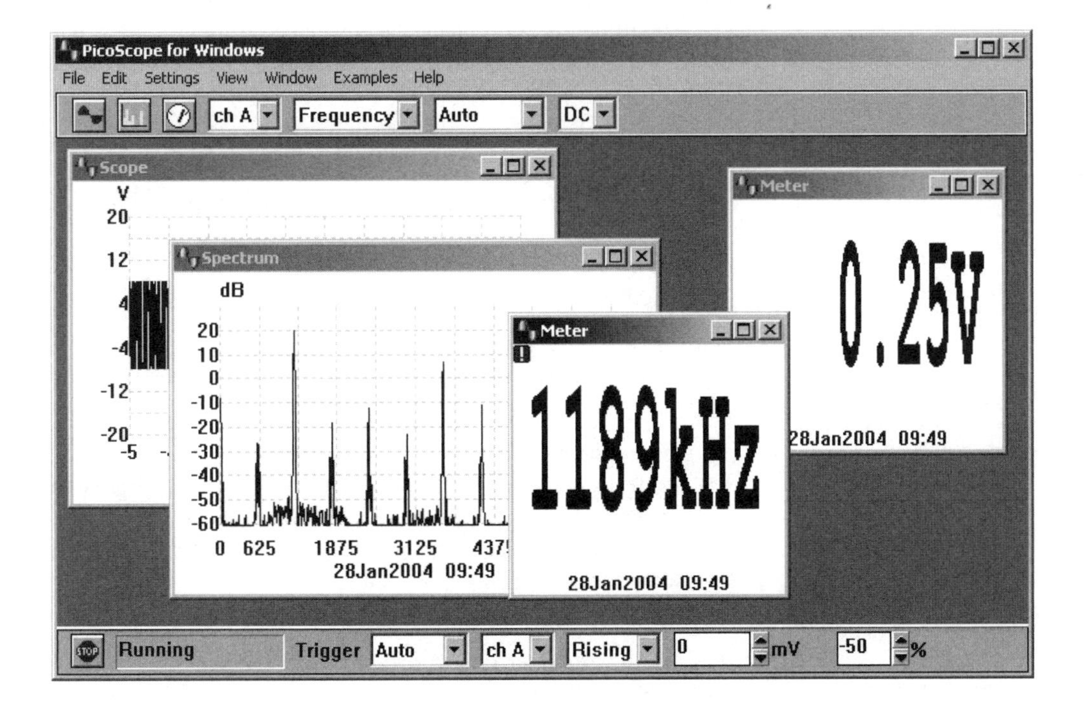

Figure 2.25 *Typical displays provided by a virtual oscilloscope instrument*

Activity 2.8

New technology is widely used in the telecommunications industry. Investigate the use of optical fibres in data communications. Present your findings in the form of a brief word processed report with appropriate diagrams.

Test your knowledge 2.11

Explain the meaning of the term 'virtual instrument'.

2.3 Legislation and standards in engineering

Engineering activities are governed by a variety of different legislation. Being aware of the implications of this legislation and conforming with the standards that underpin such legislation is vitally important. This section looks at legislation and standards and their impact on the engineering industry.

Contracts

Most engineering projects are the subject of a contract between two parties. These are variously known as a *purchaser* and a *supplier* or a *seller* and a *buyer*. For each project, a *Project Team* under the direction of a Project Manager or Contract Manager will develop a *work plan* which informs a company's overall *resource plan*. The work plan will:

- specify the required work elements that make up the project
- the level of effort and resource needed to complete each element
- a schedule for the completion of the elements.

Work on approved projects that is beyond the capacity of a company or a department's staff may be resourced through consultancy services or through partnerships with other suppliers on a sub-contract basis. Project work plans will inform the company's overall resource plan. This overall plan includes an analysis of what can be accomplished by:

- using existing department staff in regular time or by using overtime
- hiring new employees
- purchasing services from external agencies.

Senior management will produce an annual budget which will allocate financial and human resources to meet the workload on its approved projects. The allocation consists of:

- department staff (both regular and overtime)
- staff brought in from other departments (both regular and overtime)
- consultant services (for project work beyond the capacity of department staff)
- professional services (for example, specialist legal advice).

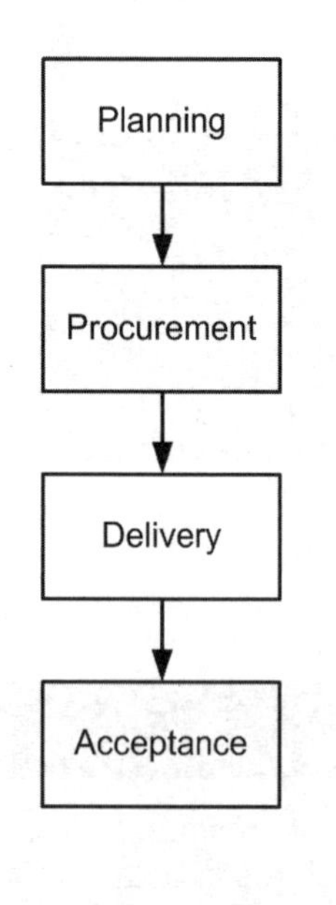

Figure 2.26 *The four different phases in a contract's life-cycle*

The *Contract Manager* (sometimes referred to as a *Project Manager*) is responsible for the administration and quality of contract products or services, and should be involved from contract initiation to contract closeout. The Contract Manager has a variety of accountabilities including:

- providing and maintaining information about the project
- coordinating the selection of project staff and any external consultants
- reviewing and evaluating contract documents
- initiating contracts and notices
- serving as the primary contact person for the project
- developing and managing the contract schedule, scope, and costing

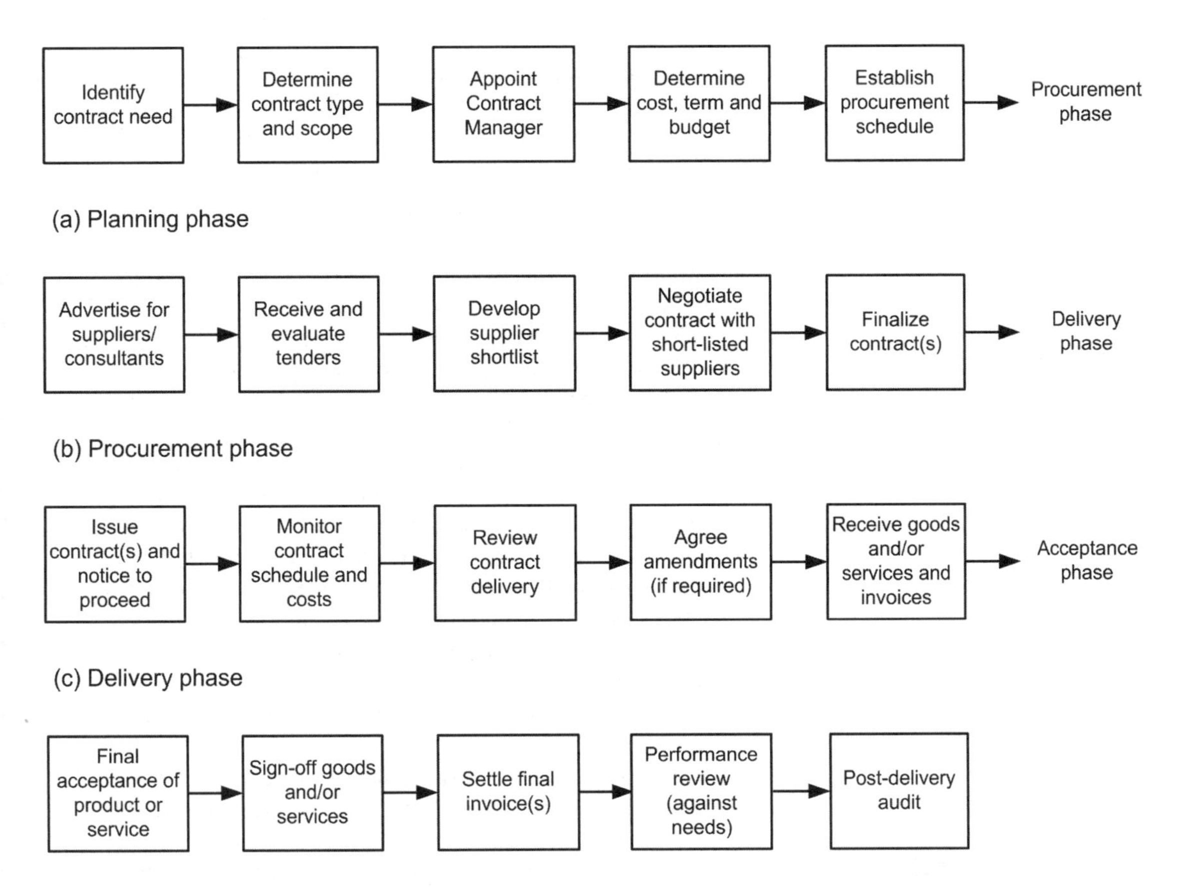

Figure 2.27 *The various stages in each phase of the execution of a contract*

- monitoring the project delivery and satisfactory completion of the project (see Figure 2.27)
- preparing periodic performance reviews and presenting these to senior management
- updating the project plan and schedule
- reviewing and approving payments
- signing-off and closing out the contract.

A Contract Manager (or Project Manager) will usually be required to develop a detailed *cost estimate*. He or she will usually do this with assistance from other functional and resource managers who have detailed knowledge of the likely costs involved.

Cost estimates may be based on several factors, including the costs of plant and equipment and the required manpower hours to complete the project. In order to simplify this complex task, the Contract Manager will usually break the scope of the work into component parts or milestones. He or she will then estimate the cost (in terms of materials, physical resources and hours) in order to achieve each milestone. Most companies use templates for costing where standard costs and rates are applied.

Test your knowledge 2.12

(a) Identify the four stages in the life-cycle of a contract.
(b) Explain the role of a Contract Manager.
(c) On what should a cost estimate be based?

Contracts and related documents are often written in legal language and can be a little intimidating at first sight, so before we take a look at the wording of a contract here is a list of some of the most important terms, together with their meanings:

- *Amendment:* An official change to the contract after contract work has begun.
- *Audit:* The formal examination and verification of project related records by an appointed person or organisation. The audit may cover a number of aspects including the appointment of consultants and other staff, billing methods, contract amendments, contract requirements, and the system of internal controls.
- *Contract Manager:* The person who is primarily responsible for coordinating, monitoring and reviewing the contract. Note that the Contract Manager may also act as a Project Manager.
- *Cost Estimate:* A document (usually forming part of a contract) which provides details of direct and indirect costs associated with the delivery of the contract.
- *Dispute:* A disagreement between the supplier and the purchaser.
- *Encumber:* To set aside, or earmark funds for a specific purpose.
- *Functional Manager:* The immediate supervisor of the staff who work on the project (often a Department or Section Head).
- *Milestone:* A point at which a particular outcome within a project is reached. A milestone can be broad, such as the date at which time a component ends, or it can be specific, such as when a particular task will be completed.
- *On-call Contract:* An on-call contract is a contract which is carried out at short-notice and when the work being requested cannot be defined adequately in advance.
- *Post Audit:* An audit completed after the project work is finished.
- *Project Schedule:* The schedule for the project, indicating the tasks, resources, milestones, completion dates, and any dependencies that exist between them.
- *Project Scope:* The work that must be done in order to deliver a product with the specified features and functions.
- *Schedule:* A chart or table that shows project tasks plotted against time. The schedule should show specific start and end dates for each task and phase of a project.

The typical format of an engineering contract is as follows:

[DATE]

[CONTRACT NAME]

- by and between -

[BUYER NAME]

BUYER

- and -

[SELLER NAME]

SELLER

TABLE OF CONTENTS

PRELIMINARY STATEMENT
1. DEFINITIONS AND INTERPRETATION
2. SCOPE OF CONTRACT
3. CONTRACT PRICE
4. PAYMENT
5. DELIVERY
6. PACKING AND SHIPPING
7. INSPECTION AND CLAIM
8. INSTALLATION, COMMISSIONING AND ACCEPTANCE
9. WARRANTY
10. SOFTWARE LICENSE
11. INTELLECTUAL PROPERTY INDEMNITY
12. TAXES
13. REPRESENTATIONS AND WARRANTIES
14. TERM
15. TERMINATION
16. CONFIDENTIALITY
17. BREACH OF CONTRACT
18. FORCE MAJEURE
19. SETTLEMENT OF DISPUTES
20. MISCELLANEOUS PROVISIONS
SCHEDULE A DEFINITIONS AND INTERPRETATION
SCHEDULE B SPECIAL TERMS
ANNEX 1 EQUIPMENT, SERVICES & DOCUMENTATION LIST AND PRICES
ANNEX 2 SPECIFICATIONS
ANNEX 3 PROJECT SCHEDULE
ANNEX 4 ENGINEERING NOTES AND INSTALLATION RESPONSIBILITIES
ANNEX 5 EQUIPMENT ACCEPTANCE TEST PLAN
ANNEX 6 SPECIMEN OF ACCEPTANCE CERTIFICATE

Activity 2.9

The first few paragraphs of a typical engineering contract are shown on page 158. Read this contract carefully and then answer the following questions:

1. Who will be responsible for delivering the contract?

2. How often should progress reports be made during the installation and commissioning phase?

3. Under what conditions can the Buyer transfer the equipment to a third party?

4. Does the contract price include tax?

Test your knowledge 2.13

Explain the meaning of the term 'milestone' in relation to the delivery of a contract.

CONTRACT FOR THE SUPPLY OF EQUIPMENT AND SERVICES

THIS CONTRACT ("Contract") is made in England on the Contract Date set out in Schedule B (Special Terms) by and between the Buyer and the Seller specified in Schedule B (Special Terms). The Buyer and the Seller shall hereinafter be referred to individually as a "Party" and collectively as the "Parties".

PRELIMINARY STATEMENT
After consultations conducted in accordance with the principles of equality and mutual benefit, the Parties have agreed to enter into a sales contract in accordance with Applicable Laws and the provisions of this Contract.

Now the Parties Hereby Agree as Follows:

1. DEFINITIONS AND INTERPRETATION
Unless the terms or context of this Contract otherwise provide, this Contract shall be interpreted in accordance with Schedule A, and each of the terms used herein shall have the meaning ascribed to it in Schedule A (Definitions and Interpretation) or Schedule B (Special Terms) as the case may be.

2. SCOPE OF CONTRACT

2.1 Supply of Equipment and Services
The Seller shall supply to the Buyer the Equipment and Services (as more specifically set out in Annex 1 hereto) [for the Project] in accordance with the terms of this Contract. The Buyer shall pay the Seller the Contract Price set out in Schedule B (Special Terms) hereto and shall perform its other responsibilities in accordance with the terms of this Contract.

2.2 Project Managers
The Seller and the Buyer shall each nominate a representative as its project supervisor for the Project (each, a "Project Manager" and jointly, the "Project Managers"). Each Project Manager shall be authorized to deal with all technical matters in connection with the Equipment and the Services during the period from the Effective Date up to the expiration of the applicable Warranty Periods. Detailed arrangements for both regularly scheduled and emergency communications between the Project Managers shall be made through consultation between the Seller and the Buyer. The Project Managers shall fully cooperate to resolve all technical issues which may arise in respect of the Equipment and the Services. If there is any dispute between the Project Managers, they shall analyse the problem, clarify the responsibilities and, to the fullest extent possible, settle it at the Installation Sites through consultation. During the period of Installation and Commissioning, the progress of work shall be reported by the Project Managers on a weekly basis.

2.3 No Transfer of Equipment
The Buyer acknowledges and agrees that the Equipment constitutes and embodies the Confidential Information and Intellectual Property of Seller. Without the prior written consent of the Seller, the Buyer may not directly or indirectly transfer the Equipment to any third party for consideration or without consideration.

3. CONTRACT PRICE

3.1 Price
The Contract Price is set out in Schedule B (Special Terms).

3.2 Consideration for Price
This Contract Price covers:
(a) the Equipment (including Hardware and Software License) and Services,
(b) the cost of shipment of the Hardware and Documentation to the Place of Delivery.

3.3 Costs and Expenses
The Contract Price does not cover costs and expenses that are for Buyer's account under this Contract, nor do they cover any taxes that may apply (including but not limited to customs duty and value added tax) payable by the Buyer in respect of this Contract.

Contracts of employment

Employment law relates to the contract that exists when one person employs another. The Contracts of Employment Act (1963) obliged employers in the UK to provide written terms of employment. This was later incorporated into the Employment Protection (Consolidation) Act 1978 and is now found in the Employment Rights Act 1996.

The law requires all contracts of employment to encompass minimum terms such as the name of the employer, the name of employee, the job title, the place of work, the hours of work, remuneration, discipline and grievance procedures, and holiday entitlement.

Contracts of employment can be written, oral, implied or a mixture of all three, but research indicates that a substantial proportion of tribunals occur where employers fail to provide no written terms of employment. From 2004, new legislation has been introduced that requires statutory disciplinary and grievance procedures should be incorporated into all contracts of employment, whether these are written or not.

Companies who fall foul of these regulations are more likely to find themselves before an *Employment Tribunal* which is a judicial body established to resolve a dispute between an employer and an employee over employment rights. Employment tribunals are able to award an employee sums in compensation where no established *disciplinary procedure* has been followed.

Disciplinary procedures

The general principles of most disciplinary procedures is that, wherever possible, problems should be resolved informally without recourse to formal procedures. Only when it is not possible for a problem to be resolved informally, or the severity of the allegation warrants it, should a formal procedure be instigated. Such a procedure usually involves several discrete stages, as described below:

Stage One: Formal Oral Warning

Normally the employee/worker will be given a formal oral warning if his/her conduct or performance failed to meet acceptable standards. He/she will be advised of the reason for the warning, the improvements required and the timescale for improvement. He/she will also be advised that this is the first stage of the Disciplinary Procedure and that he/she has the right to appeal.

Stage Two: First Written Warning

If the employee/worker's conduct or performance warrants it, or if a further offence occurs, a first written warning will be issued. The employee/worker should be advised that, if there is no satisfactory improvement, further disciplinary action will be considered under Stage Three. The written warning will be effective for a maximum of one year (from the date of the letter following the disciplinary interview). After that time it will be disregarded subject to satisfactory performance and/or conduct.

Stage Three: Final Written Warning

If there is a continuing failure to improve, and conduct or performance is not satisfactory, a final written warning will be issued to the employee/worker. This will give details of the complaint, the improvement required and the timescale for improvement. It will warn that a recommendation for dismissal will result if there is not satisfactory improvement and will advise of the right of appeal. The final written warning will normally be effective for a maximum of two years. After that time it will be disregarded subject to satisfactory performance and/or conduct.

Stage Four: Dismissal

If conduct or performance remains unsatisfactory, and the employee/worker continues to fail to reach agreed standards, he/she will be dismissed. Only named senior personnel (such as Section Heads or Department Heads) can usually take the decision to dismiss an employee/worker. The employee/worker will be provided with written reasons for dismissal, the date on which the employment is terminated, and the right of appeal. If the dismissal is 'with notice', the notice period will be as stated in the employee/worker's contract of employment, but the University may deem it appropriate to make a payment in lieu of notice. In cases of gross misconduct the employee/worker may be dismissed straight away (see below).

Gross Misconduct

Gross misconduct is generally seen as misconduct serious enough to destroy the contractual relationship between the employer and the employee/worker and make any further working relationship and trust difficult, if not impossible. Whilst it is not possible to specify all incidents which would constitute gross misconduct, examples of acts which normally would be regarded as gross misconduct include:

- fighting, or using or threatening to use physical violence
- serious negligence which causes or may cause unacceptable loss, damage or injury to persons or property
- harassment or bullying of another employee or worker
- deliberate damage to, or serious misuse of company property or the property of another employee or worker
- serious insubordination or refusal to carry out a legitimate instruction, the consequences of such a refusal being that there is a clear breach of contract between the employee and the company
- theft, fraud, or deliberate falsification of records
- incapacity at work through alcohol or use of non prescribed drugs
- irresponsible conduct of a nature likely to endanger the health and safety of the individual or others
- bringing the company into serious disrepute
- serious breach of confidence (subject to the Public Interest Disclosure Act 1998).

Test your knowledge 2.14

Under what circumstances should a disciplinary procedure be used?

European standards

The 1957 Treaty of Rome is one of the foundation stones of the European Community. One of the main aims of the Treaty was the creation of a single European market in goods and services with the objective of providing producers and consumers with the benefits of economies of scale that this offers. This unified market eliminates the confusion and cost associated with multiple local laws and standards. The ensuing harmonization of standards has been instrumental in simplifying the technical and commercial task of creating products for a European market.

Unfortunately, some of the terminology used in relation to European Directives and harmonized standards can be a little confusing so, before we begin to look at some of the directives in detail, here is a list of some of the most important terms, together with their meanings:

- *Directive:* The name given to an official EU document that defines a set of requirements
- *Apparatus:* A finished product delivering an intrinsic function, and directly usable by the end user
- *CE Marking (93/68/EEC):* A means of identifying products that comply with the Directives (and can therefore be legally offered for sale in any EU state)
- *Competent body:* A body recognized as fulfilling certain criteria and who is responsible for issuing reports or certificates under the terms of a particular Directive
- *Notified body:* A body specially identified by an EC member state as having responsibility for issuing certificates
- *Compliance:* Being able to demonstrate that a set of prescribed criteria are met (there may be several ways of demonstrating compliance with a particular Directive)
- *EMC Directive (89/336/EEC amended 92/31/EEC):* The EU specification for EMC emission and susceptibility
- *Low-Voltage Directive (LVD) (73/23/EEC):* Safety requirements for equipment normally operating at from 50 V to 1,000 V a.c. or 75 V to 1,500 V d.c.
- *New Approach:* A set of Directives intended to harmonize product safety throughout the EU.

Equipment designed for sale or use in any European Union country must comply with the relevant directives. These define a set of requirements but leave it to the standards (primarily European *harmonized standards*) to define the technical requirements. To put it simply, a manufacturer (or the manufacturer's European representative) must first ensure that a product complies with any applicable directives before CE marking the product.

It's important to understand that the CE mark (see Figure 2.28) does not indicate conformance with a particular standard. Instead, it indicates that the product complies with all of the standards that might apply to it. This means that, for example, a plastic toy and a hair-dryer will both carry the CE mark even though they meet a totally different set of standards. Now let's take a look at the directives and the CE mark in a little more detail.

CE

Figure 2.28 *The CE mark*

The EMC Directive

The Electromagnetic Compatibility (EMC) Directive 89/336/EEC is one a series of measures introduced under article 100a of the Treaty of Rome. The effect of the directive has been to introduce identical requirements for the EMC performance of electrical apparatus in every country within the European Economic Area (EEA).

As with all CE mark directives, the primary purpose of the EMC Directive is the creation of a single market for electrical goods throughout Europe. The protection requirements of the Directive are the means by which this is achieved, not the fundamental objective. In contrast to all the other CE mark directives, the EMC Directive's primary requirement is the protection of the electromagnetic spectrum, not the safety of equipment.

The EMC Directive is one of the widest in its application and all electrical products must comply. The only exceptions are for components or sub-assemblies that only function in combination with other components (we say that they have no *intrinsic function*) and certain electrical products and systems that are already covered by other directives. Examples of products which do not need to comply are plugs and sockets, capacitors, resistors and integrated circuits. More complex sub-assemblies such as thermostats, power supplies and micro-controllers do have to comply but would normally only be considered as parts of a complete product or system. Medical devices, military, automotive and certain agricultural equipment are all excluded from the scope of the EMC Directive itself because they are subject to more specific directives containing equivalent EMC provisions but almost all other apparatus, whether it is mains or battery powered, must comply. This includes the simplest hand held system (e.g. a battery powered torch) right up to the most complex installation (for example, a power station or a chemical factory!). Note that most transmitting and communications apparatus is excluded from the scope of the EMC Directive because this equipment is covered by a separate directive (the *Radio Terminal and Telecommunications Equipment Directive*).

Essentially, the EMC Directive states that products must not emit unwanted electromagnetic pollution which might otherwise cause *interference* to other appliances and services. The Directive also states that products must themselves be immune to a reasonable amount of interference. The Directive gives no figures or guidelines on what the required level of emissions or immunity are, nor does it state ant limits in terms of amplitude or frequency—this is left to the standards that are used to demonstrate *compliance* with the Directive.

In order to determine whether or not equipment complies with the EMC Directive certain tests must be applied. These tests fall into several classes including:

- *Radiated emissions:* Checks to ensure that the product does not emit unwanted radio signals
- *Conducted emissions:* Checks to ensure the product does not send out unwanted signals along its supply connections and connections to any other apparatus
- *Radiated susceptibility:* Checks that the product can withstand a typical level of electromagnetic pollution.

EC Declaration of Conformity

We: Howard Scientific Instruments
Coopers Hill Industrial Estate
Boundary Road
Little Millington
Cambridgeshire
PE10 3TY

declare that the:

PAT21 Electrical PAT Tester

meets the intent of the EMC Directive 89/336/EEC and the Low Voltage Directive 73/23/EEC. Compliance was demonstrated by conformance with the following specifications listed in the Official Journal of the European Communities:

EMC

Emissions: EN5008-1-1 (1992) Generic (Light Industrial) referring to:
(a) EN55022 Conducted, Class B
(b) EN55022 Radiated, Class B

Immunity: EN50082-1 (1992) Generic (Light Industrial) referring to:
(a) EN60801-2 (1993) Electrostatic Discharge
(b) IEC801-3 (1984) RF Field
(c) IEC801-4 (1988) Fast Transient

SAFETY EN6010-1 (1993) Installation Category II
Pollution Degree 1

S R Holmes

Steven R. Holmes
Technical Director

27th September 2001

CE

Figure 2.29 *A typical certificate of conformity for a portable appliance (note the CE mark in the bottom right corner)*

Test your knowledge 2.15

Explain the purpose of European Standards and the CE mark.

British Standards

The British Standards Institution (BSI) was the first national Standards making body in the world. Independent of government, BSI is a non-profit distributing organisation. A non-profit distributing organisation is defined as an organisation in which profits are not distributed to its directors, shareholders, employees, or any one else. Instead, profits are reinvested into the services provided.

BSI is globally recognised as an independent and impartial body serving both the private and public sectors, working with manufacturing and service industries, businesses and governments to facilitate the production of British, European and International Standards.

As well as facilitating the creation of British Standards, BSI is the UK's National Standards Body (NSB) and represents the UK interests across all of the European and international Standards committees.

Figure 2.30 *The British Standards Institute (BSI) Kitemark*

A *standard* is a published specification that establishes a common language, and contains a technical specification or other precise criteria and is designed to be used consistently, as a rule, a guideline, or a definition. Standards are applied to many materials, products, methods and services. They help to make life simpler, and increase the reliability and the effectiveness of many goods and services we use.

Standards are designed for voluntary use and do not impose any regulations. However, laws and regulations may refer to certain standards making compliance with them compulsory. British Standards are created by appropriately qualified and experienced people who are brought together (for the specific purpose of creating a standard) by BSI. They discuss and agree on the details that will form the new British Standard.

Once a consensus has been reached, a draft of the new standard is released, and anyone with an interest is invited to comment on its content. Finally, after all comments have been reviewed, the new standard is published as a British Standard.

BSI's registered certification mark, the *Kitemark* (see Figure 2.30) and its Registered Firm symbol are instantly recognised by customers and suppliers respectively, promoting a company's reliability and commitment to product quality.

Activity 2.10

Visit the BSI website at http://www.bsi-global.com and investigate the EMC testing facilities available from BSI. Use this to answer the following questions:

1. What organisation has accredited BSI's EMC testing facilities?
2. Under what circumstances can a manufacturer 'self-certify' a product under the EMC directive?
3. What other BSI product testing services are available?

Activity 2.11

(a) Visit the BSI website at http://www.bsi-global.com and investigate the BSI Kitemark scheme as it applies to printed circuit board assembly and the international standard IPC-A-610. State the three main elements of the BSI Kitemark scheme.

(b) Search the Web in order to locate a UK company that holds the BSI IPC-A-610 Kitemark. Investigate how the BSI Kitemark is used in the company's marketing and promotional literature.

Test your knowledge 2.16

Explain the purpose of the British Standards Institute (BSI).

Environmental legislation

Engineering activities can have harmful effects on the physical environment and therefore on people. In order to minimize these effects, there is a range of legislation that all engineering companies must observe. We shall look at this in more detail in Unit 5 but for now we will confine our study to the legislation and its impact on engineering activities.

The appropriate United Kingdom Acts of Parliament include: Deposit of Poisonous Wastes Act, Pollution of Rivers Act, Clean Air Act, Environmental Protection Act, Health and Safety at Work, etc and the like. Additionally, not only are there local by-laws to be observed there are also European Union (EU) directives that are activated and implemented either through existing UK legislation in the form of Acts of Parliament or mandatory instructions called *Statutory Instruments* (SI).

New Acts and directives are introduced from time to time and Industry needs to be alert to and keep abreast of these changes. Typical of these new initiatives is the European Electromagnetic Compatibility (EMC) legislation that we met earlier. The EMC Directive recognizes the well known problem of unwanted electromagnetic noise that emanates from most pieces of electrical equipment (notably computers and other digital equipment) and may cause interference to other services.

In the case of UK Acts of Parliament, the above legislation is implemented by judgement in UK Courts of Justice in the normal manner but based on EU legislation, if more appropriate, or by judgement of the European Court of Justice. The purpose of this legislation is to:

- *prevent* the environment being damaged in the first place;
- *regulate* the amount of damage by stating limits, for example, the maximum permitted amount of liquid pollutant that a factory may discharge into the sea;
- *compensate* people for damage caused, for example, from a chemical store catching fire and spreading wind borne poisonous fumes across the neighbourhood;
- *impose sanctions* on those countries or other lesser parties that choose to ignore the legislation;
- *define who is responsible* for compliance with legislation (including named persons with their precise area of responsibility documented).

Note that you are *not* expected to have a detailed understanding of the various Acts however you *should be aware of the general provisions* of the legislation and *what it is trying to achieve.* Your school, college or local library will be able to provide you with more details.

The effects of the above legislation on engineering activities has, in general, made them more difficult and more expensive to implement. A few simple examples of this follow:

- *Chemical factories* can no longer discharge their dangerous waste effluent straight into the river or sea without first passing it through some form of purification.

- *Coal fuelled power stations* must ensure that their chimney stacks do not pollute the neighbourhood with smoke containing illegal limits of grit, dust, toxic gases and other pollutants. A system of smoke filtration and purification must be (expensively) incorporated.
- *Motor vehicle* exhaust gases must be sufficiently free of oxides of nitrogen, carbon monoxide and other toxic gases. This can only be achieved by, among other things, replacing the crude petrol carburettor with a more sophisticated petrol injection system and fitting a catalyser in the exhaust pipe. All this has added to the price of the motor car and has made it more difficult for the DIY motorist to service his or her vehicle.
- All *electrical equipment* including TVs, PCs, power hand tools, electromedical machines, lighting and the like, must be tested and certified that they comply with the EMC legislation that we met earlier.

Health and Safety legislation

The *Health and Safety at Work Act (1974)* makes both the employer and the employee equally responsible for safety. Both are equally liable to be prosecuted for violations of safety regulations and procedures. It is the legal responsibility of an employee to take reasonable care of his or her own health and safety. The law expects them to act in a responsible manner so as not to endanger themselves or to endanger other workers or members of the general public. It is an offence under the Act to misuse or interfere with equipment provided for health and safety.

Human carelessness

Most accidents are caused by human carelessness. This can range from 'couldn't care less' and 'macho' attitudes, to the deliberate disregard of safety regulations and codes of practice. Carelessness can also result from fatigue and ill-health resulting from a poor working environment.

Personal habits

Personal habits such as alcohol and drug abuse can render workers a hazard not only to themselves but also to other workers. Fatigue due to a second job (moonlighting) can also be a considerable hazard, particularly when operating machines. Smoking in prohibited areas where flammable substances are used and stored can cause fatal accidents involving explosions and fire.

Supervision and training

Another cause of accidents is inadequate training. Lack of supervision can also lead to accidents if it leads to safety procedures being disregarded.

Environment

Unguarded and badly maintained plant and equipment are obvious causes of injury. However, the most common causes of accidents are falls on slippery floors, poorly maintained stairways, scaffolding and obstructed passageways in overcrowded workplaces. Noise, bad lighting, and inadequate ventilation can lead to fatigue, ill-health and carelessness. Dirty surroundings and inadequate toilet and washing facilities can lead to a lowering of personal hygiene standards.

Elimination of hazards
The workplace should be tidy with clearly defined passageways. It should be well lit and ventilated and should have well maintained non-slip flooring. Noise should be kept down to acceptable levels. Hazardous processes should be replaced with less dangerous and more environmentally acceptable alternatives.

Guards
Rotating machinery, drive belts and rotating cutters must be securely fenced to prevent accidental contact. Some machines have interlocked guards. These are guards coupled to the machine drive in such a way that the machine cannot be operated when the guard is open for loading and unloading the work. All guards must be set, checked and maintained by qualified and certified staff. They must not be removed or tampered with by operators.

Maintenance
Machines and equipment must be regularly serviced and maintained by trained fitters. Equally important is attention to such details as regularly checking the stocking and siting of first-aid cabinets and regularly checking the condition and siting of fire extinguishers. All these checks must be logged.

Personal protection
Suitable working clothes (overalls and boots or safety shows) should be worn. Some processes and working conditions demand even greater protection, such as safety helmets, earmuffs, respirators and eye protection worn singly or in combination. Such protective clothing must be provided by the employer when a process demands its use. Employees must, by law, make use of such equipment.

Safety education
This is important in producing positive attitudes towards safe working practices and habits. Warning notices and instructional posters should be displayed in prominent positions and in as many ethnic languages as necessary. Information, education and training should be provided in all aspects of health and safety, including first aid and fire procedures. Regular fire drills must be carried out to ensure that the premises can be evacuated quickly, safely and without panic.

Personal attitudes and working practices
It is important that everyone adopts a positive attitude towards safety and that all working practices are inherently safe. Adequate training is essential and personnel should be encouraged to become actively involved with ensuring that the workplace is a safe place in which to work. By taking positive steps to improve awareness of issues related to health and safety many organisations have noticed a significant reduction in accidents and the disruption to working that this inevitably causes.

Other legislation that you need to be aware of is as follows:

- Control of Substances Hazardous to Health (COSHH)
- Reporting of Injuries, Diseases and Dangerous Occurrences Regulations 1995 (RIDDOR)
- Electricity at Work Regulations 1989
- Asbestos at Work Regulations 2002
- Management of Health and Safety at Work Act 1999.

Finally, it is important to note that some industries and trade associations have introduced *Codes of Practice* which set minimum standards and govern the working practices of their members.

Activity 2.12

Consult the Health and Safety at Work Act and answer the following questions:

(a) What is an improvement notice?
(b) What is a prohibition notice?
(c) Who issues such notices?
(d) Who can be prosecuted under the Act?

Activity 2.13

Visit the COSHH section of the UK Government's Health and Safety Executive website (you will find this at www.hse.gov.uk/coshh). View or download a copy of 'COSHH: a brief guide to the regulations' and use it to identify the EIGHT steps that help to ensure that a company complies with the COSHH legislation.

Activity 2.14

Asbestos is a material that has relatively only recently been identified as a material that is dangerous to health. Use the Internet or your school or college library to answer the following questions:

1. What is asbestos?
2. Why is asbestos dangerous to health?
3. What diseases are caused by exposure to asbestos?
4. What occupations are particularly at risk from asbestos?
5. What legislation applies to the use and handling of asbestos?

Test your knowledge 2.17

Describe TWO examples of the impact of EU legislation on engineering activities.

2.4 Evaluation and modification

In the final section of this unit we will provide some brief guidance to help you with the unit assessment in which you carry out a detailed investigation of your chosen engineered product or service. We shall begin by considering an important concept, 'fitness for purpose'.

Fitness for purpose

The final *acceptance phase* (see Figure 2.28) in the execution of a contract involves an evaluation of an engineered product or service in order to confirm that it performs correctly and in accordance with the detailed specification that has been laid down (and which forms part of the contract with the supplier). In addition, as with any product or service, there is a need to assess the product as a whole in terms of its '*fitness for purpose*'. This basically means determining whether or not the product or service satisfies its intended function.

Fitness for purpose is an important concept. A product or service may be considered complete (and may conform according to its specification) but it still may not actually be fit for purpose.

As an example, consider a small family car which has four forward gears but no reverse gear. The car may perform well when driven along a highway with a full load of passengers. It may also be very economical in terms of fuel consumption and perform well in terms of emissions. However, the car could not be considered 'fit for purpose' simply because it could never be left in a parking space!

Evaluation

The evaluation of an engineered product and service involves a thorough analysis of the product or service as well as the ability to present your findings (and recommendations for changes modifications) in the form of a detailed and well-reasoned report accompanied by relevant sketches and drawings. The following questions should help you with this task:

- To which engineering sector does the product or service belong?
- What technologies are used in the design, manufacture and use of the product or service?
- Is there a published specification for the product or service?
- Does the measured performance conform to the specification?
- In which particular areas does the measured performance fail to meet specification?
- In which particular areas does the measured performance exceed specification?
- What legislation and standards apply to the product or service?
- How has the supplier or manufacturer indicated conformance with the relevant legislation and standards?
- What modifications can be made to improve the performance or cost-effectiveness of the product or service?
- Finally, is the product or service 'fit for purpose'?

Another view

Most companies have highly developed quality systems that are designed to check on the consistency and conformance of manufactured products with their specification. However, unless the specification was properly thought out in the first place, this may not actually mean that the product is 'fit for purpose'.

Product	*Sector*	*Technology*	*Legislation*	*Standards*	*Modifications*
A portable digital (DAB) radio that operates from the AC mains using an external AC adapter	Consumer electronics	Microelectronics CAD/CAM Computer-aided PCB design and manufacture	Low Voltage Directive 73/23/EEC EMC Directive 89/336/EEC	CE marking BS Kitemark	Internal batteries would eliminate the need for an external AC adapter (these would need to be appropriately rated)
A rechargeable electric toothbrush which uses two nickel-cadmium NiCd cells					
A small portable gas torch designed for use by DIY plumbers					
Motorway design and construction					
Civil aircraft maintenance provided by a specialist aircraft maintenance facility					

Table 2.1 *See Activity 2.15*

Activity 2.15

Table 2.1 (shown on page 170) provides brief details of FIVE engineered products and services. For each of these products and services complete the table showing:

(a) The engineering sector (or sectors) to which the product or service belongs

(b) The technology (or technologies) used in the design, manufacture and testing of the product or service

(c) The legislation that governs the design, manufacture and use of the product or service

(d) The standards that relate to the design, manufacture and use of the product or service

(e) Potential modifications that could be used to improve the performance and/or cost-effectiveness of the product or service.

Activity 2.16

The design modifications for an optical unit fitted to the International Space Station (ISS) are shown on the contract revision shown on page 172. Changes are shown in bold and with the original text struck through.

1. Name the sub-contractor who is supplying the optical unit.

2. Name the company that has prepared the contract.

3. How many previous revisions have been made to the contract?

4. What THREE changes have been made to the specification of the optical unit?

5. What additional equipment is to be incorporated into the optical unit?

6. What provision has been incorporated into the contract concerning electromagnetic compatibility?

7. Who is required to write the acceptance test procedure?

8. What tests are to be carried out on the first flight item?

9. What tests are to be carried out on each flight item?

10. Which USA (MIL) standards are referred to in the contract?

Test your knowledge 2.18

Explain the concept of 'fitness for purpose'.

Contract Revision for Howard Aerospace Optical Unit

Revision History: Rev. 1: 2 May 2004, Rev. 2: 24 May 2004, Rev 3: 11 September 2004

Revision 4

NorthStar Dynamics intends to purchase one Howard Aerospace CDA optical unit immediately and 4 more trackers (and associated sets of optical units) in the next fiscal year. Although the units we are purchasing are based upon the design of a standard CDA Mk.2, certain modifications must be implemented for use on the International Space Station (ISS). These modifications, and certain unique procedural requirements, are listed below; the engineering design effort as well as the procurement of unique parts (e.g. PC boards to MIL-SPEC requirements) are to be costed separately as a non-recurring expenditure.

The design modifications for compatibility with ISS usage are to be as follows:

1. Input voltage shall be ~~6V DC ± 5%~~ **12V DC ± 10%.**
2. Signal and power will be on a single MIL-C-38999 connector.
3. Complex printed circuit boards shall be fabricated to MIL-P-55110 or equivalent by a qualified vendor.
4. Parts and materials proposed for use shall be subject to NorthStar Dynamics assessment of compatibility with ISS requirements for flammability, offgassing, and fungus resistance.
5. Electronic parts shall be of the highest quality level commonly available.
6. PC assemblies shall be conformally coated using CYTEC Conap 1155 or equivalent.
7. EMC design shall be sufficient for CE certification with conformance to ISS requirements as a goal.
8. Mechanical design shall assure that the unit will conform to NorthStar Dynamics drawing numbers NS 6355-ISS-01 and NS 6355-ISS-02.
9. Thermal design shall assure that no individual electrical component on the PC boards exceeds a case temperature of ~~55C~~ **40°C** in a 25°C ground laboratory environment.
10. An integral protective cover will be provided to protect the camera windows.
11. A 640x480 pixel resolution, color Webcam will be integrated into the chassis (the USB signal to be routed through the MIL-C-38999 connector).
12. An acceptance test procedure shall be written by Howard Associates, with concurrence by NorthStar Dynamics required, which demonstrates that the flight items meet all performance requirements in a laboratory environment.
13. All non-metallic materials shall be evaluated for fungus resistance. NorthStar Dynamics will work with Howard Associates in conducting this evaluation.
14. The supplier shall furnish a Certificate of Compliance with the parts or material procured under this contract.
15. A vibration test shall be performed on each flight item. The excitation is random, ~~4.5G~~ **6.5G** r.m.s. for a duration of 60 seconds in each of three axes. The units will be unpowered during vibration.
16. Each flight item is required to operate for a minimum of 96 hours at its maximum operating temperature.
17. The first delivered flight item shall be tested to demonstrate compliance with CE EMC requirements. Testing to Space Station EMC requirements will be handled by NorthStar Dynamics after acceptance and delivery.
18. Flammability and offgassing tests will be conducted by NorthStar Dynamics after acceptance and delivery.

Date of revision: 26th January 2005

Review questions

1 List FIVE different engineering sectors and, for each sector, identify a typical product or engineering service.

2 Describe each of the following generic engineering roles:
(a) Chartered Engineer
(b) Incorporated Engineer
(c) Engineering Technician.

3 Which UK body is responsible for defining the roles of professional engineers?

4 Describe the stages in the development of a new product and explain the contribution of engineers to THREE of these stages.

5 Describe the role of the *sales engineer*. Explain how this role differs from that of a *sales representative*.

6 Explain THREE constraints under which an engineer may have to work.

7 Describe THREE different aspects of *computer aided engineering* (CAE).

8 Distinguish between 2D and 3D CAD and describe a typical application of each.

9 List FIVE features of a modern 2D CAD package.

10 In relation to a 3D CAD system, explain the terms:
(a) wire frame drawing
(b) rendering.

11 Sketch a 3D workplane and explain how the three-axis coordinate system works.

12 Explain, briefly, the role of a *Production Engineer*.

13 List the FOUR stages in the life-cycle of an engineering contract.

14 In relation to European Standards, explain the meaning of the terms:
(a) Directive
(b) New Approach.

15 Sketch the CE mark and explain what it means.

16 Sketch the BS Kitemark and explain what it means.

17 Explain why the EMC Directive is needed and give THREE examples of products to which it applies.

18 Explain why transmitters and mobile phones are not covered by the EMC Directive.

19 Describe the stages of a typical *disciplinary procedure.*

20 List FIVE examples of employee conduct that might be classed as gross misconduct.

EC Declaration of Conformity

We: Howard Scientific Instruments
Coopers Hill Industrial Estate
Boundary Road
Little Millington
Cambridgeshire
PE10 3TY

declare that the:

PAT21 Electrical PAT Tester

meets the intent of the EMC Directive 89/336/EEC and the Low Voltage Directive 73/23/EEC. Compliance was demonstrated by conformance with the following specifications listed in the Official Journal of the European Communities:

EMC

Emissions: EN5008-1-1 (1992) Generic (Light Industrial) referring to:
(a) EN55022 Conducted, Class B
(b) EN55022 Radiated, Class B

Immunity: EN50082-1 (1992) Generic (Light Industrial) referring to:
(a) EN60801-2 (1993) Electrostatic Discharge
(b) IEC801-3 (1984) RF Field
(c) IEC801-4 (1988) Fast Transient

SAFETY EN6010-1 (1993) Installation Category II
Pollution Degree 1

S R Holmes

Steven R. Holmes
Technical Director

27th September 2001

CE

Figure 2.31 *See Question 21*

21 Figure 2.31 shows a typical EC Declaration of Conformity.
(a) Which company has produced the declaration?
(b) Which European legislation is referred to in the declaration?
(c) What product is covered by the declaration?
(d) Who has signed the document and what is his/her job title?
(e) What mark appears on the declaration to show that the product conforms to the relevant EC directives?

22 Name the UK legislation that relates to the use and handling of substances that are potentially hazardous.

Unit 3 Principles of design, planning and prototyping

Summary

The unit is about the stages required in taking an initial idea and turning it into a finished engineering product or service. Starting with an identified need and a set of requirements, we shall look at how an engineer sets about the task of producing a design brief and a design specification before evaluating a range of different solutions. As you work through this chapter, you will be involved with the planning, design, manufacture and evaluation of your own engineering product or service. To get you started, we have provided you with a typical example of a product that you might want to develop. This product is of a similar complexity and standard to that which you should be working. We then continue by introducing you to engineering drawing techniques, project planning and design. We also provide you with guidance concerning the manufacture of a prototype of your product and how to present your project at the end of the unit.

3.1 Engineering products

In this unit you will study the principles of engineering design, planning and prototyping. You will design and manufacture an engineered product which can be electrical, mechanical, fluidic, electronic or any combination of these. The choice of product will be yours but will be subject to a number of constraints, including the information, materials and resources that are available to you. In order to give you an idea of what is required, we shall start by providing you with an example of an engineering product and the various stages that turned an initial design brief into a fully functional prototype. This example should give you an insight into the stages required to develop your own product before we look at engineering design, planning and prototyping in greater depth.

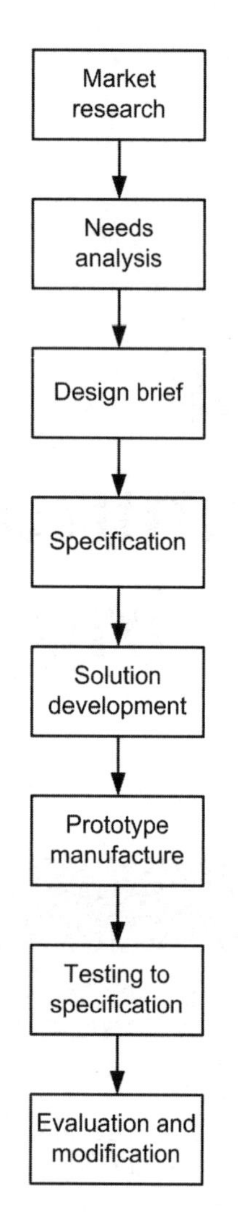

Figure 3.1 *Stages in the design and development of an engineering product*

Figure 3.1 shows the typical stages in the design and development of an engineering product or service. We will be explaining these stages in much greater detail later in this chapter but, for now, it is sufficient to say that we followed each of these steps in the development of our example product. We started with some initial market research and our analysis of this justified the need for our new product. It also provided us with some valuable pointers as to what the product must be like.

Market research and needs analysis

The market research indicated that a market existed for a small adjustable d.c. power supply that could be used in a school or college science laboratory. The research (which was based on discussions with potential users) indicated that the basic requirements for the product were as follows:

1. It should be mechanically and electrically rugged
2. It should use proven and reliable technology
3. It should be low-cost and easy to maintain
4. It should comply with appropriate European legislation (e.g. the EMC and low-voltage directives)
5. It should have one variable (3 V to 15 V) output and one fixed (+5 V) accessory output and that both outputs should be protected against a short-circuit connected to the output
6. With the exception of the accessory output and the a.c. mains input, all controls, connectors and switches should be made available on the front panel
7. It should have colour coded output terminals which will accept standard 4 mm plugs. The terminals should also allow wires that have not been fitted with plugs to be clamped directly using a screw action
8. It should operate from a standard 220 V a.c. mains supply which should be connected using a standard 3-pin IEC connector
9. It should be 'tamper-proof' (it should not be possible to remove the knobs or the enclosure without having to resort to the use of special tools)
10. It should be lightweight and portable
11. It should have LED indicators to show that the power supply is switched on and that the outputs are present.

Design brief and design specification

Having determined a need for our product and obtained a detailed list of user requirements, the next stage was that of firming up our design brief and producing a detailed specification for the power supply. This *design specification* was a detailed performance specification which included numerical values for relevant parameters (such as output voltage and output current). The detailed design specification was important because we returned when we needed to confirm that our prototype power supply met our requirements. We did this by comparing the measured performance specification with the original design specification. Taking into account the requirements listed above, we arrived at the following design specification:

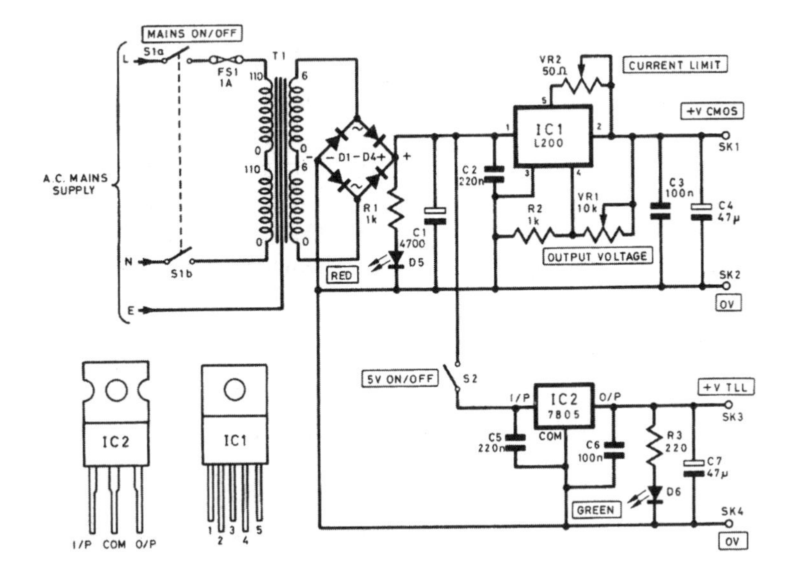

Figure 3.2 *Circuit diagram for a variable DC power supply*

Figure 3.4 *Stripboard prototype layout*

Figure 3.5 *Assembled stripboard prototype*

Figure 3.6 *Printed circuit board prototype*

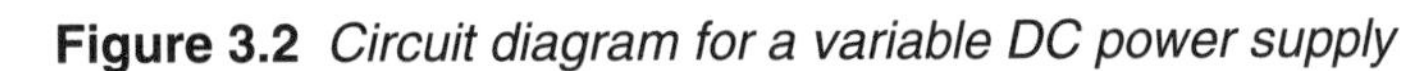

Howard Associates

DATA SHEET L200 Adjustable Voltage Regulator

MAIN FEATURES

- Adjustable output voltage down to 2.85 V
- Adjustable output current up to 2 A
- Input overload protection (up to 60 V for 10 ms)
- Thermal overload protection
- 5-pin Pentawatt® package
- Low bias current on regulation pin
- Low standby current drain
- Low cost

DESCRIPTION

The L200 is a monolithic integrated circuit voltage regulator which features variable voltage and variable current adjustment. The device is supplied in a 5-pin Pentawatt® package (a TO-3 packaged version is also available to special order). Current limiting, power limiting, thermal shutdown and input over-voltage protection (up to 60 V for 10 ms) make the L200 virtually blow-out proof. The L200 can be used in a wide range of applications wherever high-performance and adjustment of output voltage and current is required.

DIMENSIONS

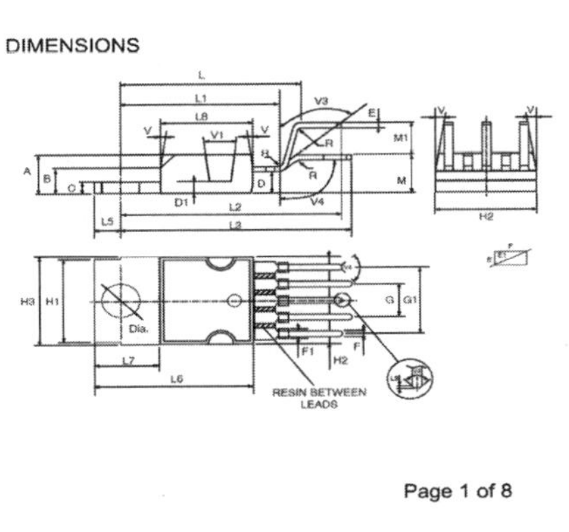

DIM.	mm			inch		
	MIN.	TYP.	MAX.	MIN.	TYP.	MAX.
A			4.8			0.189
C			1.37			0.054
D	2.4		2.8	0.094		0.110
D1	1.2		1.35	0.047		0.053
E	0.35		0.55	0.014		0.022
E1	0.76		1.19	0.030		0.047
F	0.8		1.05	0.031		0.041
F1	1		1.4	0.039		0.055
G	3.2	3.4	3.6	0.126	0.134	0.142
G1	6.6	6.8	7	0.260	0.268	0.276
H2			10.4			0.409
H3	10.05		10.4	0.396		0.409
L	17.55	17.85	18.15	0.691	0.703	0.715
L1	15.55	15.75	15.95	0.612	0.620	0.628
L2	21.2	21.4	21.6	0.831	0.843	0.850
L3	22.3	22.5	22.7	0.878	0.886	0.894
L4			1.29			0.051
L5	2.6		3	0.102		0.118
L6	15.1		15.8	0.594		0.622
L7	6		6.6	0.236		0.260
L9		0.2			0.008	
M	4.23	4.5	4.75	0.167	0.177	0.187
M1	3.75	4	4.25	0.148	0.157	0.167
V4	40° (typ.)					

Page 1 of 8

Figure 3.3 *Datasheet for the L200 voltage regulator*

Fixed output

- Output voltage adjustable from 3V to 15V
- Output current adjustable from 50 mA to 1 A max.
- 4 mm binding post connectors (red and black)

Variable output

- Output voltage fixed at +5V ±5%
- Output current 1 A max.
- 4 mm binding post connectors (yellow and black)

Input

- 200 to 240 V a.c. via IEC connector, 1 A fuse and EMC filter

Solution development and prototype manufacture

The next stage in the development of the power supply was that of examining a range of solutions before deciding on the particular solution that formed the basis of a prototype. Since the product has both electrical/electronic and mechanical aspects it was possible to consider these separately.

Research was needed to find a suitable electronic circuit and to identify the components that would be needed to build it. Various sources of information were available including books, magazines, data sheets from manufacturers and suppliers as well as the Web. After carrying out some initial investigation, a suitable circuit design was located (see Figure 3.2). This circuit was based on two low-cost readily available integrated circuit devices (see requirements 2 and 3) and the data sheets were obtained for each (see Figure 3.3).

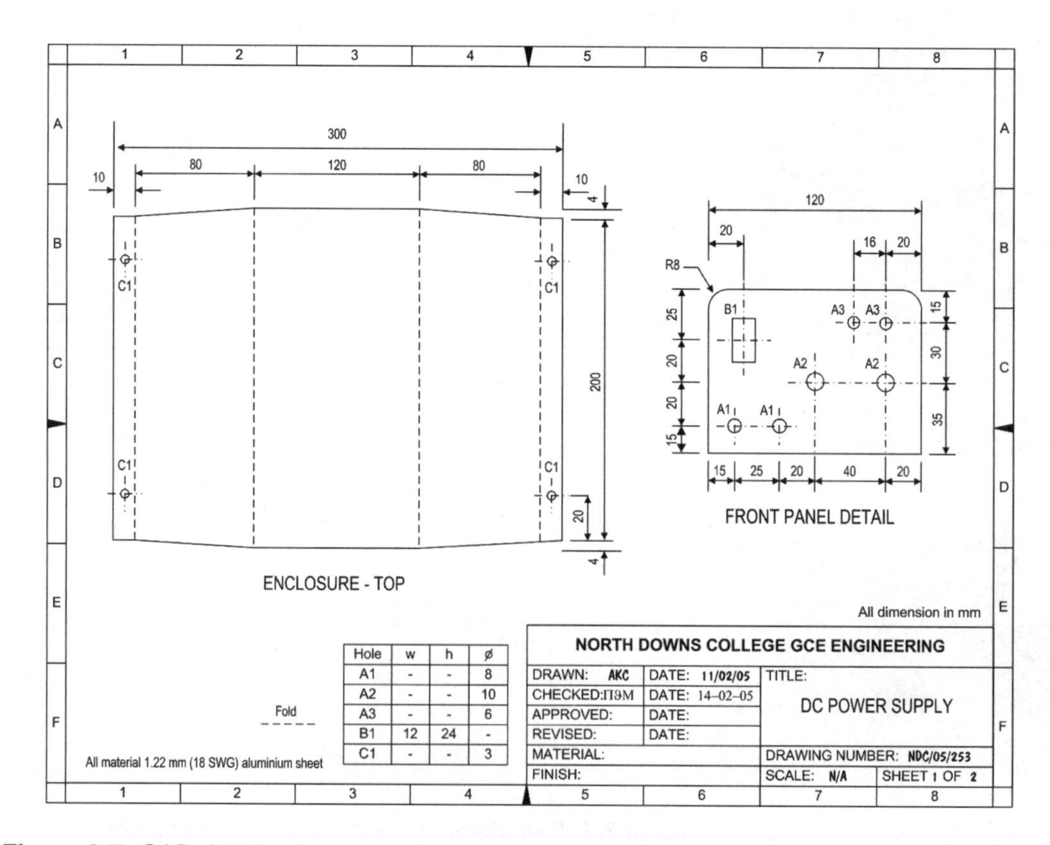

Figure 3.7 *CAD drawing for the enclosure*

The initial prototype was constructed on stripboard following the layout recommended by the circuit designer (see Figure 3.4) but modified for fitting into the enclosure (as shown in Figure 3.5). A further prototype was developed using a printed circuit board (PCB), the track layout for which was designed using a PCB CAD package which had an autorouting facility.

In order to satisfy requirements 1 and 4, a fully-screened metal enclosure was used. This was based on a simple two-part construction (see requirements 3). An internal EMC filter was fitted in order to comply with the EMC directive (see requirement 4).

Before manufacturing the enclosure it was necessary to produce some general arrangement and detail drawings showing how the sheet metal should be drilled, punched, cut and bent. These drawings had to be accurately dimensioned in order to accommodate the components used. The drawings were produced using a simple 2D CAD package. One drawing was produced to show the enclosure top and front panel detail (see Figure 3.7) whilst a second drawing was used to show the enclosure bottom and rear panel detail.

The material selected for the enclosure was 1.22 mm (18 SWG) aluminium sheet. The reasons for the choice of this material were that it was reasonably lightweight (see requirement 10), low-cost (see requirement 3), and easy to process. It could also be attractively paint finished. The assembled prototype is shown in Figures 3.8 and 3.9.

Figure 3.8 *External view of the finished prototype*

Testing to specification

Having assembled a working prototype the next stage was that of measuring its performance in order to ensure that it fully conformed with the requirements set down in the design brief. Testing was carried out by applying a variable load to the output of the power supply and measuring corresponding values of output voltage and output current. To simplify the analysis, the test data was entered into a spreadsheet and graphs of each load test were generated.

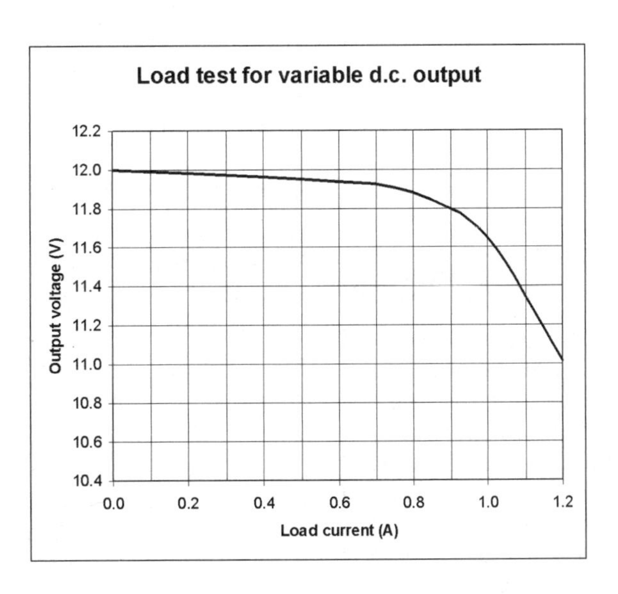

Figure 3.10 *Load test graph for the variable d.c. output*

Figure 3.9 *Internal layout of the finished prototype*

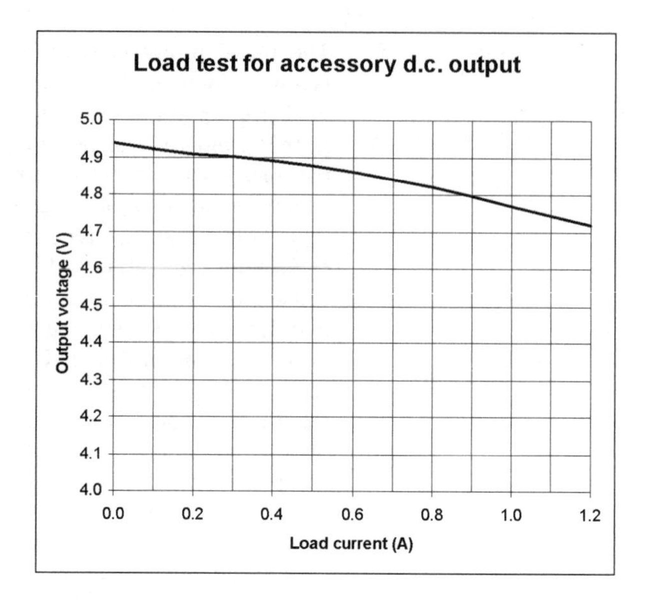

Figure 3.11 *Load test graph for the fixed accessory output*

The measured performance specifications for the power supply (obtained from the load tests) were as follows (figures from the original design specification are shown in square brackets):

Fixed output

- Minimum output voltage = 2.85V [3 V]
- Maximum output voltage = 15.23 V [15 V]
- Maximum output current = 1.2 A [1 A]

Variable output

- Output voltage (no load) = 4.94 V [5 V]
- Output voltage (1 A load) = 4.77 V [4.75 V]
- Output current = 1.2 A [1 A]

The above results provide confirmation that the prototype conformed closely with the original design specification.

Evaluation and modification

Having carried detailed testing of the prototype power supply and having verified that the design specification had been met, the next stage was that of finalising the prototype and carrying out any modifications prior to passing the design for production. Feedback was obtained from a number of 'test users' who were asked to check that each of the original requirements had been satisfied. Several recommendations were made as a result of this feedback including:

- fitting a carrying handle to the enclosure
- fitting calibrated scales to the voltage and current controls
- relocating the 5 V accessory switch to the front panel
- redrawing the circuit diagram to BS standards (for inclusion in a repair manual to be supplied with each unit).

Another view

Your tutor should be able to supply you with an example of a student portfolio based on the product described here. This will give you a better idea of what you should be aiming to produce for the unit assessment. The sample portfolio is part of the Curriculum Support Pack which accompanies this book.

3.2 Engineering drawings

Types of engineering drawing

Engineers use many different types of drawing and diagram to communicate their ideas. These drawings include block diagrams, flow diagrams, exploded diagrams and circuit schematics. It is vitally important to be able to understand each type of drawing and the circumstances in which it is used.

To avoid confusion (and to ensure that drawings are correctly interpreted) nationally and internationally recognized symbols, conventions and abbreviations are used. These are listed (and their use explained) in the appropriate British Standards (BS 8888:2004). Such standards tend to be lengthy and costly but low-cost summaries are available for students and your school or college library should have copies to which you can refer. The abridged edition for schools and colleges is 'PP 8888-1:2005 Drawing practice: a guide for schools and colleges to BS 8888:2004, Technical Product Specification'.

Block diagrams

Block diagrams show the relationship between the various elements of a system. They can be used to simplify a complex design by dividing it into a number of much smaller functional elements. Figure 3.12 shows a typical example in which the links and dependencies between various elements (inputs and outputs) can be clearly seen.

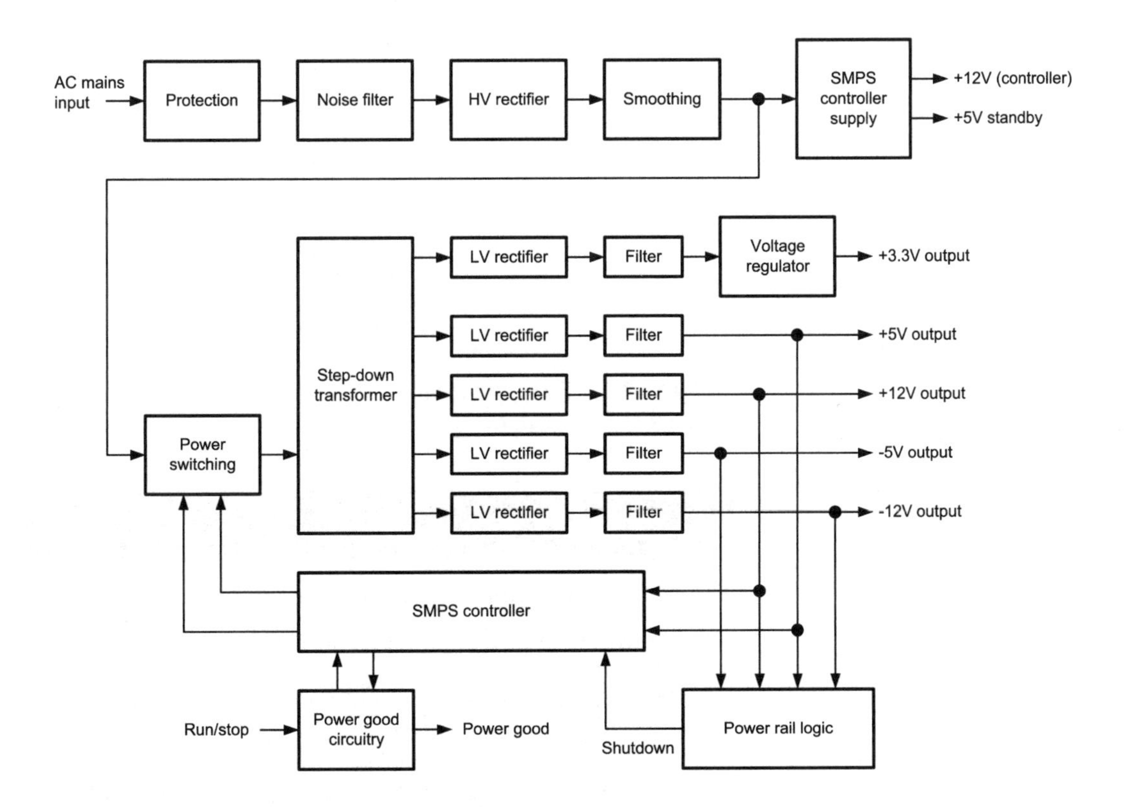

Figure 3.12 *Block diagram of a computer power supply*

Flow diagrams

Flow diagrams (or *flowcharts*) are used to illustrate the logic of a sequence of events. They are frequently used in fault-finding, computer programming (software engineering) and in process control. They are also used by production engineers when working out the best sequence of operations in which to manufacture a product or component. Figure 3.13 shows a flow chart for fault location on the power supply shown in Figure 3.12. The symbols used in diagrams are shown in Figure 3.14.

3
Slowly increase AC input voltage
Fuse blows
No
Yes
Check/replace switching transistors
Check/replace HV reservoir capacitors
All outputs missing?
No
Yes
Check SMPS controller supply
Check drive to switching transistors
Check SMPS controller
Check DC voltage at each output
Identify out of tolerance outputs
Check rectifier regulator and filter

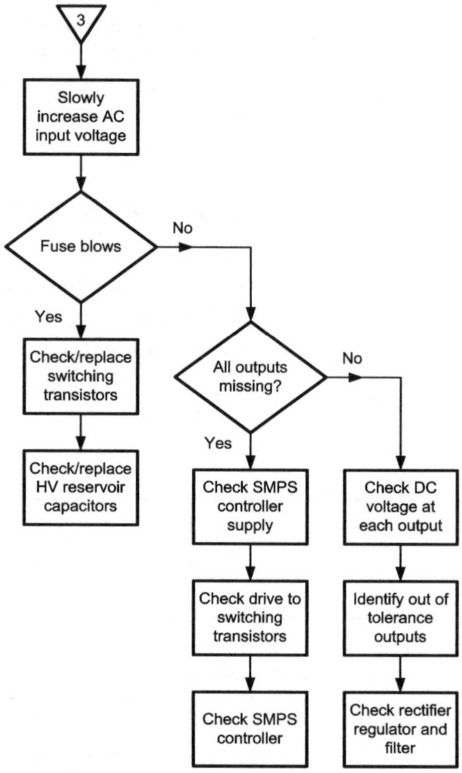

Figure 3.13 *Flow diagram for fault-finding on the power supply shown in Figure 3.12.*

Start or finish

Process

Decision

Data input/output

Annotation

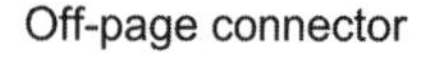
Off-page connector

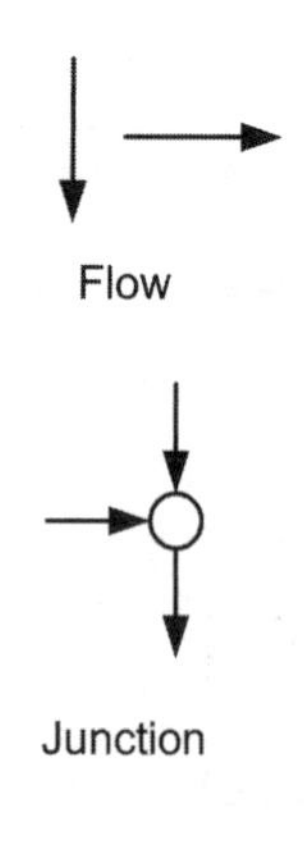

Figure 3.14 *Flow diagram symbols*

Activity 3.1

Construct a flow diagram for any ONE of the following tasks:

1. Upgrading the hard disk drive of a PC
2. Fitting a tow bar to a car
3. Replacing the cutting blade of an electric lawnmower
4. Changing the oil in a car or motorcycle
5. Installing a CD writer in a PC
6. Building and testing a radio controlled model aircraft or boat.

Test your knowledge 3.1

Your bicycle tyre is flat. Draw a flow diagram for checking the inner-tube, locating and repairing a puncture (if there is one), or replacing a faulty valve. Figure 3.15 shows you how to start the flow chart.

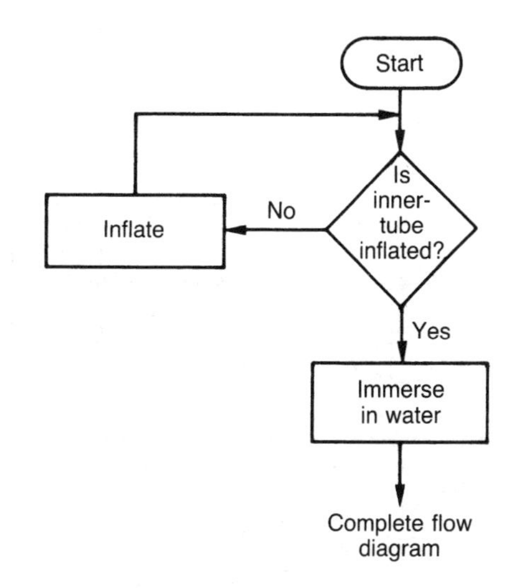

Figure 3.15 *See Test your knowledge 3.1*

Circuit diagrams

These are used to show the functional relationships between the components in a circuit. The components are represented by symbols and their position in the circuit diagram does not represent their actual position in the final assembly. Circuit diagrams are also referred to as schematic diagrams or even schematic circuit diagrams.

Figure 3.16(a) shows the circuit for an electronic filter unit using standard component symbols. Figure 3.16(b) shows a layout diagram with the components correctly positioned. Figure 3.16(c) shows the printed circuit board. This is also called a track diagram or a wiring diagram.

However, it is more usual to use the term wiring diagram where the components are hard wired, as in the wiring up of a building or the manufacture of a control cubicle. Architects use circuit diagrams to show the electrical installation of buildings. They also provide installation drawings to show where the components are to be sited. They may also provide a wiring diagram to show how the cables are to be routed to and between the components. The symbols used in architectural installation drawings and wiring diagrams are not the same as those used in circuit diagrams. Examples of architectural and topographical electrical component symbols are shown in BS 8888.

Schematic circuit diagrams are also used to represent pneumatic (compressed air) circuits and hydraulic circuits. Pneumatic circuits and hydraulic circuits share the same symbols. You can tell which circuit is which because pneumatic circuits should have open arrowheads, whilst hydraulic circuits should have solid arrowheads. Also, pneumatic circuits exhaust to the atmosphere, whilst hydraulic circuits have to have a return path to the oil reservoir. Figure 3.17 shows a typical hydraulic circuit.

Just as electrical circuit diagrams may have corresponding installation and wiring diagrams, so do hydraulic, pneumatic and plumbing circuits. Only this time the wiring diagram becomes a pipe-

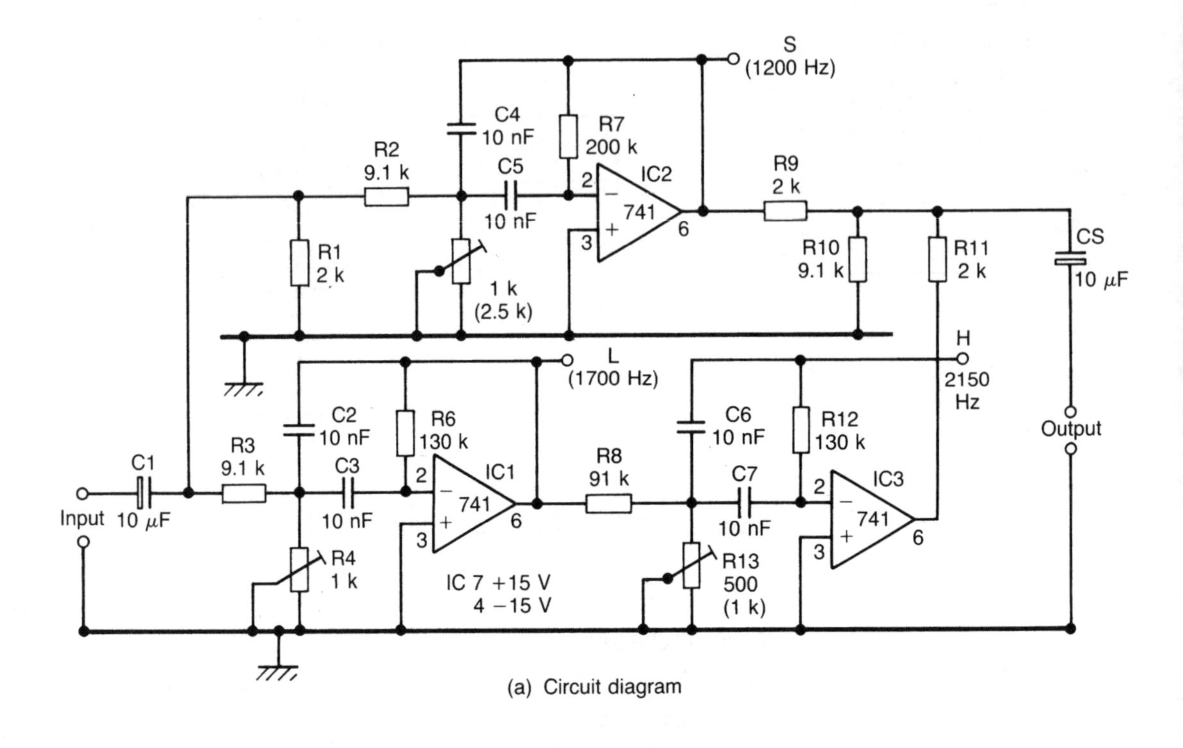

(a) Circuit diagram

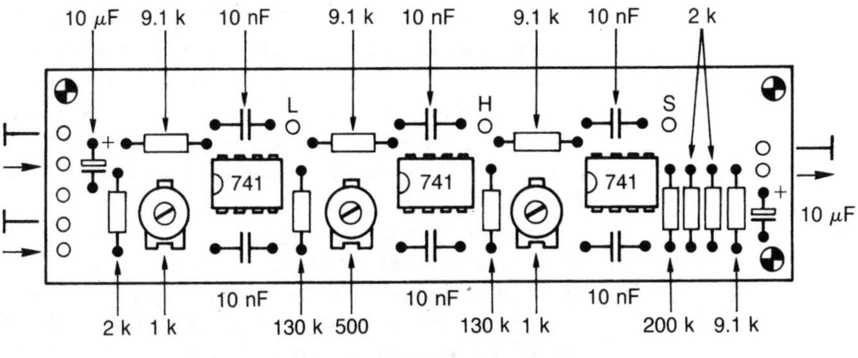

(b) Component layout diagram

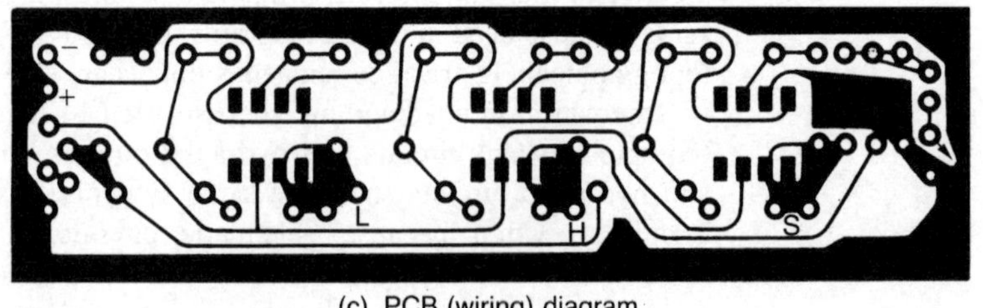

(c) PCB (wiring) diagram

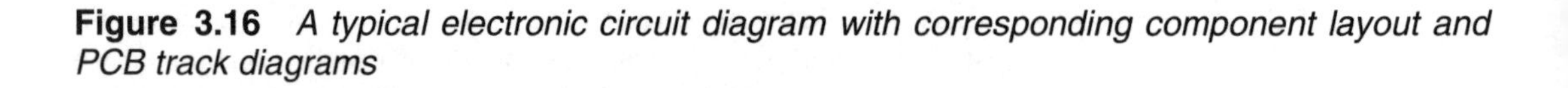
Figure 3.16 *A typical electronic circuit diagram with corresponding component layout and PCB track diagrams*

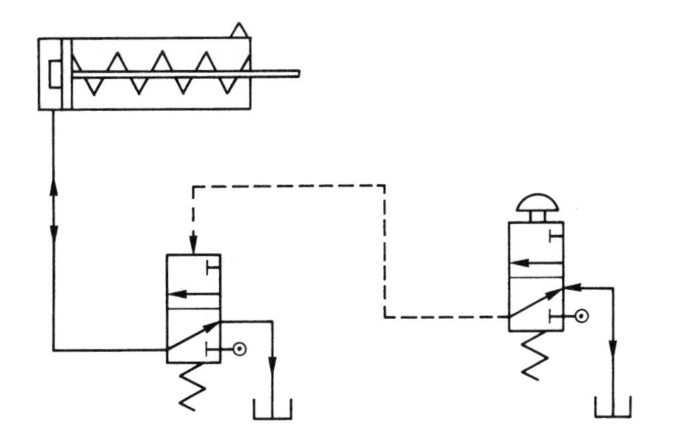

Figure 3.17 *A typical hydraulic circuit*

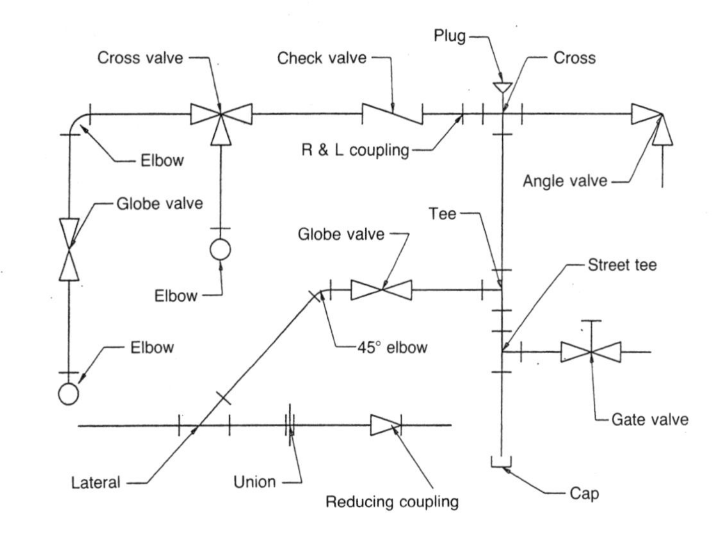

(a) Circuit diagram (schematic)

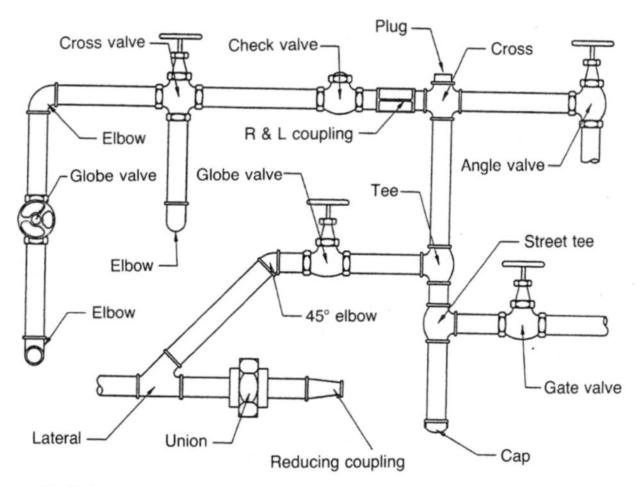

(b) Piping diagram

Figure 3.18 *A typical plumbing circuit with corresponding piping diagram*

work diagram. A plumbing example is shown in Figure 3.18. As you may not be familiar with the symbols, I have named them for you. Normally this is not necessary and the symbols are recognized by their shapes.

General Arrangement (GA) drawings

General Arrangement (GA) drawings are widely used in engineering to show the overall arrangement of an engineering assembly (such as a pump, gearbox, motor drive, or clutch). GA drawings are often supported by a number of detail drawings that provide more detailed information on the individual parts.

Figure 3.19 shows a typical GA drawing. This shows as many of the features listed above as are appropriate for this drawing. It shows all the components correctly assembled together. Dimensions are not usually given on GA drawings although, sometimes, overall dimensions will be given for reference when the GA drawing is of a large assembly drawn to a reduced scale.

The GA drawing shows all the parts. These are listed in a table together with the quantities required. Manufacturers' catalogue references are also given for bought-in components. The detail drawing numbers are also included for components that have to be manufactured as special items.

Detail drawings

As the name implies, detail drawings provide all the details required to make the component shown on the drawing. Referring to Figure 3.19, we see from the table that the detail drawing for the punch has the reference number 174/6. Figure 3.20 shows this detail drawing. In this instance, the drawing provides the following information:

- the shape of the punch
- the dimensions of the punch and the manufacturing tolerances
- the material from which the punch is to be made and its subsequent heat treatment
- the unit of measurement (millimetre)
- the projection (first angle)
- the finish
- the guidance note 'do not scale drawing'
- the name of the company
- the name of the draughtsperson
- the name of the person checking the drawing.

The amount of information given will depend upon the job. For example, drawings for a critical aircraft component will be much more fully detailed than a drawing for a wheelbarrow component.

Exploded views

The final type of specialised drawing that we shall be looking at is called an *exploded view*. Quite simply, an exploded view is a pictorial representation of a product that is taken apart. By drawing the individual component parts separately but in approximately the same physical relationship as when assembled, you can gain a very good idea of how something is put together. Exploded views can be

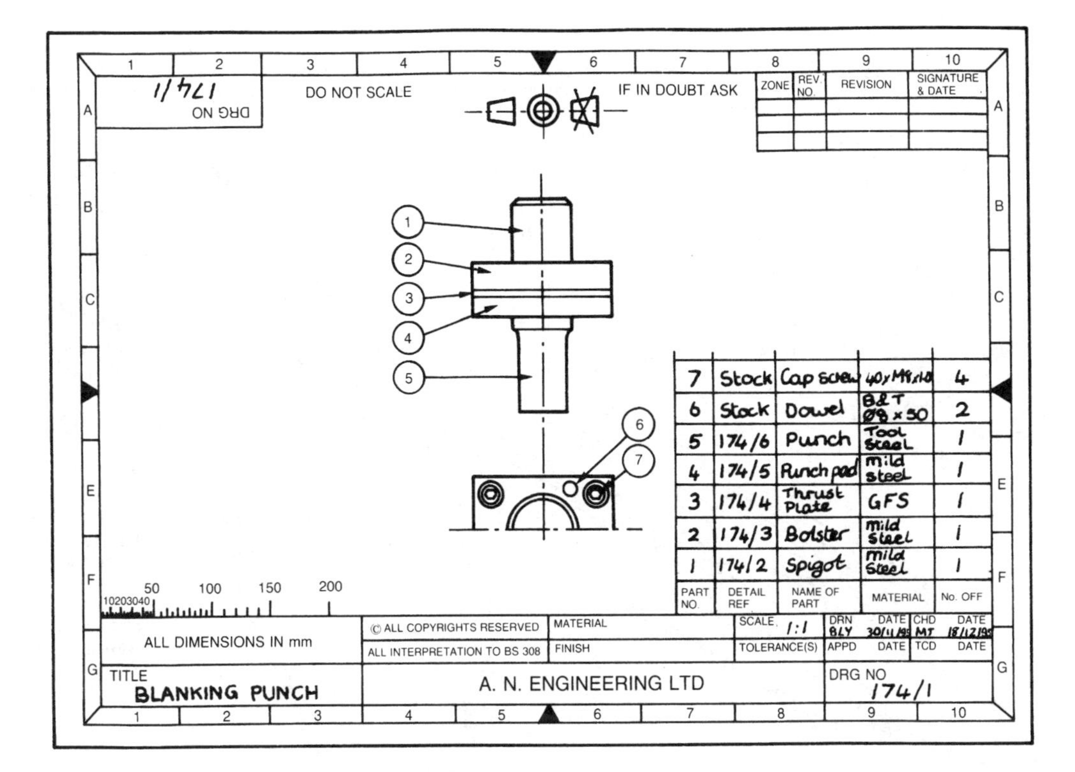

Figure 3.19 *General Arrangement (GA) drawing*

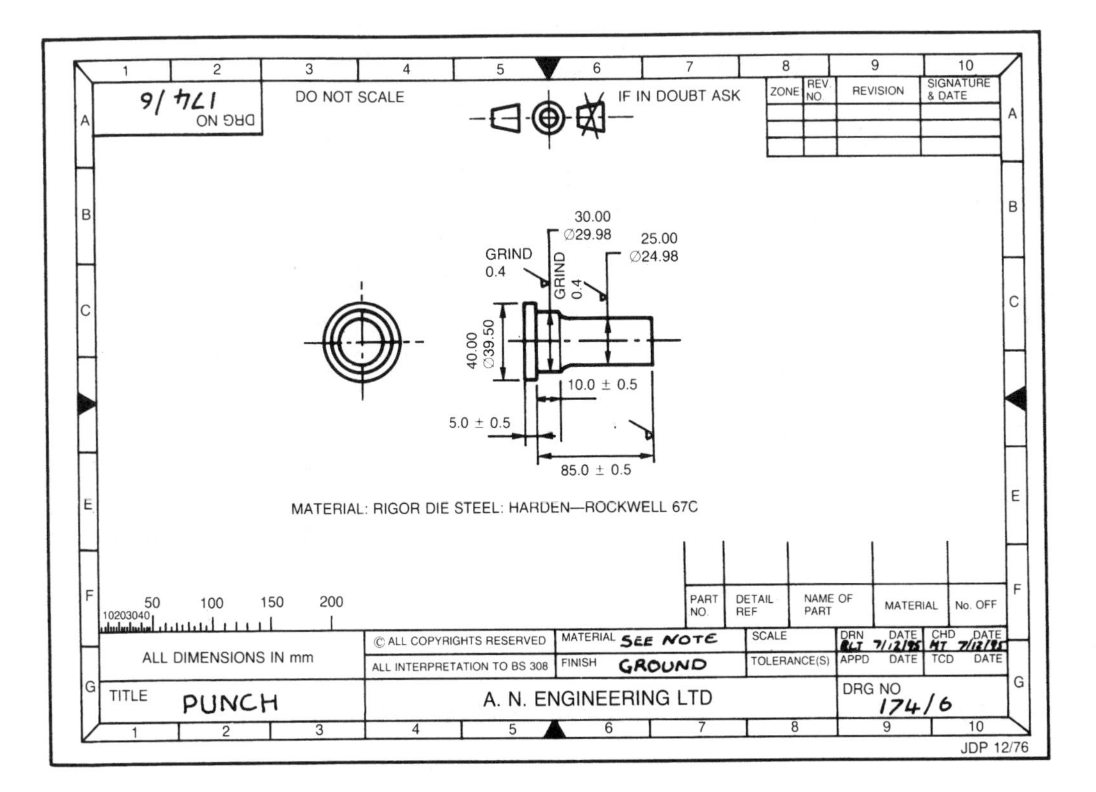

Figure 3.20 *Detail drawing*

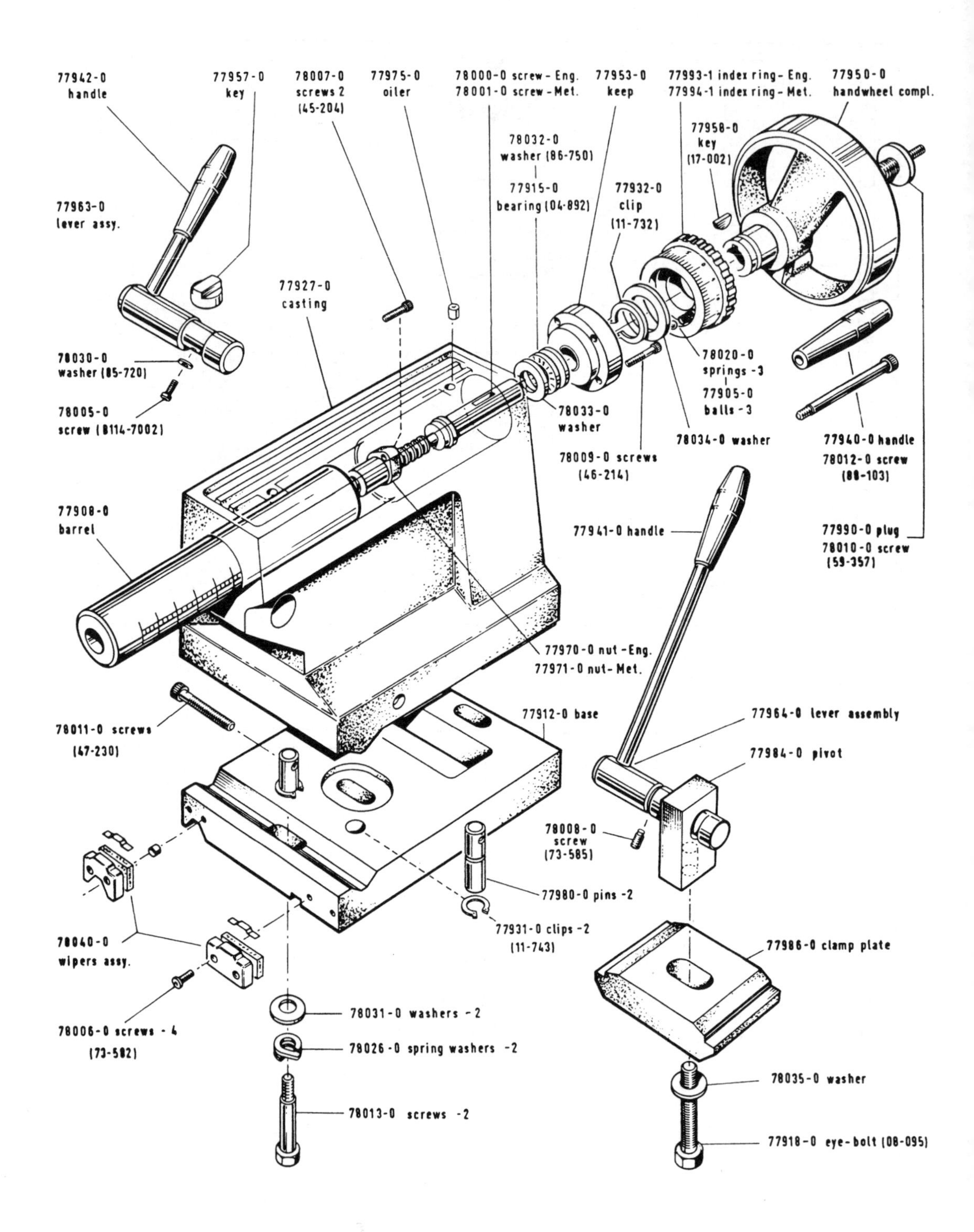

Figure 3.21 *Exploded view and parts list for the tailstock assembly*

extremely useful when a product has to be serviced or maintained. A service or maintenance engineer has only to take a look at an exploded diagram to see how the various parts fit together.

A typical exploded diagram for the tailstock assembly of a lathe is shown in Figure 3.21. Note that part numbers are included on the diagram.

Charts and graphs

In order to make it easier to understand (or when presenting information to a non-technical audience), numerical engineering data is often presented in the form of a chart or a graph. Some typical examples are shown in Figures 3.22 and 3.23. Notice how it is easy to see a *trend* when data is presented in this way.

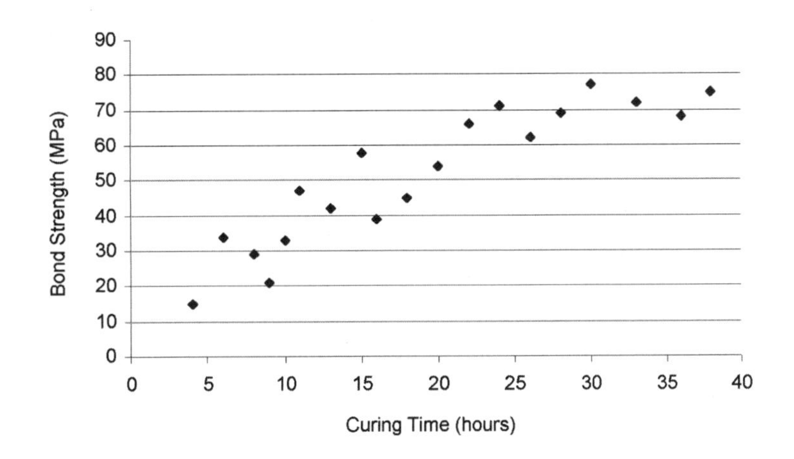

Figure 3.22 *Engineering data represented in the form of a scatter diagram*

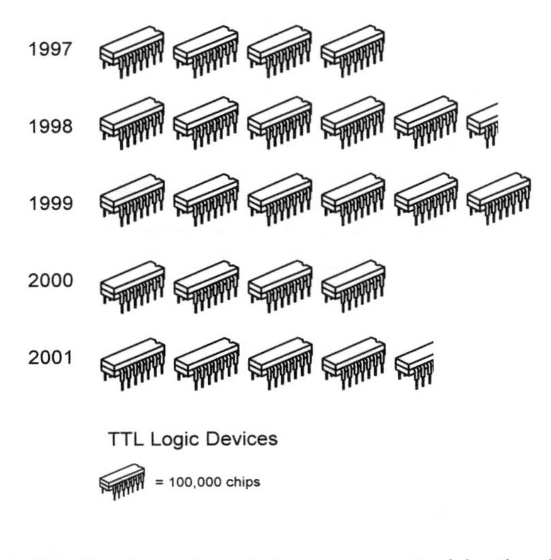

Figure 3.23 *Engineering data represented in the form of a chart*

Test your knowledge 3.2

Refer to the exploded view shown in Figure 3.21. State the function of the components with the following part numbers:

(a) 78012

(b) 78026

(c) 77927

(d) 77931

(e) 77958

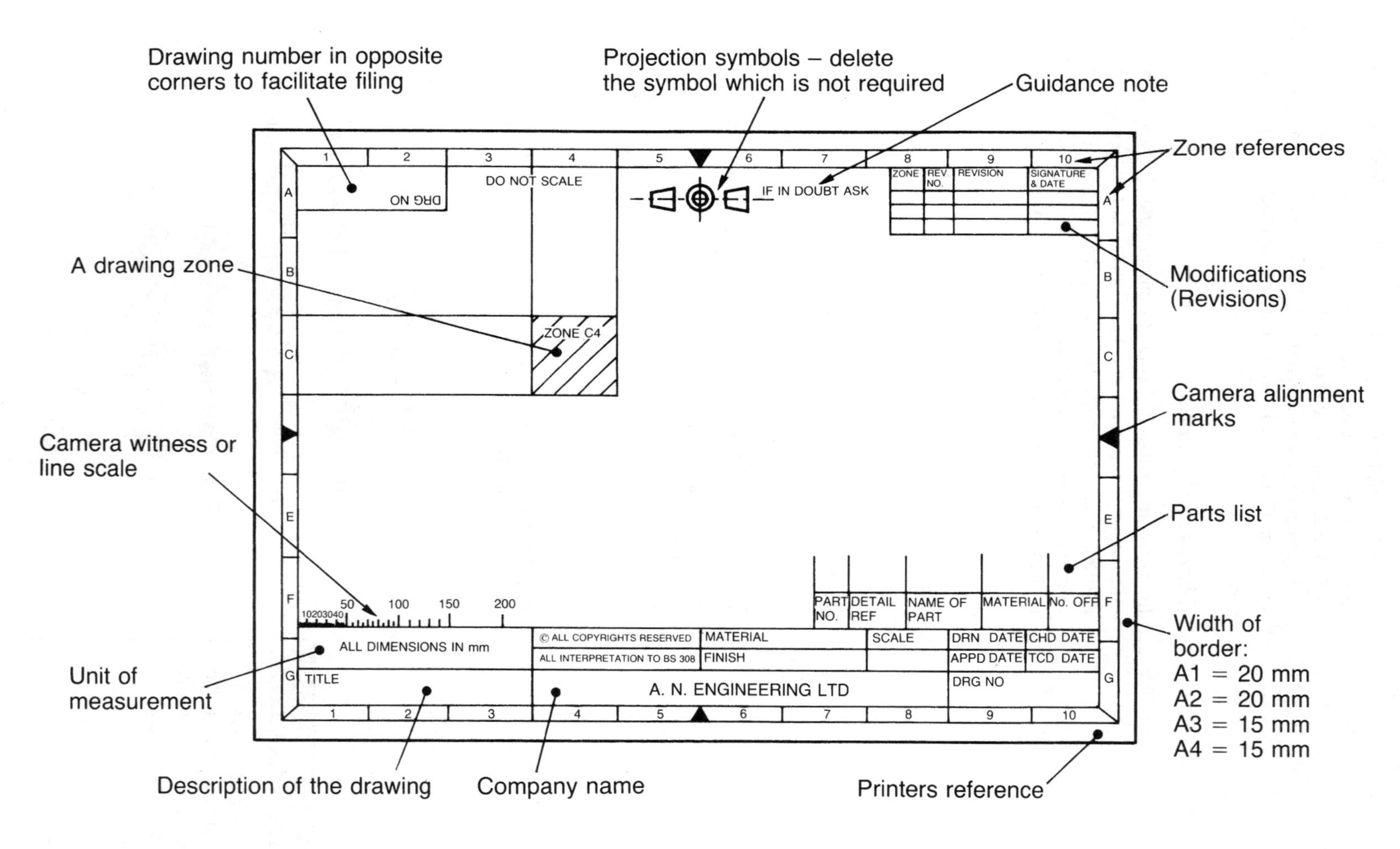

Figure 3.24 *Layout of a typical drawing sheet*

The drawing sheet

Figure 3.24 shows the layout of a typical drawing sheet. To save time these are printed to a standardized layout for a particular company, ready for the draughtsperson to add the drawing and complete the boxes and tables. The basic information found on most drawing sheets consists of:

- the drawing number and name of the company
- the title and issue details
- scale
- method of projection (first or third angle)
- initials of persons responsible for: drawing, checking, approving, tracing, etc. together with the appropriate dates
- unit(s) of measurement (inches or millimetres) and general tolerances
- material and finish
- copyright and standards reference
- guidance notes such as: 'do not scale'
- reference grids so that 'zones' on the drawing sheet can be quickly found
- modifications table for alterations which are reference related to the issue number on the drawing and identified by the means of the reference grid.

The following additional information may also be included:

- fold marks
- centre marks for camera alignment when microfilming
- line scale, so that the true size is not lost when enlarging or reducing copies
- trim marks
- orientation marks.

Activity 3.2

Use A9CAD (or any other 2D CAD package) to produce your own drawing sheet, following the guidelines shown in Figure 3.24. Save and print your drawing sheet.

Activity 3.3

Use the drawing sheet that you created in Activity 3.2 in conjunction with the diagram of the fastenings that you created earlier in Activity 2.7. Make sure that the following items are included in your drawing; title, your name, the name of your school or college, the date and your selection of materials for manufacturing the hex bolt and washer.

Engineering drawing techniques

Engineering drawings can be produced either manually or on a computer using suitable CAD software. Drawings produced manually can range from freehand sketches to formally prepared drawings produced with the aid of a drawing board and conventional drawing instruments.

The choice of technique is dependent upon a number of factors such as:

- *Speed* How much time can be allowed for producing the drawing. How soon the drawing can be commenced.
- *Media* The choice will depend upon the equipment available (e.g. CAD or conventional drawing board and instruments) and the skill of the person producing the drawing.
- *Complexity* The amount of detail required and the anticipated amount and frequency of development modifications.
- *Cost* Engineering drawings are not works of art and have no intrinsic value. They are only a means to an end and should be produced as cheaply as possible. Both initial and ongoing costs must be considered.
- *Presentation* This will depend upon who will see/use the drawings. Non-technical people can visualize pictorial representations better than orthographic drawings.

Nowadays technical drawings are increasingly produced using computer aided drawing techniques (CAD). Developments in software and personal computers have reduced the cost of CAD and made it more powerful. At the same time, it has become more 'user friendly'. Computer aided drawing does not require the high physical skill required for manual drawing which takes years of practice to achieve. It also has a number of other advantages over manual drawing. Let's consider some of these advantages:

- *Accuracy* Dimensional control does not depend upon the draughtsperson's eyesight.
- *Radii* These can be made to blend with straight lines automatically.
- *Repetitive features* For example, holes round a pitch circle do not have to be individually drawn but can be easily produced automatically by 'mirror imaging'. Again, some repeated, complex feature need only be drawn once and saved as a matrix. It can then be called up from the computer memory at each point in the drawing where it appears at the touch of a key.
- *Editing* Every time you erase and alter a manually produced drawing on tracing paper or plastic film the surface of the drawing is increasingly damaged. On a computer you can delete and redraw as often as you like with no ill effects.
- *Storage* No matter how large and complex the drawing, it can be stored digitally on floppy disk. Copies can be taken and transmitted between factories without errors or deterioration.
- *Prints* Hard copy can be produced accurately and easily on flat bed or drum plotters and to any scale. Colour prints can also be made.

Engineering drawings such as General Arrangement drawings and detail drawings are produced by a technique called orthographic drawing using the conventions set out in BS 8888. Since I will be asking you to make orthographic drawings from more easily recognized pictorial drawings, I will start by introducing you to the two pictorial techniques widely used by draughtspersons.

Oblique drawing

Test your knowledge 3.3

(a) Obtain a sheet of quadrille paper (maths paper ruled in 5 mm squares) and draw the box shown in Figure 3.25 full size. Use cabinet oblique projection.

(b) Now use your compasses to draw a 50 mm diameter hole in the centre of the front (elevation) of the box.

(c) Can you think of a way to draw the same circle on the side (receding) face of the box? Note that it will not be a true circle so you cannot use your compasses!

Figure 3.25 shows a simple oblique drawing. The front view (elevation) is drawn to true shape and size. Therefore this view should be chosen so as to include any circles or arcs so that these can be drawn with compasses. The lines forming the side views appear to travel away from you, so these are called *receders*. They are drawn at 45° to the horizontal using a 45° set-square. They may be drawn full length as in cavalier oblique drawing or they may be drawn half-length as in cabinet oblique drawing. This latter method gives a more realistic representation, and is the one you and I will be using.

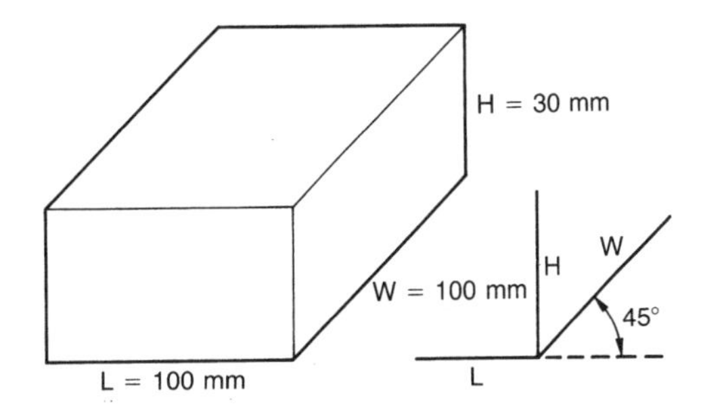

Cavalier oblique projection

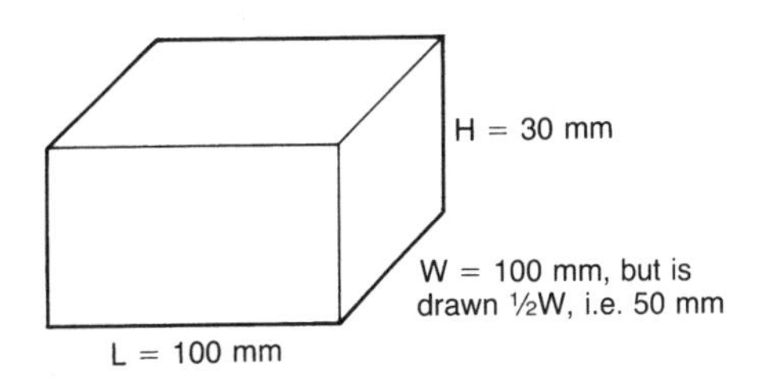

Cabinet oblique projection

Figure 3.25 *A simple oblique drawing*

Isometric drawing

Figure 3.26(a) shows an isometric drawing of our previous box. To be strictly accurate, the vertical lines should be drawn true length and the receders should be drawn to a special isometric scale. However this sort of accuracy is rarely required and, for all practical purposes, we draw all the lines full size. As you can see, the receders are drawn at 30° to the horizontal for both the elevation and the end view.

Although an isometric drawing is more pleasing to the eye, it has the disadvantage that all circles and arcs have to be constructed. They cannot be drawn with compasses. Figures 3.26(b), 3.26(c) and 3.26 (d) show you how to construct an isometric curve. You could have used this technique in Test your knowledge 3.3 to draw the circle on the side of the box drawn in oblique projection.

First we draw the required circle. Then we draw a grid over it as shown in Figure 3.26(b). Next number or letter the points where the circle cuts the grid as shown. Now draw the grid on the side elevation of the box and step off the points where the circle cuts the grid with your compasses as shown in Figure 3.26(c). All that remains is to join up the dots and you have an isometric circle as shown in Figure 3.26(d).

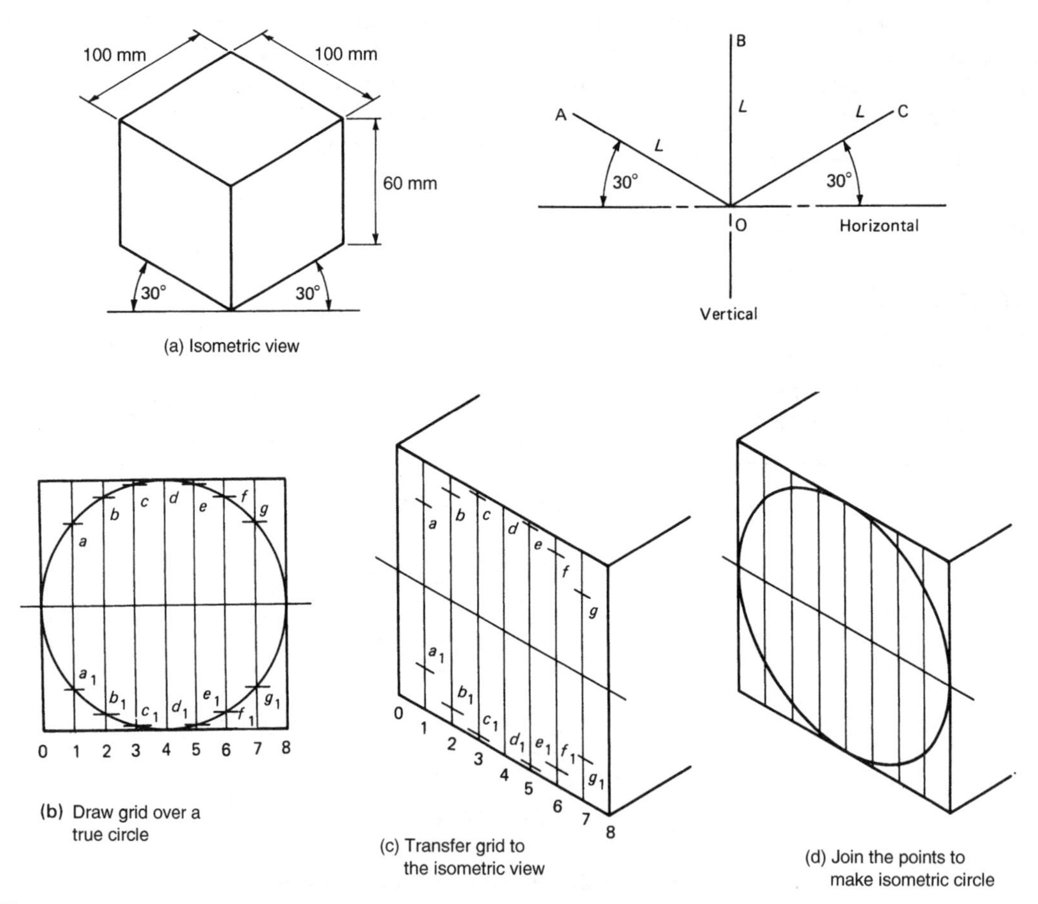

Figure 3.26 *Isometric drawing*

Another way of drawing isometric circles and curves is the 'four-arcs' method. This does not produce true curves but they are near enough for all practical purposes and quicker and easier than the previous method for constructing true curves. The steps are shown in Figure 3.27.

- Join points B and E, as shown in Figure 3.27(b). The line BE cuts the line GC at the point J. The point J is the centre of the first arc. With radius BJ set your compass to strike the first arc as shown.

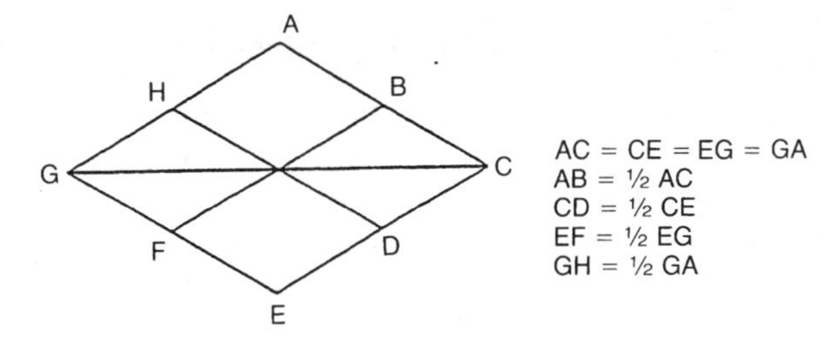

(a) Draw an isometric grid of appropriate size

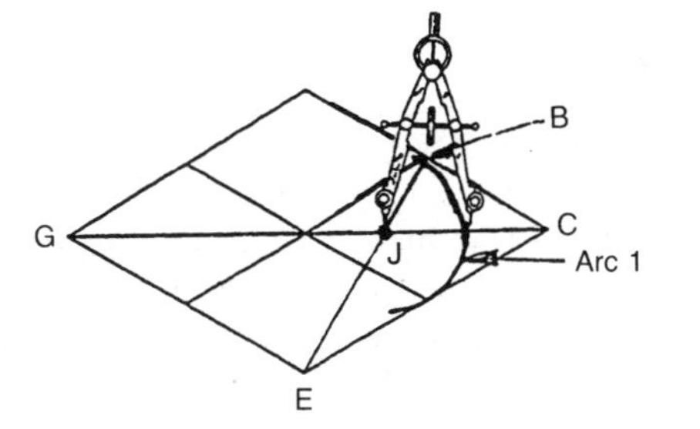

(b) Construct the 1st arc using a compass located as shown

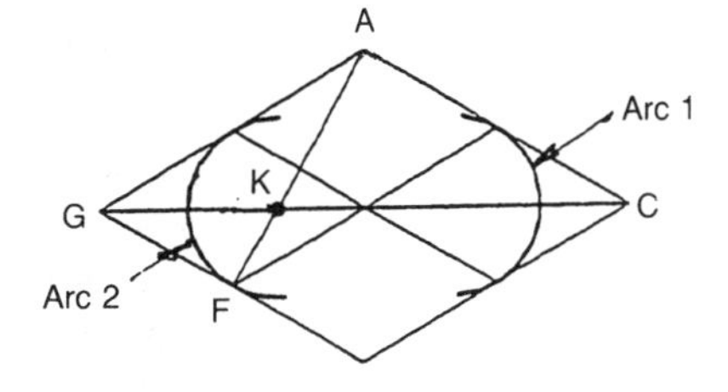

(c) Draw the 2nd arc using the construction process shown

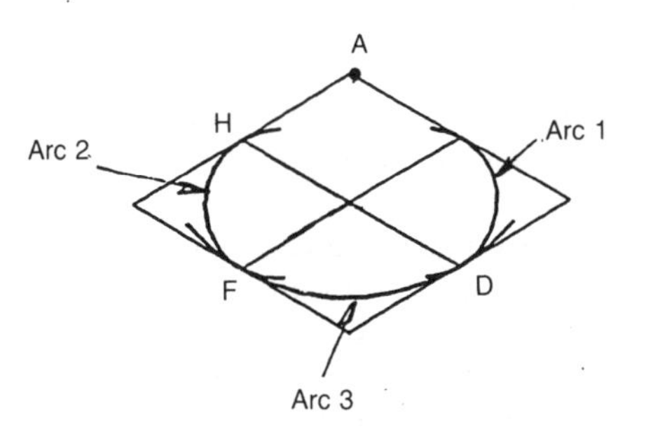

(d) Draw the 3rd arc through the points shown

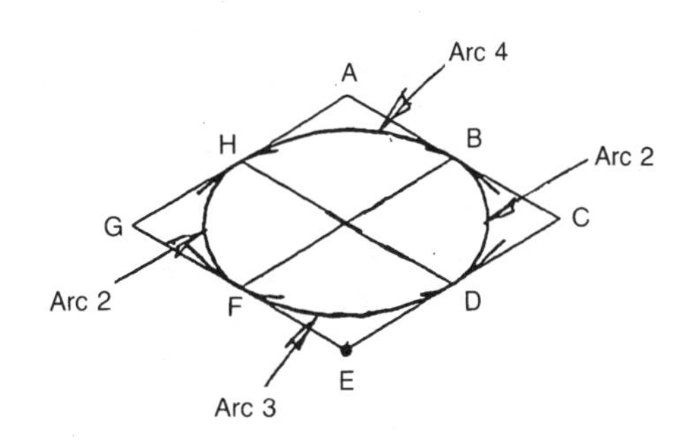

(e) Complete the process drawing the 4th arc from the opposite corner

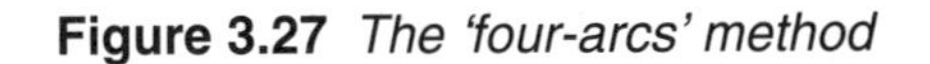
Figure 3.27 *The 'four-arcs' method*

- Join the points A and F, as shown in Figure 3.27(c). The line AF cuts the line GC at the point K. The point K is the centre of the second arc. With radius KF set your compasses to strike the second arc as shown. If your drawing is accurate both arcs should have the same radius.
- With centre A and radius AF or AD strike the third arc, as shown in Figure 3.27(d).
- With centre E and radius EH or EB strike the fourth and final arc, as shown in Figure 3.27(e). If your drawing is accurate, arcs 3 and 4 should have the same radius.

The Test your knowledge exercises will provide you with some practise with isometric drawings. You should complete these before moving on to the next section.

Test your knowledge 3.4

(a) Draw, full size, an isometric view of the box shown in Figure 3.26. Isometric ruled paper will be of great assistance if you can obtain some!

(b) Draw a 50 mm diameter isometric circle on the upper (top) face of the box. Remember that Figure 3.26 shows it on the side of the box.

Orthographic drawing

General Arrangement (GA) and detail drawings are produced by the use of a drawing technique called orthographic projection. This is used to represent three-dimensional solids on the two-dimensional surface of a sheet of drawing paper so that all the dimensions are true length and all the surfaces are true shape. To achieve this when surfaces are inclined to the vertical or the horizontal we have to use auxiliary views, but more about these later. Let's keep things simple for the moment.

Test your knowledge 3.5

Use the technique just described to draw a 40 mm diameter circle centred on the side face of the box. Start off by drawing a 40 mm isometric square in the middle of the side face.

First-angle projection

Figure 3.29(a) shows a simple component drawn in isometric projection. Figure 3.29(b) shows the same component as an orthographic drawing. This time we make no attempt to represent the component pictorially. Each view of each face is drawn out separately either full size or to the same scale. What is important is how we position the various views as this determines how we 'read' the drawing.

Engineers use two orthographic drawing techniques, either first-angle or third-angle projection. The former is called 'English' projection and the latter is known as 'American' projection. The drawing in Figure 3.29 is in first-angle projection. The views are arranged as follows:

- *Elevation* This is the main view from which all the other views are positioned. You look directly at the side of the component and draw what you see.
- *Plan* To draw this, you look directly down on the top of the component and draw what you see below the elevation.
- *End view* This is sometimes called an 'end elevation'. To draw this you look directly at the end of the component and draw what you see at the opposite end of the elevation. There may be two end views, one at each end of the elevation, or there may be only one end view if this is all that is required to completely depict the component. Figure 3.29 requires only one end view. When there is only one end view this can be placed at either end of the elevation depending upon which gives the greater clarity and ease of interpretation. Whichever end is chosen, the rules for drawing this view must be obeyed.

Test your knowledge 3.6

(a) Figure 3.28(a) shows some further examples of isometric drawings. Redraw them as cabinet oblique drawings.

(b) Figure 3.28(b) shows some further examples of cabinet oblique drawings. Redraw them as isometric drawings. Any circles and arcs on the vertical surfaces should be drawn using the grid construction method. Any arcs and circles on the horizontal (plan) surfaces should be drawn using the 'four-arcs method'.

Use faint construction lines to produce the drawing, as shown in Figure 3.29(b). When these are complete, 'line-out' the outline more heavily. Carefully remove the construction lines to leave the drawing uncluttered, thus improving the clarity. Sometimes, in examinations, you are asked to leave the construction lines so that the examiner can see how you have arrived at your answer. Figure 3.30 shows the finished drawing.

Test your knowledge 3.7

Figure 3.28 shows some components using pictorial projections. Redraw each of these drawings in first-angle orthographic projection. To start you off the first one has been drawn for you (see Figure 3.31).Note how the end view has been positioned so that you can see the web.

Third-angle projection

Figure 3.32 shows the same component, but this time I have drawn it in third-angle projection for you.

- *Elevation* Again I have started with the elevation or side view of the component and, as you can see, there is no difference.

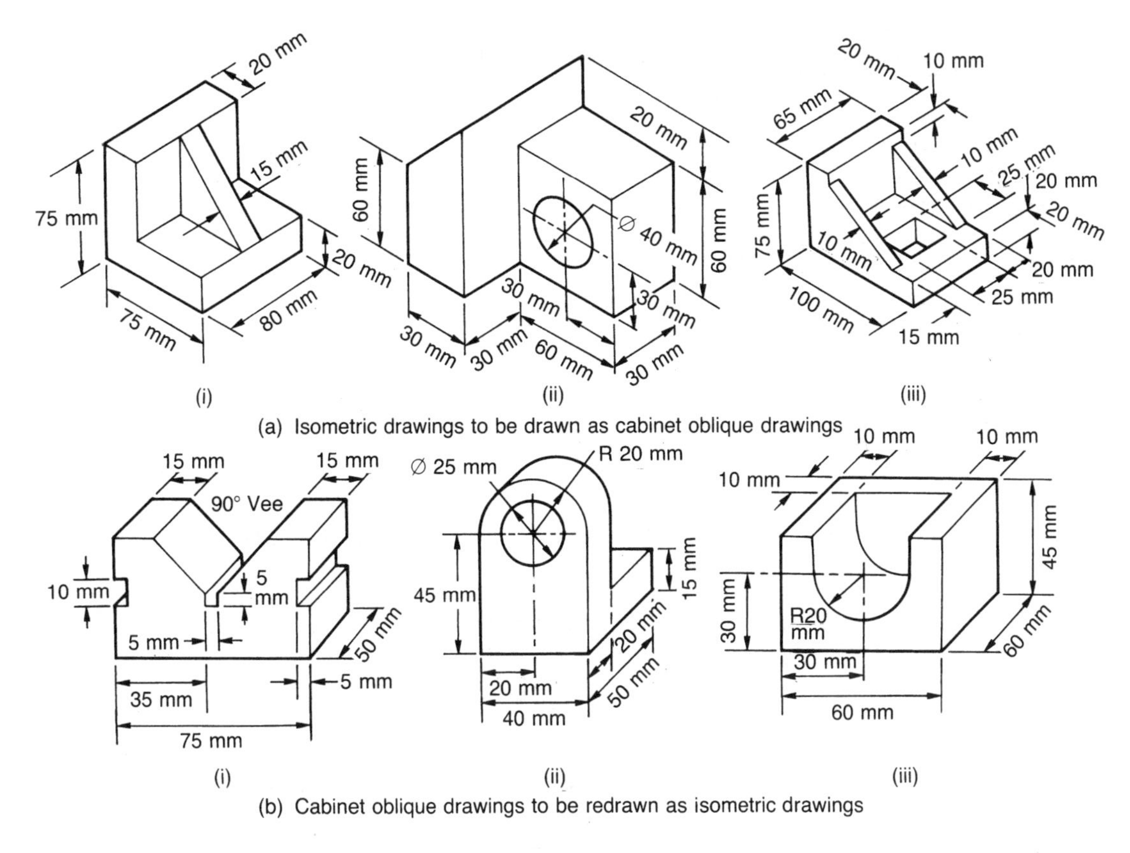

Figure 3.28 *See Test your knowledge 3.6 and 3.7*

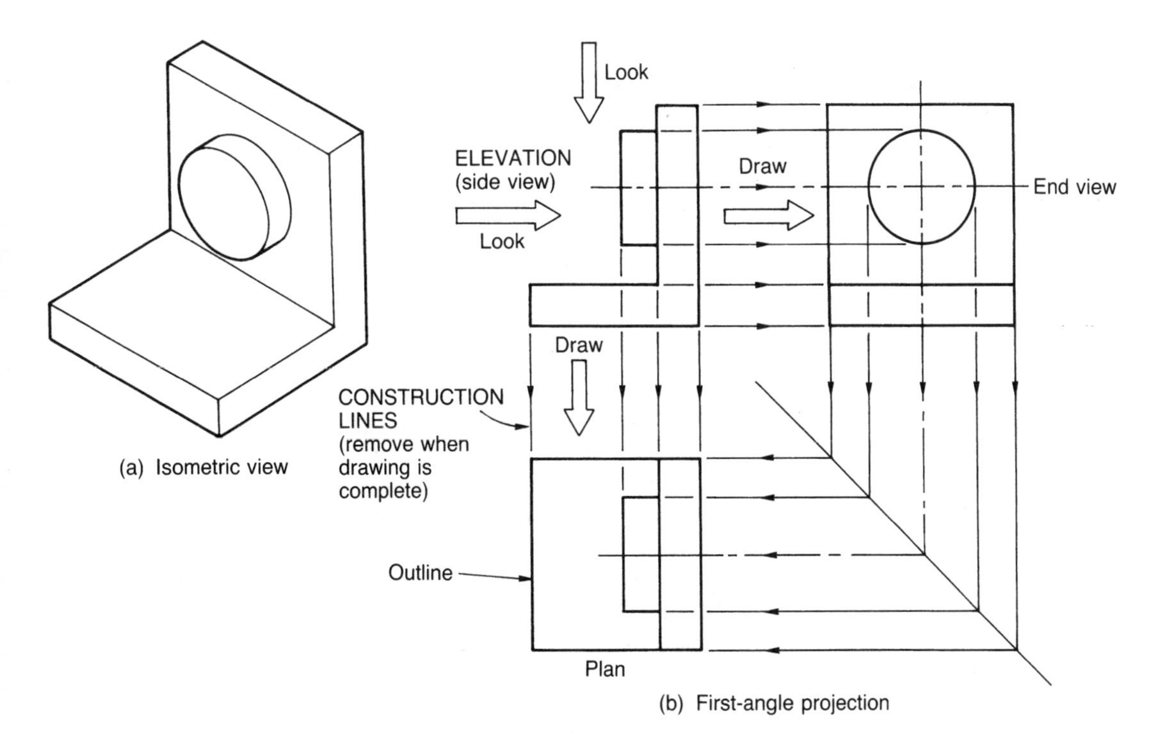

Figure 3.29 *An isometric view and its corresponding first-angle projection*

Test your knowledge 3.8

Figure 3.28 shows some components using pictorial projections. Redraw each of these drawings in third-angle orthographic projection. To start you off the first one has been drawn for you (see Figure 3.34).Note how the end view has been positioned so that you can see the web.

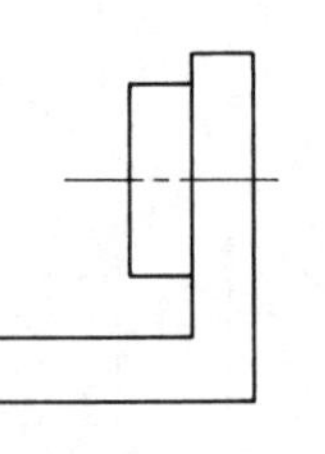

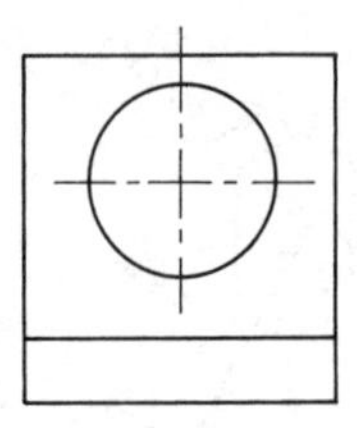

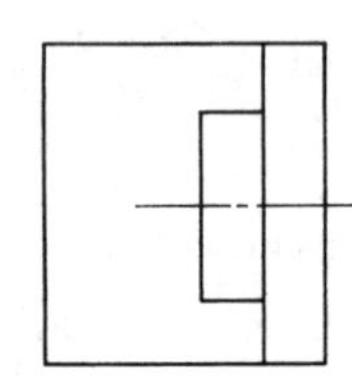

Figure 3.30 *Completed first-angle projection*

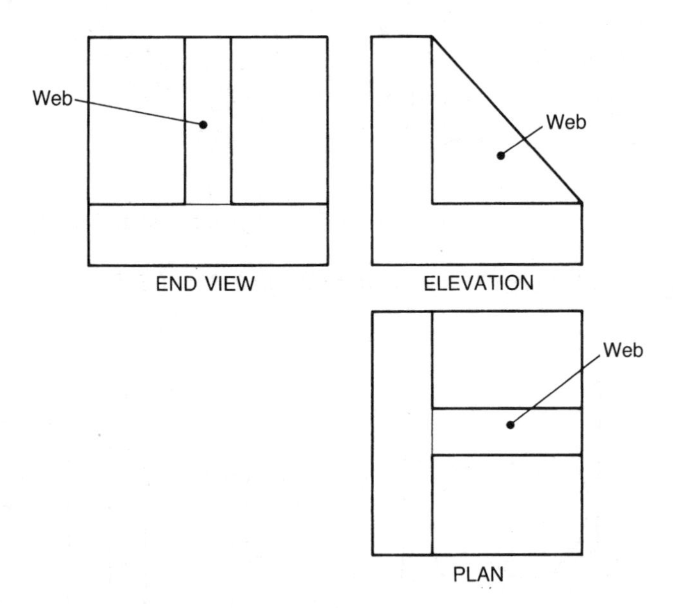

Figure 3.31 *See Test your knowledge 3.7*

- *Plan* Again we look down on top of the component to see what the plan view looks like. However, this time, we draw the plan view above the elevation. That is, in third-angle projection we draw all the views from where we look.
- *End view* Note how the position of the end view is reversed compared with first-angle projection. This is because, like the plan view, we draw the end views at the same end from which we look at the component.

Again use faint construction lines to produce the drawing as shown in Figure 3.32. Then 'line-in' the outline more heavily and carefully remove the construction lines for clarity, unless you have been

Test your knowledge 3.9

(a) Figure 3.35 shows some components drawn in first-angle projection and some in third-angle projection. Note that not all the possible views have been shown (only as many are shown as are actually needed). State which is first-angle and which is third-angle.

(b) Two of the drawings in Figure 3.35 are standard symbols for indicating whether a drawing is in first-angle or whether it is in third-angle. Which drawings do you think show these two symbols?

Test your knowledge 3.10

Figure 3.36 shows pictorial views of some more solid objects.

(a) Draw these objects in first-angle orthographic projection and label the views.

(b) Draw these objects in third-angle orthographic projection and label the views.

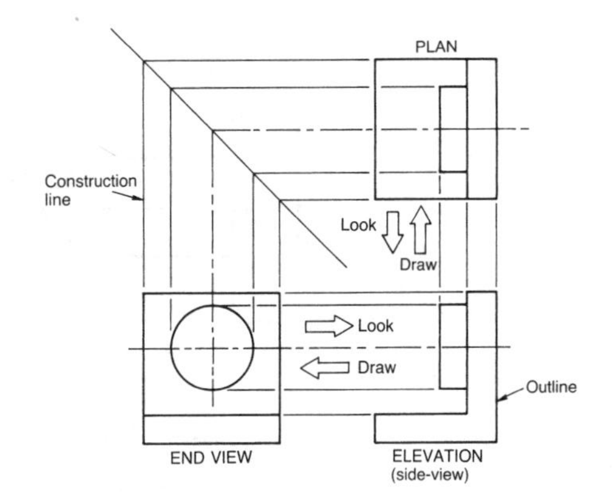

Figure 3.32 *Third-angle projection*

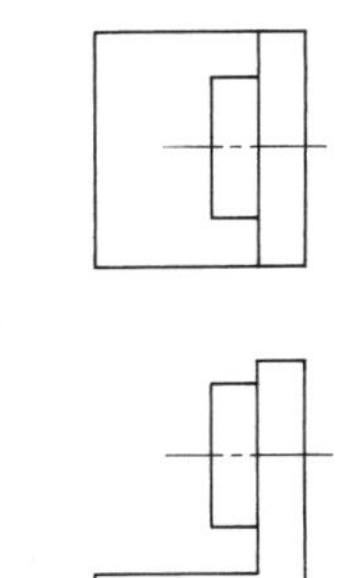

Figure 3.33 *Completed third-angle projection*

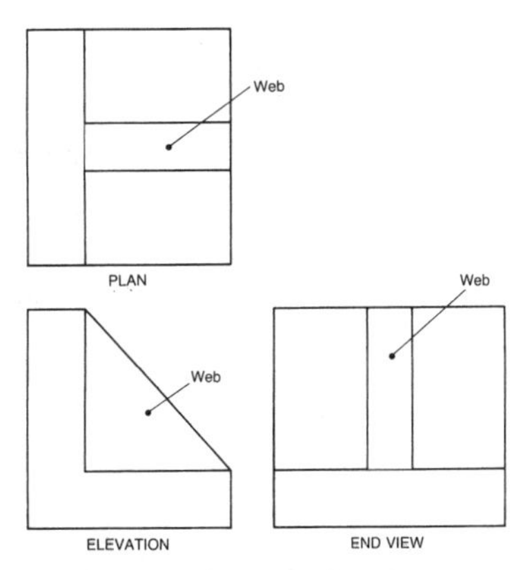

Figure 3.34 *See Test your knowledge 3.8*

instructed otherwise. Figure 3.33 shows the finished drawing in third-angle projection.

Auxiliary views

In addition to the main views on which we have just been working, we sometimes have to use auxiliary views. We use auxiliary views when we cannot show the true outline of the component or a feature of the component in one of the main views; for example, when a surface of the component is inclined as shown in Figure 3.37.

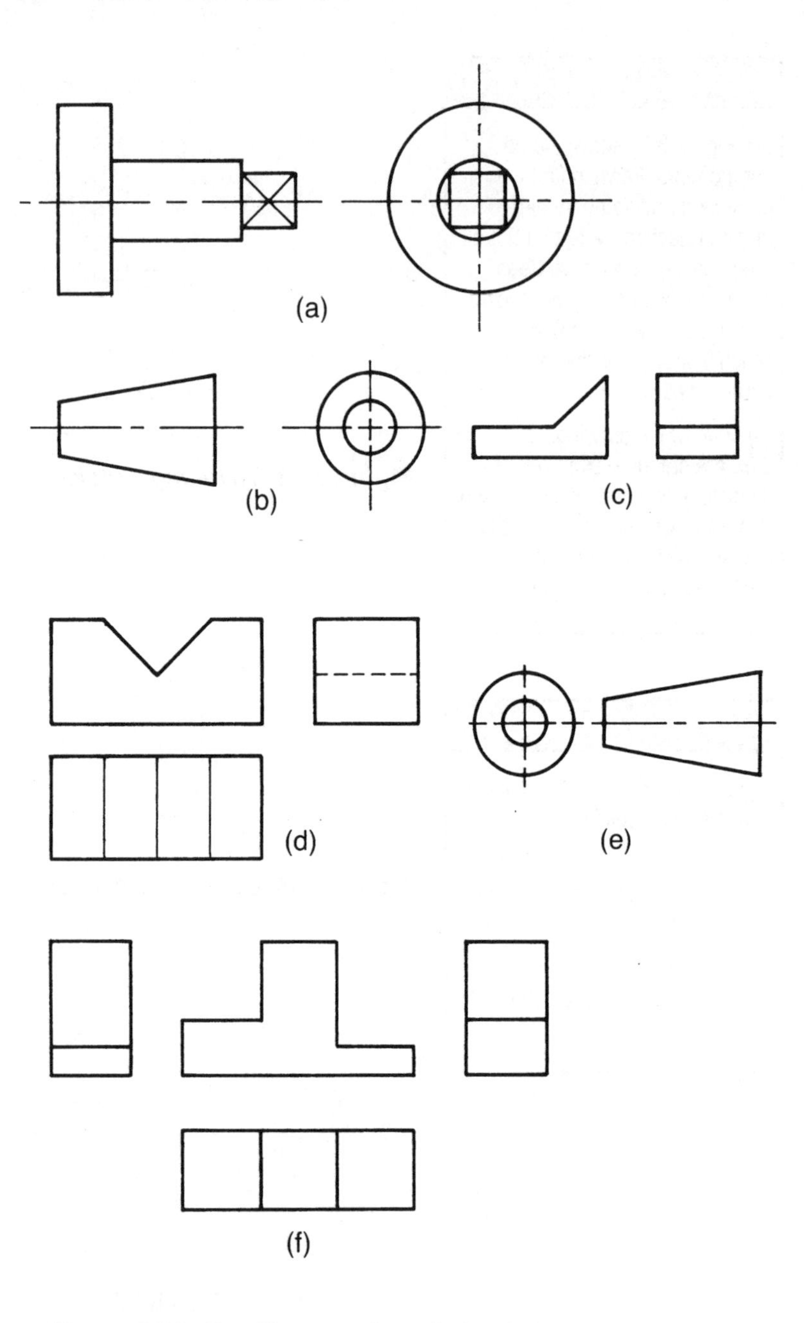

Figure 3.35 *See Test your knowledge 3.9*

Activity 3.4

Obtain a small cast or machined component from your engineering workshop. Examine this carefully and then produce a drawing of the component showing separate elevation, plan and end elevation views. Also show an isometric view of the component and include dimensions.

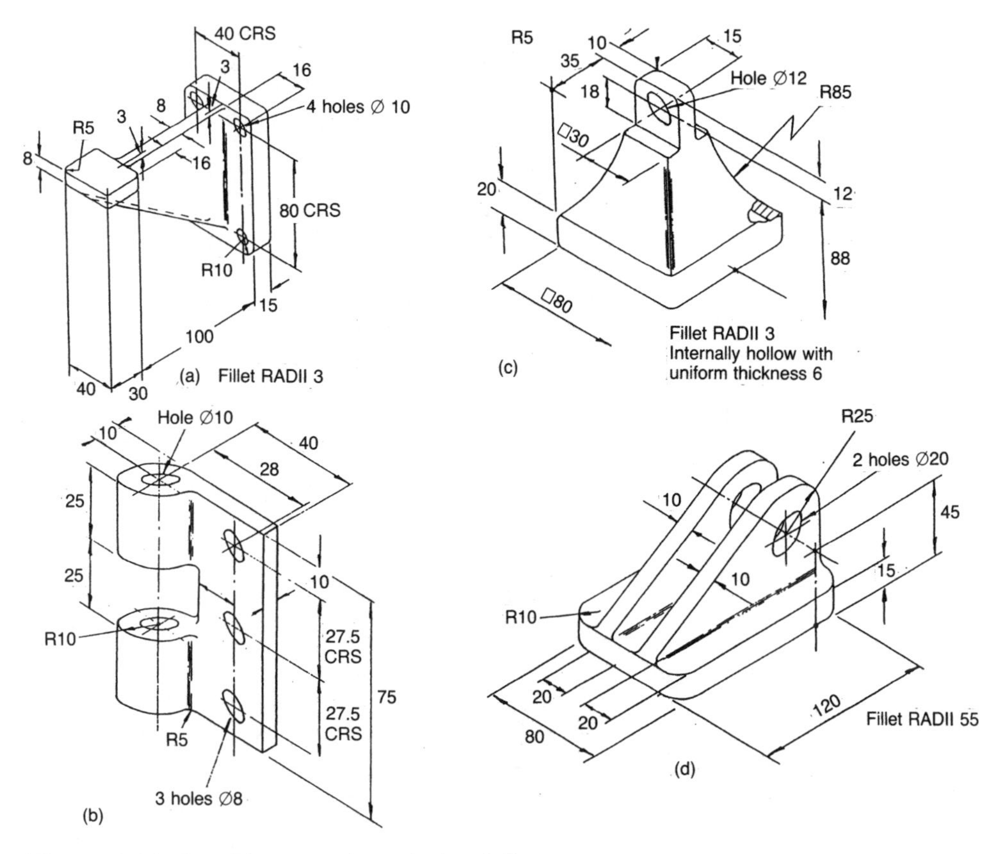

Figure 3.36 *See Test your knowledge 3.9*

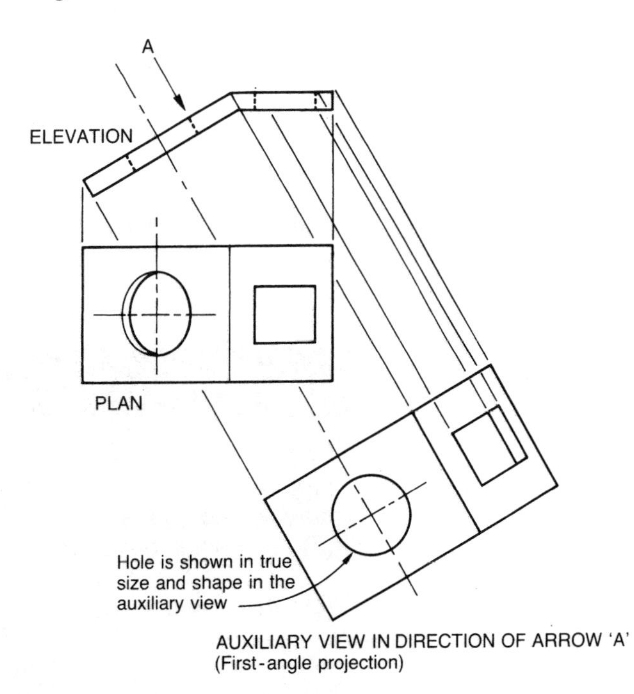

Figure 3.37 *An auxiliary view*

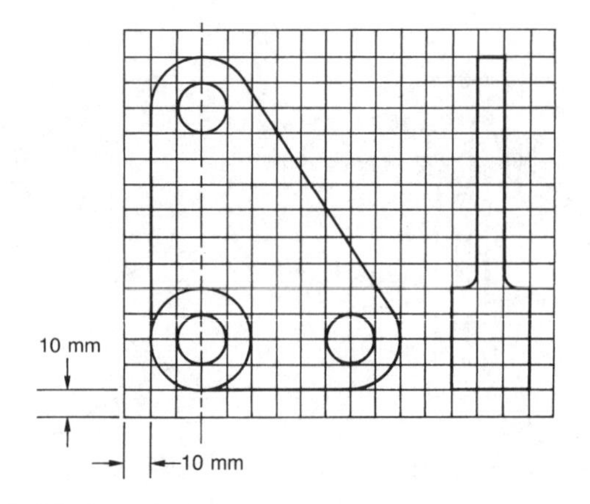

Figure 3.38 *See Activity 3.5*

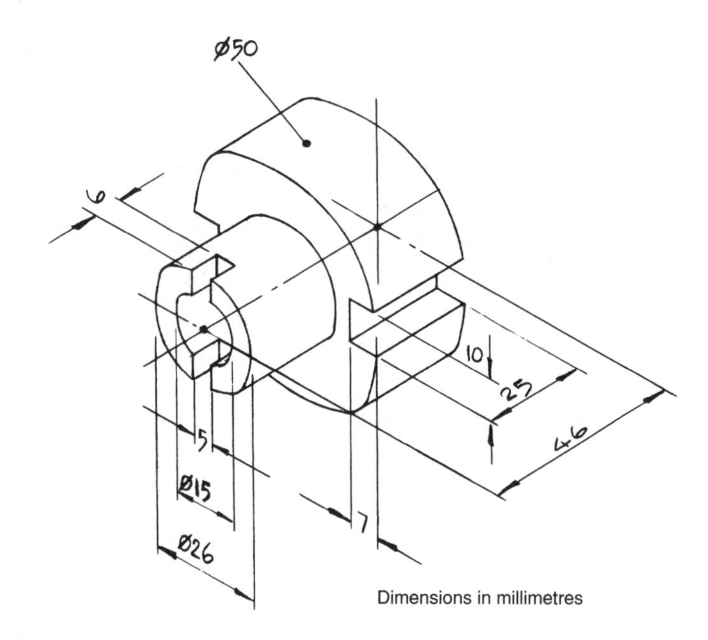

Figure 3.39 *A dimensioned drawing*

Activity 3.5

Draw the component shown in Figure 3.38 in isometric projection. Each square has a side length of 10 mm. Also draw the component in:

(a) first-angle orthographic projection (only two views required)
(b) third-angle orthographic projection (only two views required)
(c) cabinet oblique projection.

Present your results in the form of a portfolio of drawings. Clearly mark each drawing with the projection used.

Production of engineering drawings

Standard conventions are used in order to avoid having to draw out, in detail, common features in frequent use. Figure 3.39 shows a typical dimensioned engineering drawing. Figure 3.40(a) shows a pictorial representation of a screw thread. Figure 3.40(b) shows the convention for a screw thread. The convention for the screw thread is much the quicker and easier to draw.

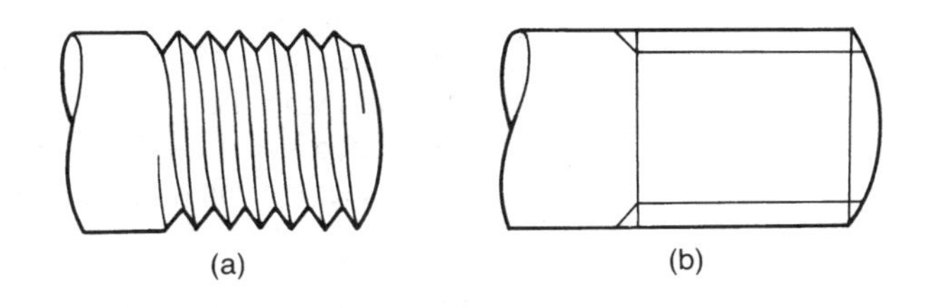

Figure 3.40 *Screw threads*

All engineering drawings should be produced using appropriate drawing standards and conventions for the following reasons:

- *Time* It speeds up the drawing process by making life easier for the draughtsperson as indicated above. This reduces costs and also reduces the 'lead-time' required to get a new product into production.
- *Appearance* The consistent use standards and conventions not only serves to make your drawings look more professional but it also improves the 'image' of yourself and your company. Badly presented drawings can send out the wrong messages and can call your competence into question.
- *Portability* Drawings produced to international standards and conventions can be read and correctly interpreted by any trained engineer, anywhere in this country or abroad. This avoids misunderstandings that could lead to expensive and complex components and assemblies being scrapped and dangerous situations arising. (Note that problems can still sometimes occur due to text and written notes that may be language dependent.)

Drawing conventions used by engineers in the UK are specified in BS 8888:2004. This is produced in several parts and is 'harmonised' with the appropriate International Standards Organization (ISO) standards.

As recommended earlier, it is well worth getting hold of a copy of 'PP 8888-1:2005 Drawing practice: a guide for schools and colleges to BS 8888:2004, Technical Product Specification'. You may also find it useful to refer to a number of other BS publications (all of which are supplied as part of BS 8888:2004 on CD-ROM). These documents include:

- BS ISO 128-20: Technical drawings. General principles of presentation. Basic conventions for lines

- BS ISO 128-21: Technical drawings. General principles of presentation. Preparation of lines by CAD systems
- BS ISO 128-22: Technical drawings. General principles of presentation. Basic conventions and application for leader lines and reference lines
- BS ISO 129-1: Technical drawings. Dimensioning. General principles, methods of execution and special indications
- BS 2917-1: Graphic symbols and circuit diagrams for fluid power systems and components. Specifications for graphic symbols
- BS EN ISO 5457: Technical drawings. Sizes and layout of drawing sheets
- BS 5070-1: Engineering diagram drawing practice. Recommendations for general principles.

Planning the drawing

Before we start the actual drawing and lay pencil to paper we should plan what we are going to do. This saves having to alter the drawing or even starting again later on. We have to decide whether the drawing is to be pictorial, orthographic or schematic. If orthographic we have to decide on the projection we are going to use. We also have to decide whether we need a formal drawing or whether a freehand sketch is all that is required. If a formal drawing is needed then we have to decide whether to use manual techniques or CAD.

Paper size

When you start to plan your drawing you have to decide on the paper size that you are going to use. Engineering drawings are usually produced on 'A' size paper. Paper size A0 is approximately one square metre in area and is the basis of the system. Size A1 is half the area of size A0, size A2 is half the area of size A1 and so on down to size A4. Smaller sizes are available but they are not used for drawing. All the 'A' size sheets have their sides in the ratio of 1:1.2. This gives the following paper sizes (see Figure 3.41):

- A0 = 841 mm × 1189 mm
- A1 = 594 mm × 841 mm
- A2 = 420 mm × 594 mm
- A3 = 297 mm × 420 mm
- A4 = 210 mm × 297 mm.

The paper size you choose will depend upon the size of the drawing and the number of views required. Be generous, nothing looks worse than a cramped up drawing, and overcrowded dimensions. It is also false economy since overcrowding invariably leads to reading errors.

As you will already have seen from some of the previous examples, the drawing should always have a border and a title block. This restricts the blank area available to draw on. Figures 3.42 and 3.43 show how the views should be positioned. These layouts are only a guide but they offer a good starting point until you become more experienced. If only one view is required then it is centred in the drawing space available.

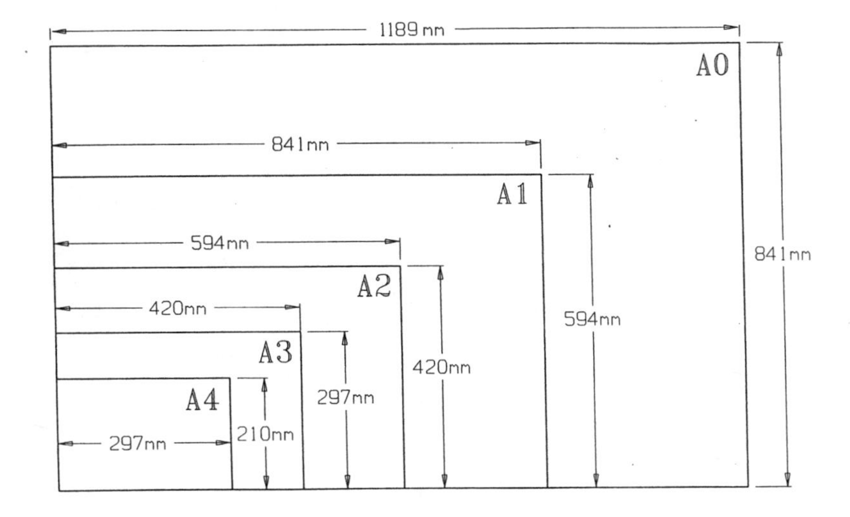

Figure 3.41 *Paper sizes*

Title block

A typical title block was shown earlier in Figure 3.24. If you refer back to this figure you will see that it is expandable vertically and horizontally to accommodate any written information that is required. The title block should contain:

- the drawing number (which should be repeated in the top left-hand corner of the drawing)
- the drawing name (title)
- the drawing scale
- the projection used (standard symbol)
- the name and signature of the draughtsperson together with the date on which the drawing was signed
- the name and signature of the person who checks and/or approves the drawing, together with the date of signing
- the issue number and its release date
- any other information as dictated by company policy.

Scale

The scale should be stated on the drawing as a ratio. The recommended scales are as follows:

- Full size = 1:1
- Reduced scales (smaller than full size) are:

  1:2 1:5 1:10
  1:20 1:50 1:100
  1:200 1:500 1:1000

  (NEVER use the words full-size, half-size, quarter-size, etc.).
- Enlarged scales (larger than full size) are:

  2:1 5:1 10:1
  20:1 50:1 100:1

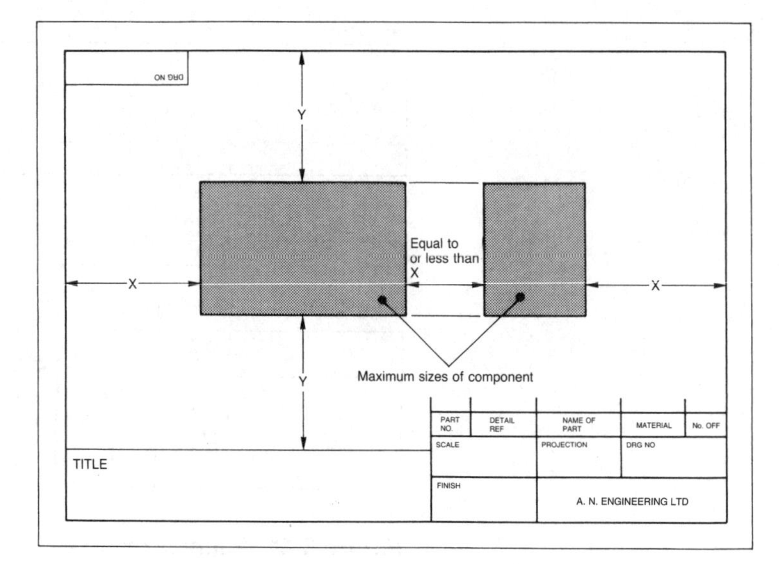

Figure 3.42 *Positioning a drawing with two component parts*

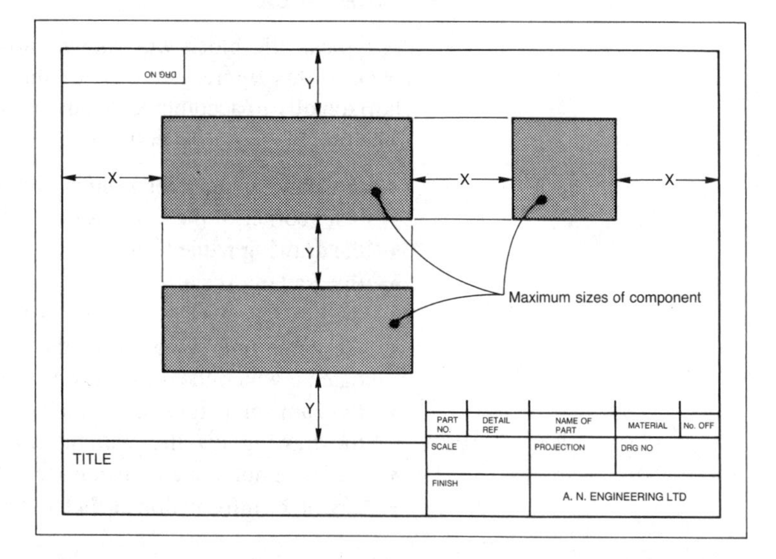

Figure 3.43 *Positioning a drawing with three component parts*

Lines and linework

The lines of a drawing should be uniformly black, dense and bold. On any one drawing they should all be in pencil or in black ink. Pencil is quicker to use but ink prints out more clearly. Lines should be thick or thin as recommended later. Thick lines should be twice as thick as thin lines. Figure 3.44 shows the types of lines recommended in BS 8888 for use in engineering drawing and how the lines should be used. This is reinforced by Table 3.1.

Sometimes the lines overlap in different views. When this happens, as shown in Figure 3.45, the following order of priority should be observed.

- Visible outlines and edges (type A) take priority over all other lines.
- Next in importance are hidden outlines and edges (type E).
- Then cutting planes (type G).
- Next come centre lines (type F and B).
- Outlines and edges of adjacent parts, etc. (type H).
- Finally, projection lines and shading lines (type B).

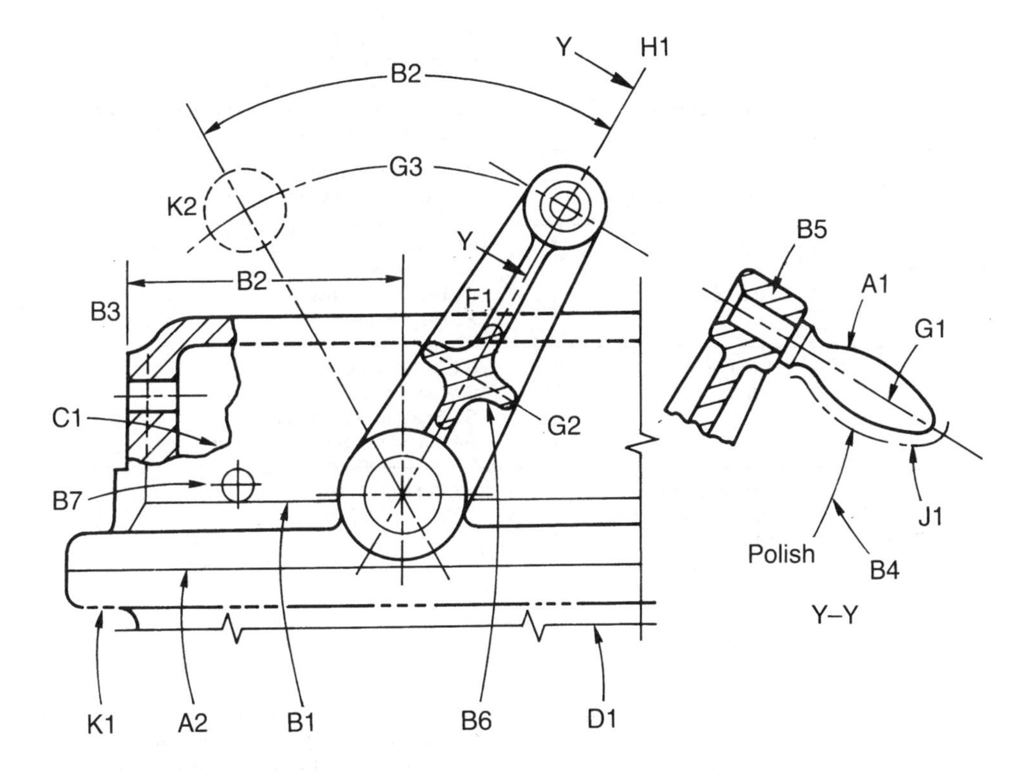

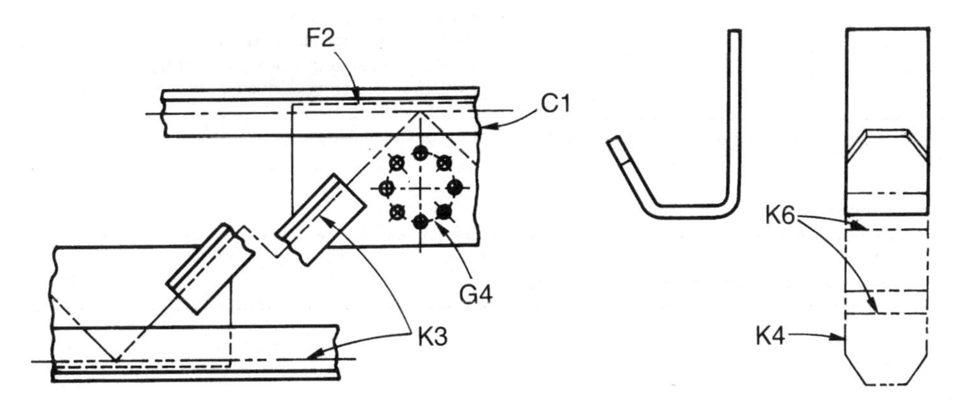

Figure 3.44 *Use of various line types*

Table 3.1 *Types of line*

Line	Description	Application
A	Continuous thick	A1 Visible outlines A2 Visible edges
B	Continuous thin	B1 Imaginary lines of intersection B2 Dimension lines B3 Projection lines B4 Leader lines B5 Hatching B6 Outlines of revolved sections B7 Short centre lines
C	Continuous thin irregular	C1 Limits of partial or interrupted views and sections, if the limit is not an axis
D	Continuous thin straight with zigzags	†D1 Limits of partial or interrupted views and sections, if the limit is not an axis
E	Dashed thick	E1 Hidden outlines E2 Hidden edges
F	Dashed thin‡	F1 Hidden outlines F2 Hidden edges
G	Chain thin	G1 Centre lines G2 Lines of symmetry G3 Trajectories and loci G4 Pitch lines and pitch circles
H	Chain thin, thick at ends and changes of direction	H1 Cutting planes
J	Chain thick	J1 Indication of lines or surfaces to which a special requirement applies (drawn adjacent to surface)
K	Chain thin double dashed	K1 Outlines and edges of adjacent parts K2 Outlines and edges of alternative and extreme positions of movable parts K3 Centroidal lines K4 Initial outlines prior to forming §K5 Parts situated in front of a cutting plane K6 Bend lines on developed blanks or patterns

NOTE. The lengths of the long dashes shown for lines G, H, J and K are not necessarily typical due to the confines of the space available.

†This type of line is suited for production of drawings by machines.

‡The thin F type line is more common in the UK, but on any one drawing or set of drawings only one type of dashed line should be used.

§Included in ISO 128-1982 and used mainly in the building industry.

Test your knowledge 3.11

Figure 3.46 shows a simple drawing using a variety of lines. List the numbers and against each number write down whether the type of line chosen is correct or incorrect. If incorrect state what type of line should have been used.

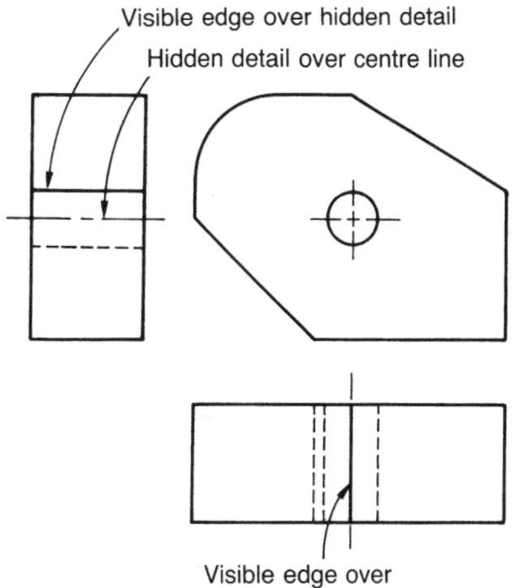

Figure 3.45 *Line priorities*

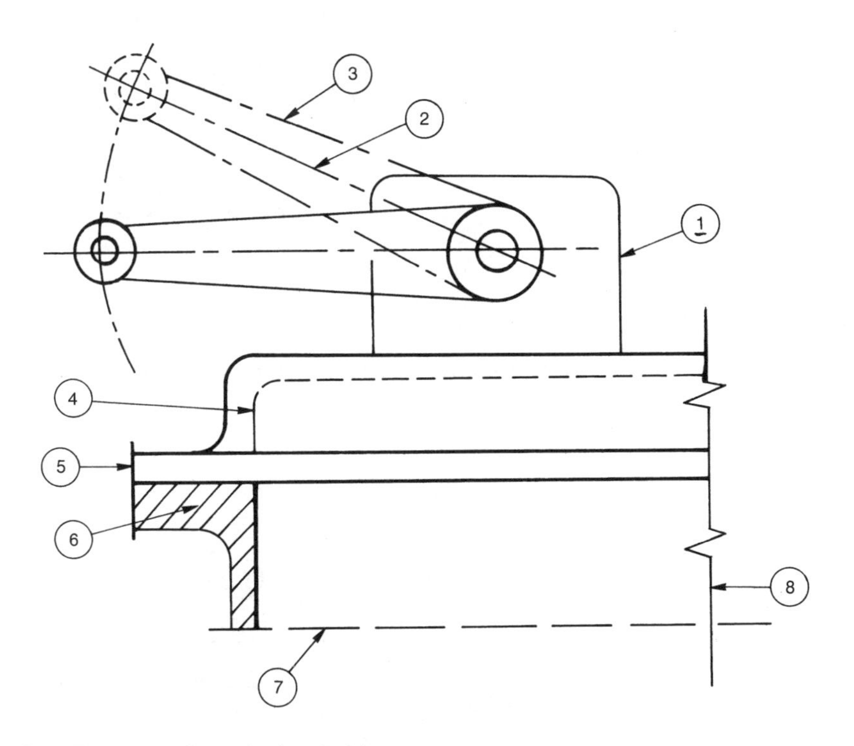

Figure 3.46 *See Test your knowledge 3.11*

Leader lines

Leader lines, as their name implies, lead written information or dimensions to the points where they apply. Leader lines are thin lines (type B) and they end in an arrowhead or in a dot, as shown in Figure 3.47(a). Arrowheads touch and stop on a line, whilst dots should always be used within an outline.

- When an arrowed leader line is applied to an arc it should be

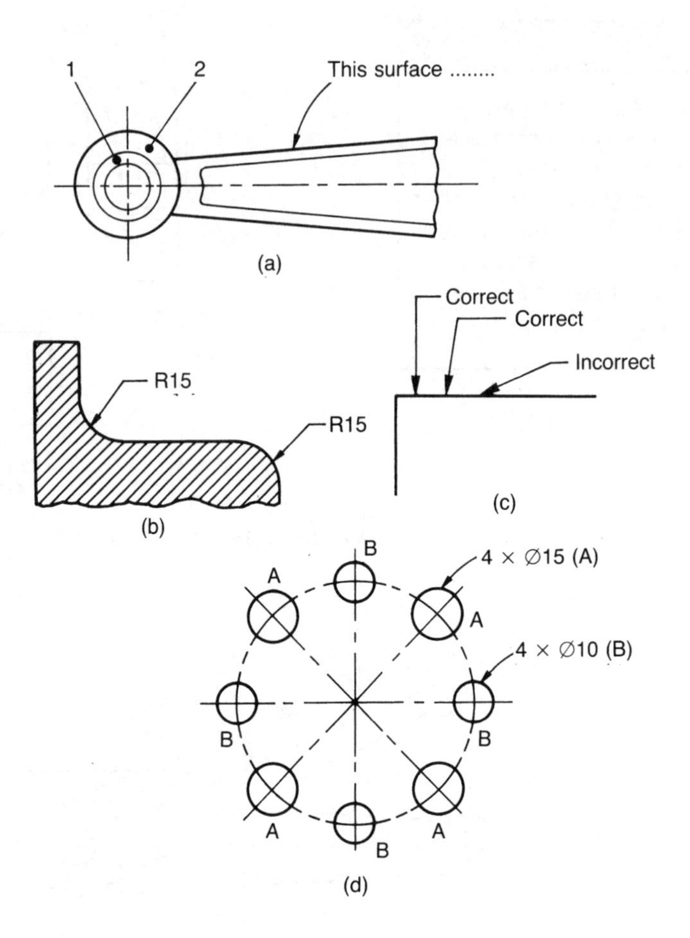

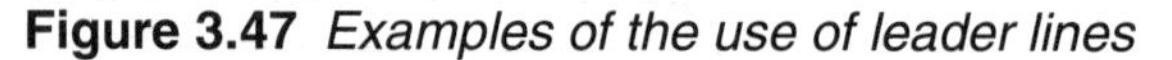

Figure 3.47 *Examples of the use of leader lines*

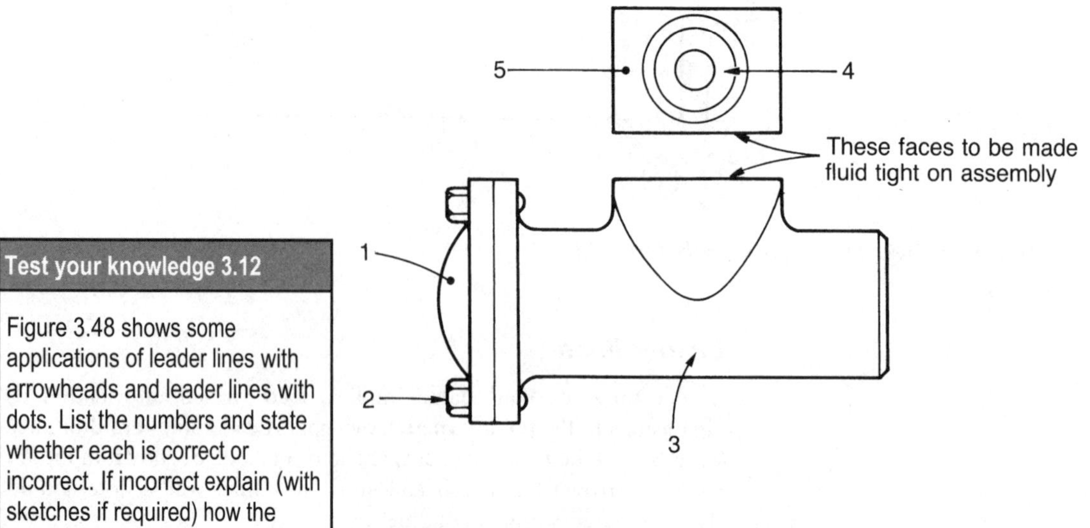

Figure 3.48 *See Test your knowledge 3.12*

Test your knowledge 3.12

Figure 3.48 shows some applications of leader lines with arrowheads and leader lines with dots. List the numbers and state whether each is correct or incorrect. If incorrect explain (with sketches if required) how the drawing should be corrected.

in line with the centre of the arc, as shown in Figure 3.47(b).
- When an arrowed leader line is applied to a flat surface, it should be nearly normal to the lines representing that surface, as shown in Figure 3.47(c).
- Long and intersecting leader lines should not be used, even if this means repeating dimensions and/or notes, as shown in Figure 3.47(d).
- Leader lines must not pass through the points where other lines intersect.
- Arrowheads should be triangular with their length some three times larger than the maximum width. They should be formed from straight lines and the arrowheads should be filled in. The arrowhead should be symmetrical about the leader line, dimension line or stem. It is recommended that arrowheads on dimension and leader lines should be some 3 to 5 mm long.
- Arrowheads showing direction of movement or direction of viewing should be some 7 to 10 mm long. The stem should be the same length as the arrowhead or slightly greater. It must never be shorter.

Letters and numerals

Style

The style should be clear and free from embellishments. In general, capital letters should be used. A suitable style could be:

ABCDEFGHIJKLMNOPQRSTUVWXYZ
1234567890

Size

The characters used for dimensions and notes on drawings should be not less than 3 mm tall. Title and drawing numbers should be at least twice as big.

Direction of lettering

Notes and captions should be positioned so that they can be read in the same direction as the information in the title block. Dimensions have special rules and will be dealt with later.

Location of notes

General notes should all be grouped together and not scattered about the drawing. Notes relating to a specific feature should be placed adjacent to that feature.

Emphasis

Characters, words and/or notes should not be emphasized by underlining. Where emphasis is required the characters should be enlarged.

Symbols and abbreviations

If all the information on a drawing was written out in full, the drawing would become very cluttered. Therefore symbols and abbreviations are used to shorten written notes. Those recommended for use on engineering drawings are listed in BS 8888, and in the corresponding student version PP 8888-1.

Test your knowledge 3.13

With reference to BS 8888 or to PP 8888-1, complete Table 3.2.

Table 3.2 *See Test your knowledge 3.13*

Abbreviation or symbol	*Term*
AF	
ASSY	
CRS	
	Centre line
	Countersink
	Counterbore
Ø	
DAG	
	Hexagon
	Internal
LH	
MATL	
	Maximum
	Minimum
	Number
PCD	
	Radius (in a note)
	Radius (preceding a dimension)
REQD	
RH	
SCR	
SH	
SK	
SPEC	
	Figure
	Diameter (in a note)
CYL	
CHAM	
CH HD	
	Equally spaced

Conventions

These are a form of 'shorthand' used to speed up the drawing of common features in regular use. The full range of conventions and examples of their use can be found in BS 8888 or PP 8888-1, so I will not waste space by listing them here. However by completing the next exercise you will use some of the more common conventions and this will help you to become familiar with them.

Test your knowledge 3.14

With reference to BS 8888 or to PP 8888-1, complete Figure 3.49. Note that you must take care to use the same types of lines as shown in the standard or the conventions will become meaningless. This applies particularly to line thickness.

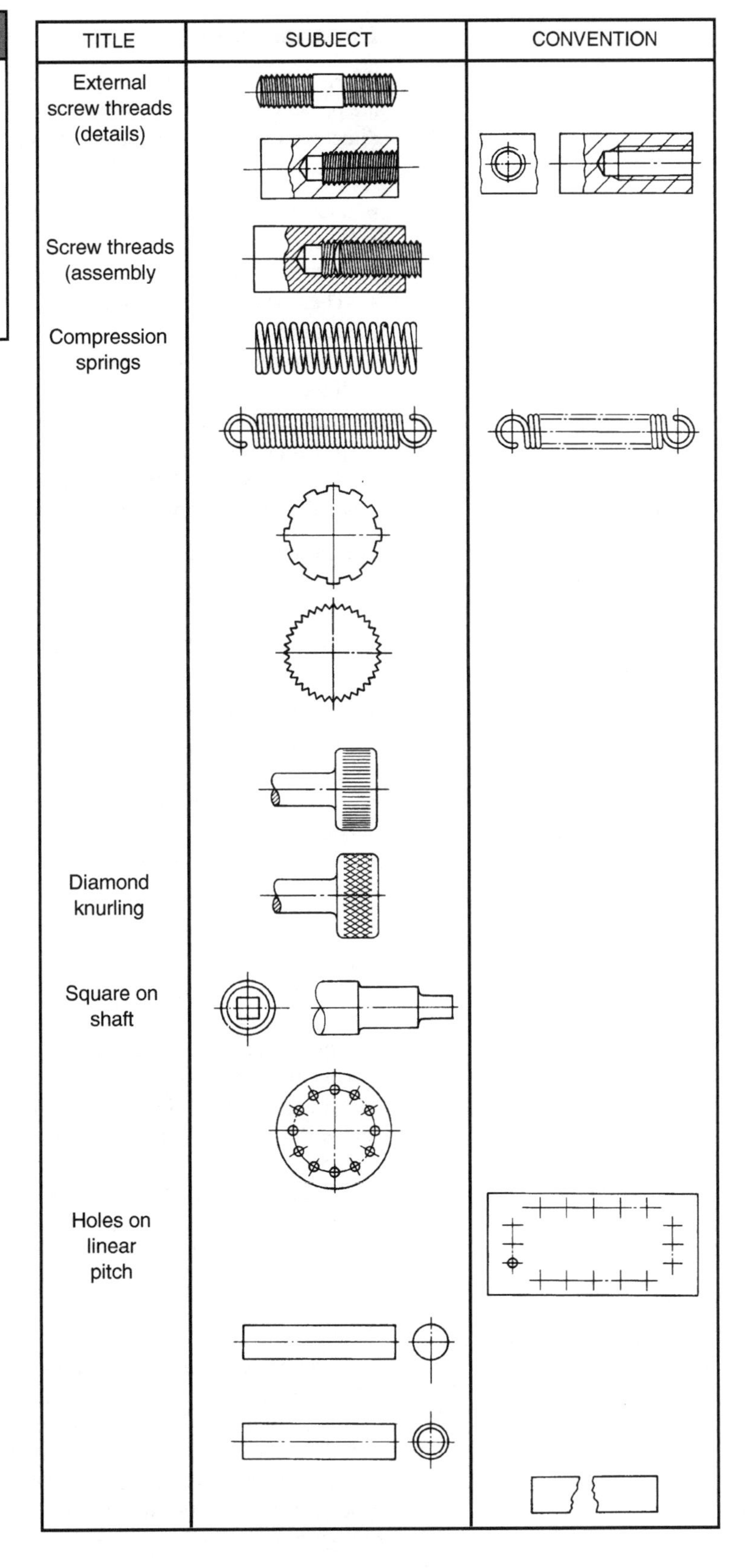

Figure 3.49 *See Test your knowledge 3.14*

Dimensioning

When a component is being dimensioned, the dimension lines and the projection lines should be thin full lines (type B). Where possible dimensions should be placed outside the outline of the object, as shown in Figure 3.50(a). The rules are:

- Outline of object to be dimensioned in thick lines (type A).
- Dimension and projection lines should be half the thickness of the outline (type B).
- There should be a small gap between the projection line and the outline.
- The projection line should extend to just beyond the dimension line.
- Dimension lines end in an arrowhead that should touch the projection line to which it refers.
- All dimensions should be placed in such a way that they can be read from the bottom right-hand corner of the drawing.

The purpose of these rules is to allow the outline of the object to stand out prominently from all the other lines and to prevent confusion.

There are three ways in which a component can be dimensioned. These are:

- Chain dimensioning, as shown in Figure 3.50(b).
- Absolute dimensioning (dimensioning from a datum) using parallel dimension lines, as shown in Figure 3.50(c).
- Absolute dimensioning (dimensioning from a datum using superimposed running dimensions, as shown in Figure 3.50(d)). Note the common origin (termination) symbol.

It is neither possible to manufacture an object to an exact size nor to measure an exact size. Therefore important dimensions have to be toleranced. That is, the dimension is given two sizes: an upper limit of size and a lower limit of size. Providing the component is made so that it lies between these limits it will function correctly. Information on Limits and Fits can be found in BS 1916.

The method of dimensioning can also affect the accuracy of a component and produce some unexpected effects. Figure 3.50(b) shows the effect of chain dimensioning on a series of holes or other features.

The designer specifies a common tolerance of ±0.2 mm. However, since this tolerance is applied to each and every dimension, the cumulative tolerance becomes ± 0.6 mm by the time you reach the final, right-hand hole, which is not what was intended. Therefore, absolute dimensioning as shown in Figure 3.50(c) and (d) is to be preferred in this example.

With absolute dimensioning, the position of each hole lies within a tolerance of ±0.2 mm and there is no cumulative error. Further examples of dimensioning techniques are shown in Figure 3.51.

It is sometimes necessary to indicate machining processes and surface finish. The machining symbol, together with examples of process notes and the surface finishes in micrometres (μm), is shown in Figure 3.52.

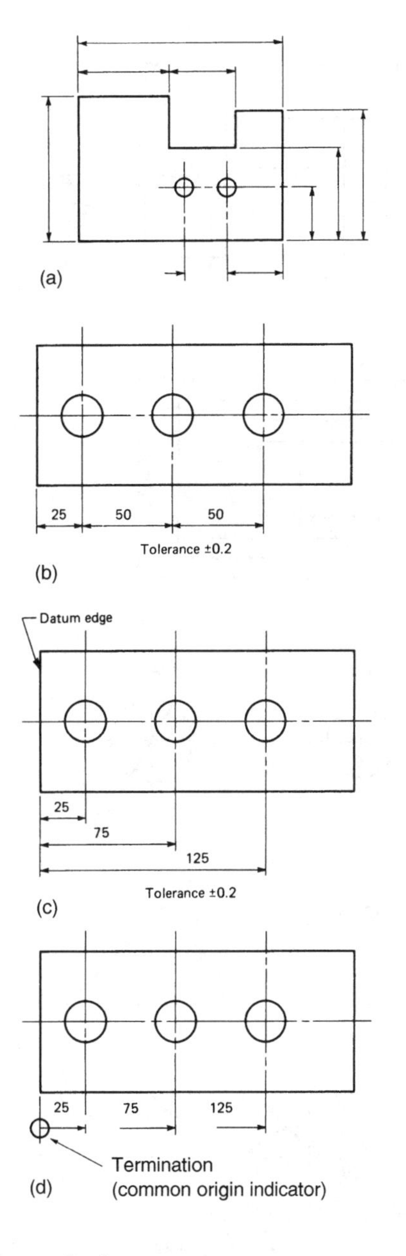

Figure 3.50 *Dimensioning*

Sectioning

Sectioning is used to show the hidden detail inside hollow objects more clearly than can be achieved using dashed thin (type E) lines. Figure 3.54(a) shows an example of a simple sectioned drawing. The cutting plane is the line A–A. In your imagination you remove everything to the left of the cutting plane, so that you only see what remains to the right of the cutting plane looking in the direction of the arrowheads. Another example is shown in Figure 3.54(b).

Figure 3.54(c) shows how to section an assembly. Note how solid shafts and the key are not sectioned. Also note that thin webs that lie on the section plane are not sectioned.

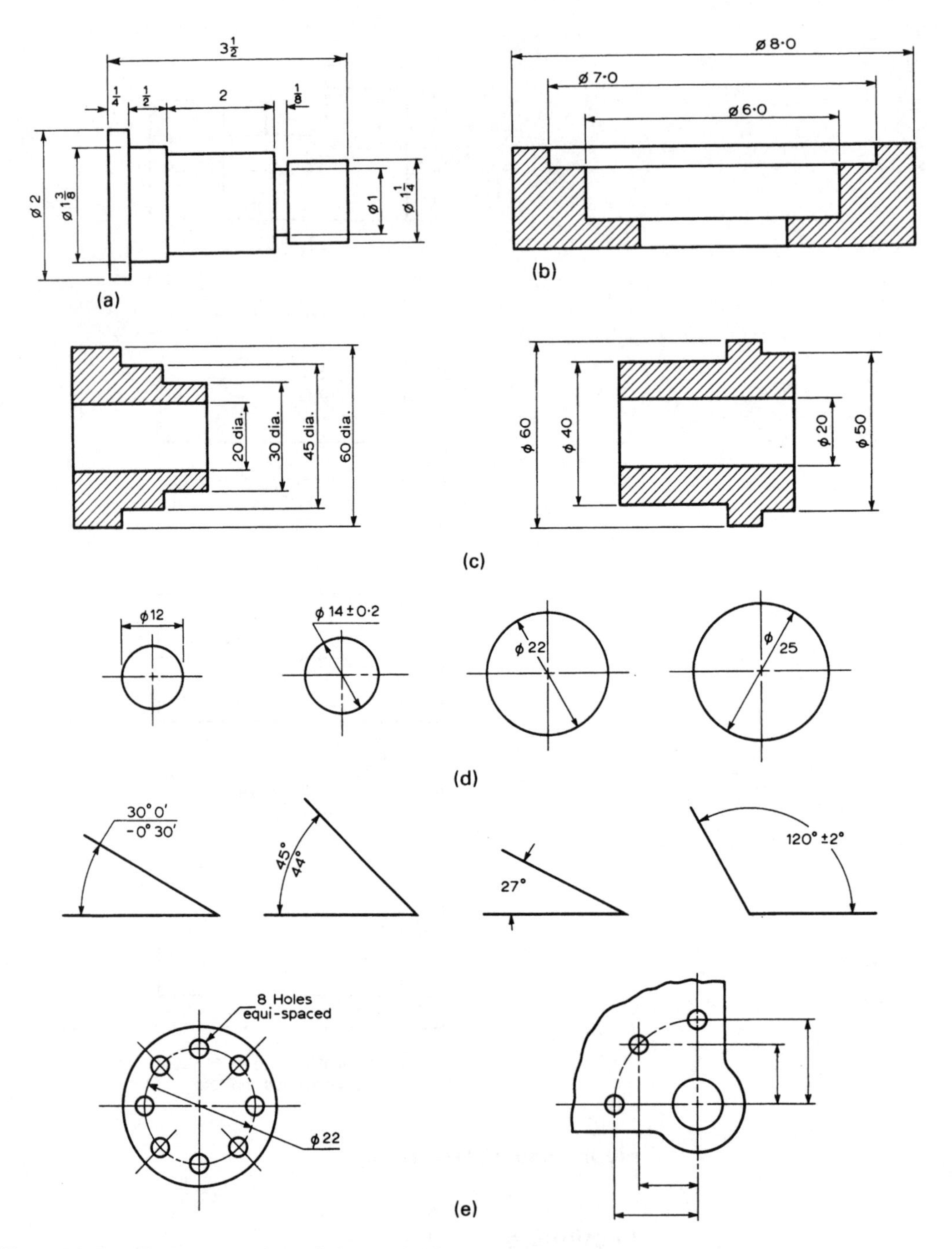

Figure 3.51 *More examples of dimensioning*

When interpreting sectioned drawings, some care is required. It is easy to confuse the terms Sectional view and Section.

Sectional view

In a sectional view you see the outline of the object at the cutting plane. You also see all the visible outlines seen beyond the cutting plane in the direction of viewing. Therefore, Figure 3.54(a) is a sectional view.

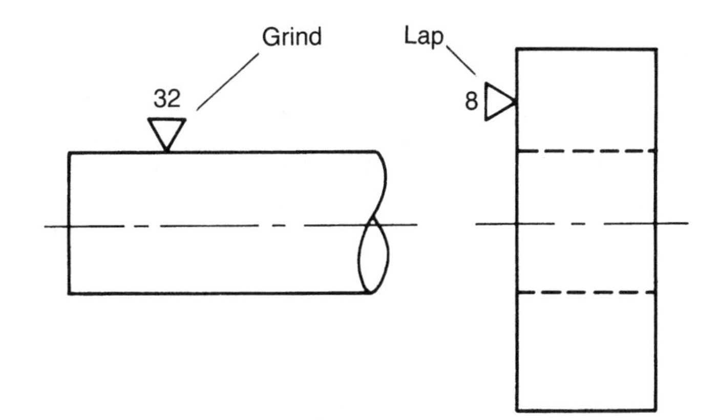

Figure 3.52 *Indicating surface finishes*

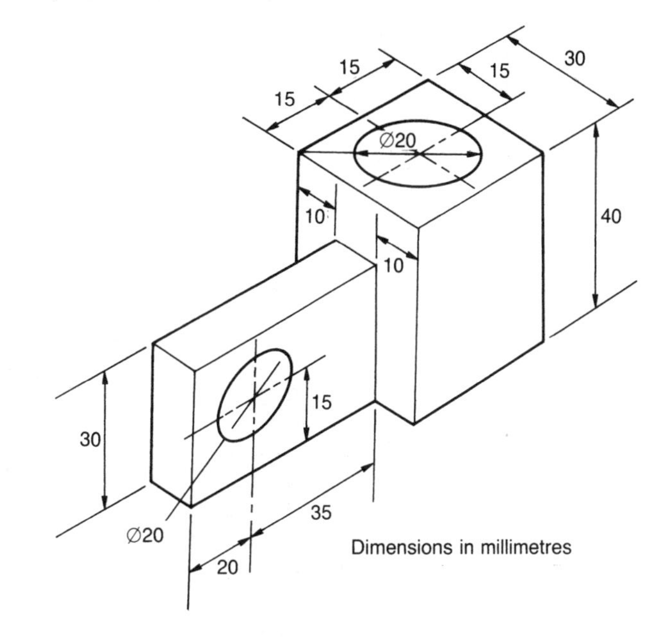

Figure 3.53 *See Test your knowledge 3.15*

Test your knowledge 3.15

Figure 3.53 shows a component drawn in isometric projection. Redraw it in first-angle orthographic projection and add the dimensions, using the following techniques:

(a) Absolute dimensioning using parallel dimension lines.

(b) Absolute dimensioning using superimposed running dimensions.

Test your knowledge 3.16

(a) Redraw Figure 3.54(a) as a section (remember that the drawing is a sectional view).

(b) Explain why Figure 3.54(b) can be a section or a sectional view.

Section

A section only shows the outline of the object at the cutting plane. Visible outlines beyond the cutting plane in the direction of viewing are not shown. Therefore a section has no thickness.

Cutting planes

You have already been introduced to cutting planes in the previous examples. They consist of type G lines; that is, a thin chain line that is thick at the ends and at changes of direction. The direction of viewing is shown by arrows with large heads. The points of the arrowheads touch the thick portion of the cutting plane. The cutting plane is labelled by placing a capital letter close to the stems of the arrows. The same letters are used to identify the corresponding section or sectional view.

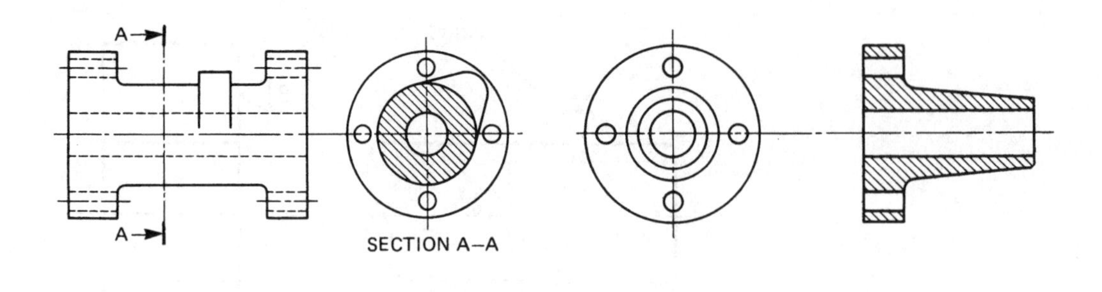

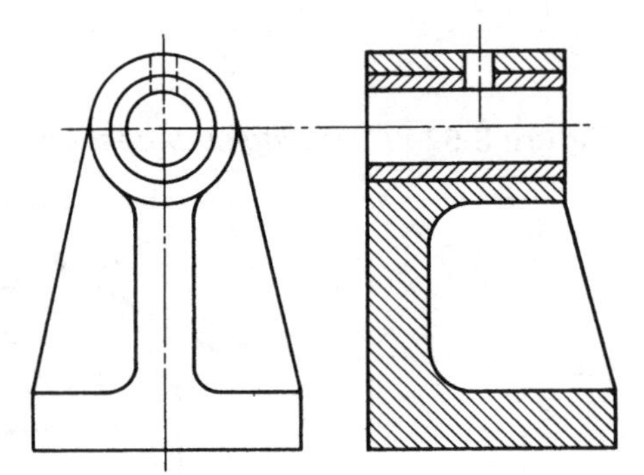

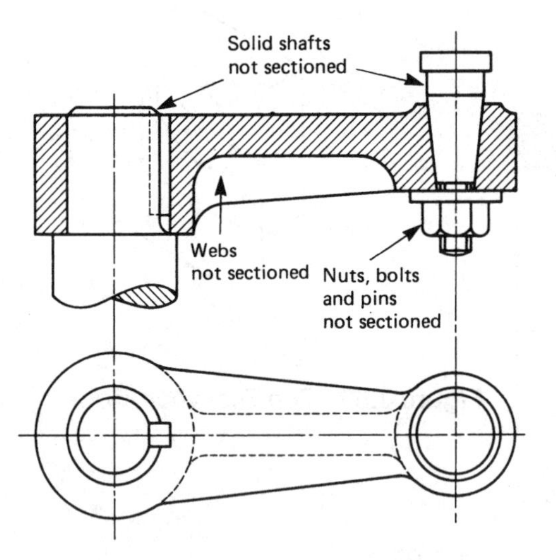

Figure 3.54 (a) to (c) *See Test your knowledge 3.16*

Hatching

You will have noticed that the shading of sections and sectional views consists of sloping, thin (type B) lines. This is called hatching. The lines are equally spaced, slope at 45° and are not usually less than 4 mm apart. However when hatching very small areas the hatching can be reduced, but never less than 1 mm. The drawings in this book may look as though they do not obey these rules. Remember that they have been reduced from much bigger drawings to fit onto the pages.

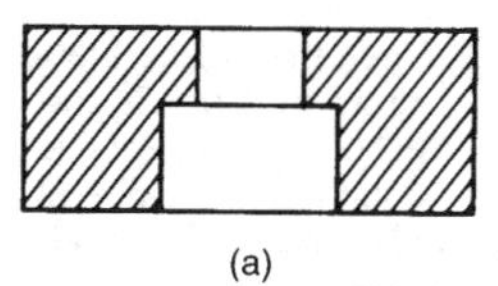

(a)

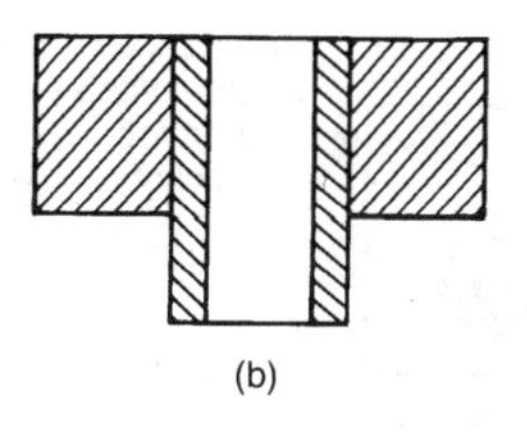

(b)

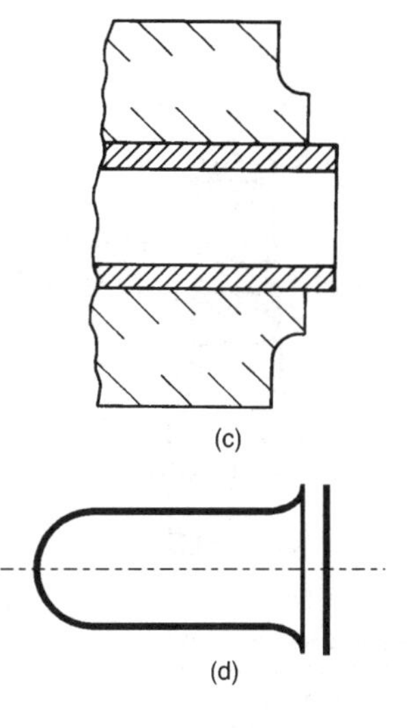

(c)

(d)

Figure 3.55 *Rules of hatching*

Figure 3.55 shows the basic rules of hatching. The hatching of separated areas is shown in Figure 3.55(a). Separate sectioned areas of the same component should be hatched in the same direction and with the same spacing.

Figure 3.55(b) shows how to hatch assembled parts. Where the different parts meet on assembly drawings, the direction of hatching should be reversed. The hatching lines should also be staggered. The spacing may also be changed.

Figure 3.55(c) shows how to hatch large areas. This saves time and avoids clutter. The hatching is limited to that part of the area that touches adjacent hatched parts or just to the outline of a large part.

Figure 3.55(d) shows how sections through thin materials can be blocked in solid rather than hatched. There should be a gap of not less than 1 mm between adjacent parts even when these are a tight fit in practice.

Finally I have included some further examples of sectioning in Figure 3.56. These include assemblies, half sections, part sections and revolved sections. Then it will be your turn to produce some engineering drawings including some or all of the features outlined in this section.

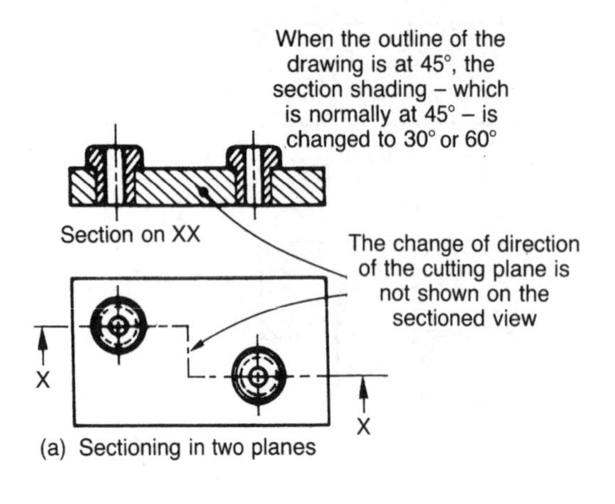

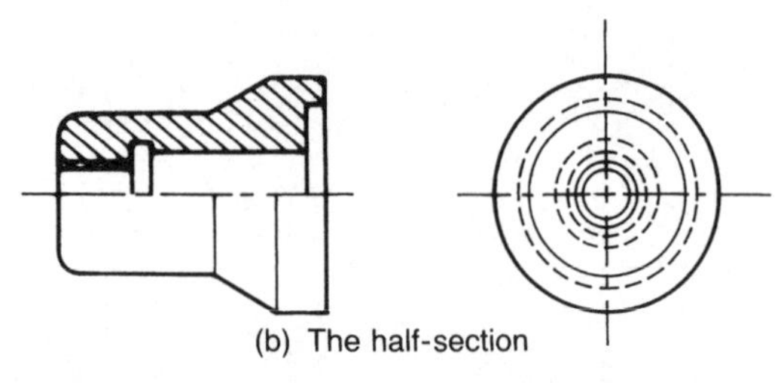

The half-section can be used with symmetrical components to show internal and external detail in the same view. The external view does not include hidden detail unless this is required for clarity or for dimensioning purposes.

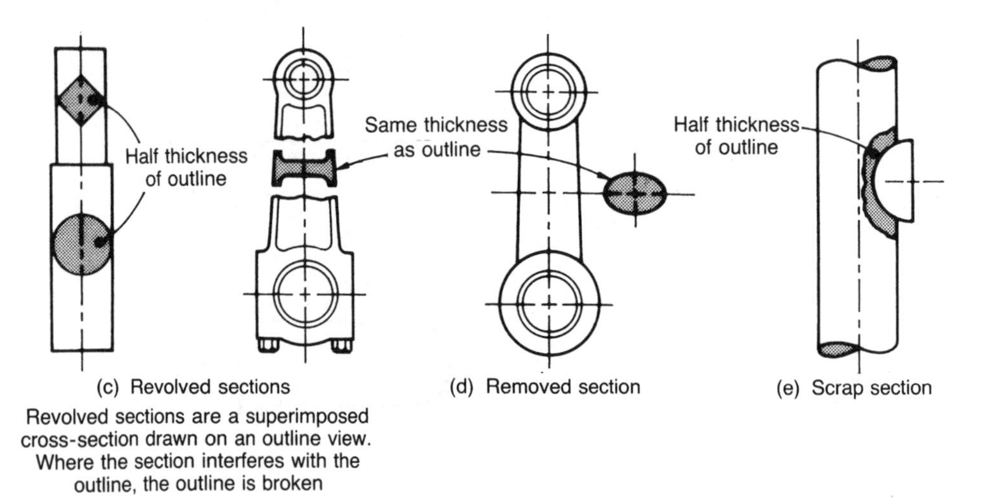

Revolved sections are a superimposed cross-section drawn on an outline view. Where the section interferes with the outline, the outline is broken

Figure 3.56 *Examples of sectioning*

Activity 3.6

Redraw Figure 3.57 in third-angle projection. Include an end view looking in the direction of arrow A, and section the elevation on the cutting plan XX.

Activity 3.7

Figure 3.58 shows a cast iron bend.

(a) Redraw and add an end view looking in the direction of arrow A.

(b) Section the elevation on the centre line.

(c) Draw an auxiliary view of flange B.

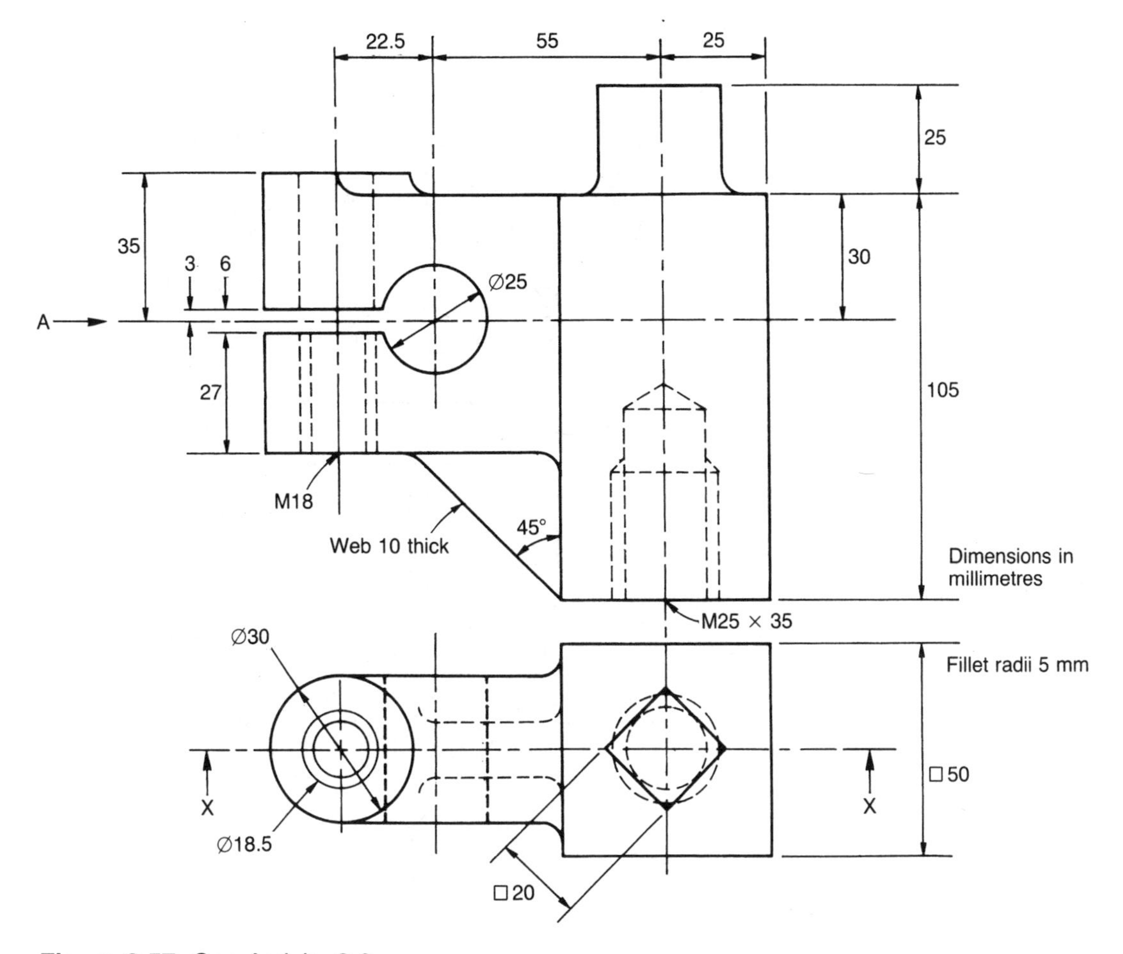

Figure 3.57 *See Activity 3.6*

Schematic diagrams

Schematic diagrams use symbols to show the connections that exist between components in an engineered product or system. Schematic diagrams are used extensively in fluid power engineering and in electronics.

Fluid power schematics

These diagrams cover both pneumatic and hydraulic circuits. The symbols that we shall use do not illustrate the physical make-up, construction or shape of the components. Neither are the symbols to scale or orientated in any particular position. They are only intended to show the 'function' of the component they portray, the connections and the fluid flow path.

Complete symbols are made up from one or more basic symbols and from one or more functional symbols. Examples of some basic symbols are shown in Figure 3.59 and some functional symbols are shown in Figure 3.60.

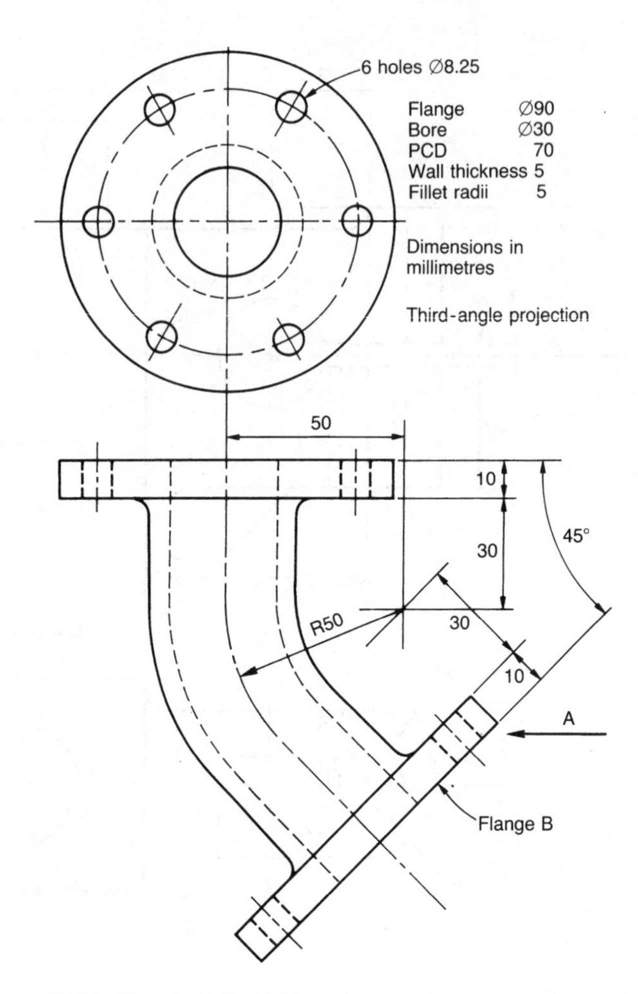

Figure 3.58 *See Activity 3.7*

Energy converters

Let's now see how we can combine some of these basic and functional symbols to produce a complete symbol representing a component. For example let's start with a motor. The complete symbol is shown in Figure 3.61.

The large circle indicates that we have an energy conversion unit such a motor or pump. Notice that the fluid flow is into the device and that it is pneumatic. The direction of the arrowhead indicates the direction of flow. The fact that the arrowhead is clear (open) indicates that the fluid is air. Therefore the device must be a motor. If it was a pump the fluid flow would be out of the circle. The single line at the bottom of the circle is the outlet (exhaust) from the motor and the double line is the mechanical output from the motor.

Now let's analyse the symbol shown in Figure 3.62.

- The circle tells us that it is an energy conversion unit.
- The arrowheads show that the flow is from the unit so it must be a pump.

Description	Symbol
Flow lines Continuous: Working line return line feed line	————
Long dashes: Pilot control lines	— — — —
Short dashes Drawn lines	- - - - - - -
Long chain enclosure line	—— - ——
Flow line connections	┴
Mechanical link, roller, etc.	o
Semi-rotary actuator	D
As a rule, control valves (valve) except for non- return valves	☐
Conditioning apparatus (filter, separator, lubricator, heat exchanger)	◇

Description	Symbol
Spring	WV
Restriction: affected by viscosity unaffected by viscosity	)(
As a rule, energy conversion units (pump, com- pressor motor)	◯
Measuring instruments	○
Non-return valve, rotary connection, etc.	o

Figure 3.59 *Basic symbols used in fluid power diagrams*

- The arrowheads are solid so it must be a hydraulic pump.
- The arrowheads point in opposite directions so the pump can deliver the hydraulic fluid in either direction depending upon its direction of rotation.
- The arrow slanting across the pump is the variability symbol, so the pump has variable displacement.
- The double lines indicate the mechanical input to the pump from some engine or motor.

Summing up, we have a variable displacement, hydraulic pump that is bi-directional.

Test your knowledge 3.17

(a) Draw the symbol for a unidirectional, fixed displacement pneumatic pump (compressor).

(b) Draw the symbol for a fixed capacity hydraulic motor.

Directional control valves

The function of a directional control valve is to open or close flow lines in a system. Control valve symbols are always drawn in square boxes or groups of square boxes to form a rectangle. This is how you recognize them. Each box indicates a discrete position for the control valve. Flow paths through a valve are known as 'ways'. Thus a 4-way valve has four flow paths through the valve. This will be the same as the number of connections. We can, therefore, use a number

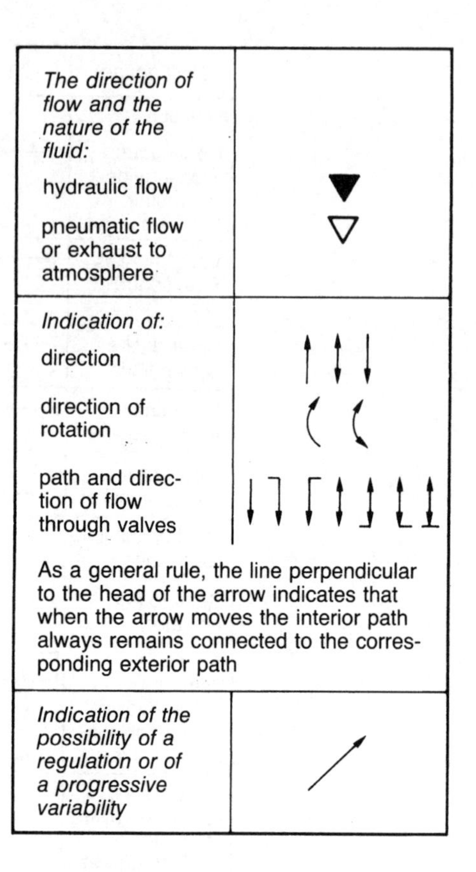

The direction of flow and the nature of the fluid:	
hydraulic flow	▼
pneumatic flow or exhaust to atmosphere	▽
Indication of:	
direction	
direction of rotation	
path and direction of flow through valves	
As a general rule, the line perpendicular to the head of the arrow indicates that when the arrow moves the interior path always remains connected to the corresponding exterior path	
Indication of the possibility of a regulation or of a progressive variability	

Figure 3.60 *Functional symbols used in fluid power diagrams*

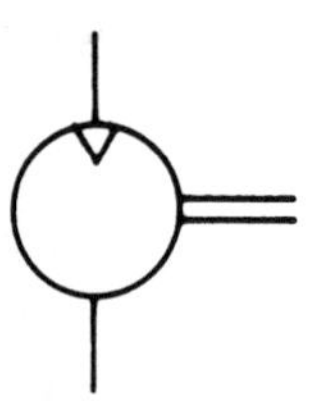

Figure 3.61 *Basic symbol for a motor*

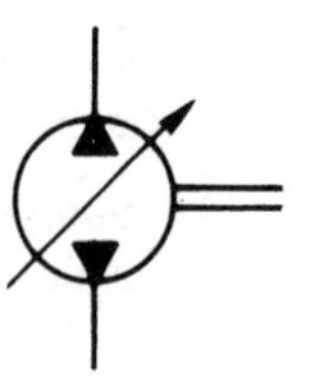

Figure 3.62 *Energy converter symbol (see text)*

code to describe the function of a valve. Figure 3.63 shows a 4/2 directional control valve. This valve has four flow paths, ports or connections and two positions. The two boxes indicate the two positions. The appropriate box is shunted from side to side so that, in your imagination, the internal flow paths line up with the connections. Connections are shown by the lines that extend

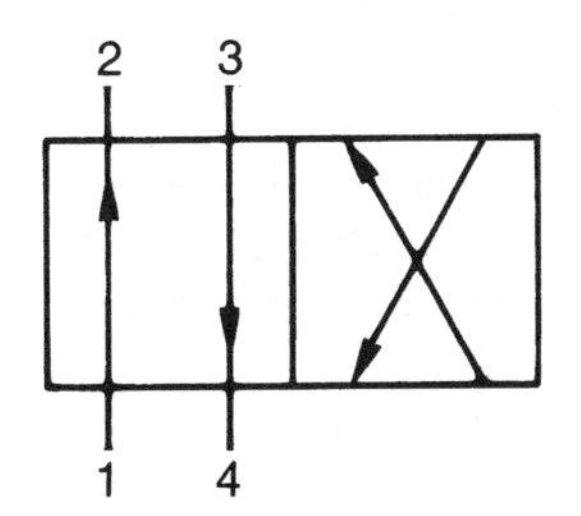

Figure 3.63 *4/2 directional control valve*

Test your knowledge 3.18

(a) State the numerical code that describes the valve shown in Figure 3.64.

(b) Describe the flow path drawn.

(c) Sketch and describe the flow path when the valve is in its alternative position.

Figure 3.64 *See Test your knowledge 3.18*

'outside' the perimeters of the boxes.

As drawn, the fluid can flow into port 1 and out of port 2. Fluid can also flow into port 3 and out of port 4. In the second position, The fluid flows into port 3 and out of port 1. Fluid can also flow into port 4 and out of port 2.

Valve control methods

Before we look at other examples of directional control valves, let's see how we can control the positions of a valve. There are four basic methods of control, these are:

- Manual control of the valve position.
- Mechanical control of the valve position.
- Electromagnetic control of the valve position.
- Pressure control of the valve positions (direct and indirect).
- Combined control methods.

The methods of control are shown in Figure 3.65. With simple electrical or pressure control, it is only possible to move the valve to one, two or three discrete positions. The valve spool may be located in such positions by a spring loaded detent.

Combinations of the above control methods are possible. For example a single solenoid with spring return for a two position valve. Let's now look at some further directional control valves (DCVs).

- Figure 3.66(a) shows a 4/2 DCV controlled by a single solenoid with a spring return.
- Figure 3.66(b) shows a 4/3 DCV. That is, a directional control valve with 4 ports (connections) and 3 positions. It is operated manually by a lever with spring return to the centre. The service ports are isolated in the centre position. An application of this valve will be shown later.

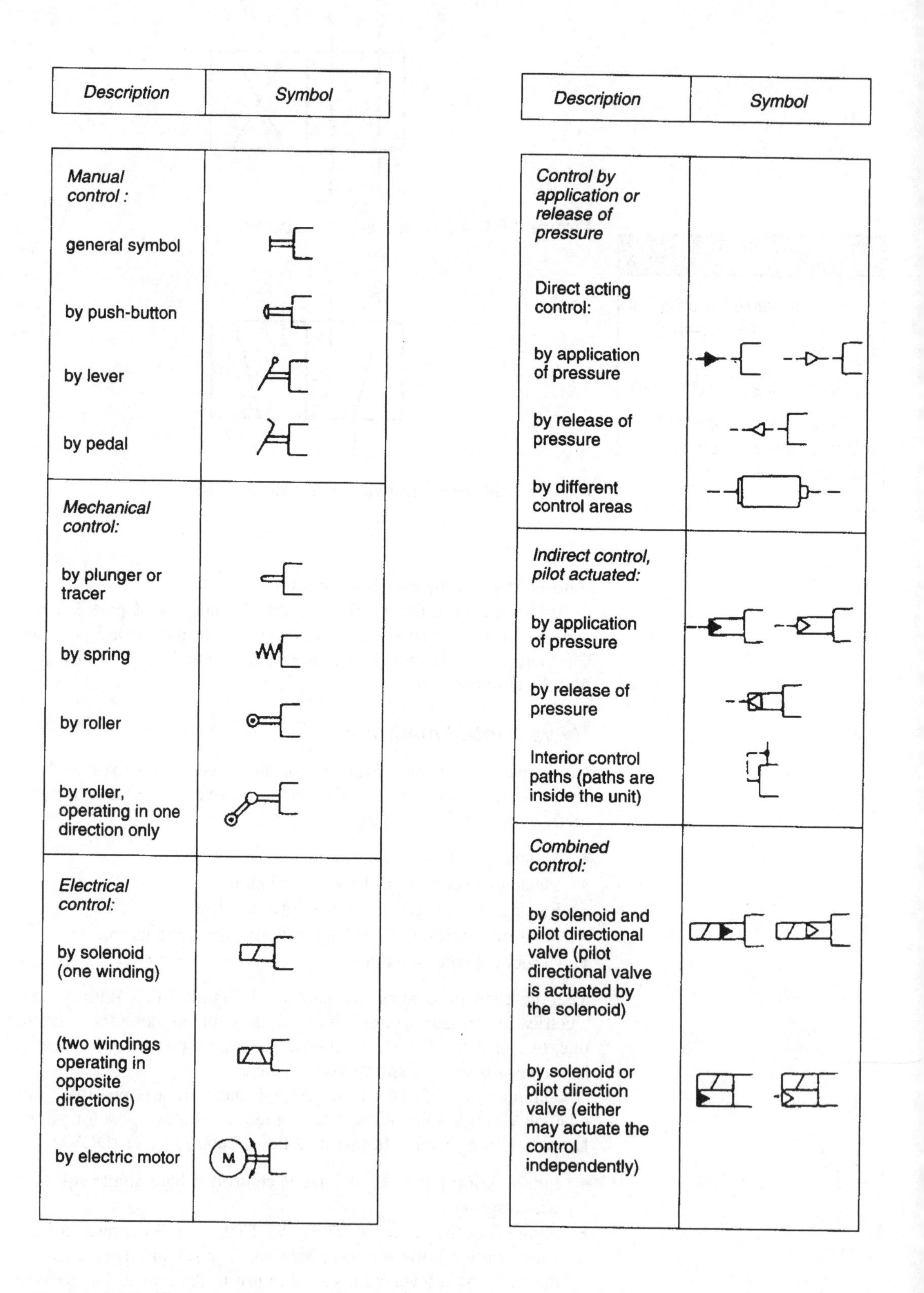

Description	Symbol
Manual control:	
general symbol	
by push-button	
by lever	
by pedal	
Mechanical control:	
by plunger or tracer	
by spring	
by roller	
by roller, operating in one direction only	
Electrical control:	
by solenoid (one winding)	
(two windings operating in opposite directions)	
by electric motor	M

Description	Symbol
Control by application or release of pressure	
Direct acting control:	
by application of pressure	
by release of pressure	
by different control areas	
Indirect control, pilot actuated:	
by application of pressure	
by release of pressure	
Interior control paths (paths are inside the unit)	
Combined control:	
by solenoid and pilot directional valve (pilot directional valve is actuated by the solenoid)	
by solenoid or pilot direction valve (either may actuate the control independently)	

Figure 3.65 *Methods of control*

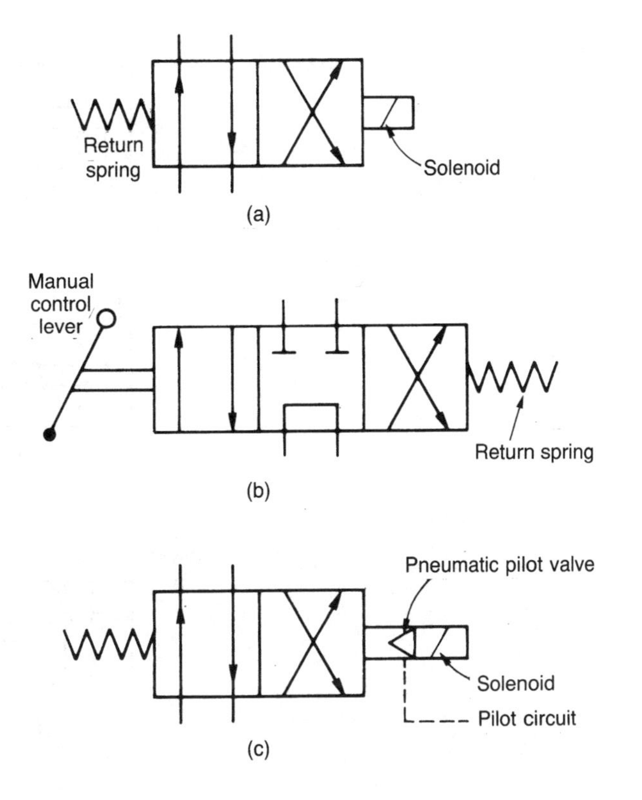

Figure 3.66 *Various types of DCV*

- Figure 3.66(c) shows a 4/2 DCV controlled by pneumatic pressure by means of a pilot valve. The pilot valve is actuated by a single solenoid and a return spring.

Linear actuators

A linear actuator is a device for converting fluid pressure into a mechanical force capable of doing useful work and combining this force with limited linear movement. Put more simply, a piston in a cylinder. The symbols for linear actuators (also known as 'jacks' and 'rams') are simple to understand and some examples are shown in Figure 3.68.

- Figure 3.68(a) shows a single-ended, double-acting actuator. That is, the piston is connected by a piston rod to some external mechanism through one end of the cylinder only. It is double acting because fluid pressure can be applied to either side of the piston.

Test your knowledge 3.19

Describe the DCV whose symbol is shown in Figure 3.67.

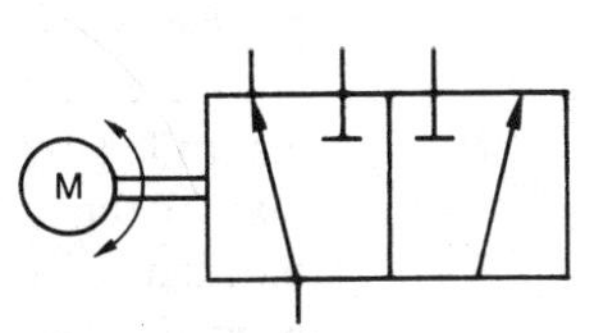

Figure 3.67 *See Test your knowledge 3.19*

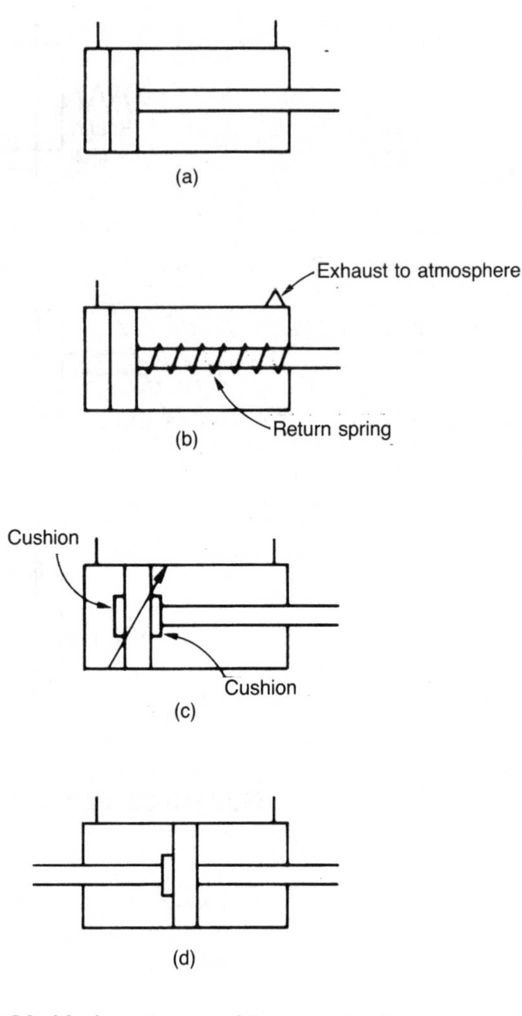

Figure 3.68 *Various types of linear actuator*

- Figure 3.68(b) shows a single-ended, single-acting actuator with spring return. Here the fluid pressure is applied only to one side of the piston. Note the pneumatic exhaust to atmosphere so that the air behind the piston will not cause a fluid lock.
- Figure 3.68(c) shows a single-ended, single-acting actuator, with double variable cushion damping. The cushion damping prevents the piston impacting on the ends of the cylinder and causing damage.
- Figure 3.68(d) shows a double-ended, double-acting actuator fitted with single, fixed cushion damping.

We are now in a position to use the previous component symbols to produce some simple fluid power circuits.

Figure 3.69 shows a single-ended, double-acting actuator controlled by a 4/3 tandem centre, manually operated DCV. Note that in the neutral position both sides of the actuator piston are blocked off, forming a hydraulic lock. In this position the pump flow is being returned directly to the tank. Note the tank symbol. This system is being supplied by a single direction fixed displacement hydraulic pump.

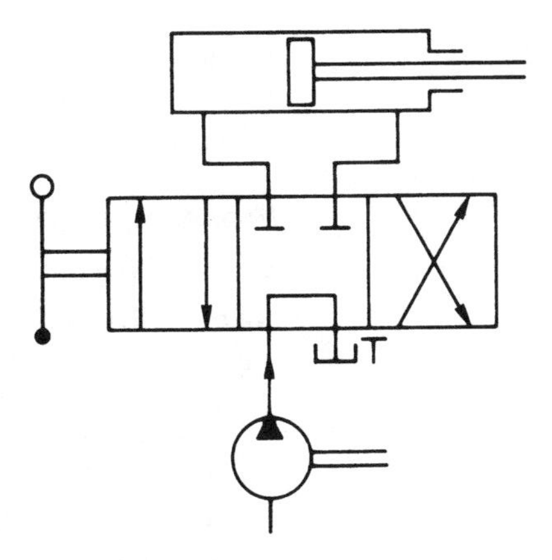

Figure 3.69 *Actuator controlled by a DCV*

Figure 3.70 shows a simple pneumatic hoist capable of raising a load. The circuit uses two 2-port manually operated push-button valves connected to a single-ended, single-acting actuator. Supply pressure is indicated by the circular symbol with a black dot in its centre. Valve 'b' has a threaded exhaust port indicated by the extended arrow. When valve 'a' is operated, compressed air from the air line is admitted to the underside of the piston in the cylinder. This causes the piston to rise and to raise the load. Any air above the piston is exhausted to the atmosphere through the threaded exhaust port at the top of the cylinder. Again this is indicated by a long arrow. When valve 'b' is operated, it connects the cylinder to the exhaust and the actuator is vented to the atmosphere. The load is lowered by gravity.

Both these circuits are functional, but they do not have protection against over-pressurization, neither do they have any other safety devices fitted. Therefore, we need to increase our vocabulary of

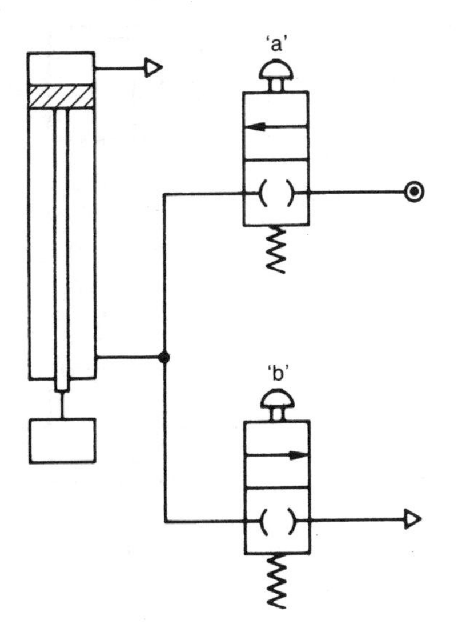

Figure 3.70 *A simple pneumatic hoist*

components before we can design a safe, practical circuit. We will now consider the function and use of pressure and flow control valves.

Pressure relief and sequence valves

Figure 3.71 shows an example of a pressure relief (safety) valve. In Figure 3.71(a) the valve is being used in a hydraulic circuit. Pressure is controlled by opening the exhaust port to the reservoir tank against an opposing force such as a spring. In Figure 3.71(b) the valve is being used in a pneumatic circuit so it exhausts to the atmosphere.

Figures 3.71(c) and 3.71(d) show the same valves except that this time the relief pressure is variable, as indicated by the arrow drawn across the spring. If the relief valve setting is used to control the normal system pressure as well as acting as an emergency safety valve, the adjustment mechanism for the valve must be designed so that the maximum safe working pressure for the circuit cannot be exceeded.

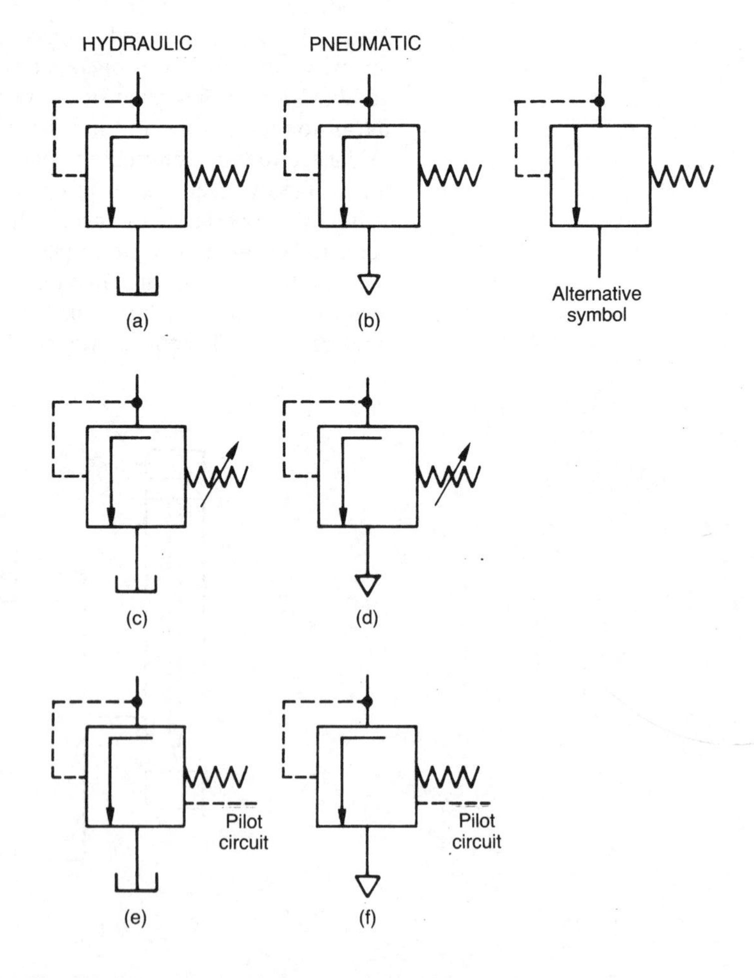

Figure 3.71 *Use of a pressure relief valve*

Figures 3.71(e) and 3.71(f) show the same valves with the addition of pilot control. This time the pressure at the inlet port is not only limited by the spring but also by the pressure of the pilot circuit superimposed on the spring. The spring offers a minimum pressure setting and this can be increased by increasing the pilot circuit pressure up to some pre-determined safe maximum. Sometimes the spring is omitted and only pilot pressure is used to control the valve.

Sequence valves are closely related to relief valves in both design and function and are represented by very similar symbols. They permit the hydraulic fluid to flow into a sub-circuit, instead of back to the reservoir, when the main circuit pressure reaches the setting of the sequence valve. You can see that Figure 3.72 is very similar to a pressure relief valve (PRV) except that, when it opens, the fluid is directed to the next circuit in the sequence instead of being exhausted to the reservoir tank or allowed to escape to the atmosphere.

Flow control valves

Flow control valves, as their name implies, are used in systems to control the rate of flow of fluid from one part of the system to another. The simplest valve is merely a fixed restrictor. For operational reasons this type of flow control valve is inefficient, so the restriction is made variable as shown in Figure 3.73(a). This is a throttling valve. The full symbol is shown in Figure 3.73(b). In this example the valve setting is being adjusted mechanically. The valve rod ends in a roller follower in contact with a cam plate.

Sometimes it is necessary to ensure that the variation in inlet pressure to the valve does not affect the flow rate from the valve. Under these circumstances we use a pressure compensated flow control valve (PCFCV). The symbol for this type of valve is shown in Figure 3.74. This symbol suggests that the valve is a combination of a variable restrictor and a pilot operated relief valve. The enclosing box is drawn using a long-chain line. This signifies that the components making up the valve are assembled as a single unit.

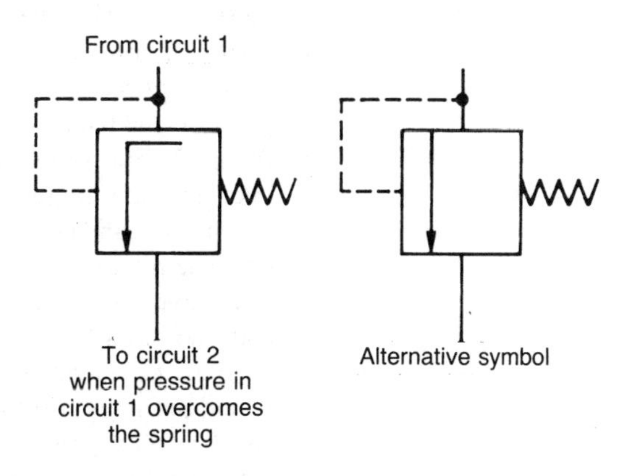

Figure 3.72 *Sequence valve*

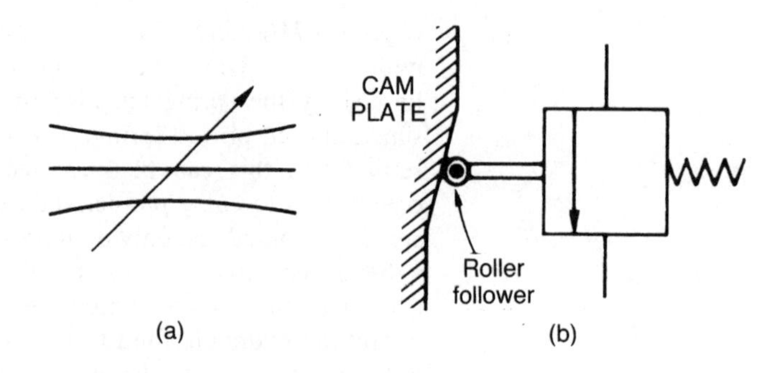

Figure 3.73 *Fluid control valves*

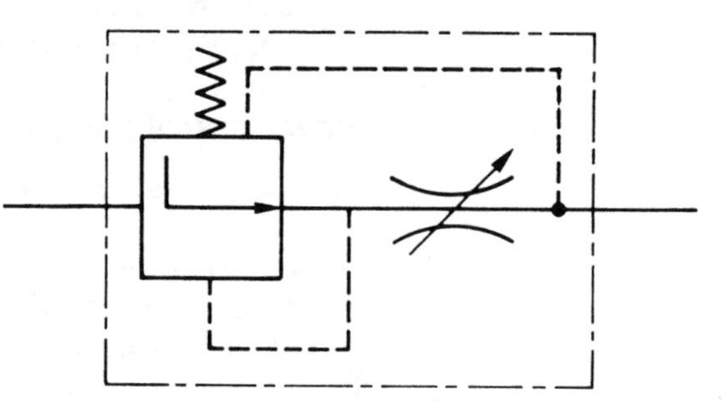

Figure 3.74 *Pressure compensated flow control valve*

Non-return valves and shuttle valves

The non-return valve (NRV), or check valve as it is sometimes known, is a special type of directional control valve. It only allows the fluid to flow in one direction and it blocks the flow in the reverse direction. These valves may be operated directly or by a pilot circuit. Some examples are shown in Figure 3.75.

- Figure 3.75(a) shows a valve that opens (is free) when the inlet pressure is higher than the outlet pressure (back-pressure).
- Figure 3.75(b) shows a spring-loaded valve that only opens when the inlet pressure can overcome the combined effects of the outlet pressure and the force exerted by the spring.
- Figure 3.75(c) shows a pilot controlled NRV. It only opens if the inlet pressure is greater than the outlet pressure. However, these pressures can be augmented by the pilot circuit pressure.
  (i) The pilot pressure is applied to the inlet side of the NRV. We now have the combined pressures of the main (primary) circuit and the pilot circuit acting against the outlet pressure. This enables the valve to open at a lower main circuit pressure than would normally be possible.
  (ii) The pilot pressure is applied to the outlet side of the NRV. This assists the outlet or back-pressure in holding the valve closed. Therefore it requires a greater main circuit pressure to open the valve. By adjusting the pilot pressure in these two examples we can control the circumstances under which the NRV opens.

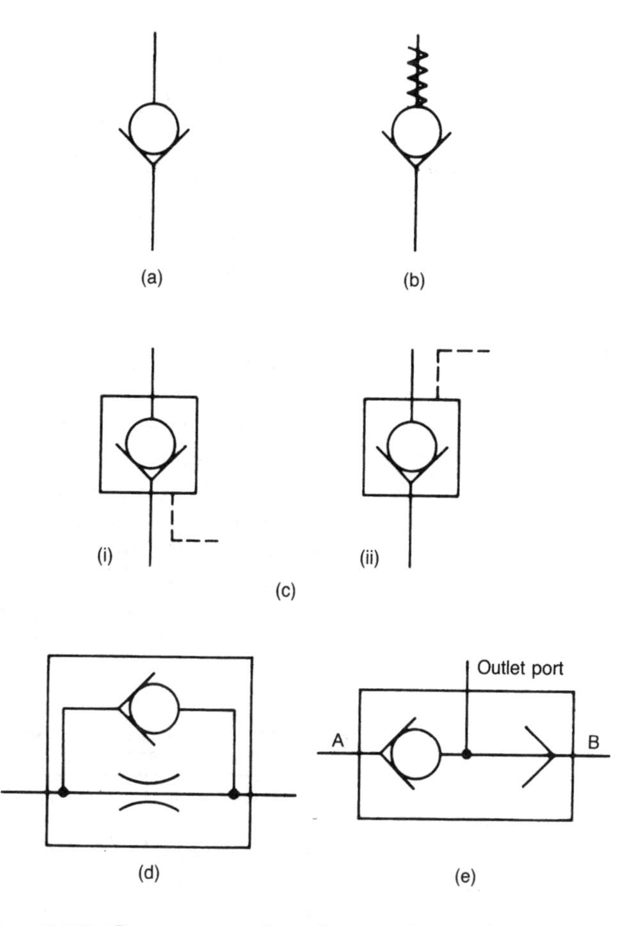

Figure 3.75 *Some examples of non-return valves*

- Figure 3.75(d) shows a valve that allows normal full flow in the forward direction but restricted flow in the reverse direction. The valves previously discussed did not allow any flow in the reverse direction.
- Figure 3.75(e) shows a simple shuttle valve. As its name implies, the valve is able to shuttle backwards and forwards. There are two inlet ports and one outlet port. Imagine that inlet port A has the higher pressure. This pressure overcomes the inlet pressure at B and moves the shuttle valve to the right. The valve closes inlet port B and connects inlet port A to the outlet port. If the pressure at inlet port B rises, or that at A falls, the shuttle will move back to the left. This will close inlet port A and connect inlet port B to the outlet. Thus, the inlet port with the higher pressure is automatically connected to the outlet port.

Conditioning equipment

The working fluid, be it oil or air, has to operate in a variety of environments and it can become overheated and/or contaminated. As its name implies, conditioning equipment is used to maintain the fluid in its most efficient operating condition. A selection of conditioning equipment symbols are shown in Figure 3.76. Note that all conditioning device symbols are diamond shaped.

Filters and *strainers* have the same symbol. They are normally identified within the system by their position. The filter element (dotted line) is always positioned at 90° to the fluid path.

Water traps are easily distinguished from filters since they have a drain connection and an indication of trapped water. Water traps are particularly important in pneumatic systems because of the humidity of the air being compressed.

Lubricators are particularly important in pneumatic systems. Hydraulic systems using oil are self-lubricating. Pneumatic systems use air, which has no lubricating properties so oil, in the form of a mist, has to be added to the compressed air line.

Heat exchangers can be either heaters or coolers. If the hydraulic oil becomes too cool it becomes thicker (more viscous) and the system becomes sluggish. If the oil becomes too hot it will become too thin (less viscous) and not function properly. The direction of the arrows in the symbol indicates whether heat energy is taken from the fluid (cooler) or given to the fluid (heater). Notice that the cooler can show the flow lines of the coolant.

Filters, water traps, lubricators and miscellaneous apparatus

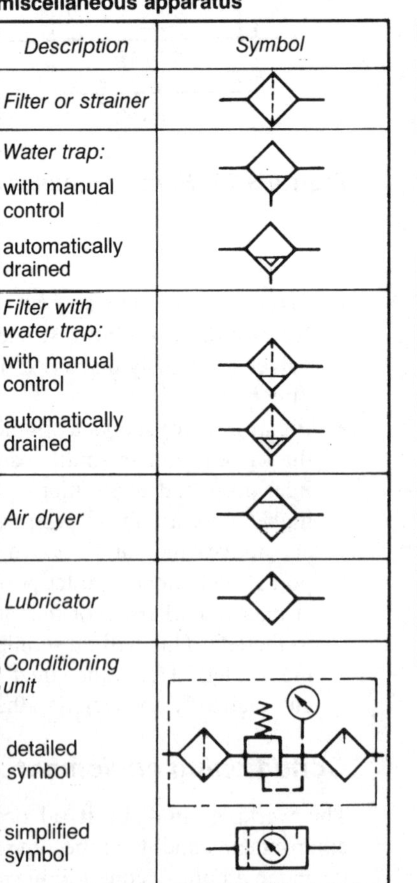

Description	*Symbol*
Filter or strainer	
Water trap:	
with manual control	
automatically drained	
Filter with water trap:	
with manual control	
automatically drained	
Air dryer	
Lubricator	
Conditioning unit	
detailed symbol	
simplified symbol	

Heat exchangers

Description	*Symbol*
Temperature controller (arrows indicate that heat may be either introduced or dissipated)	
Cooler (arrows indicate the extraction of heat)	
without representation of the flow lines of the coolant	
with representation of the flow lines of the coolant	
Heater (arrows indicate the introduction of heat)	

Figure 3.76 *Symbols for conditioning devices*

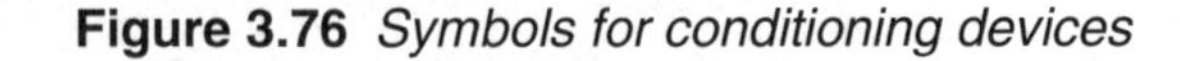

There is one final matter to be considered before you can try your hand at designing a circuit, and that is the pipework circuit to connect the various components together. The correct way of representing pipelines is shown in Figure 3.78.

- Figure 3.78(a) shows pipelines that are crossing each other but are not connected.
- Figure 3.78(b) shows three pipes connected at a junction. The junction (connection) is indicated by the solid circle (or large dot, if you prefer).

Figure 3.78(c) shows four pipes connected at a junction. On no account can the connection be drawn as shown in Figure 3.78(d). This is because there is always a chance of the ink running where lines cross on a drawing. The resulting 'blob' could then be misinterpreted as a connection symbol with disastrous results.

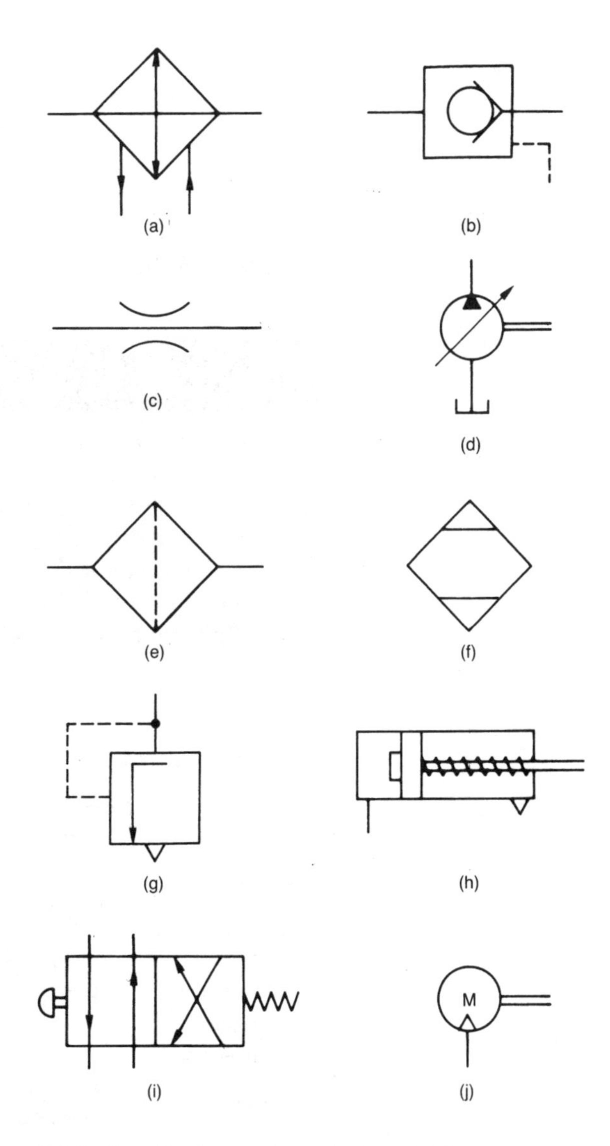

Figure 3.77 *See Test your knowledge 3.20*

Test your knowledge 3.20

Figure 3.77 shows a range of fluid circuit symbols. Name the symbol and explain the purpose of the device that it represents.

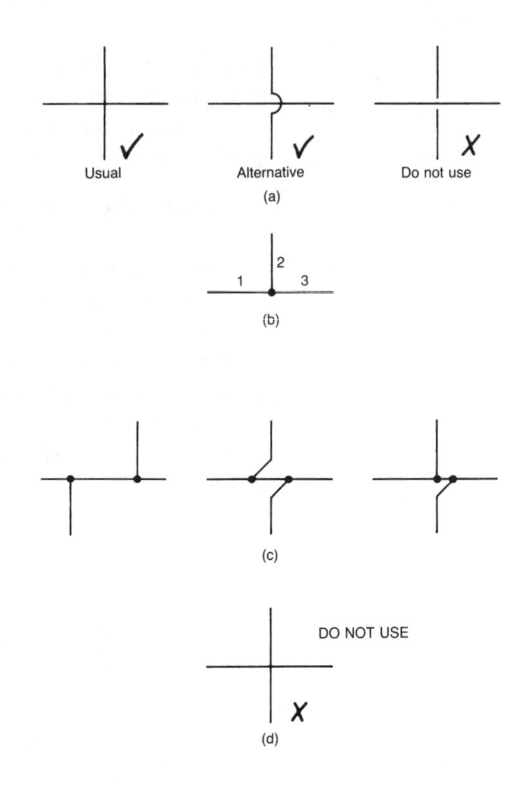

Figure 3.78 *Representing pipelines*

Activity 3.8

Figure 3.79 shows the general principles for the hydraulic drive to the ram of a shaping machine. The ram is moved backwards and forwards by a double-acting single-ended hydraulic actuator. The drawing was made many years ago and it uses outdated symbols. Using current symbols and practices (as set out in BS PP 8888) you are to draw a schematic hydraulic circuit diagram for the machine.

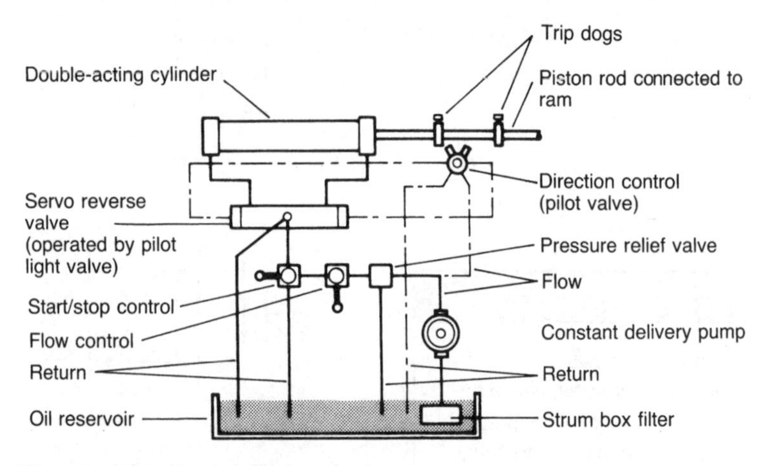

Figure 3.79 *See Activity 3.8*

Electrical and electronic schematics

Electrical and electronic schematics (or *circuits*) can also be drawn using schematic symbols to represent the various components. The full range of symbols and their usage can be found in BS 8888:2004. This is a very extensive standard and well beyond the needs of this book. However, for our immediate requirements you should refer to PP 8888-1:2005 Drawing practice: a guide for schools and colleges to BS 8888:2004. Figure 3.80 shows a selection of symbols that will be used in the following examples.

- A *cell* is a source of direct current (d.c.) electrical energy. Primary cells have a nominal potential of 1.5 volt seach. They cannot be recharged and are disposable. Secondary cells are rechargeable. Lead–acid cells have a nominal potential of 2 volts and nickel cadmium (NiCd) cells have a nominal potential of 1.2 volts.
- *Batteries* consist of a number of cells connected in series to increase the overall potential. A 12 volt car battery consists of six lead–acid secondary cells of 2 volts each.
- *Fuses* protect the circuit in which they are connected from excess current flow. This can result from a fault in the circuit, from a fault in an appliance connected to the circuit or from too many appliances being connected to the same circuit. The current flowing in the circuit tends to heat up the fuse wire. When the current reaches some pre-determined value the fuse wire melts and breaks the circuit so the current can no longer flow. Without a fuse the circuit wiring could overheat and cause a fire.
- *Resistors* are used to control the magnitude of the current flowing in a circuit. The resistance value of the resistor may be fixed or it may be variable. Variable resistors may be pre-set or they may be adjustable by the user. The electric current does work in flowing through the resistor and this heats up the resistor. The resistor must be chosen so that it can withstand this heating effect and sited so that it has adequate ventilation.
- *Capacitors*, like resistors, may be fixed in value or they may be preset or variable. Capacitors store electrical energy but, unlike secondary cells, they may be charged or discharged almost instantaneously. The stored charge is much smaller than the charge stored by a secondary cell. Large value capacitors are used to smooth the residual ripple from the rectifier in a power pack. Medium value capacitors are used for coupling and decoupling the stages of audio frequency amplifiers. Small value capacitors are used for coupling and decoupling radio frequency signals and they are also used in tuned (resonant) circuits.
- *Inductors* act like electrical 'flywheels'. They limit the build up of current in a circuit and try to keep the circuit running by putting energy back into it when the supply is turned off. They are used as current limiting devices in fluorescent lamp units. They are used as chokes in telecommunications equipment. Inductors are also used together with capacitors to make up resonant (tuned) circuits in telecommunications equipment.

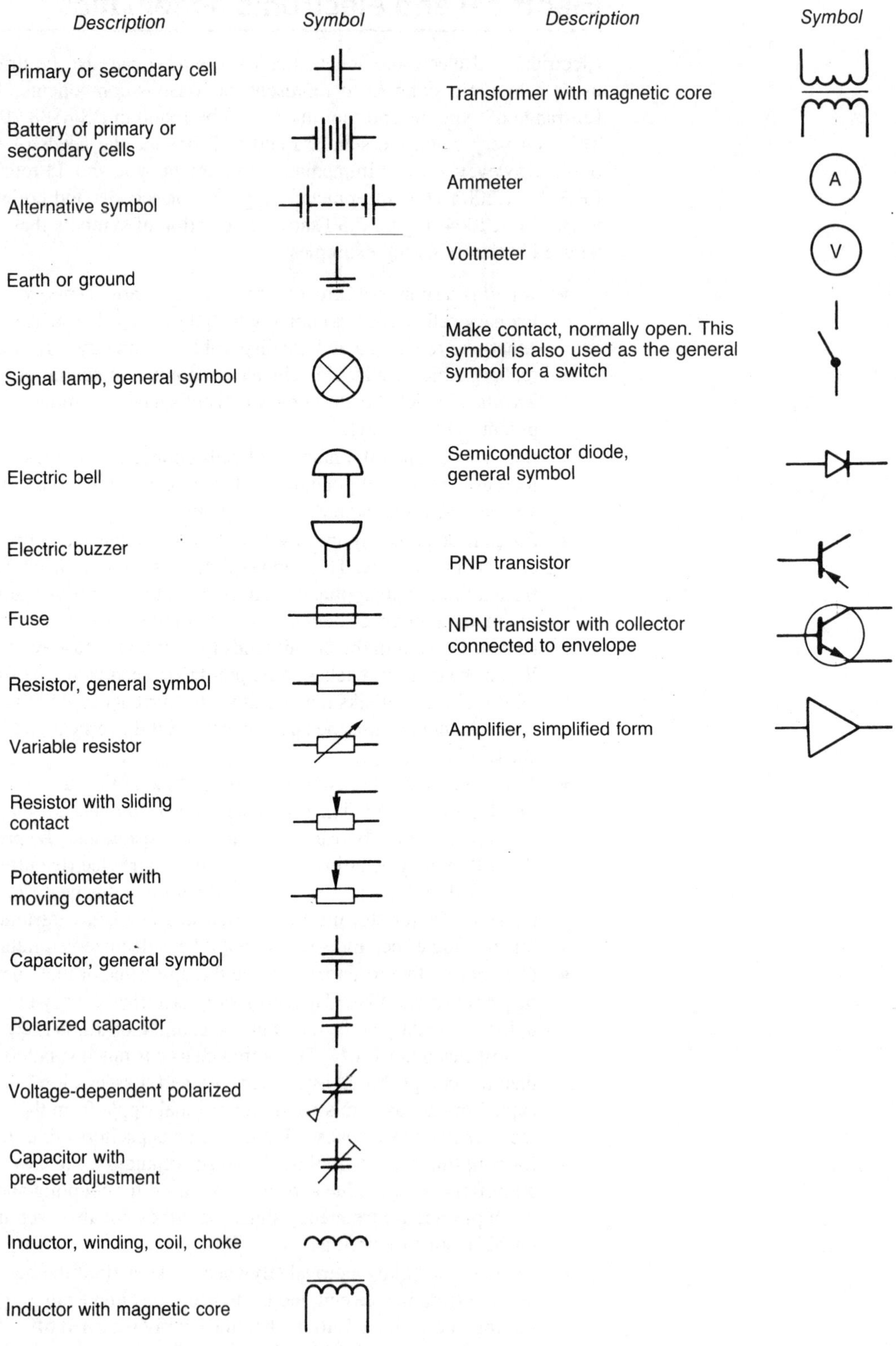

Description	*Symbol*
Primary or secondary cell	
Battery of primary or secondary cells	
Alternative symbol	
Earth or ground	
Signal lamp, general symbol	
Electric bell	
Electric buzzer	
Fuse	
Resistor, general symbol	
Variable resistor	
Resistor with sliding contact	
Potentiometer with moving contact	
Capacitor, general symbol	
Polarized capacitor	
Voltage-dependent polarized	
Capacitor with pre-set adjustment	
Inductor, winding, coil, choke	
Inductor with magnetic core	

Description	*Symbol*
Transformer with magnetic core	
Ammeter	A
Voltmeter	V
Make contact, normally open. This symbol is also used as the general symbol for a switch	
Semiconductor diode, general symbol	
PNP transistor	
NPN transistor with collector connected to envelope	
Amplifier, simplified form	

Figure 3.80 *Electronic symbols*

- *Transformers* are used to raise or lower the voltage of alternating currents. Inductors and transformers cannot be used in direct current circuits. You cannot get something for nothing, so if you increase the voltage you decrease the current accordingly so that (neglecting losses), $V \times I = k$ where k is a constant for the primary and secondary circuits of any given transformer.
- *Ammeters* measure the current flowing in a circuit. They are always wired in series with the circuit so that the current being measured can flow through the meter.
- *Voltmeters* measure the potential difference (voltage) between two points in a circuit. To do this they are always connected in parallel across that part of the circuit where the potential is to be measured.
- *Switches* are used to control the flow of current in a circuit. They can only open or close the circuit. So the current either flows or it doesn't.
- *Diodes* are like the non-return valves in hydraulic circuits. They allow the current to flow in one direction only as indicated by the arrowhead of the symbol. They are used to rectify alternating current (a.c.) and convert it into d.c.
- *Transistors* are used in high-speed switching circuits and to magnify radio and audio frequency signals.
- *Integrated circuits* consist of all the components necessary to produce amplifiers, oscillators, central processor units, computer memories and a host of other devices fabricated onto a single slice of silicon; each chip being housed in a single compact package.

Let's look at some examples of schematic circuit diagrams using these symbols. All electric circuits consist of:

- A source of electrical energy (e.g. a battery or a generator).
- A means of controlling the flow of electric current (e.g. a switch or a variable resistor).
- An appliance to convert the electrical energy into useful work (e.g. a heater, a lamp, or a motor).
- Except for low-power battery operated circuits, an over-current protection device (fuse or circuit breaker).
- Conductors (wires) to connect these various circuit elements together. Note that the rules for drawing conductors that are connected and conductors that are crossing but not connected are the same as for drawing pipework (see Figure 3.78).

Figure 3.81 shows a very simple circuit that satisfies the above requirements. In Figure 3.81(a) the switch is 'closed' therefore the circuit as a whole is also a closed loop. This enables the electrons that make up the electric current to flow from the source of electrical energy through the appliance (lamp) and back to the source of energy ready to circulate again. Rather like the fluid in our earlier hydraulic circuits. In Figure 3.81(b) the switch is 'open' and the circuit is no longer a closed loop. The circuit is broken. The electrons can no longer circulate. The circuit ceases to function. We normally draw our circuits with the switches in the 'open' position so that the circuit is not functioning and is 'safe'.

Another view

There are some notable differences between the symbols used in the USA and those use in the UK for electronic schematics. It's also important to be aware that many European-based companies use the USA symbols (not the BS symbols). The most notable differences are in the symbols used for resistors, capacitors, and logic devices. Watch out for this! (Figure 3.2 on page 177 is a good example).

Figure 3.82 shows a simple battery operated circuit for determining the resistance of a fixed value resistor. The resistance value is obtained by substituting the values of current and potential into the formula, $R = V/I$. The current in amperes is read from the ammeter and the potential in volts is read from the voltmeter. Note that the ammeter is wired in series with the resistor so that the current can flow through it. The voltmeter is wired in parallel with the resistor so that the potential can be read across it. This is always the way these instruments are connected.

Figure 3.83 shows a circuit for operating the light over the stairs in a house. The light can be operated either by the switch at the bottom of the stairs or by the switch at the top of the stairs. Can you work out how this is achieved? The switches are of a type called 'two-way, single-pole'. The circuit is connected to the mains supply. It is protected by a fuse in the 'consumer unit'. This unit contains the main switch and all the fuses for the house and is situated adjacent to the supply company's meter and main fuse. Figure 3.84 shows a two-stage transistorised amplifier. It also shows a suitable power supply. Table 3.3 lists and names the components.

Figure 3.85 shows a similar amplifier using a single chip. Such an amplifier would have the same performance but fewer components are required. Therefore it is cheaper and quicker to make.

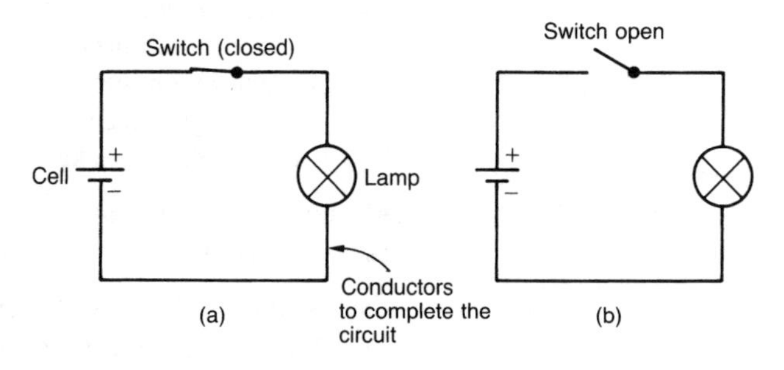

Figure 3.81 *A simple electric circuit*

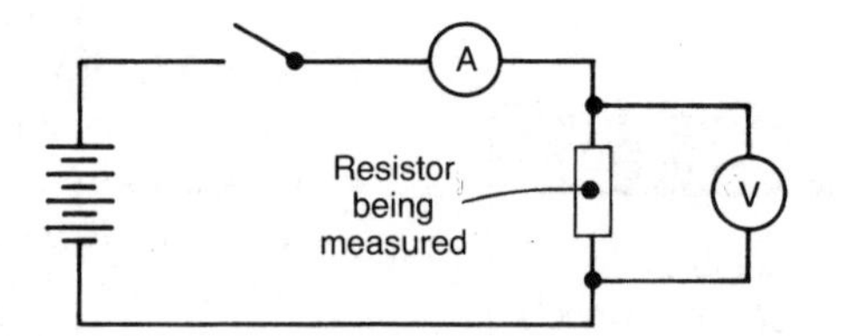

Figure 3.82 *Circuit for determining resistance*

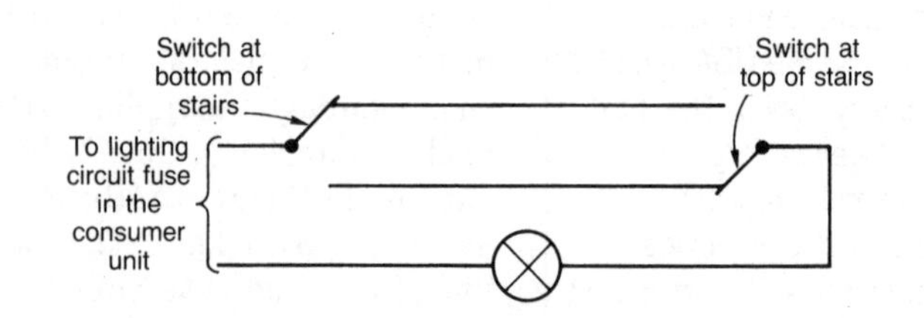

Figure 3.83 *Two-way lighting switch*

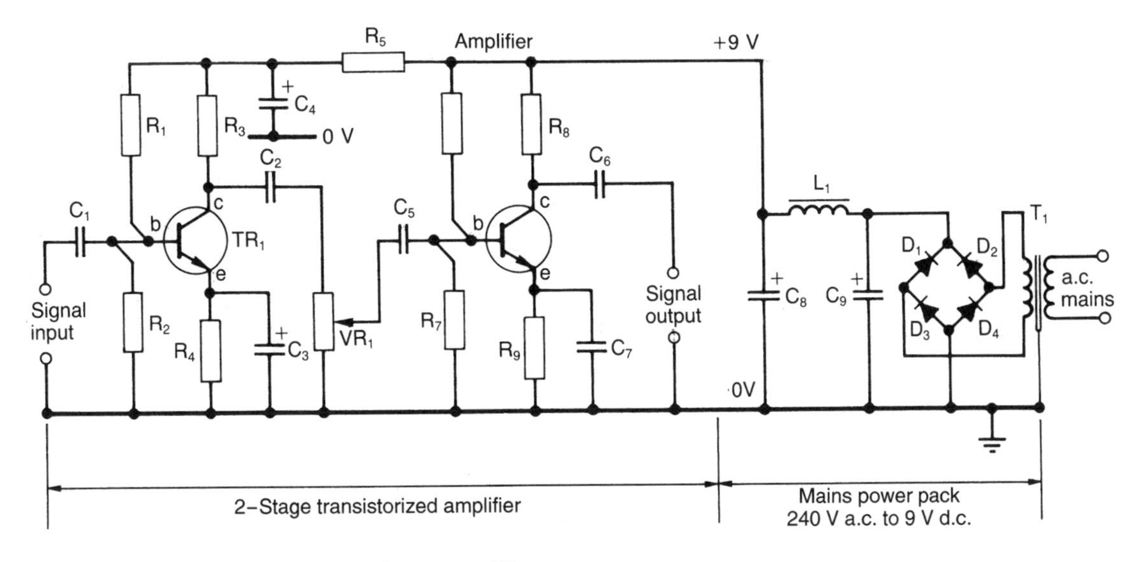

Figure 3.84 *Two-stage transistor amplifier*

Table 3.3 *Component list for the two-stage transistor amplifier*

Component	*Description*
R_1 to R_9	Fixed resistors
VR_1	Variable resistor
C_1 to C_9	Capacitors
D_1 to D_4	Diodes
TR_1 and TR_2	Transistors
T_1	Mains transformer
L_1	Inductor (choke)

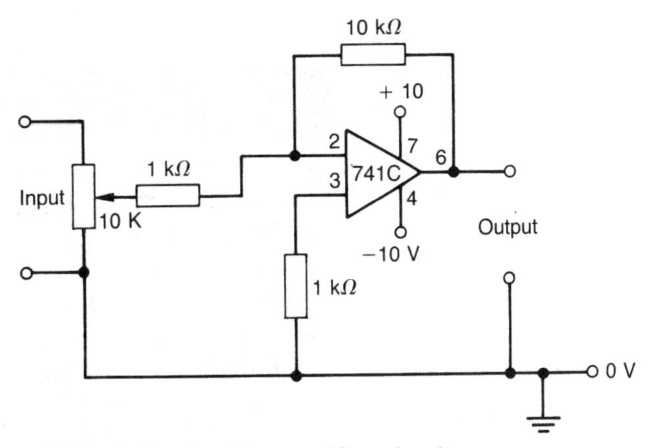

Figure 3.85 *A single-chip amplifier circuit*

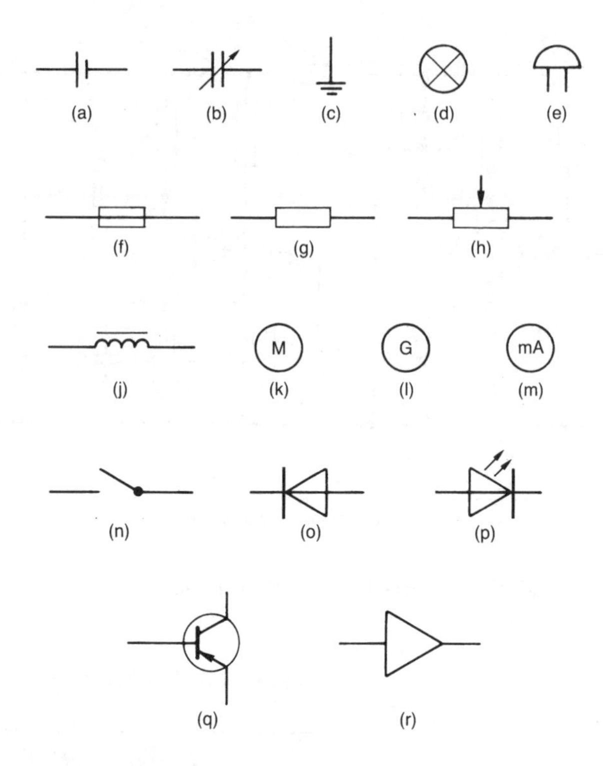

Figure 3.86 *See Test your knowledge 3.21*

Test your knowledge 3.21

Figure 3.86 shows some typical electrical and electronic component symbols. Name them and briefly explain what they do.

Activity 3.9

Draw a simple circuit for powering four electric lamps from a battery. It must be possible to turn each lamp on or off independently. A master switch must also be provided so that the whole circuit can be turned on or off. A fuse must be included in the circuit in order to prevent the battery being overloaded.

Activity 3.10

Draw a schematic circuit diagram for a battery charger having the following features:

- the primary circuit of the transformer (i.e. the side that is connected to the a.c. mains supply) is to have an on/off switch, a fuse and an indicator lamp
- the secondary circuit of the transformer is to have a bridge rectifier, a variable resistor to control the charging current, a fuse and an ammeter to indicate the charging current.

Activity 3.11

Figure 3.87 shows an electronic circuit.

(a) Draw up a component list that numbers and names each of the components (include values where given).

(b) Use manufacturer's data to obtain package outline and dimensions for the BF115 transistor.

(c) Draw the circuit diagram using a 2D CAD (such as A9CAD) package and label each component. Add a component list to the drawing.

(d) Draw a dimensioned package outline for the transistor and add this to the drawing.

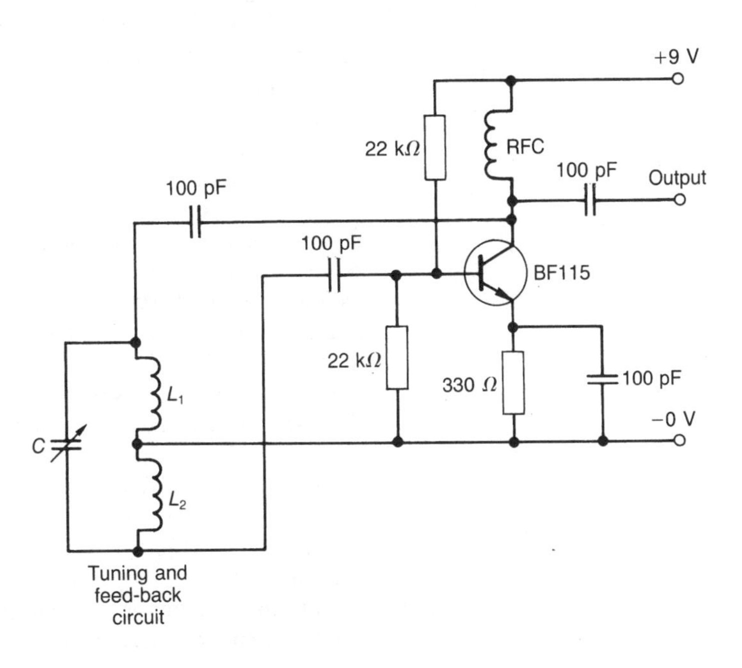

Figure 3.87 *See Activity 3.11*

3.3 Project planning

Having dealt at some length with the way in which engineers communicate their ideas graphically it is worth introducing the techniques that engineers use when they plan a project.

Planning is an important task and it needs to be carried out in detail and also before any of the design work or prototype manufacture is carried out. An effective plan will save both time and money and it will ensure that resources are not wasted.

A plan defines a strategy for a project in terms of the individual tasks and activities that take place and the resources that will be required. It is also important to share this plan with others that are involved and obtain their agreement to supply the required input and resources within the timescale defined in the plan.

Design strategies

What is a design strategy? It is basically having a design method and it consists of two things:

1 A *framework of intended actions* within which to work
2 Some form of *management control function* to enable you to adapt your actions as the problem unfolds.

Using a design strategy may seem to divert effort and time from the main task of designing, but this may not be a bad thing, as the purpose of a design strategy will make you think of the way design problems will be dealt with. It also provides you with an awareness as to where the design team is going and how it intends to get there.

The purpose then of having a strategy is to ensure that design activities are realistic with respect to the constraints of both time and resources within which the design team has to work. In a manufacturing company the most used strategy is a sequence of actions that have previously been applied to an already existing product.

For instance, to design a variation of an already designed electromechanical device or engineering service, the strategy most likely to be used for the new variation will be the same tactics and design methods used for the previous design. This would therefore be making use of a 'pre-established' strategy. The relevant tactics would be drawn from conventional techniques and rational methods already familiar to the design team. This type of strategy applies to innovative designs.

It is not always possible to have a Design Strategy, as would be the case in research design situations but having no particular plan of action would be a type of strategy in itself. This could be referred to as an 'inventive' strategy. The type of final design may be purely inventive, where no previous market exists.

Often the designers may not know when or what the final outcome may be, although hopefully they may achieve some degree of success in designing a material, product or engineering service that can be commercially exploited. The relevant tactics would be mainly creative.

The two strategies mentioned are extreme forms. In all probability, most designs require a compromise between the two, certain parts of the project design may need the inventive strategy if it calls for unknown areas of engineering design.

The 'pre-established' strategy is predominantly a convergent design approach, whereas the 'inventive' strategy is predominantly a 'divergent' approach. Usually the aim of a design strategy is to converge onto a final, evaluated and detailed design.

Sometimes in reaching that final design it may be necessary in some areas of the design to diverge, so as to widen the search for new ideas and solutions. Therefore the overall design process is mainly convergent, but has elements of divergent thinking.

Convergent thinkers are usually good at detail design, and evaluating and choosing the most suitable solution from a range of options. On the other hand, divergent thinkers are best at conceptual design problems and are able to produce a wide range of alternative solutions.

Project planning

Having established a design strategy we can begin to construct an outline plan for the project. There are typically six stages in most design projects and they are as follows:

- Clarifying objectives.
- Establishing functions.
- Setting requirements.
- Generating alternatives.
- Evaluating alternatives.
- Improving details.

Now let us look in more detail at the stages that make up each stage of the planning process and expand the meaning of each:

1 Clarifying objectives

This stage would in all probability have been dealt with during the design brief and design proposals stages, when the objectives of the design would be stated. But you would still need to clarify them before design can commence.

Aim
To understand the client's needs and to clarify the design objectives and sub-objectives, and the relationships between them.

Method
This is best achieved by carrying out a needs analysis in conjunction with the client and potential users.

2 Establishing functions

Although the client may have specified the functions expected from the design the designer may find a more radical or innovative solution by reconsidering the level of the problem definition. He or she may be able to offer the customer a better solution to the functional problems of the design in excess to the expected at no extra cost.

Aim
To establish the functions required.

Method
Break down the overall function into sub-functions. The sub-functions will comprise of all the functions expected within the product.

3 Setting requirements

Design problems are usually all set within certain limits, these limits may be cost, weight, size, safety or performance, etc. or any combination of them. Quantify the requirements in the form of a design specification.

Aim
To produce an accurate specification of the performance of the designed product.

Method
Identify the required performance attributes, these may well have been considered at the design feasibility study stage or may have been specified by the client.

4 Generating alternatives

Even if you think you have a good design solution (or can simply modify or adapt an existing product) always look further, if time and costs permit, for alternative solutions. A radical approach can often generate new and better solutions than the one that may result from making small changes to an existing product.

Aim
To have a choice of solutions to allow comparisons of ideas in solving the design requirements.

Method
It would help to draw several design layout drawings or models available to enable discussions to take place with the design team and the client.

5 Evaluating alternatives

When some alternative design proposals have been thought about and maybe some design layouts have been produced, the evaluation of the alternatives can be discussed.

Aim
To evaluate the alternatives, choose the ones that fully satisfy the client's needs and meet all of the identified requirements. Note, also, that you will normally be looking for the most cost-effective solution and your client may be willing to compromise on performance if the cost can be reduced. The final decision will normally rest with the client.

Method
Compare the value of the alternative design proposals against the original proposal agreed with the customer on the basis of performance and cost.

6 Improving details

There are two main reasons for improving details, they are either aimed at increasing the product value to the customer or reducing the cost to the producer.

Aim
To increase or maintain the value of the product to the customer at the same time reducing its cost to the producer. This may follow directly from the adoption of an alternative solution but may also result from making minor changes to components, or finish.

Method
This is normally approached using two methods, one is called *value engineering* and the other is known as *value analysis*.

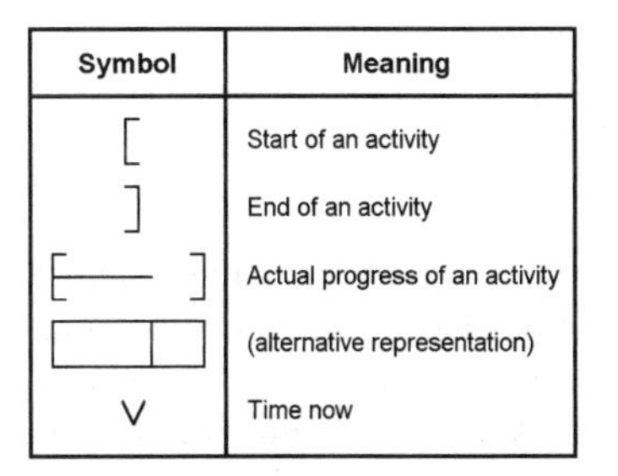

Symbol	Meaning
[	Start of an activity
]	End of an activity
[——]	Actual progress of an activity
▭▭	(alternative representation)
V	Time now

Figure 3.88 *Gantt chart symbols*

Gantt charts

A Gantt chart is simply a bar chart that shows the relationship of the activities that make up a project over a period of time. When constructing a Gantt chart, activities are listed down the page whilst time runs along the horizontal axis. The standard symbols used to denote the start and end of activities, and the progress towards their completion, are shown in Figure 3.88.

A simple Gantt chart is shown in Figure 3.89. This chart depicts the relationship between four activities A to D that make up a project. The horizontal scale is marked off in intervals of 1 day, with the whole project completed by day 14. At the start of the sixth day (see *time now*) the following situation is evident:

- Activity A has been completed
- Activity B has been partly completed and is on schedule
- Activity C has not yet started and is behind schedule
- Activity D is yet to start.

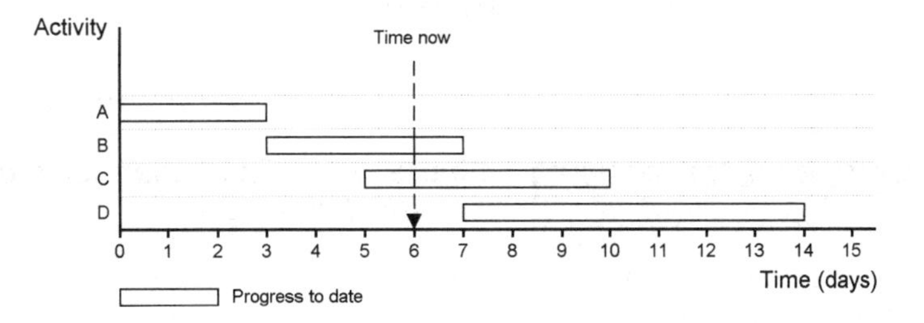

Figure 3.89 *A simple Gantt chart*

Another example is shown in Figure 3.90. This chart depicts the relationship between six activities A to F that make up a project. The horizontal scale is marked off in intervals of 1 day, with the whole project completed by day 18. At the start of the eighth day (again marked *time now*) the following situation is evident:

- Activity A has been completed
- Activity B has been partly completed but is running behind schedule by two days
- Activity C has been partly completed and is running ahead of schedule by one day

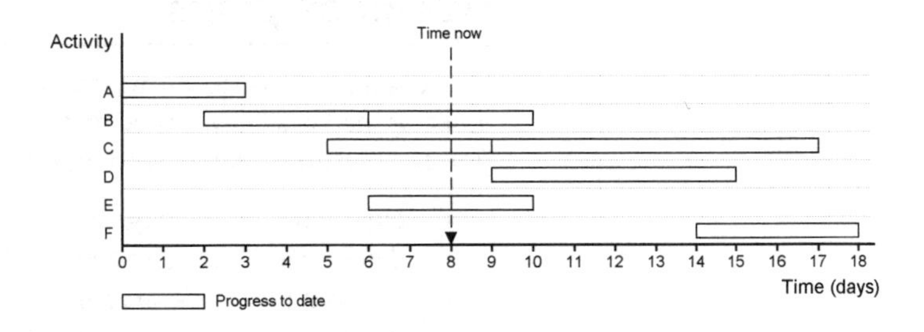

Figure 3.90 *Another simple Gantt chart*

- Activity D is yet to start
- Activity E has started and is on schedule
- Activity F is yet to start.

As an example of how a Gantt chart is used in practical situation take a look at Figure 3.91. This shows how a loudspeaker manufacturer might use this technique to track the progress of a project to produce a new loudspeaker design. The chart shows the situation at the beginning of Day 4 with all activities running to schedule.

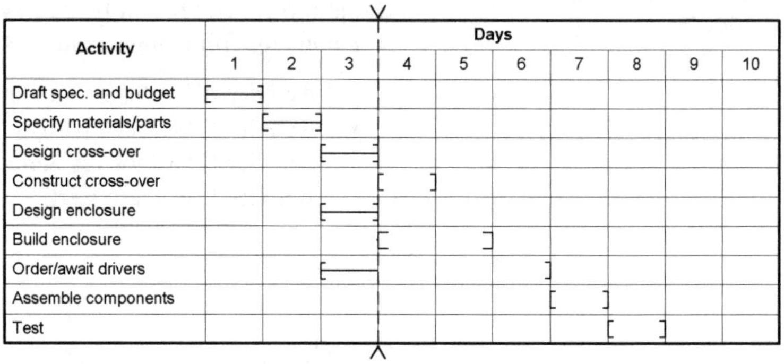

Activity	Days									
	1	2	3	4	5	6	7	8	9	10
Draft spec. and budget										
Specify materials/parts										
Design cross-over										
Construct cross-over										
Design enclosure										
Build enclosure										
Order/await drivers										
Assemble components										
Test										

Figure 3.91 *Gantt chart for producing a prototype of a new loudspeaker design*

Test your knowledge 3.22

Figure 3.92 shows the Gantt chart for the design and manufacture of a loudspeaker. Assuming that the project is at the start of day 6, which activities are:

(a) on schedule

(b) behind schedule by one day, or less

(c) behind schedule by more than one day

(d) ahead of schedule.

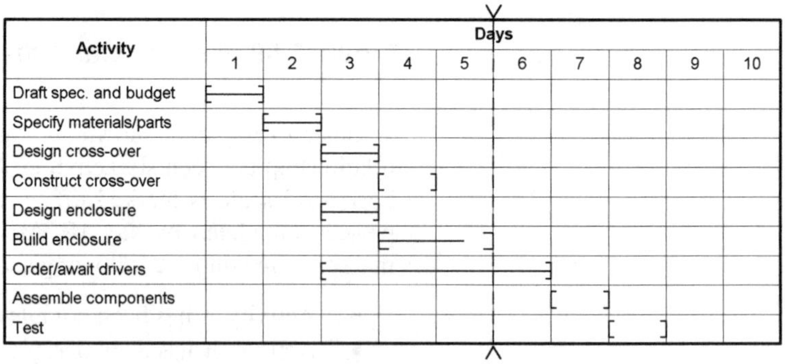

Activity	Days									
	1	2	3	4	5	6	7	8	9	10
Draft spec. and budget										
Specify materials/parts										
Design cross-over										
Construct cross-over										
Design enclosure										
Build enclosure										
Order/await drivers										
Assemble components										
Test										

Figure 3.92 *See Test your knowledge 3.22*

Business plans

On occasions, it is necessary to provide a detailed business plan in order to make a case for a particular business venture or project. Before a business plan is written, it is necessary to:

- clearly define the target audience for the business plan
- determine the content and level of detail in the plan
- map out the plan's structure (contents)
- decide on the likely length of the plan
- identify all the main issues to be addressed in the plan.

Shortcomings in the concept and gaps in supporting evidence and proposals need to be identified. This will facilitate an assessment of research to be undertaken before any drafting commences. It is also important to bear in mind that a business plan should be the end result of a careful and extensive research and development project that must be completed before any serious writing should be started.

A typical business plan comprises the following main elements:

- An *introduction* which sets out the background and structure of the plan
- A *summary* consisting of a few pages that highlight the main issues and proposals.
- A *main body* containing sections or chapters divided into numbered sections and subsections. The main body should include financial information (including a *profitability forecast*).
- *Market and sales projections* should be supported by valid market research. It is particularly important to ensure that there is a direct relationship between market analysis, sales forecasts and financial projections. It may also be important to make an assessment of competitors' positions and their possible response to the appearance of a rival product.
- *Appendices* should be used for additional information, tabulated data, and other background material. These (and their sources) should be clearly referenced in the text.

The financial section of the plan is of crucial importance and, since it is likely to be read in some detail, it needs to be realistic about sales expectations, profit margins and funding requirements ensuring that financial ratios are in line with industry norms. It is also essential to make realistic estimates of the cost and time required for product development and the need to secure external sources of funding.

When preparing a plan it is often useful to include a number of 'what-if' scenarios. These can help you to plan for the effects of escalating costs, reduction in sales, or essential resources becoming scarce. During a *what-if analysis*, you may also wish to consider the *halve-double* scenario in which you examine the financial viability of the project in the event that sales projections are halved and costs and time are doubled. The results can be sobering!

When writing a business plan it is necessary to:

- avoid unnecessary jargon
- economise on words
- use short crisp sentences and bullet points
- check spelling, punctuation and grammar
- concentrate on relevant and significant issues
- break the text into numbered paragraphs, sections etc.
- relegate detail to appendices
- provide a contents page and number pages
- write the summary last.

Finally, it can be useful to ask a consultant or other qualified outsider to review your plan in draft form and be prepared to adjust the plan in the light of comments secured and experiences gained.

3.4 Design

The design process

The design process is the name given to the various stages that we go through when we design something. Each stage in the process follows the one that goes before it and each marks a particular *phase* in a project. The various phases in a design project involve:

- understanding and describing the problem
- carrying out a needs analysis
- developing a design brief and a design specification
- carrying out research
- generating ideas and investigating potential solutions
- evaluating potential solutions
- applying scientific principles to the design
- producing and reading engineering drawings
- selecting appropriate drawing techniques
- developing and testing a prototype
- communicating the chosen design solution.

The design brief

In order to begin developing ideas you must identify the key requirements from the information supplied by the client brief (this is also sometimes referred to as a *client brief*). The design brief allows you to develop the product in terms of:

- the function of the product
- who will be using the product
- what the product should look like
- the materials it should be made from
- the technology necessary to produce it
- the production cost (or service delivery cost)
- the number required (and thus the scale of production)
- the timescale for the product
- health and safety issues
- issues relating to quality and 'fitness for purpose'.

The design brief needs to be developed into a formal *design specification*. The specification should be *quantified* (in terms of parameters such as size, weight, energy consumption, efficiency, etc) and it should be *measurable*. The design specification needs to address the function of the product, its form, size and shape and size, ergonomic features, and so on. Note that the design specification may specify particular parameters in different ways such as:

- input: 13.8 V d.c.
- minimum output: 40 W
- max. weight: 25 kg
- minimum turning circle: 3.75 m
- speed range: 270 rev/s to 350 rev/s
- height: 1.5 m ± 0.05 m
- frequency: 1.75 kHz ± 2.5%.
- supply voltage: 200 to 240 V a.c.

The design brief may also provide an indication of what materials and processes should be used in the manufacture of the product but this is not always the case. Furthermore, there may be good reasons for using different materials in the production of a prototype from those that will actually be used when the product goes into production.

Developing ideas

In everyday life, ideas often seem to flow naturally. When designing an engineered product or service this is not always the case. Furthermore, if you can only come up with a limited number of ideas (say one or two) you might need to generate more ideas to provide you with a wider range of alternatives or *options*.

In order to generate ideas we can make use of one or two tried and tested ideas. The first of these is called *brainstorming*. In brainstorming a group of people sit around and fire ideas at one another. There are several basic rules for brainstorming:

- Everyone in the group must contribute and has an equal right to be heard
- All ideas (however unlikely or preposterous) must be treated with equal respect
- Everything should be written down so that no ideas are lost (usually one member of the group is made responsible for this and ideas are recorded on a flip chart so that all can see what has been written down)
- Adequate time should be set aside for the exercise and there should be no interruptions
- It's important to avoid probing ideas too deeply. This can be left until a later stage.
- Agree, at the end of the session a selection (typically three or four) of ideas that should be considered as *candidates* for carrying forward to the next stage of the process. These are the ideas that the group considers (by poll, if necessary) to be the most feasible in terms of satisfying the design brief. Do not, at this stage, reject the other ideas because you might need to come back to these later!
- At first sight, some ideas may be considered less credible or less serious than others by some of the members of the group (we often describe such ideas as being *off-the-wall*). Nobody in the group should be made to feel bad or inferior if other members of the group consider their ideas strange or unworkable. Some of the most innovative engineering projects have resulted from brainstorming sessions that have unearthed ideas that, at first sight, have been considered unworkable by the majority of those involved.

Another technique that is used to generate ideas is *mind mapping*. A mind map is a sketch or drawing that allows you to identify all the factors that need to be taken into account when developing a solution to a design brief. The name of the product or service appears at the centre of the mind map and each of the solutions and other factors are placed around it. The map can then be progressively expanded as more detail is added.

There are a number of advantages of using a mind map to generate ideas and to understand the relationship between them. These include:

- The design brief (or design problem) appears in the centre of the map and it's thus very easy to see how all of the potential solutions and any other factors relate to it
- The links that exist between solutions and other factors can be immediately recognised
- The map can be easily grasped without having to read a lot of words
- It's easy to extend a map or add more information to it
- A mind map can help to stimulate thought and aid understanding
- It's often easier and faster to create a mind map than spend a lot of time putting your ideas in writing!

A further technique is that of carrying out *research* into similar or competing products or engineering services in order to determine how the product or service works and how it can be improved. This exercise is sometimes also carried out as part of the initial market research.

The next stage in the design process requires a number of potential solutions to the design problem to be identified and investigated. You would normally produce a range of different solutions including those that may rely on different technology, different materials or different operating principles.

At this point you will be able to identify *production constraints* for each of your design solutions including availability (and cost) of materials, components and resources and the means of processing or manufacturing the product. You should also consider issues relating to Health and Safety and any legislation that may have an impact on the manufacture, use or disposal of the product.

Having developed a range of solutions you will be in a position to make an informed decision about which one should be your final design solution. When carrying out this *formative evaluation* you will sometimes find that the choice will be obvious (for example, the alternative solutions may be prohibitively expensive or use dangerous materials or processes in their manufacture) however you may sometimes find that it is quite difficult to arrive at a final choice. In such cases you might wish to assemble a *design review panel* of colleagues and/or potential users in order to help you arrive at a final solution.

Once the final design solution has been agreed, you will require a detailed set of drawings, diagrams, component lists, and assembly instructions. Your drawings will incorporate, or be supplemented by, design notes that explain how the design operates, and how it is to be manufactured. Where numerical values are included as part of the proposed solution, you will need to demonstrate how you applied your scientific and mathematical understanding to arrive at the design values. For example, if you are designing an electric water pump, you will need to relate the energy input and motor efficiency to the head of water that the pump will raise. Your design calculations need to be checked carefully for accuracy and validity.

Finally, you will need to communicate your final design solution using the drawings and diagrams that you have prepared together with the component and parts lists and other written information. Depending on your design solution you may need to produce any or all of the following types of document:

- freehand sketches
- general arrangement drawings
- detail drawings
- circuit diagrams
- flow diagrams
- schematic diagrams
- exploded views
- component/parts lists.

3.5 Manufacturing a prototype

Manufacturing a prototype involves interpreting the design information in order to produce a working model of the engineered product or service. The manufacture of a prototype usually involves a number of discrete stages, as follows:

- identify the main parts, sub-assemblies and components used in the product and check these against the component lists or parts lists
- identify materials and processes for each component part
- identify those component parts that will need to be manufactured and those that can be purchased
- identify sources for all component parts that will be purchased and arrange for their purchase or supply
- decide upon the sequence of manufacture/assembly
- ensure that the necessary tools and processes are available for manufacture
- manufacture and check each part for dimension, value, tolerance, etc.
- assemble and test the product
- carry out a detailed performance measurement and compare the measured specifications with the original design specification.

The materials, processes and techniques likely to be used for manufacturing the prototype were described in detail in Unit 1. Whilst manufacturing the prototype it is well worth maintaining a record of the processes used and the sequence of operations. This information may prove to be valuable later on when the product goes into production.

Finally, it is important to stress that the prototype is a 'first attempt' at producing a representative working product. In most practical engineering situations a number of prototypes will usually be developed and each one may incorporate different (and hopefully improved) features. When several prototypes are produced, it is important to ensure that any differences in performance can be accounted for. The aim is usually that of achieving a consistency of performance and this means that the product should be repeatable (in other words, the design specification should be achieved each and every time).

3.6 Project presentation

At the end of this unit you will have to explain your final design solution to other people. Your presentation must:

- give reasons for your final choice that refer to the key features in the design brief and your design specification
- show details of your final design idea
- give an explanation of how your final design solution meets the client design brief
- respond to feedback, checking against the design criteria and suitability for the user, and modify your proposed solution, if necessary.

When you work on a design project it's a good idea to keep a *design portfolio* (see Activity 3.12) containing all of the notes, sketches and drawings that you use at each phase of the design process. A design portfolio should contain notes, sketches and drawings showing what was done, why it was done, when it was done, and who did it. The design portfolio will become invaluable when it comes to presenting your design solution to other people and it should normally contain:

A description of the problem
The problem is the task set (or the need identified) by the client (or that you have identified for yourself. This section should be quite brief and will normally just be a few sentences that describe the problem. However, in some cases you may wish to add some sketches or drawings to enhance or clarify your description.

A statement of the design brief
The design brief (or *client brief*) states what needs to be done to solve the problem. It should be written from the perspective of the client. In other words, it should say what the client is looking for. You will normally wish to agree the design brief with your client so it's important to get the wording right! Once again, this section is usually just text but here again, there may be occasions when you might wish to use a drawing or a sketch to clarify and enhance your wording.

Results of the research and investigation
You will need to summarise the results of your research and investigation using charts, diagrams, tables and any other method that helps convey the results to your readers. Since charts and diagrams can usually be easily understood this is often the best way to represent the outcome of your research and investigation. One important point is that you should always quote the source of any data that you use. You should also provide a copy of any market research that you used.

The investigation may also consider other aspects such as identifying primary and secondary markets for the product or service, existing products and service that may be similar or competing in the same market, materials and manufacturing constraints, ergonomics and aesthetics, issues relating to standards and Health and Safety.

A summary of the candidate solutions and the process that you went through to generate them
List each of the candidate solutions that you arrived at and explain how you arrived at them. Give brief details of any brainstorming session that you held or include a mind map if you used one.

The design criteria and the design specification
Include the design criteria that you established as well as the detailed design specification.

The candidate solutions
You should describe the design ideas for each of our candidate solutions and the process by which you arrived at the final design solution. If you used an evaluation matrix this should be included here. You should include details and sketches of the ideas that you rejected. Although these ideas may not be taken forward, they give an idea of the process that you went through and, at some time in the future, you or your client may want to return to them!

The final design solution
You will need to describe your final design solution in detail. To do this, you will need to use detailed sketches and drawings. These will help to convey your ideas to the client and they may also assist with marketing the product or service that you have designed.

Your working and presentation drawings
Your working drawings must be precise and drawn to scale. They must give specific information about dimensions and the materials that should be used. You may wish to include drawings of individual parts as well as details showing how your product should be assembled. Dimensions should be clearly indicated and all other relevant information should be included. One or more presentation drawings should be included to show what the finished product will look like. You should choose an appropriate drawing technique for this (perspective, isometric, oblique, 3D rendered CAD etc).

Information relating to manufacturing the product or supplying the service
You will need to provide details showing how you envisage that your product will be manufactured or how your service will be delivered. For example, you may need to supply a parts list, a component list or a cutting diagram. Much will depend on the individual product or service. Finally, you should provide a simple step-by-step explanation of what needs to be done to assemble or construct your designed product. A series of sketches might be useful here!

A final evaluation of your work
Your design folder will not be complete without a full evaluation of the work that you have done. You should also include comments and feedback that you received from your client as well as information on any modifications that were incorporated as the work went on. In many design projects your client will insist on holding a regular review meeting in order to discuss progress with the project. If your client does not ask for this, you might find it useful to organise a presentation of your work or provide a brief written report at various stages of the project.

You need to justify your final design solution by demonstrating that you have been able to comply with each of the essential design criteria as well as meeting the design specification that you agreed with your client. You also need to say how you complied with any relevant standards or legislation.

You need to evaluate your work with a critical eye and indicate where there are any particular strengths or weaknesses. You should also provide full details of any testing or measurement that you carried out. If your measurements did not confirm that you have met the design specification you need to suggest why this is and what should be done to ensure that the design specification is met.

Finally, you need to ask yourself whether your final design solution meets your client's needs and expectations. You should also comment on any improvements or modifications that could be made to improve the product or service or make it more cost effective.

Delivering the presentation

Your formal presentation should be a verbal presentation supported by appropriate visual aids (for example, a PowerPoint presentation or overhead projector transparencies) or a brief *technical report* (included sketches and other visual aids). In either case, your presentation should be delivered in a way that is appropriate to your audience. This will invariably mean that you should keep your presentation brief and to the point. At the same time, you should ensure that you have covered all the main points that make up your design solution. In any event, your presentation must be interesting and appropriately paced so that the attention of the audience does not wander. If you are delivering a verbal presentation it's also important to be a good listener and be able to respond to any questions or queries that are raised by your audience.

Fortunately, much of the material that you need to include in your verbal presentation will already exist. All you need to do is to summarise it and present it in a way that your audience can quickly and easily grasp. In most cases, you will wish to supply your audience with notes or printed *handouts*. These can be based on PowerPoint screens of overhead projector transparencies and can be augmented with sketches and presentation drawings, as appropriate.

If you choose to include a brief technical report, this can be assembled from the material that you have collected as you worked through the various stages of the project. Typical section headings might include:

Summary
A brief overview for busy readers who need to quickly find out what the report is about

Introduction
This sets the context and background and provides a brief description of the problem that you have solved and may also include a statement of the design brief

Main body
A comprehensive description of the design solution including how and why it was chosen together with details of the research and investigation that you carried out. You should also include (and comment on) the design criteria and the final design specification. Your design solution should be presented together with sketches and drawings.

Evaluation
A detailed evaluation of the work that you carried out including any

problems that you encountered and how you solved them.

Recommendations
This section should provide information on how the design solution should be implemented (including, for example, information on manufacture or assembly). You may also wish to include any modifications or changes that you would recommend.

Conclusions
You should end your report with a few concluding remarks about your design solution and how effective you think it's likely to be at solving your client's problem.

References
This section should provide readers with a list of sources for further information relating to the scientific principles or technology used, including (where appropriate) relevant standards and legislation.

Finally, it's important to take care when you present your work in the form of a written report. To avoid confusion, the normal conventions of grammar and punctuation must be used. Words must be correctly spelt so do make use of a dictionary if you are uncertain about spelling. Never use jargon terms and acronyms unless you are sure that your audience will understand them. Layout is important so use numbered sections and paragraphs and try to keep your sentences short and to the point.

Activity 3.12

The assessment of this unit is based on the design, development, prototype manufacture and testing, and evaluation of your engineered product. Your work should be submitted in the form of a portfolio which contains evidence of the work that you have carried out. In addition, your teacher or tutor will provide you with a 'witness statement' detailing the quality of your oral presentation. This final activity takes the form of a 'checklist' for the contents of your portfolio. You should check that each item is present before you submit your design portfolio:

1. Details of the original client brief.
2. The needs analysis and list of requirements.
3. The final design specification.
4. Alternative solutions considered.
5. The final design solution (and why it was chosen).
6. Detailed engineering drawings, diagrams and schematics.
7. Evidence of project planning (e.g. a Gantt chart).
8. Details of prototype manufacture and assembly.
9. Photographs of the completed prototype.
10. Details of testing, evaluation and recommendations for any modifications necessary prior to production.
11. A copy of your presentation materials (including any notes or technical report).
12. This list presented as a 'contents' sheet.

Review questions

1 Summarise each of the main stages in the design and development of an engineering product.

2 Explain the function of a *design specification.*

3 Give an example of a typical design specification for an engineered product. Include typical figures in your answer.

4 Explain the function of a *block diagram.*

5 Give an example of a typical block diagram for an engineered product and label your drawing clearly.

6 Explain the function of a *flow diagram.*

7 Give an example of a typical flow diagram and label your drawing clearly.

8 Explain, with the aid of a sketch, the relationship that exists between A5, A4 , A3 and A2 paper.

9 List THREE examples of schematic diagrams and explain where each is used.

10 With the aid of a sketch, explain how drawing zones are identified on a formal engineering drawing.

11 List FIVE items that usually appear in the title block of a formal engineering drawing.

12 Explain briefly what is meant by:

(a) a General Arrangement (GA) drawing
(b) a detail drawing.

13 Sketch the conventional line style used in an engineering drawing to represent:

(a) the limit of a partial view
(b) hidden edges and outlines
(c) centre lines
(d) extreme positions of movable parts.

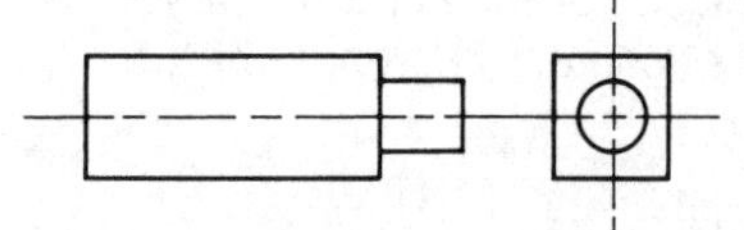

Figure 3.93 *See Question 14*

14 Identify the projections shown in Figures 3.93 and 3.94.

15 Identify the line types marked A and B in Figure 3.95.

16 Identify each of the components shown in Figure 3.96.

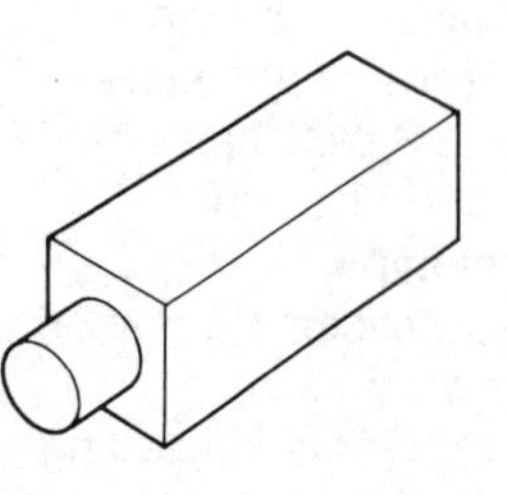

Figure 3.94 *See Question 14*

17 A flat square steel plate has sides measuring 220 mm. The plate has a round hole with a diameter of 110 mm cut in its exact centre. Use standard drawing conventions to show an orthographic view of this component. Label your drawing and include dimensions.

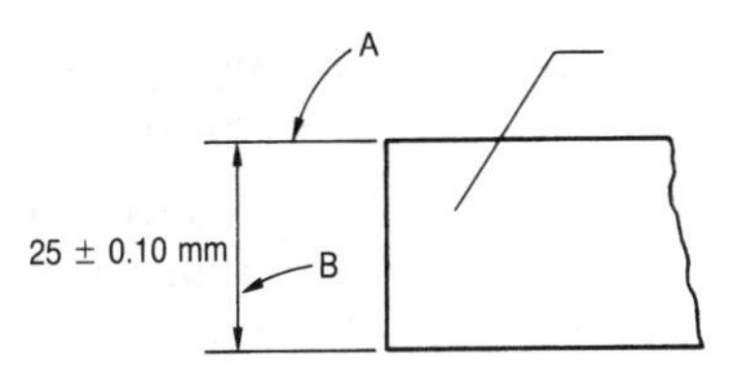

Figure 3.95 *See Question 15*

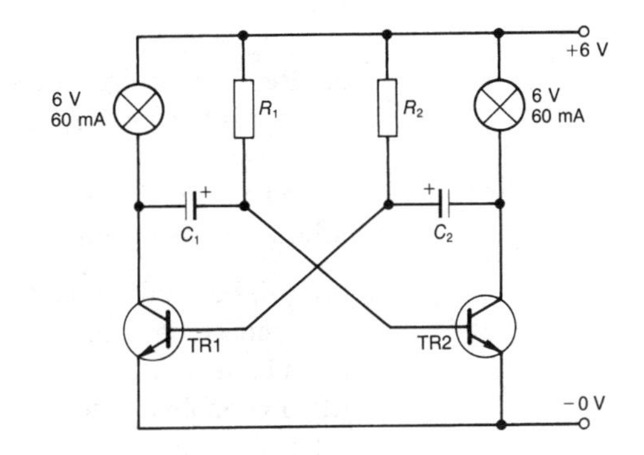

Figure 3.96 *See Question 16*

18 Refer to the drawing shown in Figure 3.97. What kind of drawing is this and what does the hatched area show?

19 Refer to the drawing shown in Figure 3.98. Which one of the styles shown is correct for use in a formal engineering drawing?

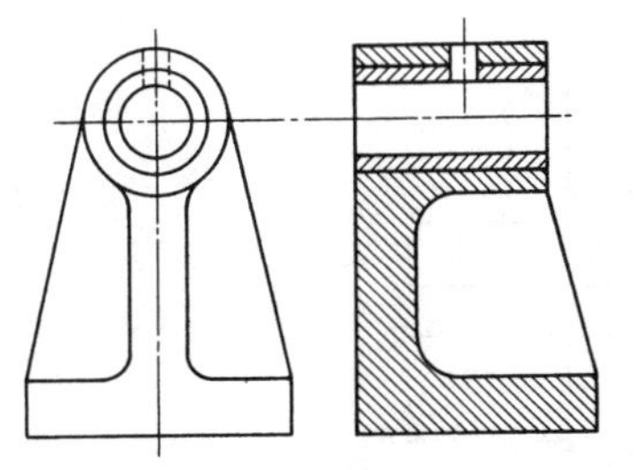

Figure 3.97 *See Question 18*

(a) Material: Steel
Grind 0.5

(b) MATERIAL: STEEL
GRIND 0.5

(c) *Material: Steel*
Grind 0.5

(d) MATERIAL: STEEL
GRIND 0.5

Figure 3.98 *See Question 19*

20 An engineering component comprises a metal alloy cube having sides measuring 100 mm. Draw this component using:

(a) cavalier oblique projection
(b) cabinet oblique projection
(c) isometric projection.

21 Refer to the drawing shown in Figure 3.99. What type of projection has been used for the drawing and explain, briefly, how it was constructed.

22 Refer to the drawing shown in Figure 3.100. Redraw the component using third-angle projection.

23 Sketch engineering drawing symbols that are used to indicate the following components:

(a) a 4/2 directional control valve
(b) a non-return valve
(c) a battery
(d) a variable resistor
(e) a semiconductor diode
(f) an iron-cored transformer.

24 Describe, with the aid of an example, the contents of a typical *design brief.*

25 Describe TWO methods of generating ideas. In what circumstances might these methods be useful as part of the design process?

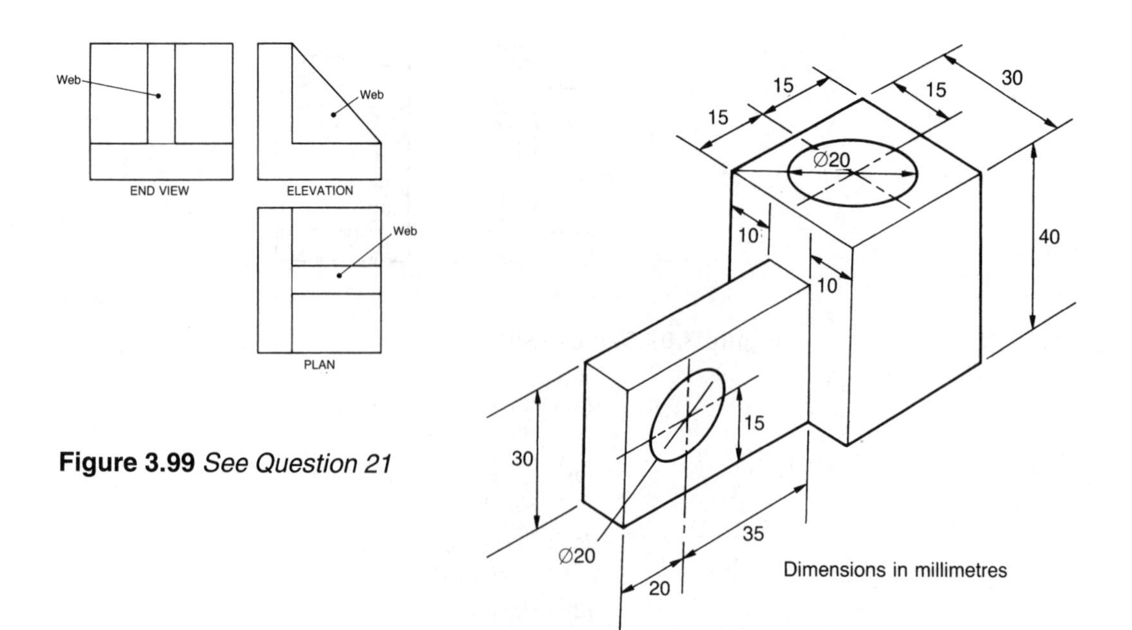

Figure 3.99 *See Question 21*

Figure 3.100 *See Question 22*

Unit 4 Applied engineering systems

Summary

In this unit you will look at ways in which engineering techniques and principles are applied in engineering systems and how a systems approach can be used to solve a wide variety of engineering problems. The systems that you will investigate in this unit range from the structural systems that you might find in civil and mechanical engineering to the highly sophisticated electronic and microelectronic systems used in control and instrumentation. The unit is externally assessed by a series of practical activities. These will involve you putting into practice what you have learned from the unit as well as from the additional research and investigation that you will carry out in preparation for the external assessment.

4.1 Static structural systems

Systems are all around us. Indeed, we use the term 'system' so liberally and in such a wide range of contexts that it is difficult to explain what the term means in just a few words. However, at the risk of over-simplification, we can say that every system has:

- a function or purpose
- a number of inputs and outputs
- a boundary which defines what is inside the system and what lies outside it
- a number of smaller components or elements linked together in a particular way.

Inside a system, energy may be converted from one form to another and signals or commands passed from one element to another. We start our study of engineering systems by investigating the systems that are used to build framed structures like bridges, roofs, and cranes.

Force, mass and weight

Force, mass, weight and density are important in many engineering applications and it's quite likely that you already have some idea of what these terms mean. However, do you understand the difference between mass and weight? And, how would you measure a force? This section is designed to help you get up to speed with these important concepts.

Force

A force is a push or pull exerted by one object on another. If the object remains in *equilibrium* (i.e. if it doesn't move or change in any way) then, for each force acting on the object there is another equal and opposite force that acts against it. The force that is applied to an object (or a *body*) is often called an *action* whilst the opposing force is referred to as a *reaction*. As long as the object (or *body*) doesn't move or change in any way, action and reaction will be equal and opposite.

It also follows that, if the forces of action and reaction acting on an object are not equal and opposite, the object will move or change in some way. You can test this theory out very easily by finding a wall and pushing against it. If the wall doesn't move (hopefully it won't!) then you will experience a force pushing back. If you increase the force that you apply to the wall (the *action*) the force pushing back (the *reaction*) will also increase by the same amount.

Now try the same experiment by pushing against a door that is partially open. There will still be some force exerted back by the door but this will be much less than the force that you apply. Because of this imbalance of forces (action being greater than reaction) the door will move and will swing open. This simple experiment leads us to the following conclusions:

- When a body is at rest (or in *equilibrium*) the *action* and *reaction* forces acting on it will be equal and opposite.
- If the action and reaction forces acting on a body are not equal and opposite a change (in this case *motion*) will be produced.

Mass and weight

The mass of a body is defined as the *quantity of matter* in the body. It's important to be aware that the mass of a body remains the same regardless of where the body is. So, for example, a mass of 50 kg will be the same on the surface of the Earth as it will be in outer space (where there is 'zero gravity').

The weight of a body is determined by its mass and the gravitational force acting on the body. So, if there is no gravitational force (for example, in outer space) then a body will have no weight! However, in most practical cases we are concerned with what things weigh on the surface of the Earth in which case the relationship between mass and weight is given by:

$$W = m\,g$$

where W is the weight in Newton (N), m is the weight in kg, and g is the gravitational acceleration (in m/s^2). On the surface of the Earth g is a constant equal to 9.81 m/s^2.

The weight of a body decreases as the body is moved away from the centre of the Earth. Weight obeys the inverse square law. This simply means that weight is inversely proportional to the square of the distance from the centre of the Earth. In other words:

$$W \propto \frac{1}{d^2}$$

where W is the weight (in N) and d is the distance from the centre of the Earth (in m).

Finally, it is essential to remember that mass and weight are not the same thing! The mass of an object remains the same wherever it is whereas, the weight of a body is determined by the product of its mass and the gravitational force that is acting on it.

Example 4.1

A light alloy beam has a mass of 17.5 kg. Determine the weight of the beam.

Here we will assume that the beam is being used at the Earth's surface. In which case:

$W = m\,g = 17.5 \times 9.81 =$ **171.68 N**

Example 4.2

A lunar lander weighing 8.25 kN on Earth, is to be used on a mission to explore the surface of the Moon. Given that the gravitational acceleration on the moon is one sixth (0.16) of that on Earth, determine the weight of the lander on the surface of the Moon.

The weight of the lander will be reduced in direct proportion to the reduction in gravitational acceleration. Hence the lander will weigh 0.16×8.25 kN = **1.38 kN** on the surface of the Moon.

Density

The density of a body is defined as the mass per unit volume. In other words, the density of an object is found by dividing its mass by its volume. The density of a particular material is a fundamental property of that material. Expressing this as a formula gives:

$$\rho = \frac{m}{V}$$

where ρ is the density in kg/m^3, m is the mass in kg, and V is the volume in m^3.

We sometimes express the density of an object relative to that of pure water (at 4°C). The density of water under these conditions is 1,000 kg/m^3. The densities (and relative densities) of various engineering materials are shown in Table 4.1.

Table 4.1 *Density of various materials*

Material	*Density (kg/m^3)*	*Relative density*
Aluminium	2,700	2.7
Brass	8,500	8.5
Cast iron	7,350	7.35
Concrete	2,400	2.4
Copper	8,960	8.96
Glass	2,600	2.6
Mild steel	7,850	7.85
Wood (oak)	690	0.69

Example 4.3

Determine the mass of an aluminium block which has dimensions 50 mm × 110 mm × 275 mm.

The total volume of the aluminium block will be given by:

$V = (50 \times 10^{-3}) \times (110 \times 10^{-3}) \times (275 \times 10^{-3}) = (1.5125 \times 10^{6}) \times 10^{-9}$

From which, $V = \mathbf{1.5125 \times 10^{-3}\ m^3}$

Re-arranging the formula for density to make m the subject gives:

$m = \rho \times V$

From the table, the value of ρ for aluminium is 2,700 kg/m^3 hence the mass of the block will be given by:

$m = 2{,}700 \times (1.5125 \times 10^{-3}) = 4{,}083.75 \times 10^{-3}$ = **4.08 kg**

Test your knowledge 4.1

A sample of a metal alloy consists of a cube with sides of length 50 mm that weighs 4.5 N. Determine the density of the material.

Force diagrams

Every force has three important properties that are used to describe it. These properties are:

- size (or *magnitude*) (see Figure 4.1a)
- direction (see Figure 4.1b)
- point of application (see Figure 4.1c).

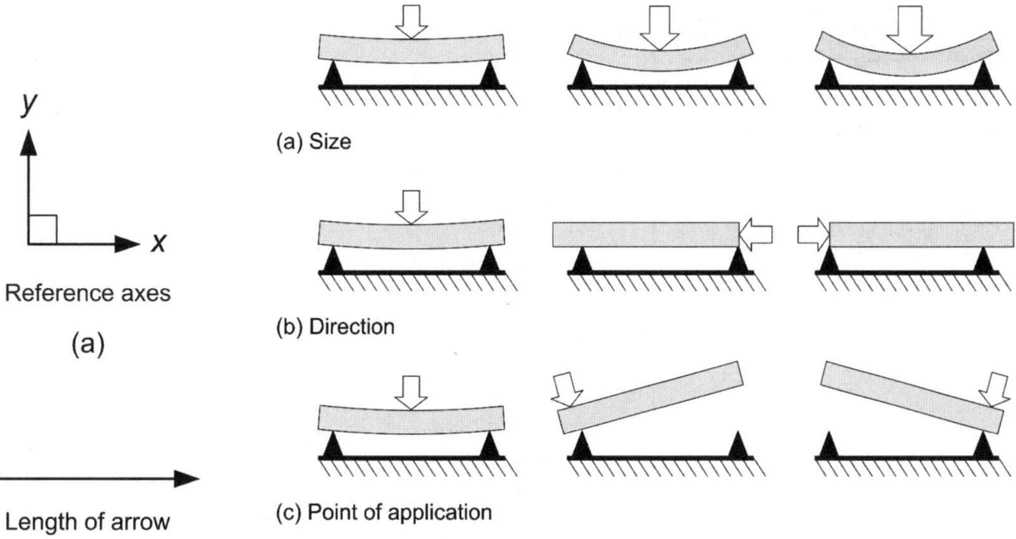

Figure 4.1 *Properties of any force*

Length of arrow indicates magnitude

(b)

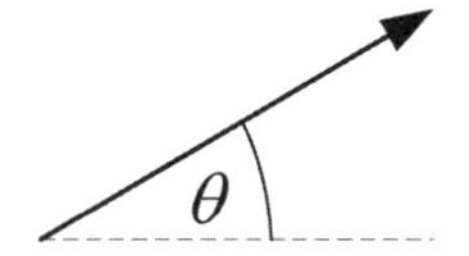

Direction, θ (relative to the x-axis)

(c)

Figure 4.2 *Diagrammatic representation of a force*

Engineers frequently use diagrams to show the effect of forces and also to help solve problems involving a number of forces acting at the same time. In order to specify the direction of a force we use a set of references axes (see Figure 4.2a). The horizontal direction is generally referred to as the x-axis whilst the vertical direction is generally known as the y-axis. Note, however, that reference axes are something that we have introduced for our own convenience and thus we need not be constrained to any particular orientation.

The size (or *magnitude*) of force is indicated by its length (see Figure 4.2b) whilst its direction (usually specified relative to the x-axis) is indicated by the angle, θ, as shown in Figure 4.2(c).

A single force, F, can be resolved into two components acting at right angles, $F \sin \theta$ and $F \cos \theta$ (as shown in Figure 4.3). Note that we have included the reference axes in this diagram although these are usually not shown in force diagrams.

Just as we can resolve a single force into two components acting at right angles we can also find the one single force (the *resultant*) of two forces acting at right angles, as shown in Figure 4.5

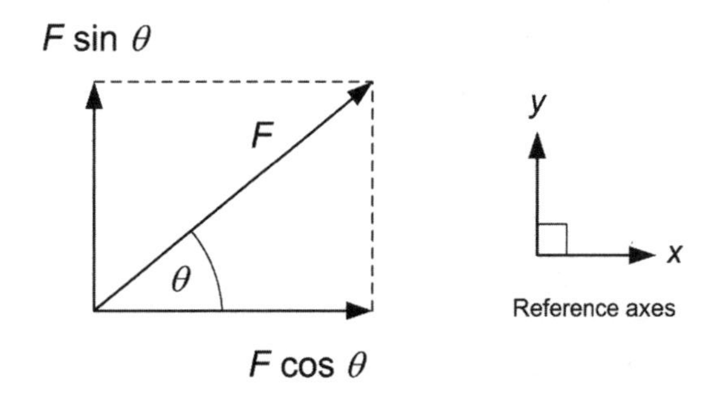

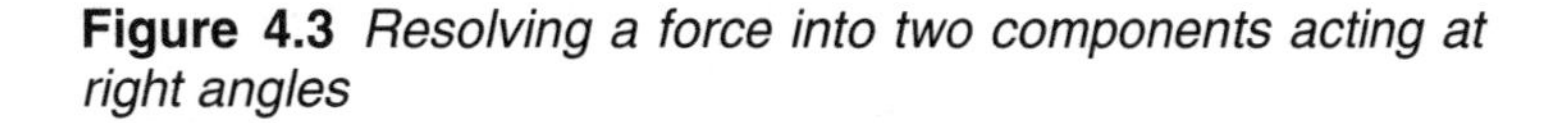

Figure 4.3 *Resolving a force into two components acting at right angles*

Another view

The book's companion website (see page ix) has a spreadsheet tool for solving simple force diagrams. You may find this a useful means of checking your results or experimenting with other values!

Example 4.4

Determine the horizontal and vertical components of a 20 N force acting at a direction of 60° to the horizontal.

The force diagram is shown in Figure 4.4. The 20 N force has been shown acting at 60° to the *x*-axis.

The horizontal component of the force, *Q*, will be given by:

$Q = 20 \times \cos 60° = 20 \times 0.5 =$ **10 N**

Similarly, the vertical component of the force, *P*, will be given by:

$P = 20 \times \sin 60° = 20 \times 0.866 =$ **17.32 N**

Hence a 20 N force acting at 60° to the horizontal can be replaced by two forces of 10 N and 17.32 N acting at right angles.

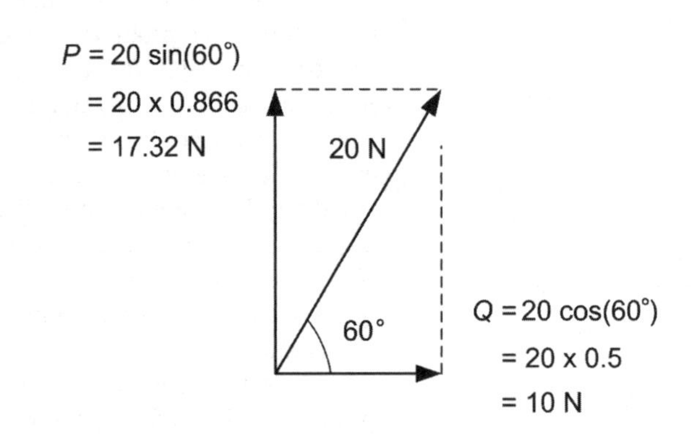

Figure 4.4 *See Example 4.4*

Test your knowledge 4.2

A force, *F*, of 20 N acts at 27° to the horizontal. Find the horizontal and vertical components of this force.

Test your knowledge 4.3

A force, *F*, acts at 75° to the horizontal. If the horizontal component of this force is 1.5 kN, determine the value of *F*.

Test your knowledge 4.4

A force, *F*, acts at 15° to the horizontal. If the vertical component of this force is 75 N, determine the value of *F*.

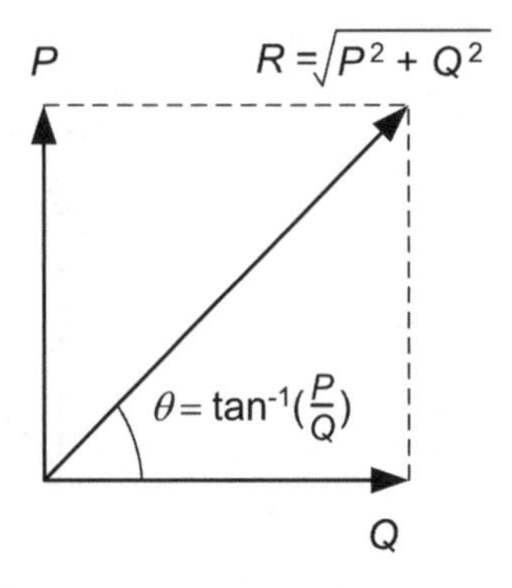

Figure 4.5 *Determining the resultant of two forces acting at right angles*

Test your knowledge 4.5

Forces of 5 N and 12 N act at right angles to one another. Determine the magnitude and direction of the resultant relative to the 5 N force.

Example 4.5

Determine the resultant of two forces, 3 N and 4 N acting at right angles to one another.

The force diagram is shown in Figure 4.6. The 4 N force has been shown acting along the *x*-axis.

The magnitude of the resultant can be calculated from:

$$R = \sqrt{3^2 + 4^2} = \sqrt{25} = 5\ \text{N}$$

The angle between the resultant and the *x*-axis can be found from:

$$\theta = \tan^{-1}\left(\frac{3}{4}\right) = \tan^{-1}(0.75) = 36.7°$$

Hence the resultant is 5 N acting at 36.7° to the 4 N force, as shown in Figure 4.6.

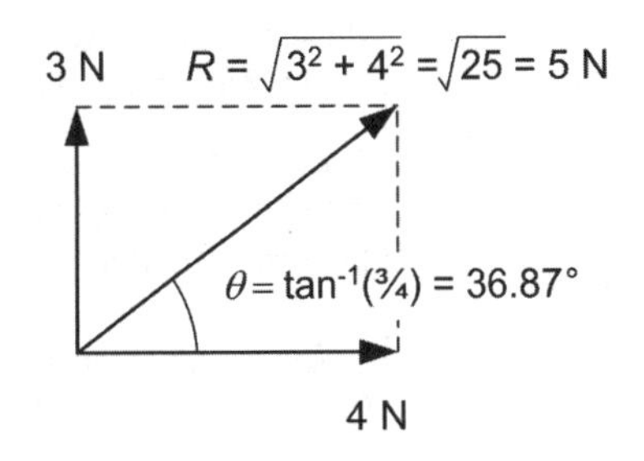

Figure 4.6 *See Example 4.5*

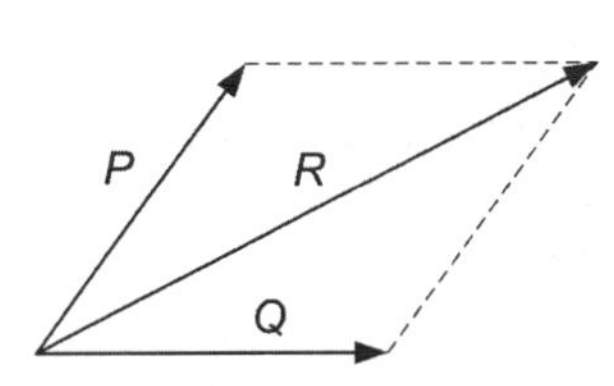

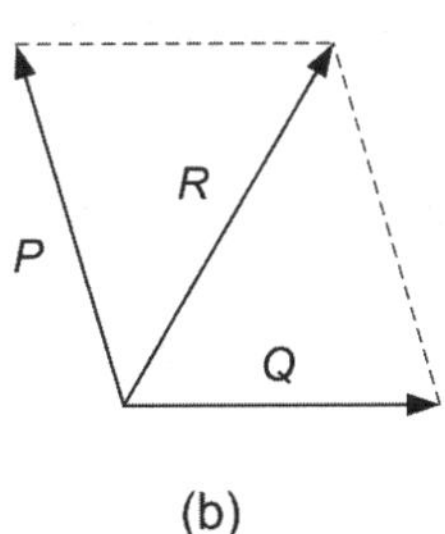

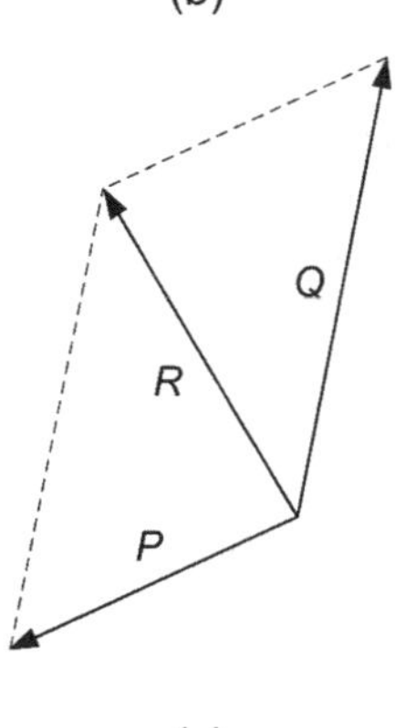

Figure 4.7 *Examples of the parallelogram of forces*

So far, we have considered the case when two forces are at right angles to one another. When this is not the case we can determine the resultant force by constructing a *parallelogram of forces*, as shown in Figure 4.7.

In Figure 4.7, the resultant, R, of the two forces P and Q is determined by constructing the diagonal, R. The magnitude and direction of R (relative to one of the other two forces) can be found by scale drawing or by calculation as follows:

1. Resolve each of the known forces (P and Q) into their horizontal and vertical components (see Example 4.4)
2. Find the total horizontal and total vertical components (taking into account their direction)
3. Determine the magnitude and direction of the total horizontal and total vertical force components (see Example 4.5).

This method may seem a little long-winded but it will usually produce a more precise answer than can be achieved by means of scale drawing.

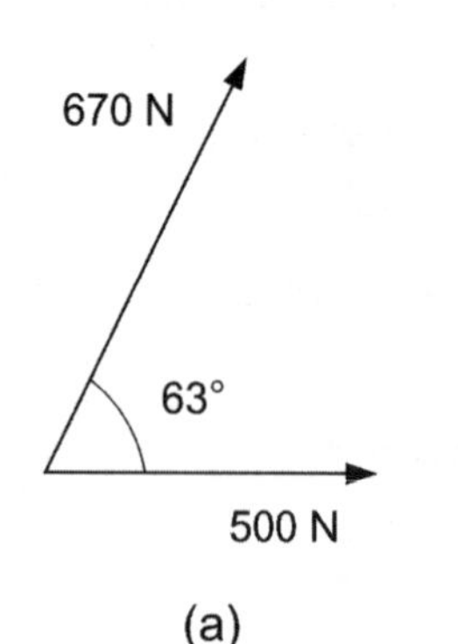

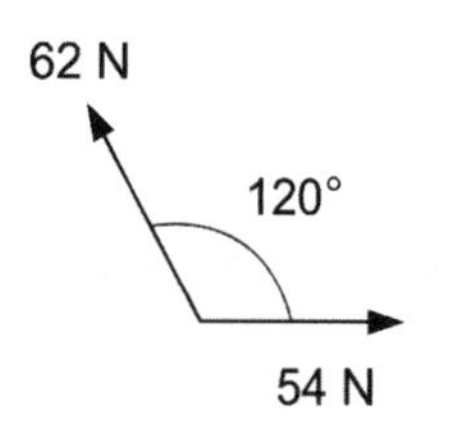

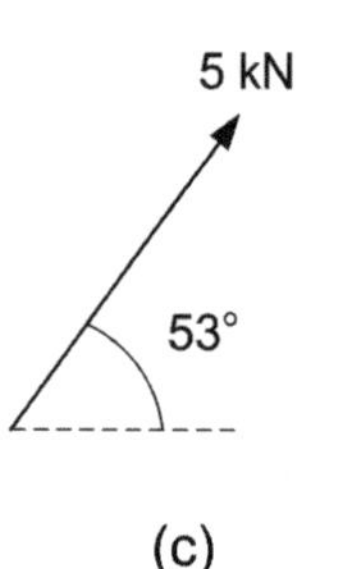

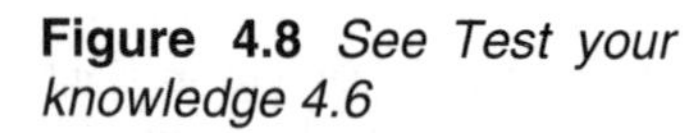

Figure 4.8 *See Test your knowledge 4.6*

Example 4.6

Figure 4.9 shows the arrangement of a small crane with a jib, BC, and a steel cable, AC. Determine the forces acting in the cable and in the jib when the crane carries a load of 20 kN.

The force diagram (i.e. *triangle of forces*) is shown in Figure 4.10 where *W* is the weight of the load, *T* is the thrust in the jib, and *P* is the tension in the cable.

In order to solve this problem we shall use a graphical method (see Figure 4.11). This requires the use of some squared paper, a protractor, a rule, and a drawing pencil. The drawing is constructed to a convenient scale (in this case 1 cm = 2 kN) and the steps are as follows:

1. First draw a line to represent the load, *W* (20 kN). This line must be 10 cm in length and it must be aligned vertically.

2. Construct an angle of 60° at the base of the vertical line and then draw a straight line to represent the force in the jib, *T*. Project this line towards the top edge of the drawing paper.

3. Construct an angle of 33° at the top of the base of the vertical line and then draw a straight line to represent the force in the steel cable, *P*. Project this line towards the top edge of the drawing paper until it meets the line that represents the force in the jib.

4. Locate the point of intersection between the two lines and then measure their lengths. Convert these lengths (using the scaling factor) to force in Newton.

5. The length of the line representing *P* is 11 cm. This indicates that the force in the cable is **22 N**.

6. The length of the line representing *T* is 18.5 cm. This indicates that the force in the jib is **37 N**.

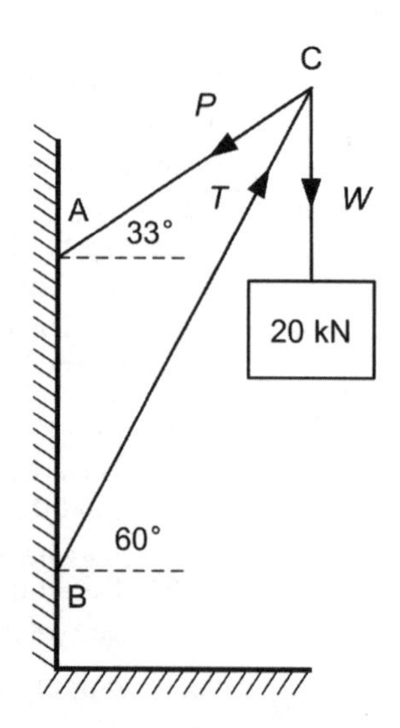

Figure 4.9 *See Example 4.6*

Test your knowledge 4.6

1. Determine the resultant of the two forces shown in Figures 4.8 (a) and 4.8(b).

2. Determine the horizontal and vertical components of the force shown in Figure 4.8(c).

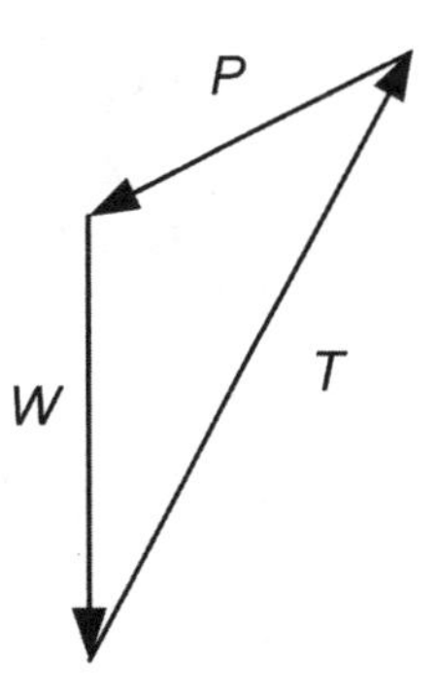

Figure 4.10 *Triangle of forces, see Example 4.6*

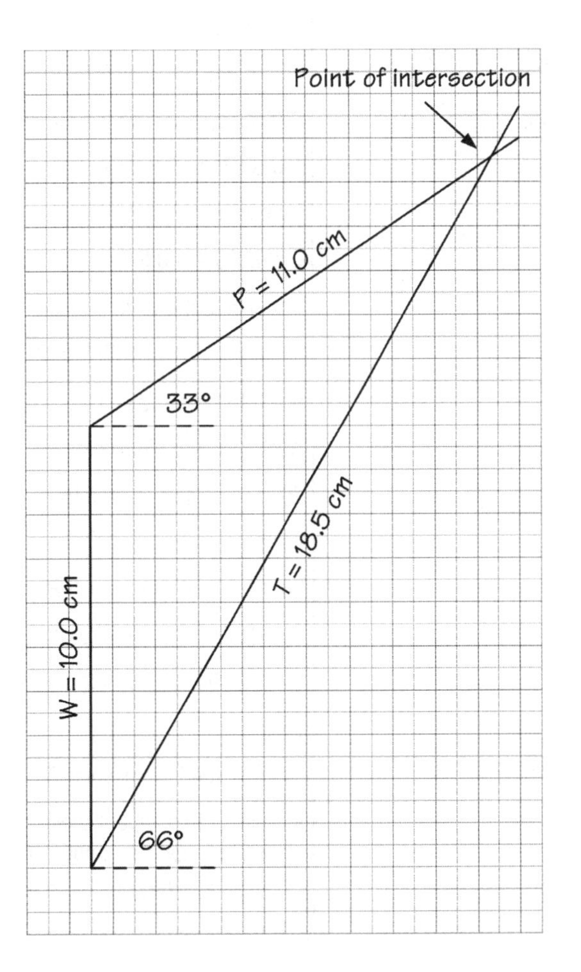

Figure 4.11 *Scale drawing, see Example 4.6*

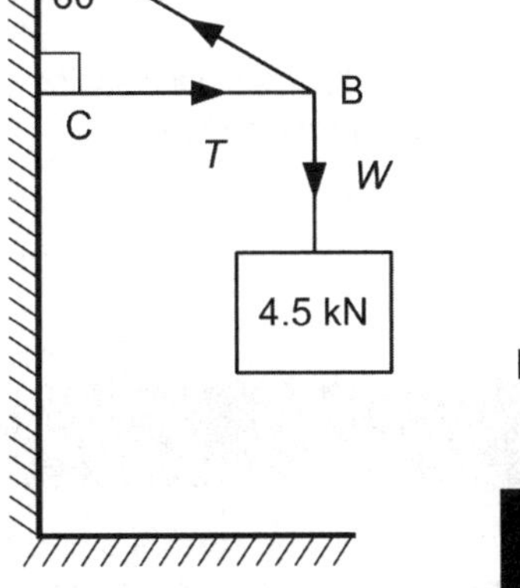

Figure 4.12 *See Activity 4.1*

Activity 4.1

Use a graphical method to determine the forces shown in Figure 4.12.

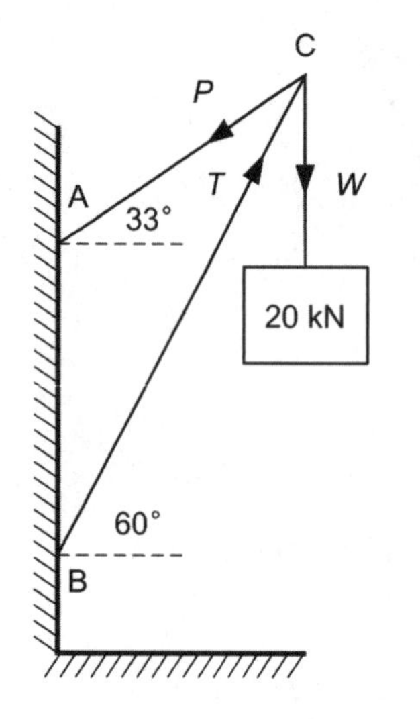

Figure 4.13 *See Example 4.7*

Example 4.7

Figure 4.13 shows the arrangement of a small crane with a jib, BC, and a steel cable, AC. Determine the forces acting in the cable and in the jib when the crane carries a load of 20 kN.

This time we will use a non-graphical method based on resolving the forces in the vertical and horizontal directions. The resultant of the forces in both the horizontal and vertical directions must be zero because the system of forces is in *static equilibrium* (i.e. it isn't in motion).

Firstly, resolving the forces into their vertical components gives:

$20 = T \cos 30° - P \cos 57°$ (downward force = upward force)

from which:

$20 = 0.866\ T - 0.545\ T$ (i)

Secondly, resolving the forces into their horizontal components gives:

$P \cos 33° = T \cos 60°$ (leftward force = rightward force)

from which:

$0 = 0.839\ T - 0.5\ T$ (ii)

Equations (i) and (ii) need to be solved simultaneously, as follows:

Multiplying (ii) by 1.732 (or 2×0.866) gives:

$0 = 1.453\ P - 0.866\ T$ (iii)

Adding (i) and (iii) gives:

$20 + 0 = 1.453\ P - 0.545\ P$ (the terms in T are eliminated)

hence:

$20 = 0.908\ P$ from which $P =$ **22.06 N**

Substituting for P in equation (iii) gives:

$0 = (0.839 \times 22.06) - 0.5\ T$

thus:

$0.5\ T = 18.48$ from which $T =$ **36.9 N**

It is useful to compare these calculated values with those found earlier using the graphical method. Which method do you feel is more accurate and why is it more accurate?

Activity 4.2

Use a non-graphical method to determine the forces shown in Figure 4.12. Compare your answers with those that you obtained in Activity 4.1. Give reasons for any significant discrepancy in your answers.

Pressure

Pressure is exerted whenever a force is applied to an object such as a floor, wall or the inside surfaces of a container. Pressure is defined as the ratio of force (or load) applied perpendicular (i.e. at right angles) to the surface, to the area over which the force (or load) acts. Thus:

$$P = \frac{F}{A}$$

where P is the pressure in Pascal (Pa), F is the force (in N), and A is the area (in m^2).

Test your knowledge 4.7

1. A vehicle chassis has a mass of 475 kg. Determine the weight of the chassis.
2. A mild steel girder has a mass of 560 kg and is supported by two concrete pillars each having a cross-sectional area of 0.044 m^2. Assuming that the load is distributed evenly, determine the pressure exerted on each of the two pillars.

Example 4.8

A lathe weighs 425 kN. Determine the pressure exerted on the workshop floor if the load is distributed over a surface area of 0.875m^2.

Now

$$P = \frac{F}{A} = \frac{425 \times 10^3}{0.875} = \mathbf{485.7\ kPa}$$

Stress and strain

No solid body is perfectly rigid and, when forces are applied to it, changes in dimensions occur. Because they are usually very small, such changes are usually imperceptible. For example, you would not usually expect to be able to see the deflection of a girder bridge when a heavy lorry crosses it. However, deflection does occur due to the weight of the lorry and the forces that this imposes on the structure of the bridge. For engineers, this deflection can be important and steps must be taken when designing structures like bridges in order to ensure that the material is appropriate and will not change its shape or, even worse, fail due to the presence of a load.

The three types of mechanical force that can act on a body are compression (*compressive force*), tension (*tensile force*), and shear (*shear force*), as shown in Figure 4.14.

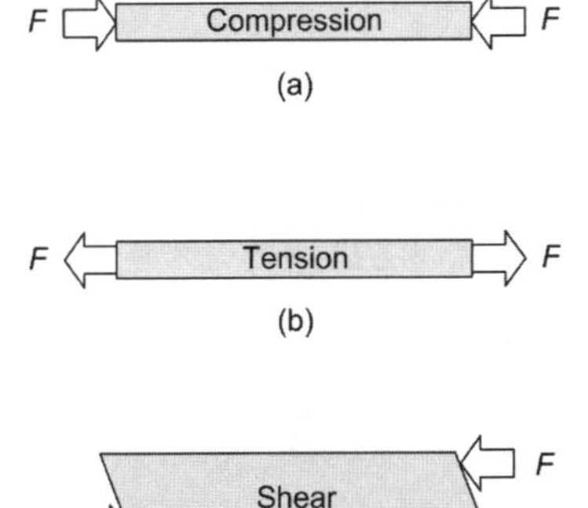

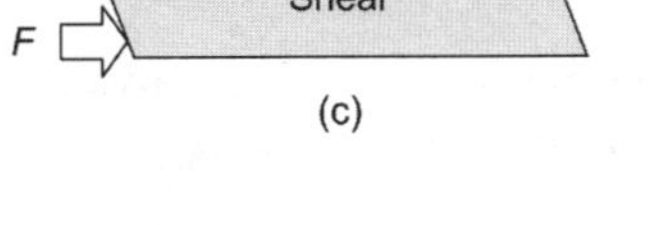

Figure 4.14 *Three types of mechanical force*

Tensile force

Tensile force is a force that tends to stretch (or extend) an object, as shown in Figure 4.14(b). Examples of tensile forces are:

- the force that acts on a rope or cable when carrying a suspended load
- the force that acts on a rubber band when it is stretched
- the force that acts on a guy rope supporting a tent or a flagpole
- the force that acts on a bolt when a nut is tightened
- the force that acts on a tow bar that connects a car to a trailer.

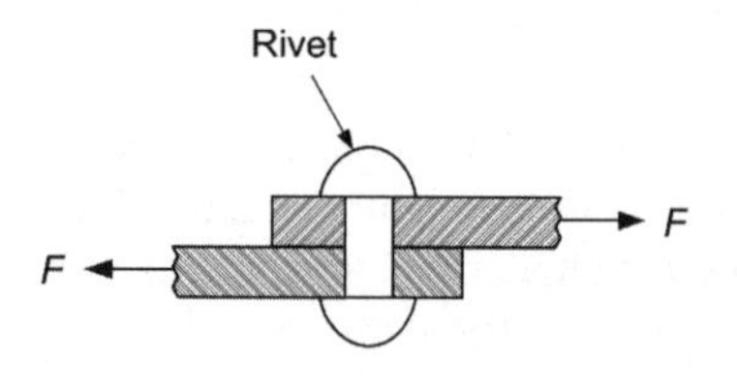

Figure 4.15 *Shear force acting on a rivet*

Test your knowledge 4.8

Give THREE typical examples of each of the following types of force:

(a) Tensile force

(b) Compressive force

(c) Shear force.

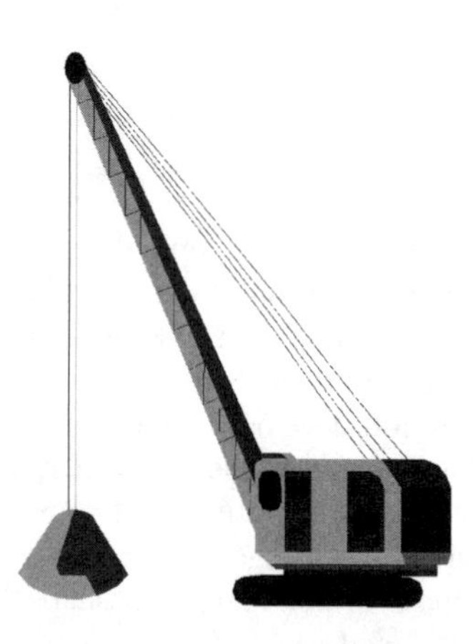

Figure 4.16 *See Test your knowledge 4.9*

Test your knowledge 4.9

Identify TWO examples of forces present in Figure 4.16 that are examples of:

(a) Tensile forces

(b) Compressive forces.

Compressive force

Compressive force is a force that tends to squeeze or crush an object, as shown in Figure 4.14(a). Examples of compressive forces are:

- the force that acts on a pillar supporting a bridge
- the force that acts on the front forks of a bicycle
- the force that acts on the jib of a crane
- the force that acts on a pallet when carrying a load.

Shear force

Shear force is the force that tends to move one face of an object relative to an opposite face, as shown in Figure 4.14(c). Examples of shear force are:

- the force that acts on a rivet (see Figure 4.15)
- the force that acts on a sheet of metal being cut in a guillotine
- the force that acts on a branch being cut by garden shears
- the force that acts on a bolt when struck radially with a mallet.

Stress

When a force acting on an object causes changes in its dimensions the object is said to be under stress. Stress is the ratio of applied force, F, to cross-sectional area, a (see Figure 4.17a). Hence:

$$\text{Stress, } \sigma = \frac{F}{a} \text{ Pa}$$

Note that, for tensile and compressive forces, the area concerned is that which is at right angles to the direction of the force.

Strain

Strain is defined as the fractional change in dimension that occurs when an object is under stress. For an object that has a length, l, that experiences a changes of length, x (see Figure 4.17b), the strain will be given by:

$$\text{Strain, } \varepsilon = \frac{x}{l}$$

Note that strain has no units and is sometimes expressed as a percentage:

$$\text{Percentage strain, } \varepsilon = \frac{x}{l} \times 100$$

Shear stress

For a shear force, the shear strain is determined by the force divided by the area parallel to the direction of the force (bd in Figure 4.17c). Hence:

$$\text{Sheer stress, } \tau = \frac{F}{bd} \text{ Pa}$$

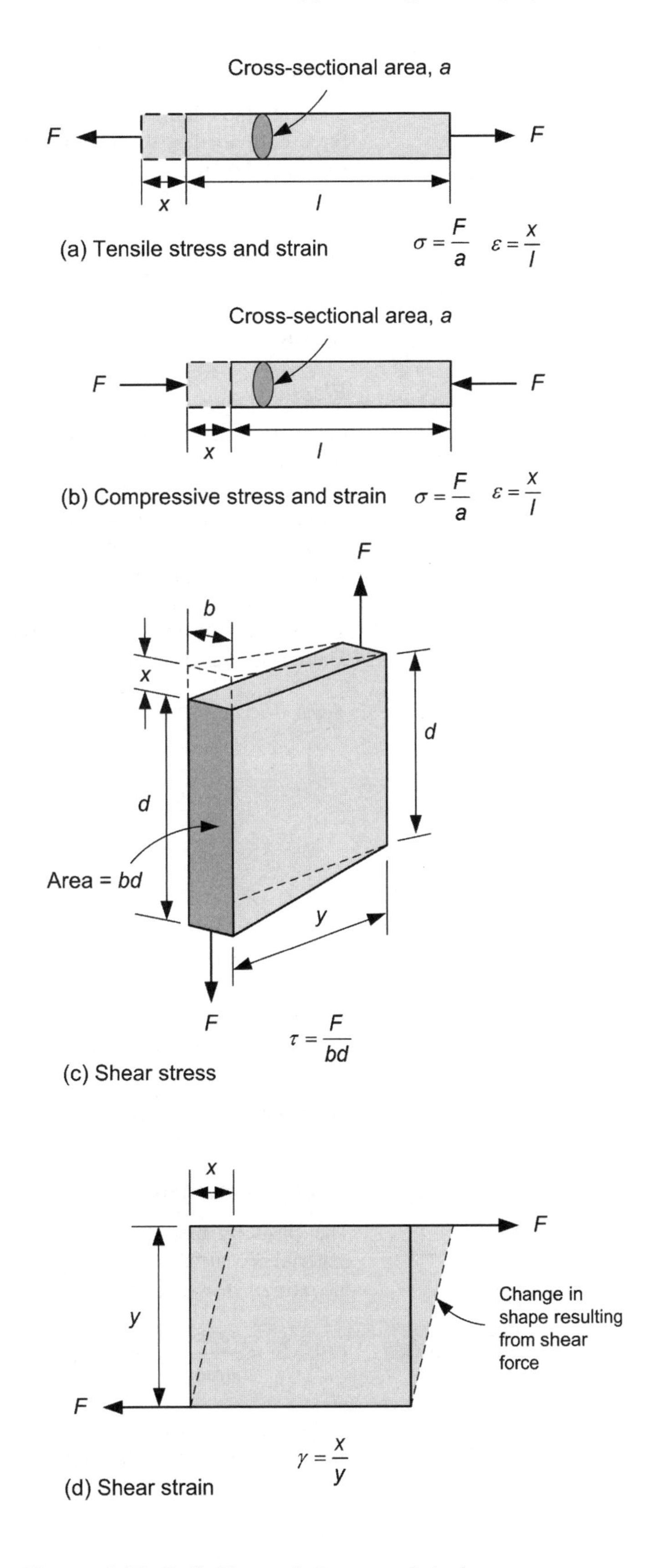

Figure 4.17 *Definitions of stress and strain*

Shear strain

Shear strain is defined as the change in length in the direction of the force, x, divided by the length, l, at right angles to the direction of the force (as shown in Figure 4.17d). Hence:

$$\text{Shear strain,} \quad \gamma = \frac{x}{l}$$

As with the other forms of strain, shear strain has no units.

Elasticity

Elasticity is the ability of a material to return to its original shape and size when any externally applied forces are removed. Consider a material that is subject to a tensile force. Provided that the force applied to a material is within the limits for which the material behaves as an elastic material, any increase in length of the material will be directly proportional to the force applied. This is known as *Hooke's Law* and it follows that, within the elastic limit of a material, the strain produced is directly proportional to the stress producing it. These important relationships are illustrated in Figure 4.18.

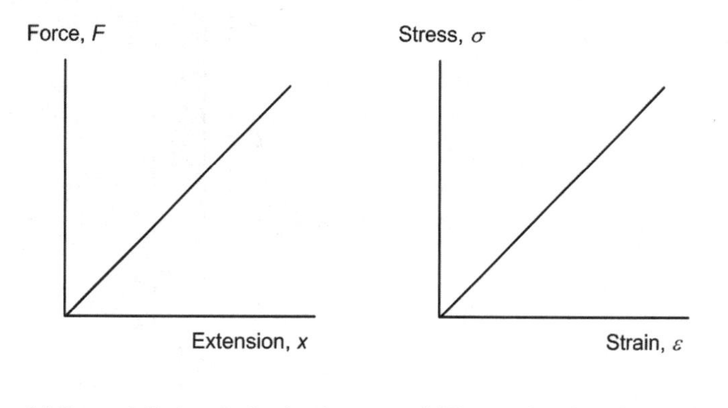

(a) Force plotted against extension (b) Stress plotted against strain

Figure 4.18 *Behaviour of elastic materials*

Elasticity modulus

The elasticity modulus, E (also known as *Young's modulus*), of a material is a measure of the *stiffness* of a material and it is defined as the ratio of stress, σ, to strain, ε. Hence:

$$E = \frac{\text{stress}}{\text{strain}} = \frac{\sigma}{\varepsilon}$$

In the case of shear stress, τ, and shear strain, γ, the elasticity modulus is known as the *modulus of rigidity*, G, and is defined as follows:

$$G = \frac{\text{shear stress}}{\text{shear strain}} = \frac{\tau}{\gamma}$$

Tensile testing

Tensile tests are often carried out in order to determine the strength of engineering materials. During a tensile test, a force (or *load*) is applied to a specimen of the material and the resulting extension is measured. Testing is usually carried out using specialised apparatus (a *tensometer*) and specimens that conform to a standard specification (see Figure 4.19). A typical load/extension graph produced during a tensile test is shown in Figure 4.20. The regions on this graph can be explained as follows:

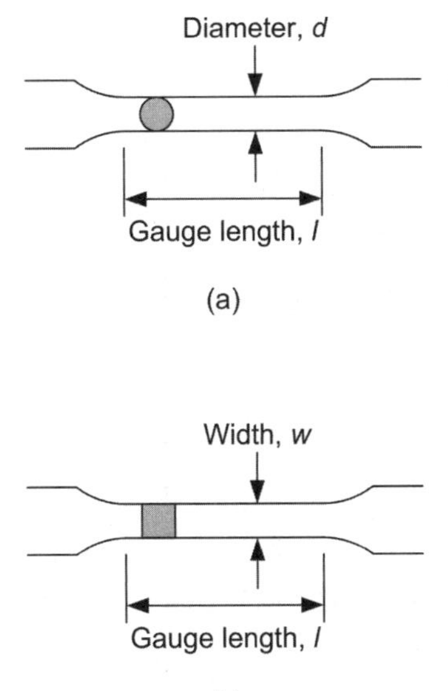

Figure 4.19 *Typical gauge specimens for tensile testing*

- Between A and B stress is directly proportional strain (i.e. Hooke's Law applies). Note that the slope of the graph in this area is used to determine the value of *elasticity modulus* for the material.
- Point B is the limit of proportionality and is the point at which stress is no longer proportional to strain for any further increase in load.
- Point C is the *elastic limit* and a specimen loaded beyond this point will not return to its original length when the load is removed.
- Point D is the point at which there is a rapid increase in extension for no further increase in load. This point is known as the *yield point*.
- Between points D and E further extension of the material occurs until point E is reached.
- Point E is the ultimate tensile strength of the material. It represents the maximum load that can be carried by a sample with a given cross-sectional area.
- Between points E and F a waist or neck appears in the material which precedes a fracture failure of the sample.
- Point F is the point at which the sample becomes fractured.

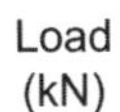

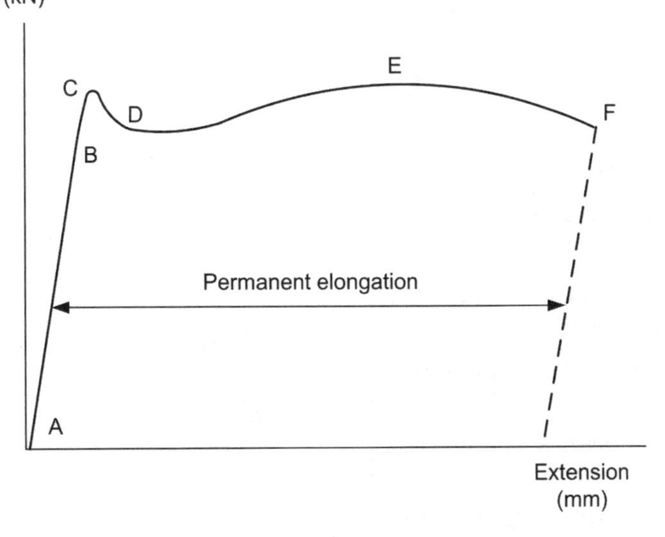

Figure 4.20 *A typical load/extension graph for an elastic material*

Another view

In most material tests (where a sample is tested to destruction) the final point on the stress-strain (or load-extension) graph is the point that occurs immediately before fracture (point F in Figure 4.20).

Test your knowledge 4.10

Figure 4.21 shows the results of a tensile test carried out on a test piece which has a length of 100 mm and a diameter of 20 mm. For this sample, use the graph to determine:

1. The elastic limit
2. The yield point
3. The elasticity modulus.

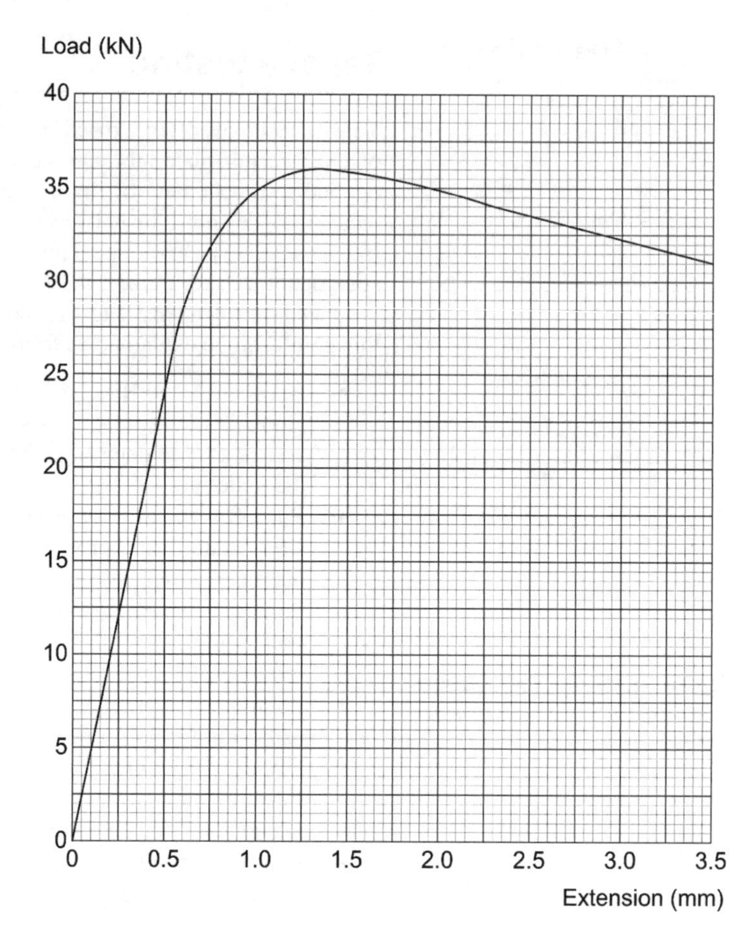

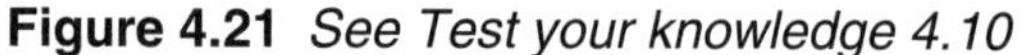

Figure 4.21 *See Test your knowledge 4.10*

Test your knowledge 4.11

For the materials shown in Figure 4.22:

1. Which material is the strongest?
2. Which material stretches the most?
3. Which material fractures after the least extension?
4. Which material fractures after the greatest extension?
5. Which material has the highest value of elasticity modulus?
6. Which material has the lowest value of elasticity modulus?

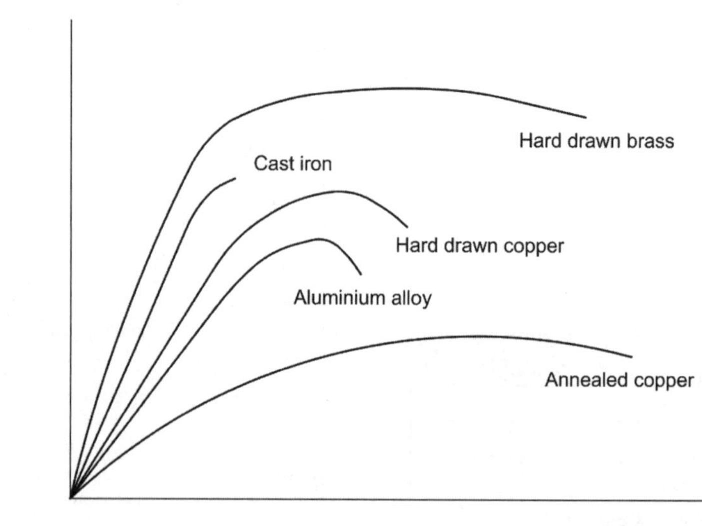

Figure 4.22 *Typical force/extension curves for various engineering materials*

Another view

The book's companion website (see page ix) has a spreadsheet tool for solving simple stress/strain problems. You may find this a useful means of checking your results or experimenting with other values!

Example 4.9

A sample of brass used in a tensile test consists of a 1 m length of wire with a diameter of 0.8 mm. The following results were obtained from the load test:

Load (N)	0	50	100	150	200	250
Extension (mm)	0	1.10	2.20	3.35	4.40	5.51

Plot the load/extension graph and use it to determine the value of elasticity modulus for the material.

The graph showing load plotted against extension is shown in Figure 4.23. The value of elasticity modulus is determined as follows:

$$\text{Elasticity modulus, E} = \frac{\text{stress}}{\text{strain}} = \frac{\text{force}}{\text{area}} \times \frac{\text{original length}}{\text{extension}}$$

$$\text{Thus, E} = \frac{\text{force}}{\text{extension}} \times \frac{\text{original length}}{\text{area}} = \text{slope of graph} \times \frac{\text{original length}}{\text{area}}$$

Now the cross-section area of the sample will be given by:

$$a = \pi r^2 = \pi\left(\frac{d}{2}\right)^2 = \pi\left(\frac{0.8\times10^{-3}}{2}\right)^2 = \pi\left(0.4\times10^{-3}\right)^2 \text{ m}^2$$

From which:

$$a = \pi \times 0.16\times10^{-6} = 0.5\times10^{-6} \text{ m}^2$$

$$\text{The slope of the graph} = \frac{270}{6\times10^{-3}} \text{ N/m}$$

$$\text{Hence the elasticity modulus, E} = \frac{270}{6\times10^{-3}} \times \frac{1}{0.5\times10^{-3}} \text{ Pa}$$

$$\text{From which, E} = \frac{270}{3} \times 10^9 = 90 \times 10^9 = \mathbf{90\ GPa}$$

Activity 4.3

The following results were obtained from a load test on a 500 mm sample of wire having a diameter of 1.2 mm:

Load (N)	0	100	200	300	400	500
Extension (mm)	0	0.85	1.75	2.50	3.35	4.20

Plot the load/extension graph and use it to determine the value of elasticity modulus for the material.

> **Another view**
>
> Often, the transition from a straight line to a curve on the stress-strain diagram is very gradual. In such cases it may be rather difficult to specify a precise value for the proportional limit. Furthermore, it may be inappropriate to quote the proportional limit as a practical elastic limit for design purposes. For these reasons, a yield stress (also called yield strength) can be established using an offset method. These methods define the yield stress by constructing a line having the same slope as the linear portion of the stress-strain curve that passes through the horizontal axis at a staring of typically, 0.2%, 0.1% or 0.05%. The point where this offset line intersects with the stress-strain curve then defines the relevant yield stress.

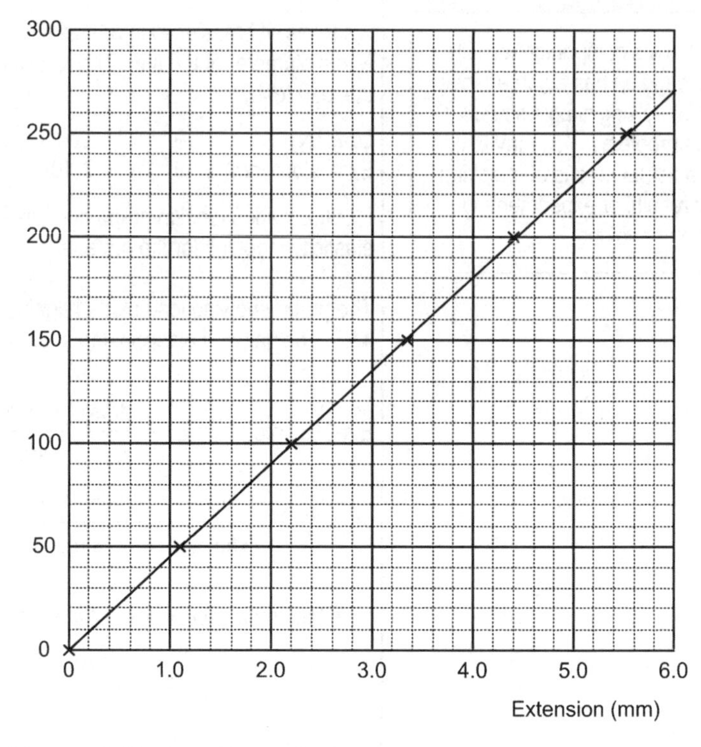

Figure 4.23 *See Example 4.9*

Table 4.2 *See Activity 4.4*

Material	*Elasticity modulus or Young's modulus (GPa)*	*Yield strength (MPa)*	*Tensile strength (MPa)*
Naval brass	100	206	
Phosphor bronze		193	455
Cast iron	117		
Carbon steel	200		
Chrome steel	210		
Stainless steel			770
Titanium	112		792

Activity 4.4

Table 4.2 shows information relating to some common metals. Use the matSdata database (see Activity 1.2 on page 4) to complete the table and then explain the significance of the data in terms of the characteristics of the material and its performance under load. Present your findings in the form of a brief written report.

Framed structures

Framed structures (or *trusses*) are commonly used in engineering and Figure 4.24 shows two examples. Figure 4.24(a) shows a *bridge truss* whilst Figure 4.24(b) shows a *roof truss*. A truss is a frame where the joints are assumed to be pin-jointed and frictionless and where the actions (i.e. loads) and reactions are only applied at the joints in the frame. The members of the truss (i.e. the individual parts of the framework) are either in compression (in which case they are referred to as *struts*) or they are in tension (in which case they are referred to as *ties*). The difference between ties and struts is illustrated in Figure 4.25.

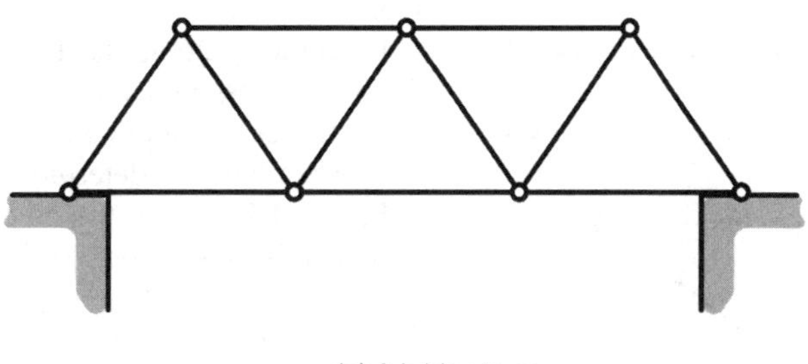

(a) A bridge truss

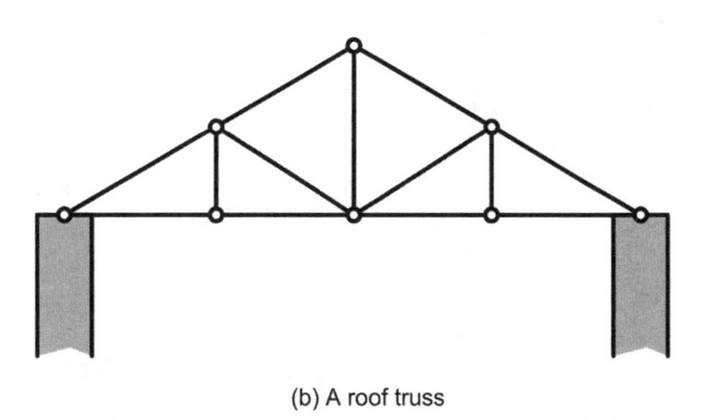

(b) A roof truss

Figure 4.24 *Examples of framed structures*

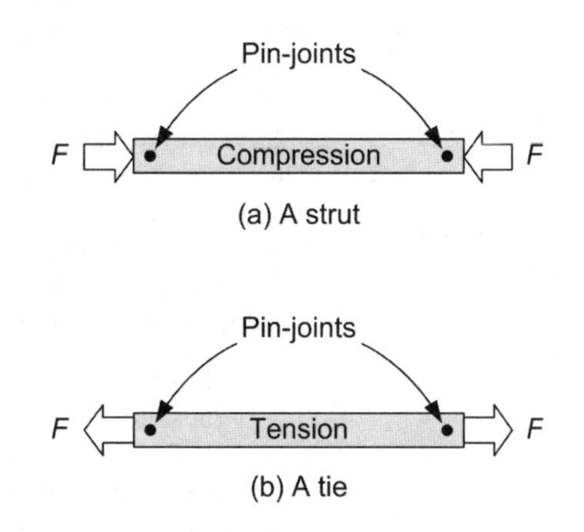

(a) A strut

(b) A tie

Figure 4.25 *Forces in the members of a framework*

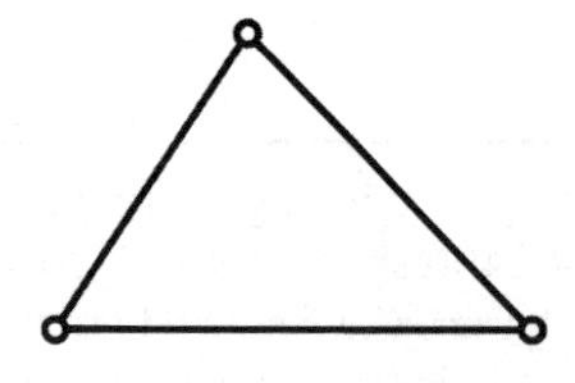

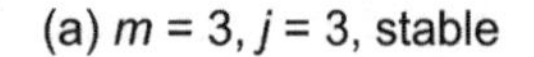

(a) $m = 3, j = 3$, stable

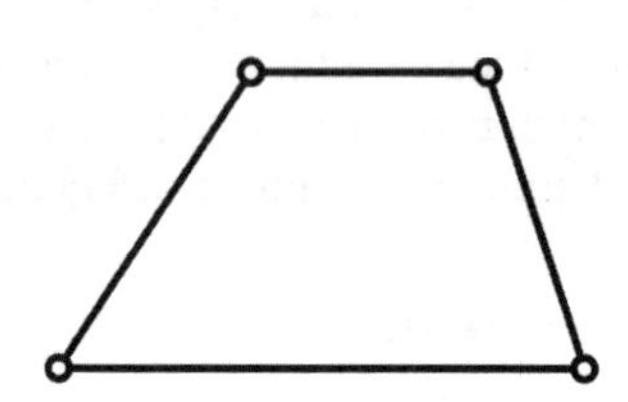

(b) $m = 4, j = 4$, unstable

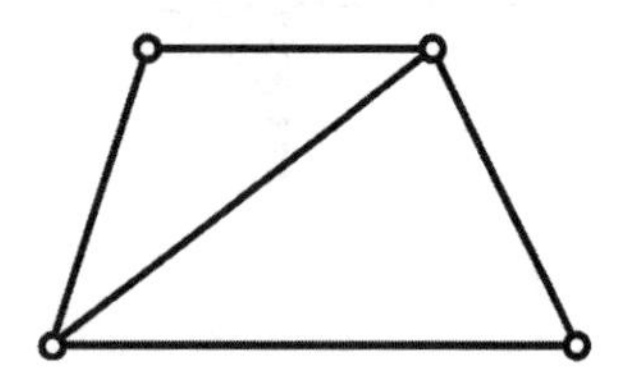

(c) $m = 5, j = 4$, stable

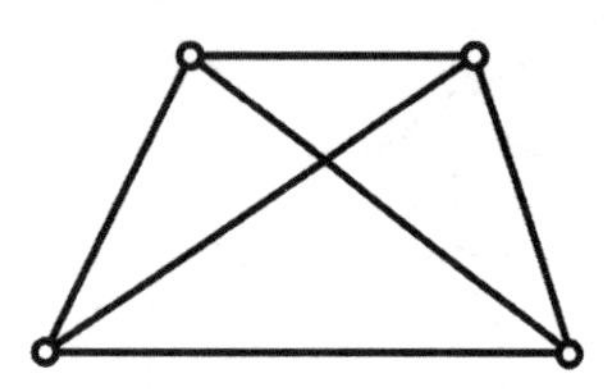

(d) $m = 4, j = 4$, stable

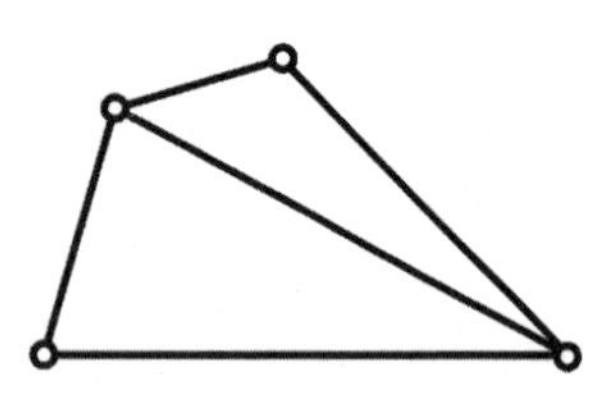

(e) $m = 5, j = 4$, stable

Figure 4.26 *Example of various frameworks*

The trusses that we shall examine in this unit all have members in the same plane and they are therefore referred to as *plane trusses*. The structural members of these trusses can be flat strips, rods, bars, channels or I-beams which may be bolted together at their ends. Note that, because they allow forces to act in any direction, bolts and welded joints can usually be assumed to provide effective pin-jointed connections.

In order to analyse a pin-jointed structure and be able to determine the forces in the individual members of which it is made, the structure needs to be *statically determinate*. This sounds complex but it simply means that it should be stable. To illustrate this important point take a look at each of the frameworks shown in Figure 4.26.

In Figure 4.26(a) the framework is that of a triangle and it has three members and three joints. The structure will retain its shape when forces are applied at the joints and therefore the framework is said to be statically determinate.

Figure 4.26(b), on the other hand, shows a framework which has four members and four joints. If forces are applied at the pin joints the framework will change shape in response. The structure is therefore unpredictable and is described as being *statically indeterminate*. Figure 4.27 shows the effect of applying a small force on the shape of the framework shown in Figure 4.26(b).

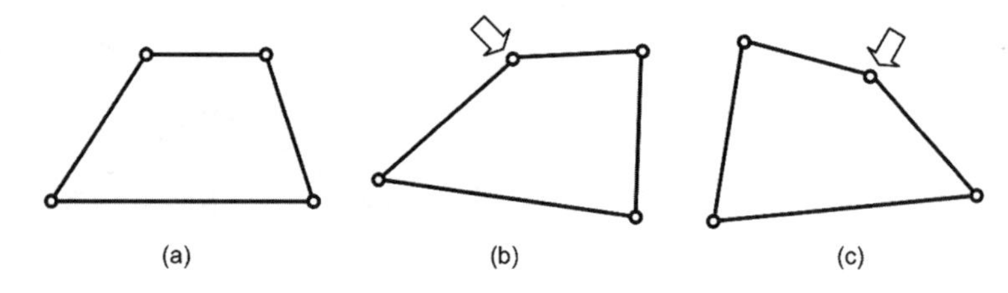

Figure 4.27 *Effect of applying a force to an unstable framework*

The framework shown in Figure 4.26(b) can be made statically determinate by simply adding one more member (a diagonal). This new structure has five members and four joints and is shown in Figure 4.26(c).

If we add a further member (the remaining diagonal) to Figure 4.26(c) we would arrive at the structure shown in Figure 4.26(d). This framework is also statically determinate so our extra member has not actually improved the stability of the framework. Because of this, the added member is said to be *redundant*.

The relationship between the number of members, m, the number of joints, j, and the stability of a framework is as follows:

- If $(m + 3) < 2j$ the framework is statically indeterminate (unstable).
- If $(m + 3) = 2j$ the framework is statically determinate (stable) and there are no redundant members present.
- If $(m + 3) > 2j$ the framework is statically determinate (stable) and there are one or more redundant members present.

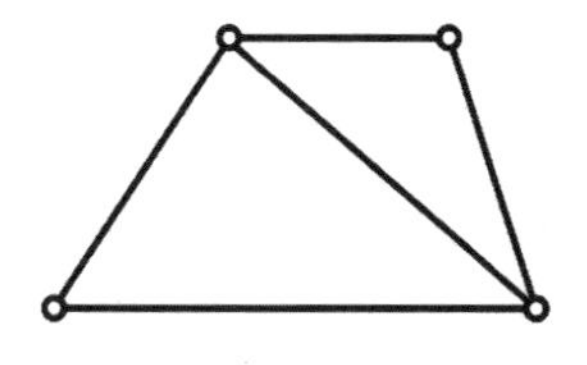

(a)

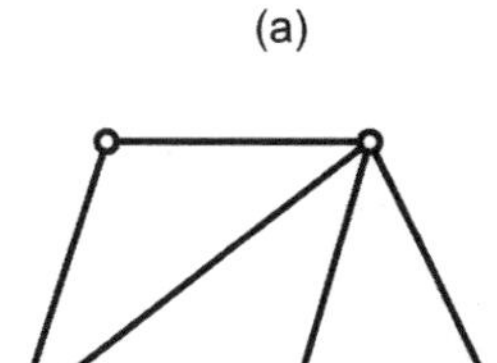

(b)

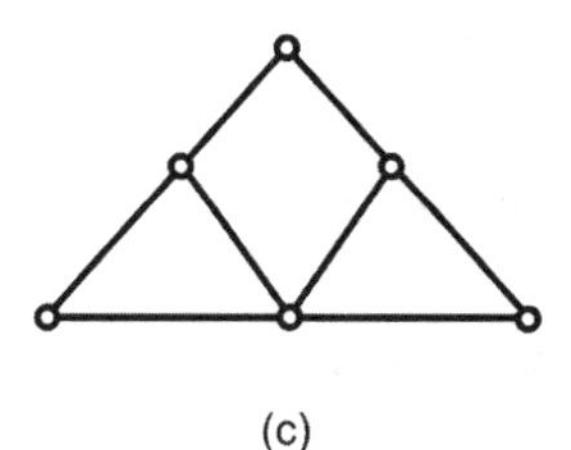

(c)

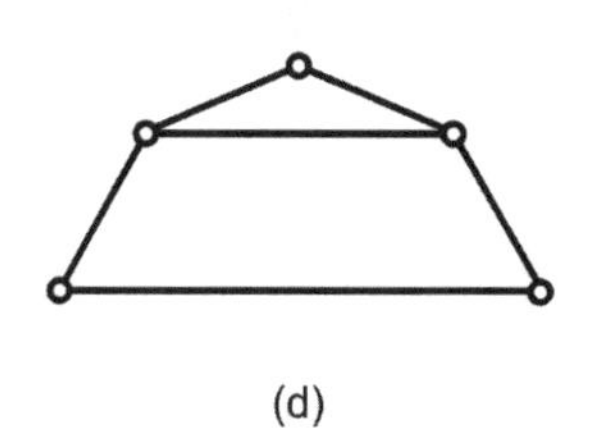

(d)

Figure 4.29 *See Test your knowledge 4.12*

Analysis of frameworks

In order to analyse a framework it needs to be pin-jointed, statically determinate, and any actions or reactions need to be present at the joints (and not part way along any of the members). The framework can be supported at one or more of the joints either by means of a support that only allows a reaction in one direction (at right angles to the support) or by means of a support that allows reactions in any direction (but normally described in terms of two components at right angles). This is shown in Figure 4.28.

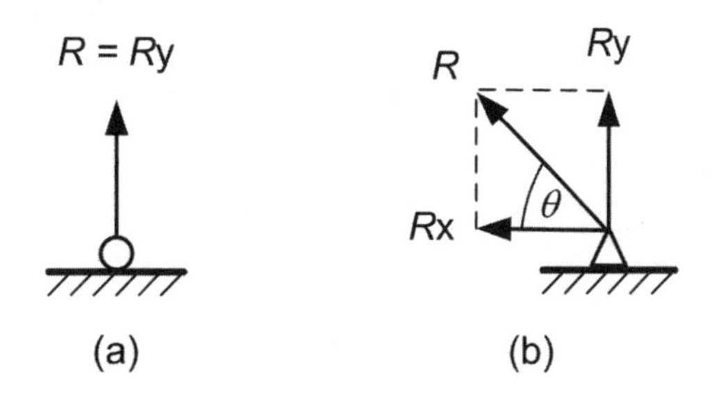

Figure 4.28 *Framework supports. A rolling support is shown in (a) whilst a fixed support is shown in (b).*

A truss can be analysed by considering the forces at each of its joints and supports. This is known as the *method of joints* and it involves resolving the forces at each joint independently into horizontal and vertical components. This generates a series of equations which can then be solved provided that we know the magnitude and direction of the applied forces. Since the framework is in equilibrium, we can also resolve the total forces acting on it in both the vertical and horizontal directions. It is thus possible to write down equations for the reaction forces in terms of the applied forces, as shown in Figure 4.30.

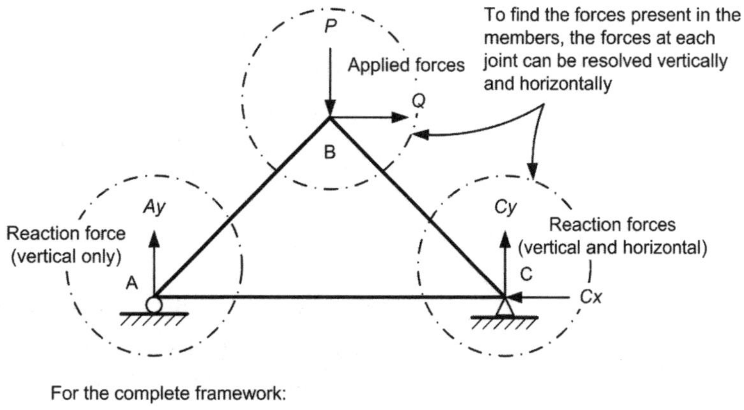

Figure 4.30 *The method of joints*

Test your knowledge 4.12

Figure 4.29 shows various frameworks. Determine whether these are stable or not and, if the framework is unstable, suggest the placement of an additional member that will ensure stability.

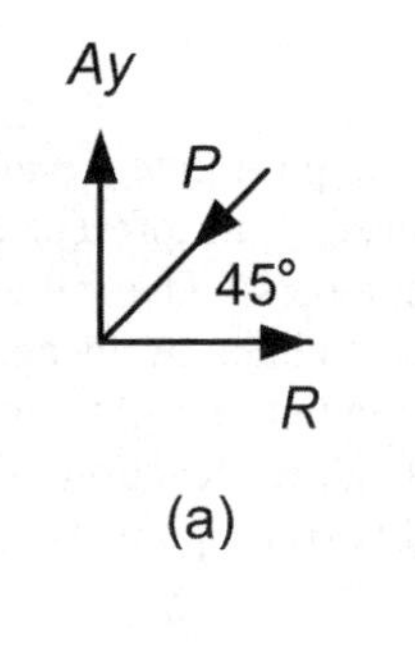

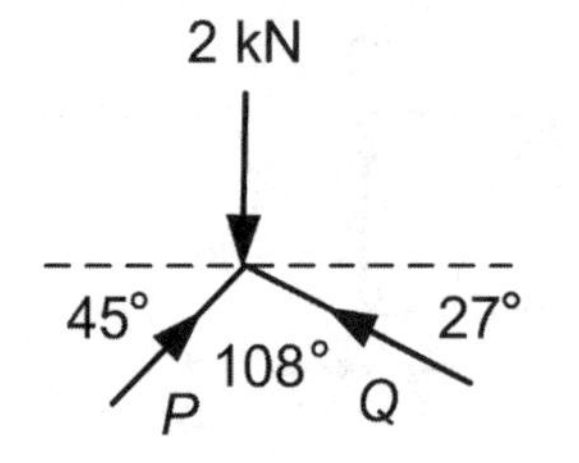

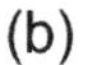

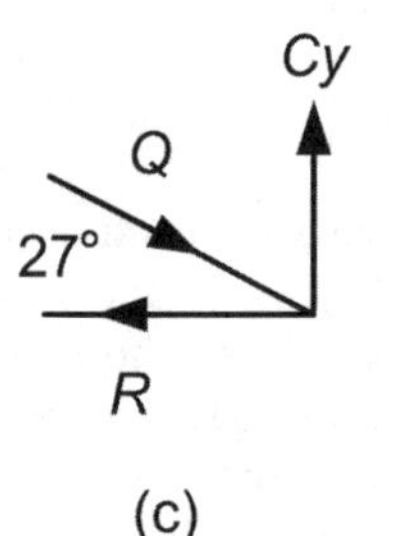

Figure 4.31 *See Example 4.10*

Example 4.10

Figure 4.32 shows a pin-jointed structure. Determine the reactions at the two supports and the force in each of the members.

The first step in solving this problem is to mark the diagram with letters to represent each joint and each force present. In most cases it should also be possible to reason the direction of the forces and thus to identify which members are in tension and which are in compression.

Figure 4.33 shows the forces in the members, *P*, *Q* and *R*. By inspection, *P* and *Q* are both compressive forces whilst *R* is a tension force. Since the load is applied vertically downwards, the reaction forces present at the two supports, *A* and *C*, must act in the opposite (vertical) direction. There can be no horizontal reaction force.

Resolving forces at point B in the horizontal direction (see Figure 4.31b) gives:

$P \cos 30° = Q \cos 27°$ therefore $0.707\,P = 0.891Q$ (i)

Resolving forces at point B in the vertical direction (see Figure 4.31b) gives:

$2 = Q \cos 63° + P \cos 45°$ therefore $2 = 0.454\,Q + 0.707\,P$ (ii)

Combining (i) and (ii) gives:

$2 = 0.454\,Q + 0.891\,Q = 1.345\,Q$ from which $Q =$ **1.487 kN**

Rearranging equation (i) gives:

$$P = \frac{0.891}{0.707} \times 1.487 = \mathbf{1.874\ kN}$$

Resolving forces at point C in the horizontal direction (see Figure 4.31c) gives:

$R = Q \cos 27° = 1.487 \times 0.891 =$ **1.325 kN**

Resolving forces at point C in the vertical direction (see Figure 4.31c) gives:

$Cy = Q \cos 63° = 1.487 \times 0.454 =$ **0.675 kN**

Resolving forces at point A in the horizontal direction (see Figure 4.31a) gives:

$R = P \cos 45°$ so $P = \dfrac{1.325}{0.707} =$ **1.874 kN**

Resolving forces at point A in the vertical direction (see Figure 4.31a) gives:

$Ay = P \cos 45° = 1.874 \times 0.707 =$ **1.325 kN**

Finally, it is worth checking that the total upward force is equal to the total downward force, as follows:

Total upward force = Total downward force

$$2 = Ay + Cy$$
$$2 = 1.325 + 0.675 = 2\text{ kN}$$

Figure 4.34 shows the structure with all forces marked.

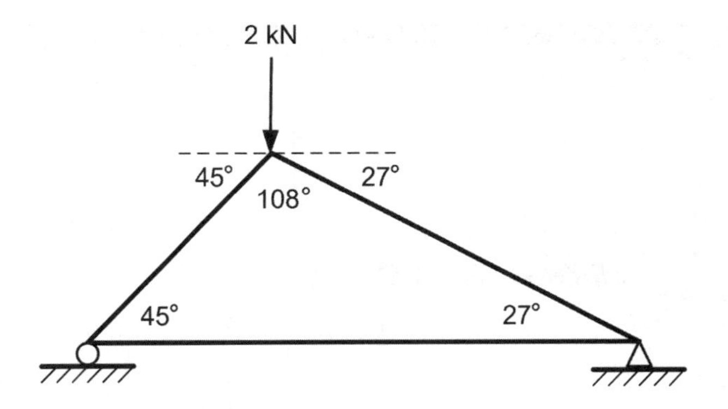

Figure 4.32 *See Example 4.10*

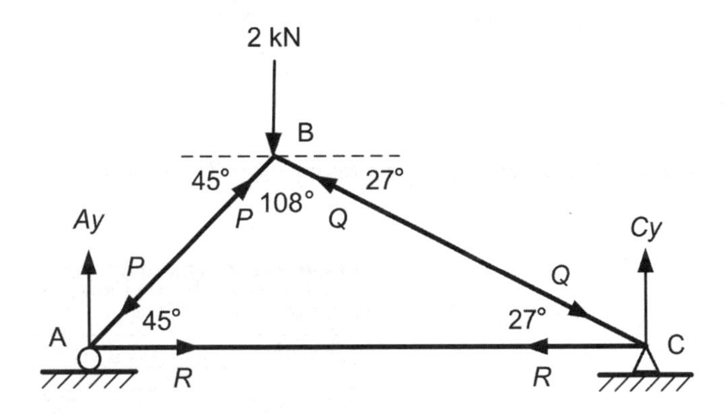

Figure 4.33 *See Example 4.10*

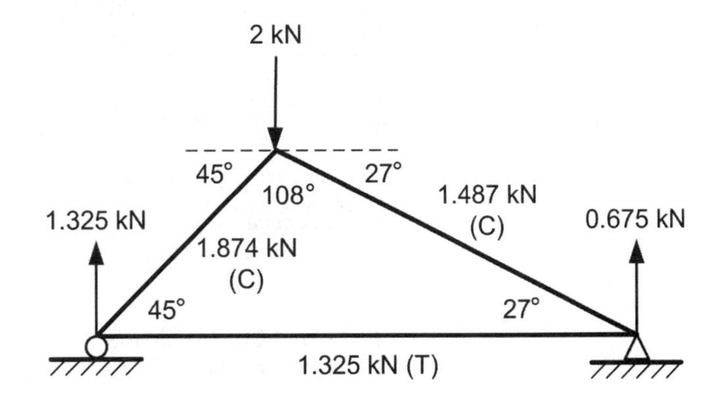

Figure 4.34 *See Example 4.10*

Finally, it is worth mentioning that more complex frameworks are often analysed using structural analysis software. Figures 4.35 and 4.36 show screens produced by MDSolids during the analysis of a framework with nine members. This package (which is ideal for student use—see the website for details) is capable of determining the stress in each member and can display its results using a variety of different units.

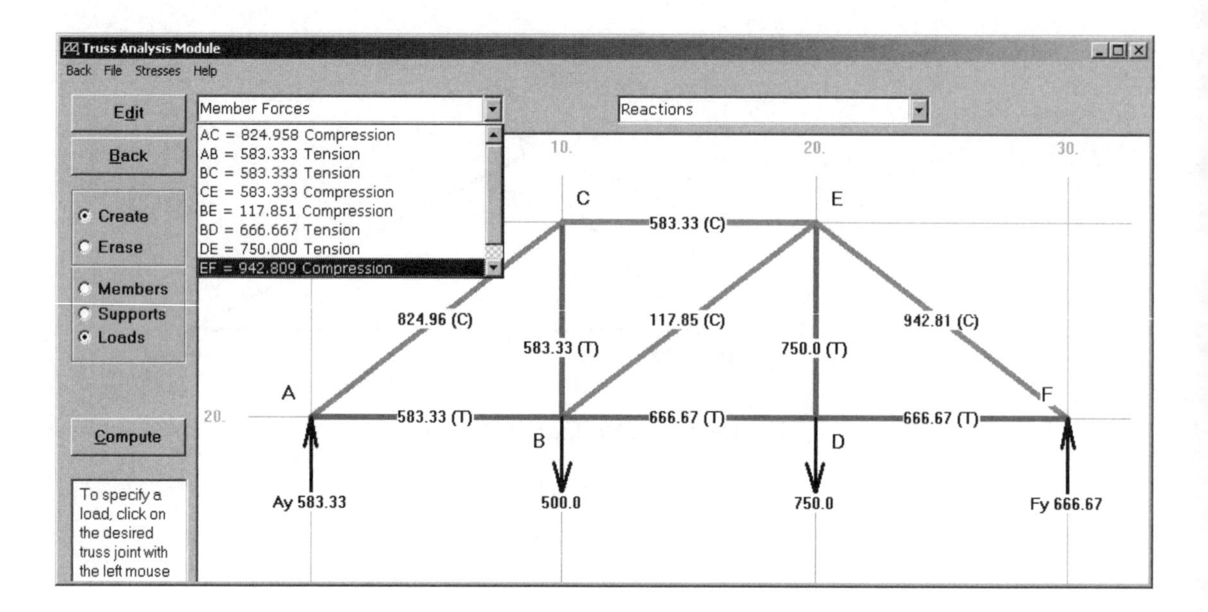

Figure 4.35 *Framework analysis using the MDSolids software package*

Normal Stresses in Truss Members

Member	Force (kN)	Area (mm²)	Stress (MPa)
BC	583.333	1,200.00	486.111
CE	-583.333	1,200.00	-486.111
BE	-117.851	640.00	-184.142
BD	666.667	1,200.00	555.556
DE	750.000	1,200.00	625.000
EF	-942.809	1,200.00	-785.674
DF	666.667	1,200.00	555.556

Force Units: kN — Area Units: mm² — Stress Units: MPa — Compute

Figure 4.36 *MDSolids can also determine the stress present in each member of a framework*

Activity 4.5

Use MDSolids (or a similar framework analysis package) to analyse the problem in Example 4.10. Check that the answers agree with those obtained by calculation and also determine the stress present in each member of the framework if all members have a cross-sectional area of 25 mm^2.

Activity 4.6

Use the method of joints to determine the reaction forces and the force in each member of the frameworks shown in Figures 4.37 to 4.40.

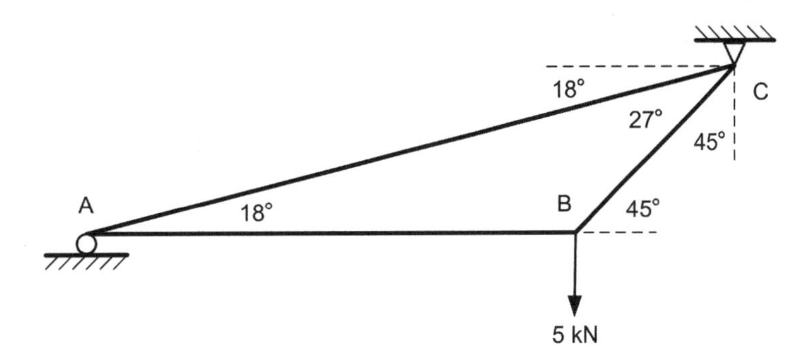

Figure 4.37 *See Activity 4.6*

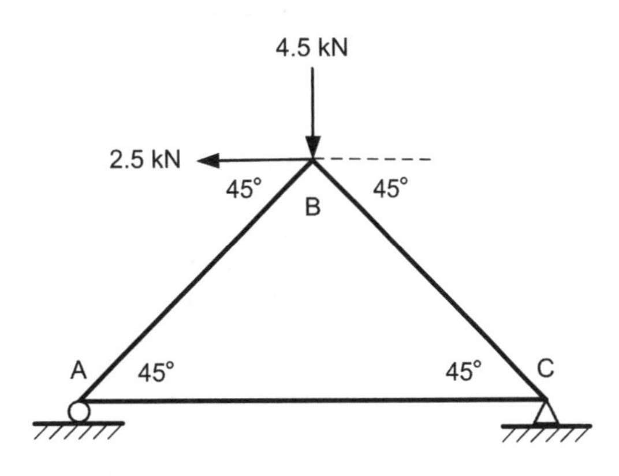

Figure 4.38 *See Activity 4.6*

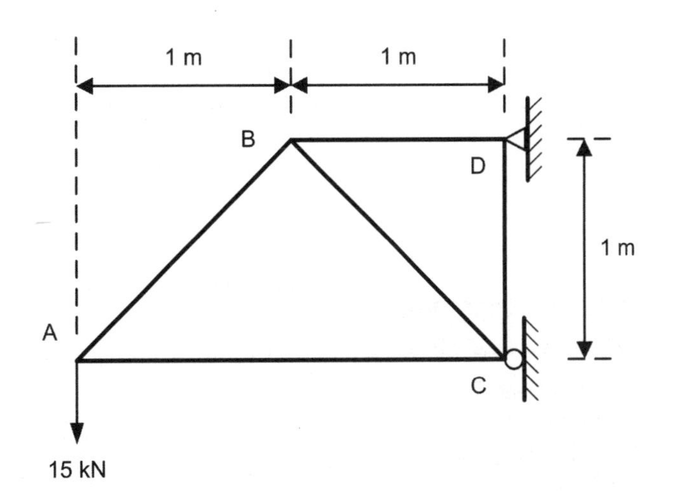

Figure 4.39 *See Activity 4.6*

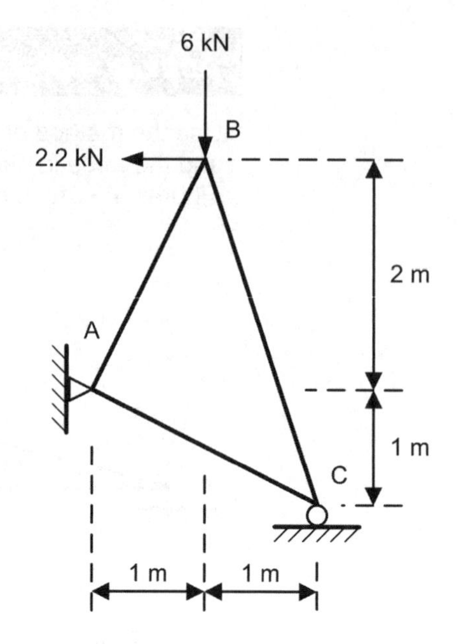

Figure 4.40 *See Activity 4.6*

Factor of safety

It should be obvious that the stress that appears in a mechanical component must be kept well below its ultimate tensile stress for all conditions of loading. In order to ensure that these values of stress are never approached, a *factor of safety* is often introduced into mechanical design calculations. Factor of safety is usually defined as follows:

$$\text{Factor of safety} = \frac{\text{Ultimate tensile stress}}{\text{Allowable working stress}}$$

However, since plastic deformation is usually unacceptable, we would normally wish to ensure that stresses are restricted to those within the elastic range. Hence we often use yield stress (or proof stress) rather than ultimate tensile stress in our calculations of allowable working stress. Hence a better definition would be:

$$\text{Factor of safety} = \frac{\text{Yield stress}}{\text{Allowable working stress}}$$

From which:

$$\text{Allowable working stress} = \frac{\text{Yield stress}}{\text{Factor of safety}}$$

Note that practical values of factor of safety range from about 2.5 (for uncritical statically-loaded systems) to 10 (for systems that are critical or may have sudden shock loads applied).

Test your knowledge 4.13

Explain what is meant by *factor of safety* and explain why this is important in the design of structural frameworks.

4.2 Electro-mechanical systems

Having dealt with simple structural systems and frameworks, we shall now move on to investigate electromechanical systems. In these *energy converting systems*, the inputs and outputs are electrical power and mechanical movement. Examples of electro-mechanical systems include:

- domestic appliances (such as washing machines)
- generators and alternators
- electrical power tools (such as electric drills)
- pneumatic systems (where compressed air is used to transmit power between components).

Electrical circuits

You will already be familiar with the most basic electric circuit which uses only two components; a cell (or battery) acting as a source of e.m.f., and a resistor (or *load*) through which a current is passing. These two components are connected together with wire conductors in order to form a completely closed circuit as shown in Figure 4.41.

Current, voltage and resistance

Electric current is the name given to the flow of electrons (or negative *charge carriers*). The ability of an energy source (e.g. a *battery*) to produce a current within a conductor may be expressed in terms of electromotive force (e.m.f.). Whenever an e.m.f. is applied to a circuit a potential difference (p.d.) exists. Both e.m.f. and p.d. are measured in volts (V). In many practical circuits there is *only one* e.m.f. present (the battery or supply) whereas a p.d. will be developed across *each* component present in the circuit.

The *conventional flow* of current in a circuit is from the point of more positive potential to the point of greatest negative potential (note that electrons move in the opposite direction!). *Direct current* result from the application of a direct e.m.f. (derived from batteries or a DC supply, such as a generator or a power supply. An essential characteristic of such supplies is that the applied e.m.f. does not change its polarity (even though its magnitude might vary).

For any conductor, the current flowing is directly proportional to the e.m.f. applied. The current flowing will also be dependent on the physical dimensions (length and cross-sectional area) and material of which the conductor is composed. The amount of current that will flow in a conductor when a given e.m.f. is applied is inversely proportional to its resistance. Resistance, therefore, may be thought of as an 'opposition to current flow'; the higher the resistance the lower the current that will flow (assuming that the applied e.m.f. remains constant).

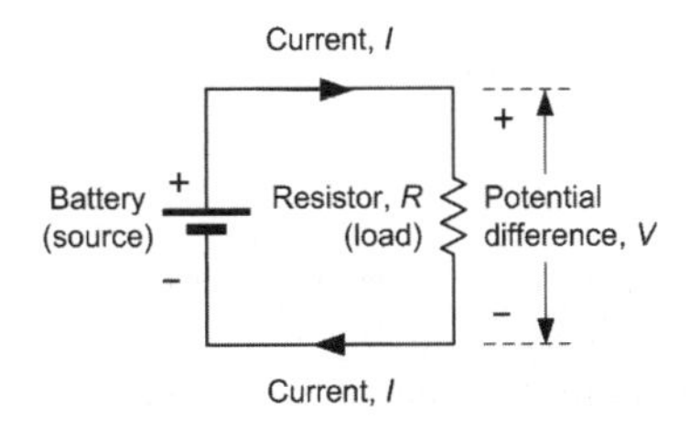

Figure 4.41 *A simple DC circuit consisting of a battery (source) and resistor (load)*

Ohm's Law

Provided that temperature does not vary, the ratio of p.d. across the ends of a conductor to the current flowing in the conductor is a constant. This relationship is known as Ohm's Law and it leads to the relationship:

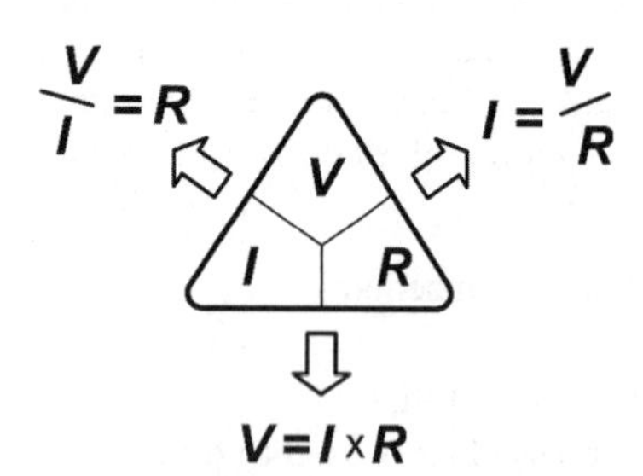

Figure 4.42 *V, I and R triangle*

$$\frac{V}{I} = \text{a constant} = R$$

where V is the potential difference (or voltage drop) in volts (V), I is the current in amps (A), and R is the resistance in ohms (Ω).

The formula may be arranged to make V, I or R the subject, as follows:

$$V = I \times R, \quad I = \frac{V}{R}, \quad \text{and} \quad R = \frac{V}{I}$$

The triangle shown in Figure 4.42 should help you remember these three important relationships.

Another view

The book's companion website (see page ix) has a spreadsheet tool for solving simple electrical cable problems. You may find this a useful means of checking your results or experimenting with other values!

Example 4.11

A current of 100 mA flows in a 56 Ω resistor. What voltage drop (potential difference) will be developed across the resistor?

Here we must use $V = I\,R$ and ensure that we work in units of volts (V), amps (A), and ohms (Ω).

$V = I \times R = 0.1 \times 56 = \mathbf{5.6\ V}$

(Note that 100 mA is the same as 0.1 A)

Test your knowledge 4.14

1. Sketch a simple electric circuit and label the voltage and current present in it.
2. State Ohm's law.

Example 4.12

A 18 Ω resistor is connected to a 9 V battery. What current will flow in the resistor?

Here we must use:

$I = \frac{V}{R} = \frac{9}{18} = \frac{1}{2} = 0.5\text{ A} = \mathbf{500\,mA}$

Power, work and energy

Energy exists in many forms including kinetic energy, potential energy, heat energy, light energy etc. Kinetic energy is concerned with the movement of a body whilst potential energy is the energy that a body possesses due to its position. Energy can be defined as 'the ability to do work' whilst power can be defined as 'the rate at which work is done'.

In electrical circuits, energy is supplied by batteries or generators. It may also be stored in components such as capacitors and inductors. Electrical energy is converted into various other forms of energy by components such as resistors (producing heat), loudspeakers (producing sound), light emitting diodes (producing light).

Power, P, is the rate at which energy is converted from one form to another and it is measured in *Watts*. Thus:

$$\text{Power, } P = \frac{\text{Energy, } W}{\text{Time, } t}$$

The unit of energy is the joule (J). Power is the rate of use of energy and it is measured in watts (W). A power of 1W results from energy being used at the rate of 1J per second. Thus Joules are equivalent to *Watt-seconds*. If the power was to be measured in kilowatts (kW) and the time in hours (rather than seconds) then the unit of electrical energy would be expressed in terms of *kilowatt-hours kW-h* (commonly knows as a *unit of electricity*). The electricity meter in your home records the *amount of energy* that you have used in kW-h.

The power in a circuit is equivalent to the product of voltage and current. Hence:

$$P = I \times V$$

where P is the power in watts (W), I is the current in amps (A), and V is the voltage in volts (V).

The formula may be arranged to make P, I or V the subject, as follows:

$$P = I \times V, \quad I = \frac{P}{V}, \quad \text{and} \quad V = \frac{P}{I}$$

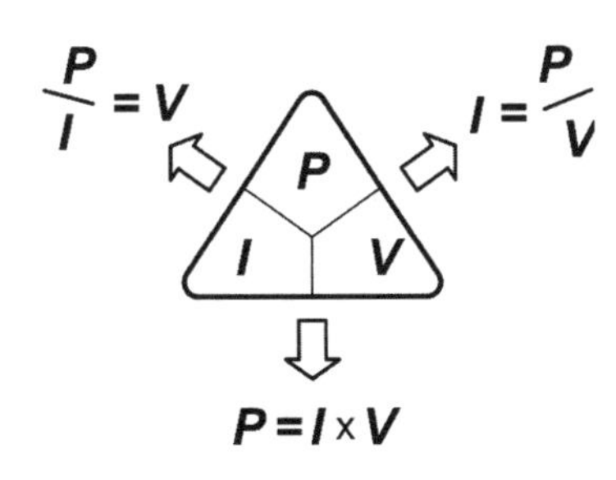

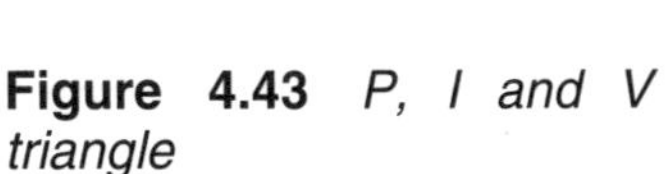

Figure 4.43 *P, I and V triangle*

The triangle shown in Figure 4.43 should help you remember these three important relationships.

Finally, we can combine the relationship that we met earlier from Ohm's Law with the relationships involving power to obtain the following useful relationships:

$$P = I^2 \times R \quad \text{and} \quad P = \frac{V^2}{R}$$

Test your knowledge 4.15

Give THREE examples of energy conversion in electric circuits.

Example 4.13

A generator provides an output of 1.5 kW for 20 minutes. How much energy has it supplied?

Here we will use $W = P t$

where P = 1.5 kW = 1,500 W and t = 20 minutes = 20 × 60 = 1,200 s.

Thus:

W = 1,500 × 1,200 = 1,800,000 J = **1.8 MJ**

Another view

The book's companion website (see page ix) has a spreadsheet tool for solving simple electrical power and energy problems. You may find this a useful means of checking your results or experimenting with other values!

Example 4.14

An electric fire is rated at 2 kW and it is used for 12 hours per day for four weeks. How many units of electricity (kW-h) is used by the fire?

Once again we must use $W = Pt$

where P = 2 kW and t = 12 x 7 x 4 = 336 hours

Thus:

W = 2 × 336 = **672 kW-h**

Example 4.15

If electricity costs 4.5 p per unit, determine the cost of operating the electric fire in Example 4.14.

The fire uses 336 units (kW-h) of electricity. The cost is thus given by:

336 x 4.5 = 1512 p = **£15.12**

Example 4.16

A current of 1.5 A is drawn from a 3 V battery. What power is supplied?

Here we must use $P = I \times V$ (where I = 1.5 A and V = 3 V):

$P = I \times V$ = 1.5 A × 3 V = **4.5 W**

Hence a power of 4.5 W is supplied.

Example 4.17

A current of 20 mA flows in a 1 kΩ resistor. What power is dissipated in the resistor and what energy is used if the current flows for 10 minutes?

Here we must use $P = I^2R$ (where I = 200 mA and R = 1,000 Ω):

$$P = I^2R = (0.2 \times 0.2) \times 1{,}000 = 0.04 \times 1{,}000 = \mathbf{40\ W}$$

Hence the resistor dissipates a power of 40 W.

To find the energy we need to use $W = P \times t$ (where P = 40 W and t = 10 minutes):

$$W = P \times t = 40 \times (10 \times 60) = 24{,}000\ \text{J} = \mathbf{24\ kJ}$$

Mechanical work and energy

Work done

Earlier we defined energy in terms of "the ability to do work". Since work is required to change the energy possessed by a body (either by virtue of its position or its velocity) *work* is actually synonymous with *energy*.

Mechanical work is done when a force overcomes the resistance of a load and moves it through a distance.

Work done = force × distance or $W = F\,d$

Hence, mechanical work can be measured in terms of the product of force and distance (in Newton-metres). Thus we arrive at another definition of the Joule:

1 Joule (J) = 1 Newton-metre (Nm)

Work can also be defined in terms of the change in energy (expressed in Joules). Mechanical energy can take one of three forms:

- Potential energy (PE)
- Strain energy (a form of PE)
- Kinetic energy (KE).

Potential energy

Potential energy (PE) is the energy of a body by virtue of its position relative to some datum. The PE of a body is the product of its weight (i.e. the force acting downwards, $m \times g$) and the height above the datum point. Thus:

$\mathrm{PE} = m\,g\,h$

From this we can conclude that the work done in raising a load of mass, m, through a distance, d, is given by:

Work done = force × distance = $m\,g\,h$

In other words, the work done in raising the object is equal to the PE that the object gains.

Strain energy

Strain energy is a form of potential energy (PE) possessed by an elastic body when deformed within its elastic range. An example of strain energy is the energy stored in a spring when it is stretched or compressed.

Kinetic energy

Kinetic energy (KE) is the energy of a body by virtue of its motion. When an object having a mass, *m*, is travelling in a straight line with a velocity, *v*, its KE will be given by:

$$\mathrm{KE} = \frac{1}{2} m v^2$$

Example 4.18

A load having a mass of 100 kg is to be raised through a vertical height of 10 m using the hoist arrangement shown in Figure 4.44 which is driven by a motor rated at 1 kW input. Assuming that only 40% of the electrical power input is converted into useful mechanical energy, determine the time taken to raise the load.

The energy required to lift the load = $m\,g\,h = 100 \times 9.81 \times 10 = 9.81$ kJ

The rated motor input power = 1 kW and the mechanical output power = 40% × 1 kW = 400 W

Since power = energy/time, we can infer that, time = energy/power

Thus time = 9.81 kJ / 400 W = 9,810/400 = **24.53 s**

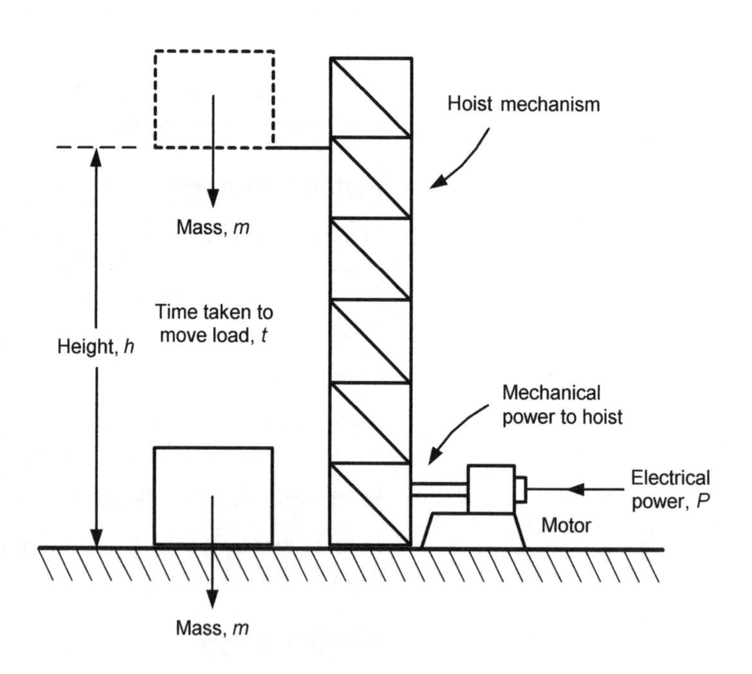

Figure 4.44 *See Example 4.18*

The ability to convert energy in one form to another is vitally important in a large number of practical engineering systems. However, each time a conversion takes place, some energy is lost. Consider, for example, the hoist arrangement shown in Figure 4.44. Here, electrical energy (taken from the supply) is converted into mechanical potential energy. Losses occur at each stage of this process (for example, in the motor due to the resistance of the electrical conductors) and minimising them can be an important task for the engineer. We shall look at this topic in a little more detail when we consider the efficiency of electro-mechanical systems.

Test your knowledge 4.16

Distinguish between *potential energy* (PE) and *kinetic energy* (KE). Give ONE example of each form of energy.

Efficiency

In Example 4.18 you may have noticed that, we took into account the losses that occur in an electro-mechanical system. In this case we were told that only 40% of the input power was converted into useful mechanical power output. The losses can be mainly accounted for in terms of friction and losses in the electrical conductors, both of which manifest themselves as heat.

In general, the efficiency of an electro-mechanical system is given by:

$$\text{Efficiency} = \frac{\text{Output}}{\text{Input}} \times 100\%$$

We can also express this another way:

$$\text{Efficiency} = \frac{\text{Input} - \text{losses}}{\text{Input}} \times 100\%$$

Example 4.19

A generator produces an output of 7.5 kW from a mechanical input of 9.25 kW. Determine the power loss and the efficiency of the generator.

The power loss will be given by:

Lost power = Input power – output power = 9.25 – 7.5 = 1.75 kW

$$\text{Efficiency} = \frac{\text{Output}}{\text{Input}} \times 100 = \frac{7.5}{9.25} \times 100 = \mathbf{81.1\%}$$

Electro-mechanical system elements

Various elements are used in electro-mechanical systems including motors, mechanical linkages, gear trains, and power transmissions. We shall briefly look at a number of these system elements, starting with DC and AC motors which are the prime movers in many practical engineering systems.

DC motors

DC motors are used in a wide variety of electro-mechanical systems where motion is to be produced from a direct current (DC) electrical supply (e.g. a battery). The basic principle of the DC motor is illustrated in Figure 4.45. A loop of wire that is free to rotate is placed inside a permanent magnetic field (see Figure 4.45). When a DC current is applied to the loop of wire, two equal and opposite forces are set up which act on the conductor in the directions indicated in Figure 4.45.

The direction of the forces acting on each arm of the conductor can be established by again using the right-hand grip rule and *Fleming's left-hand rule*. Now because the conductors are equidistant from their

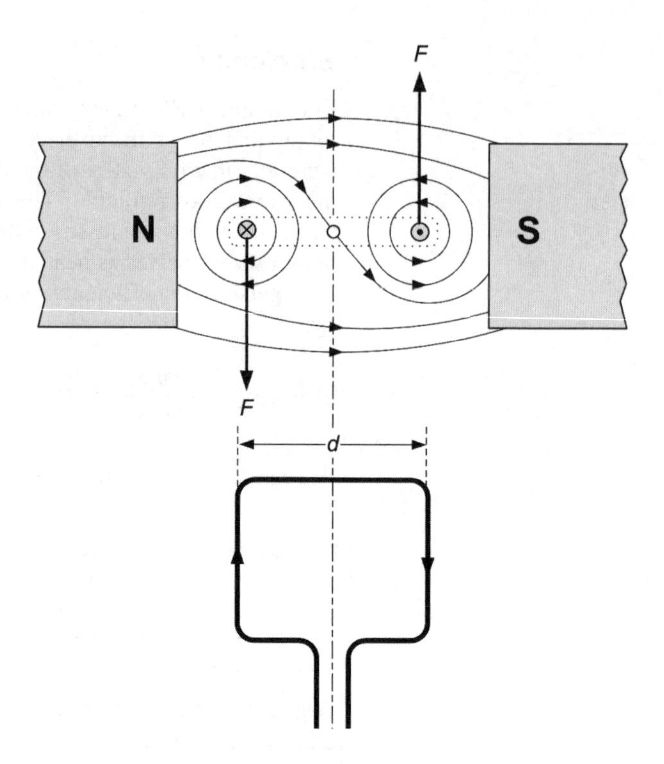

Figure 4.45 *Torque on a current carrying loop suspended within a permanent magnetic field*

pivot point and the forces acting on them are *equal and opposite,* then they form a *couple*. The *moment* of this couple is equal to the magnitude of a single force multiplied by the distance between them and this moment is known as *torque, T*. Now,

$$T = F\,d$$

where T is the torque (in Newton-metres, Nm), F is the force (N) and d is the distance (m).

The torque produces a *turning moment* such that the coil or loop rotates within the magnetic field. This rotation continues for as long as a current is applied. A more practical form of DC motor consists of a rectangular coil of wire (instead of a single turn loop of wire) mounted on a former and free to rotate about a shaft in a permanent magnetic field, as shown in Figure 4.46.

In real motors, this rotating coil is known as the *armature* and consists of many hundreds of turns of conducting wire. This arrangement is needed in order to maximize the force imposed on the conductor by introducing the *longest possible* conductor into the magnetic field. The force used to provide the torque in a motor is directly proportional to the magnitude of the magnetic flux, B. Instead of using a permanent magnet to produce this flux, in a real motor, an electromagnet is used. Here an electromagnetic field is set up using the *solenoid* principle (Figure 4.47). A long length of conductor is wound into a coil consisting of many turns and a current

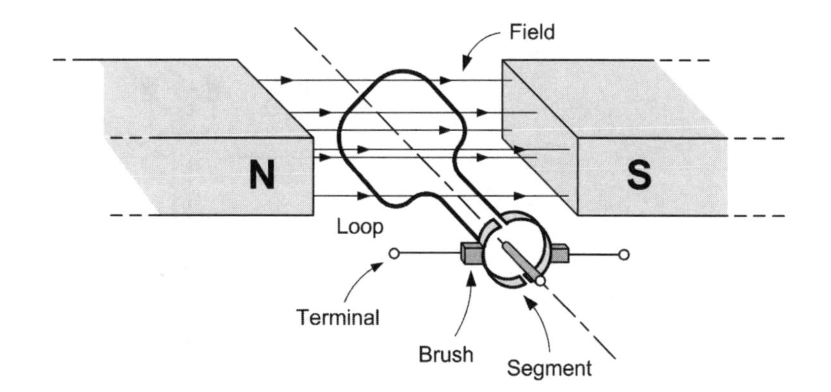

Figure 4.46 *Basic DC motor with commutator arrangement*

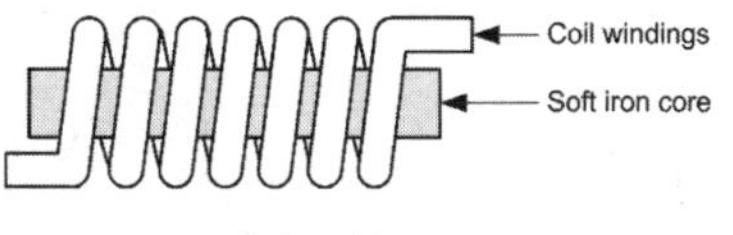

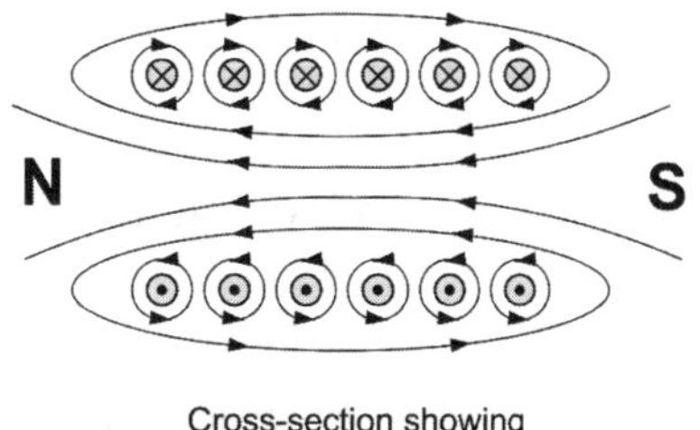

Figure 4.47 *Magnetic field produced by a solenoid*

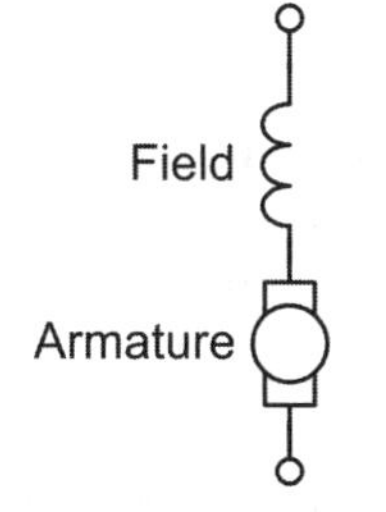

Figure 4.48 *DC motor with a series field winding*

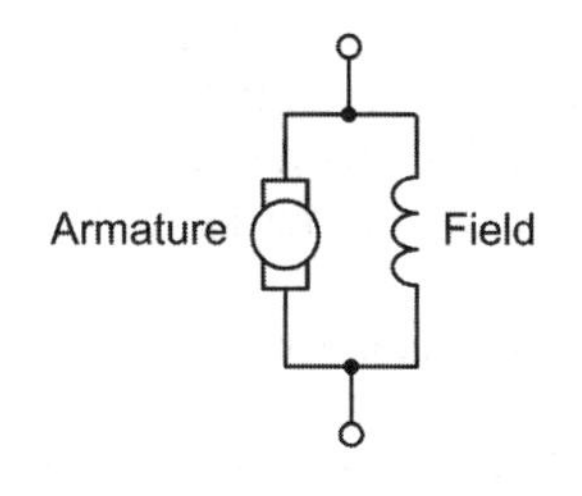

Figure 4.49 *DC motor with a parallel field winding*

passed through it. This arrangement constitutes a *field winding* and each of the turns in the field winding assists each of the other turns in order to produce a strong magnetic field, as shown in Figure 4.47. The field winding may be connected in series or in parallel with the armature as shown in Figures 4.48 and 4.49.

Now returning to the simple motor illustrated in Figure 4.46, we know that when current is supplied to the armature (*rotor*) a torque is produced. In order to produce continuous rotary motion, this torque (turning moment) must always act in the same direction.

Therefore, the current in each of the armature conductors must be reversed as the conductor passes between the North and South magnetic field poles. The *commutator* acts like a rotating switch, reversing the current in each armature conductor at the appropriate time to achieve this continuous rotary motion.

In Figure 4.50(a) the rotation of the armature conductor is given by Fleming's left-band rule. When the coil reaches a position mid-way way between the poles (Figure 4.50b), no rotational torque is produced in the coil. At this stage the commutator reverses the current in the coil. Finally (Figure 4.50c) with the current reversed,

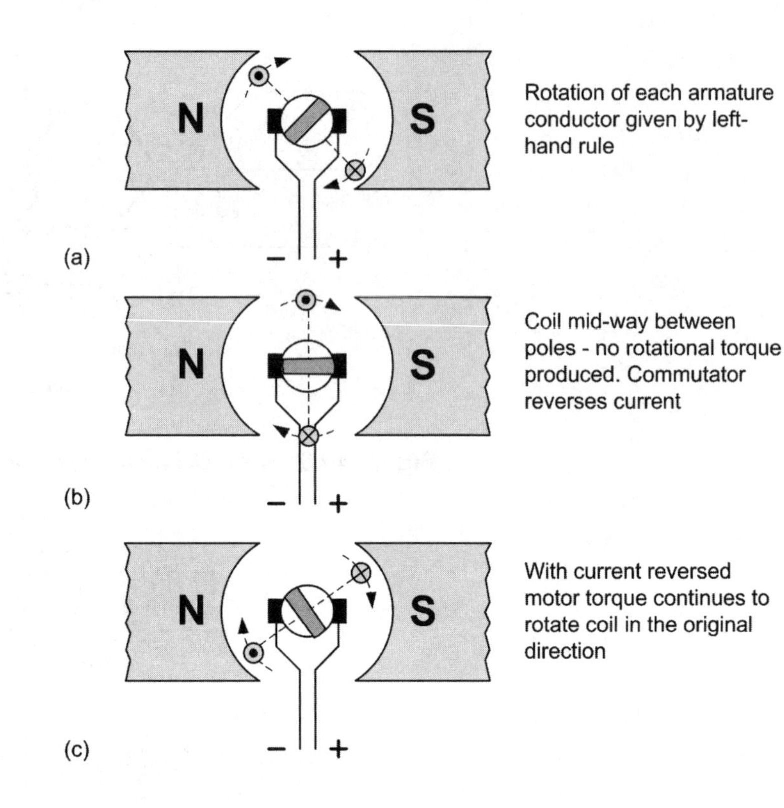

Figure 4.50 *Action of the commutator*

the motor torque now continues to rotate the coil in its original direction.

The field winding of a DC motor can be connected in various different ways according to the application envisaged for the motor in question. The following configurations are possible:

- series-wound (Figure 4.48)
- shunt-wound (Figure 4.49)
- compound-wound (where both series and shunt windings are present as shown in Figure 4.51)

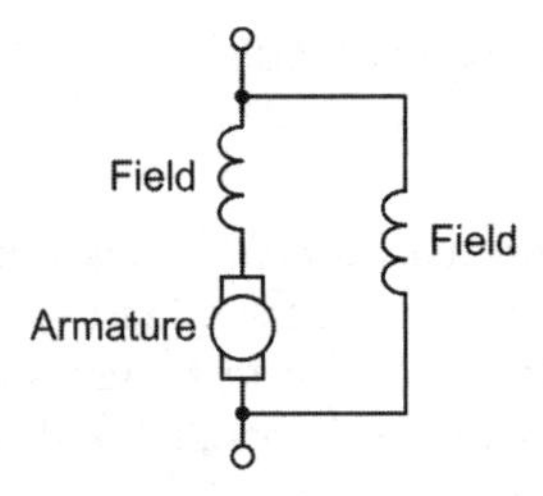

Figure 4.51 *Compound wound DC motor*

In the series-wound DC motor the field winding is connected in series with the armature and the full armature current flows through the field winding (see Figure 4.48). This arrangement results in a DC motor that produces a large starting torque at slow speeds. This type of motor is ideal for applications where a heavy load is applied from rest. The disadvantage of this type of motor is that on light loads the motor speed may become excessively high. For this reason this type of motor should not be used in situations where the load may be accidentally removed.

In the shunt-wound DC motor the field winding is connected in parallel with the armature and thus the supply current is divided between the armature and the field winding (see Figure 4.51). This arrangement results in a DC motor that runs at a reasonably constant speed over a wide variation of load but does not perform well when heavily loaded.

The compound-wound DC motor has both series and shunt field windings (see Figure 4.51) and therefore combines some of the properties of each type of motor.

AC motors

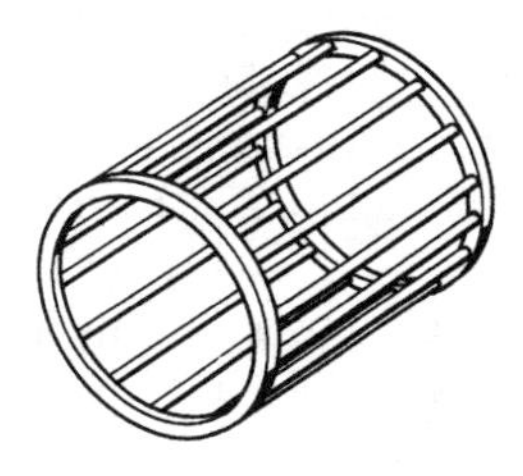

Figure 4.52 *Squirrel cage rotor of an AC induction motor*

AC motors are used to produce motion from alternating current (AC) rather than direct current (DC). AC motors offer significant advantages over their DC counterparts. AC motors can, in most cases, duplicate the operation of DC motors and they are significantly more reliable. The main reason for this is that the commutator arrangements (i.e. brushes and slip rings) fitted to DC motors are inherently troublesome. Because the speed of an AC motor is determined by the frequency of the AC supply that is applied it, AC motors are well suited to constant-speed applications.

The principle of all AC motors is based on the generation of a rotating magnetic field. It is this rotating field that causes the motor's rotor to turn. AC motors are generally classified into two types:

- synchronous motors
- induction motors.

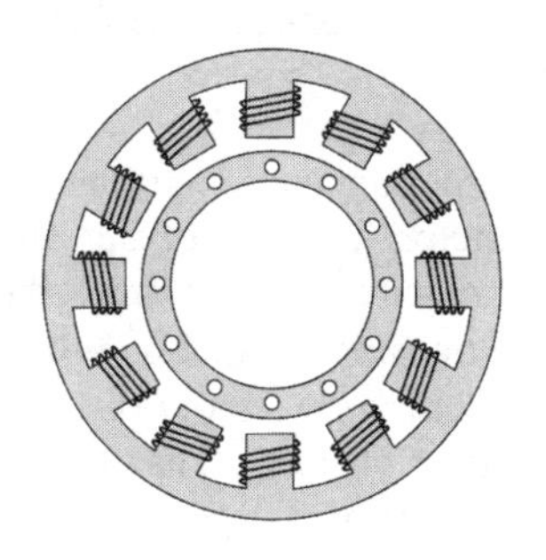

Figure 4.53 *Typical stator construction for an AC induction motor*

The *synchronous motor* is effectively an AC generator (i.e. an alternator) operated as a motor. In this machine, AC is applied to the stator and DC is applied to the rotor.

The *induction motor* is different in that no source of AC or DC power is connected to the rotor. Of these two types of AC motor, the induction motor is by far the most commonly used. Figures 4.52 and 4.53 show the basic construction of an AC induction motor whilst Figure 4.54 shows the force acting on the rotor. Regardless of the type of motor, the basic principle of operation of an induction motor is the same. The rotating magnetic field generated in the stator induces an e.m.f. in the rotor. The current in the rotor circuit caused by this induced e.m.f. sets up a magnetic field. The two fields interact, and cause the rotor to turn in the same direction as the rotating magnetic flux generated by the stator, as shown in Figure 4.54.

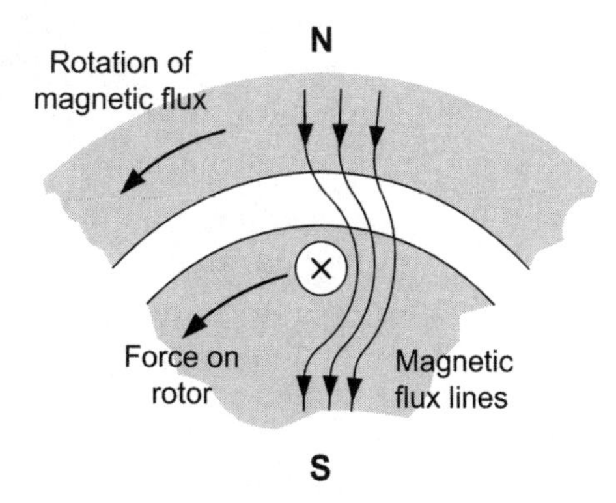

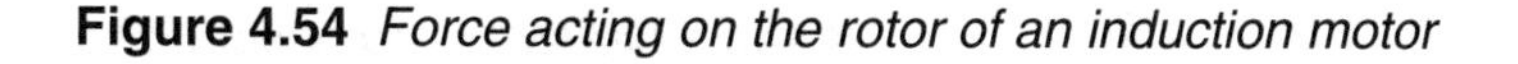

Figure 4.54 *Force acting on the rotor of an induction motor*

Stepper motors

Stepper motors provide an accurate means of controlling and positioning a variety of different mechanisms, both linear and rotary. Stepper motors can be easily interfaced to digital circuitry and this makes them ideal for use in microprocessor or microcomputer-based systems. A wide variety of different types and styles of stepper motor are currently available to the engineer suitable for applications that can vary from feeding paper into a printer to rotating a platform on a battleship.

Stepper motors are a form of synchronous motor in which the magnetic field is electronically switched in order to cause the armature to rotate to the required position and/or at the required speed. Stepper motors have no brushes or contacts. They are thus highly reliable.

A complete stepper motor system consists of several elements, usually combined with some form of command signal interface, as shown in Figure 4.56. The individual command signals (direction, step, reset, etc) usually appear at conventional TTL logic levels (i.e. 0V to +5V) and they may be derived from conventional switches or from more a central logic control system used to control a number of devices (such as other stepper motors, actuators, heaters, warning devices, etc).

The indexing logic (sometimes also referred to as a 'controller' or 'indexer') is a circuit that generates the required sequence of step and direction pulses in order to produce the required motion from the stepper motor. Each pulse is a digital logic transition, either 0–1–0 or 1–0–1. The indexing logic requires a standard logic supply (usually +5V) at a modest current (typically less than 50 mA). The stepper motor, on the other hand, requires a high current (frequently more

Figure 4.55 *A typical stepper motor. This motor operates from a 24V DC supply. The winding resistance of the stepper motor is 80 Ω and it requires a current of 0.3 A. The motor generates a step angle of 1.8° and thus 200 steps (in the same direction) are needed to perform a full revolution.*

than 250 mA) and thus some form of power amplifier (or *driver*) is required between the indexing logic output and the stepper motor coils.

In the most simple of applications, the input to the stepper motor controller board can consist solely of the three basic command signals, STEP, DIRECTION and RESET but in more sophisticated applications some form of 'intelligent control' is incorporated so that the stepper motor controller is able to accept high-level language commands.

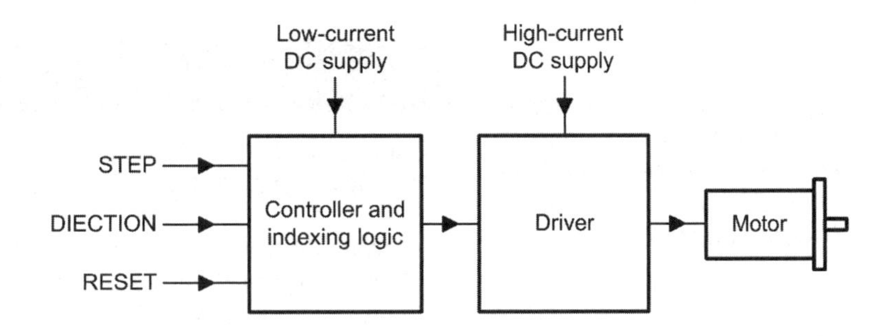

Figure 4.56 *A typical stepper motor controller*

The construction of a simple variable reluctance stepper motor is shown in Figure 4.57. The motor has three windings which are used to energise opposite pairs of poles. The windings and their respective pole pairs are spaced by an angle of 60°. The stator consists of a low-reluctance high-permeability material which supports a magnetic flux when any one of the windings is energised by applying a current to it. The rotor has four 'teeth' and is free to rotate inside the stator and its six poles. The rotor is supported by bearings (not shown in Figure 4.57). Like the stator, the rotor is made from low-reluctance high-permeability material. One end of each of the three windings is connected to a common supply rail whilst the other ends are supplied from the output of a driver stage. In most practical applications, the common wire typically goes to the positive supply.

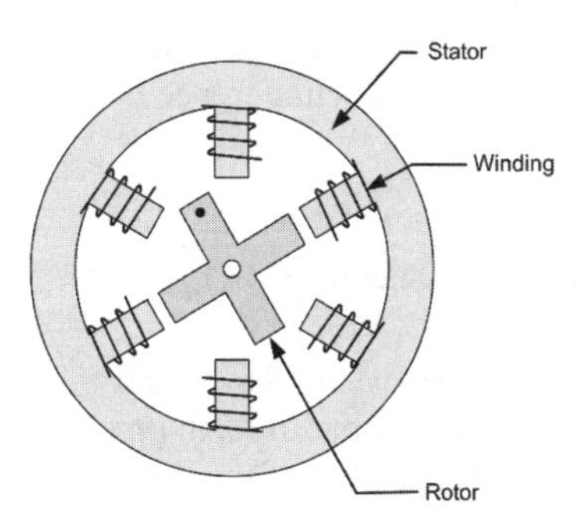

Figure 4.57 *Construction of a simple stepper motor*

In operation, the windings are energized in sequence. Depending upon which winding is energized, the rotor will move due to the magnetic forces present in the space between the rotor and stator. Assume that the two windings are energized as shown in Figure 4.58. The rotor will move from its initial position (as shown in Figure 4.58) so that it aligns with the north (N) pole and south (S) pole produced by the induced magnetic flux. The rotor will be held in this position for as long as the two windings remain energized.

Now assume that the next pair of windings (moving in a clockwise direction) becomes energized. The *nearest* pair of rotor teeth will move in order to support the passage of the induced magnetic flux. This will result in a 30° anti-clockwise movement of the rotor as shown in Figure 4.59.

Next assume that the third pair of windings (continuing to move in a clockwise direction) becomes energized. The rotor teeth will move

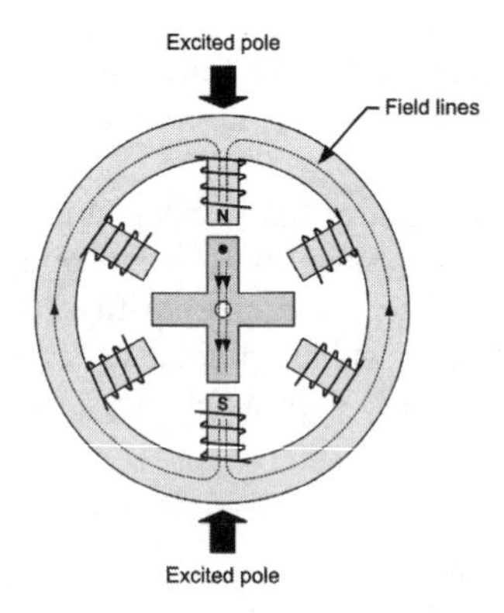

Figure 4.58 *The stepper motor's rotor will move from its initial position when one of the pairs of windings become energized*

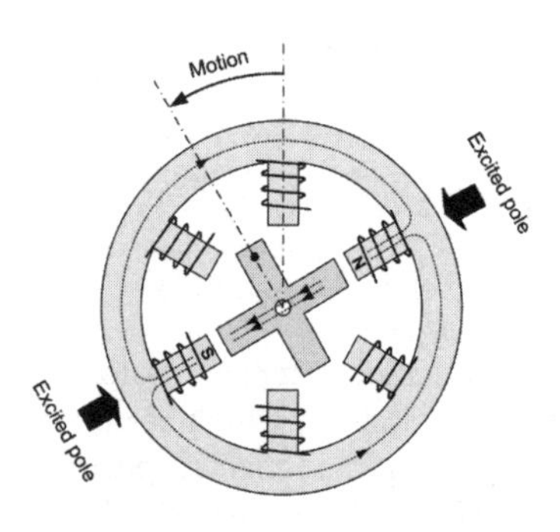

Figure 4.59 *When the next pair of windings becomes energized the rotor will move one step, as shown. Note that the rotor has moved through an angle of 30º*

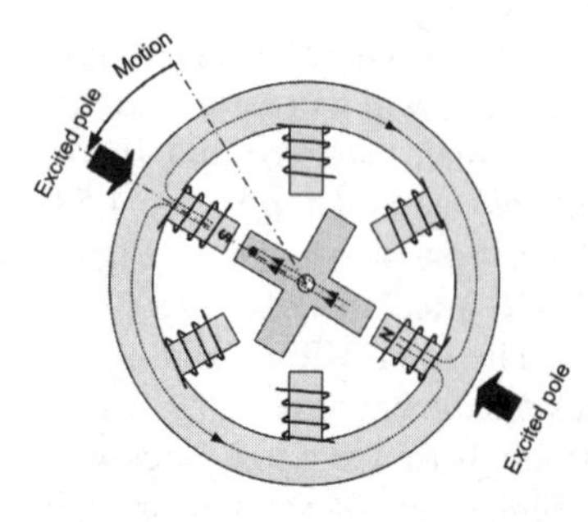

Figure 4.60 *When the third pair of windings becomes energized the rotor will move a further 30º step, as shown*

through another step in order to support the passage of the induced magnetic flux. This will result in a further 30º anti-clockwise movement of the rotor as shown in Figure 4.60. If the windings are energised in turn in a similar manner, continuous rotation will occur. A full-revolution of the rotor will be produced as a result of a total of twelve steps. The fundamental stepping angle of this simple motor will thus be 360°/12 = 30°. In practice stepper motors have many more poles and windings and they produce much smaller step angles as a result.

Activity 4.7

Obtain a data sheet for a stepper motor and use it to answer the following questions:

1. What is the fundamental step angle of the motor?
2. How many steps are there in one complete revolution of the motor's output shaft?
3. Sketch a diagram to show how the windings are arranged on the motor.
4. What is the DC resistance of the motor's windings?
5. What is the typical operating voltage and current for the motor?

Gears and gear trains

Gears and arrangements of multiple gears (known as *gear trains*) are used to transmit rotary motion (such as that produced by a motor). Since they are able to change both the magnitude and line of action of a force they comprise a simple *machine*. A simple gear train consists of two *spur gears* (as shown in Figure 4.61a). One of these gears is driven (and known as the *driver*) and the other one is connected to the load (and is known as the *follower*). The teeth of both the driver and follower are spaced so that they exactly fill the circumference of each wheel and also that they mesh together without any interference. The number of teeth on the driver and follower will thus be in direct proportion to the circumference of the two wheels. Hence:

$$\frac{\text{Number of teeth on driver, } T_A}{\text{Number of teeth on follower, } T_B} = \frac{\text{Circumference of driver, } l_A}{\text{Circumference of follower, } l_B}$$

The ratio of the number of revolutions made by the driver and the follower will be inversely proportional to the number of teeth present (the fewer the number of teeth the greater the number of revolutions). Hence:

$$\frac{\text{Number of revolutions made by driver}}{\text{Number of revolutions made by follower}} = \frac{\text{Number of teeth on follower, } T_B}{\text{Number of teeth on driver, } T_A}$$

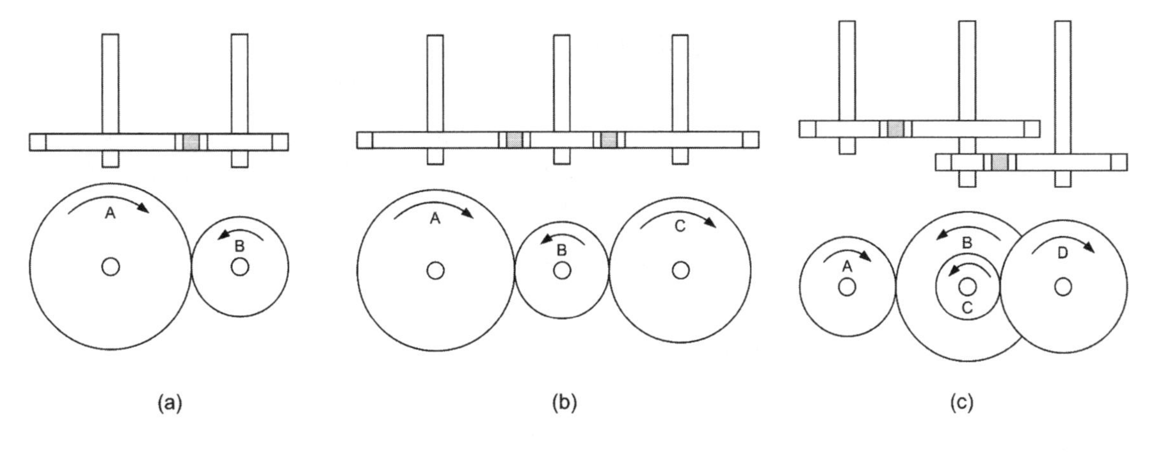

Figure 4.61 *Gear train arrangements*

It follows from this that the speed of rotation of the driver and follower will also be inversely proportional to the ratio of the number of teeth. Hence:

$$\frac{N_B}{N_A} = \frac{T_A}{T_B}$$

where N_B and N_A are the speed of rotation of the follower and driver respectively, and T_A and T_B are the number of teeth on the driver and follower respectively.

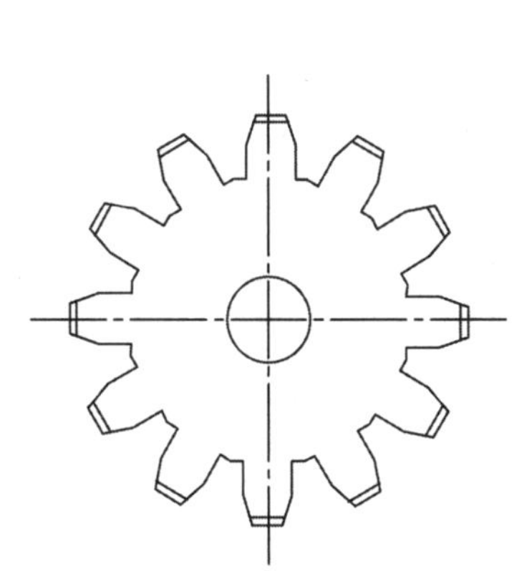

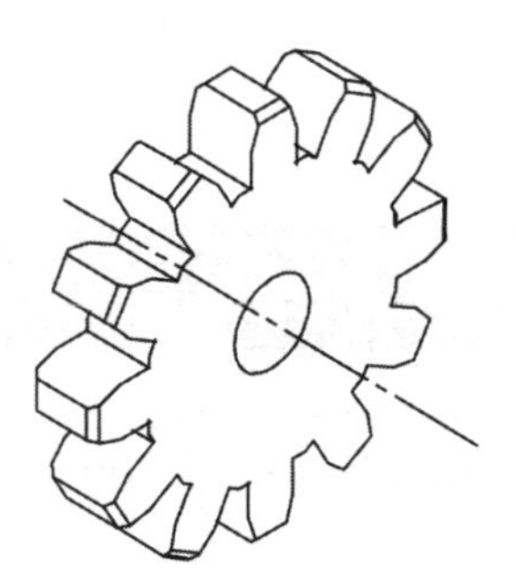

Figure 4.62 *A gear wheel*

Figure 4.63 *A typical gear train fitted to a DC motor. The driver (upper left, centre) is attached to the motor whilst the final follower (upper right) drives the load*

A *compound gear* arrangement is shown in Figure 4.61(b). This arrangement uses three gear wheels and the intermediate gear (B) is known as an *idler*. It is also important to note that, with an odd number of gears in the train, the output from the follower will be in the same direction, either clockwise (CW) or anti-clockwise (ACW), as that of the driver. Now:

$$\frac{N_B}{N_A} = \frac{T_A}{T_B} \quad \text{and} \quad \frac{N_C}{N_B} = \frac{T_B}{T_C}$$

Hence:

$$\frac{N_C}{N_A} = \frac{N_B \dfrac{T_B}{T_C}}{N_B \dfrac{T_B}{T_A}} = \frac{T_B}{T_C} \times \frac{T_A}{T_B} = \frac{T_A}{T_C}$$

This result shows that the ratio of driver speed to follower speed is independent of the idler (and so the idler affects the direction of the output but *not* the speed).

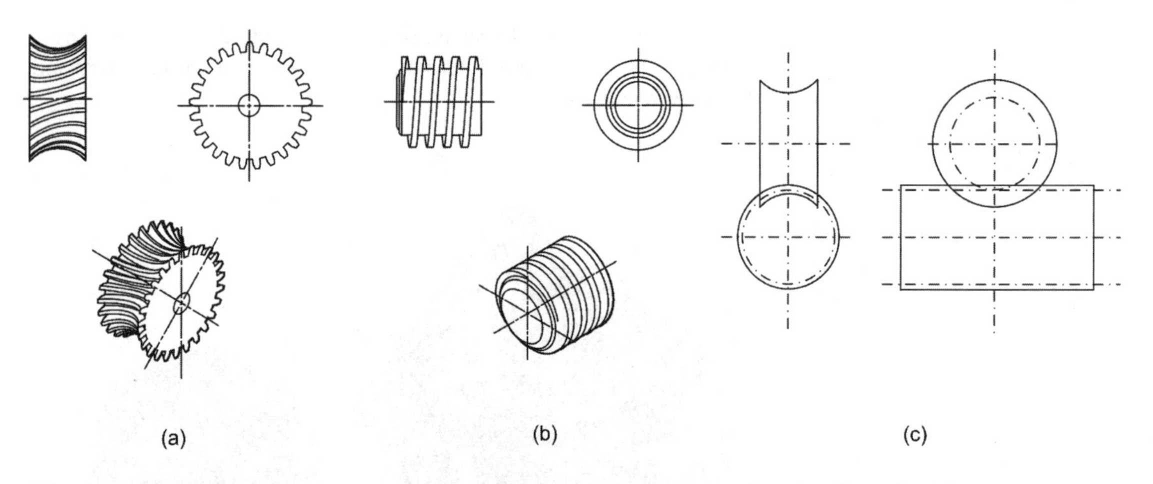

Figure 4.64 *A worm drive arrangement can be used for changing the line of action*

Activity 4.8

A more complex compound gear train arrangement is shown in Figure 4.61(c). In this arrangement gear wheels B and C are *coaxial.*

1. Obtain an expression for the output shaft speed in terms of the input shaft speed.

2. If the number of teeth on A, B, C and D are respectively 24, 42, 18, and 36 respectively, determine the output shaft speed if the input shaft is driven at a speed of 50 revolutions per second.

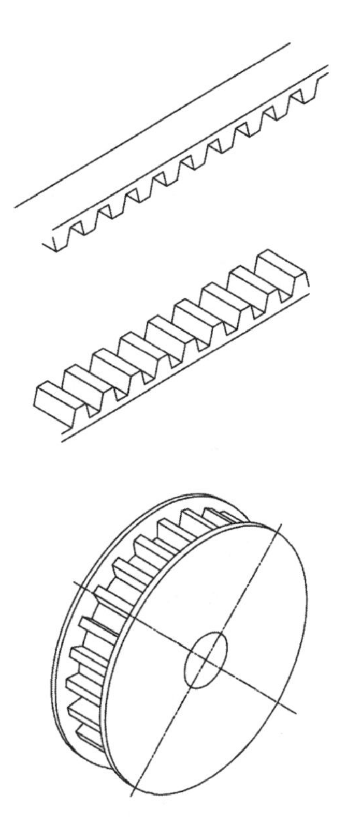

Figure 4.65 *A typical toothed belt drive*

Belt and chain drives

Belt and chain drives provide alternatives to using gear trains as a means of transmitting power from one shaft to another. A belt drive is based on a belt which passes over a series of pulley wheels. The belt may be flat or V-shaped and either keyed to the pulley wheels or it may be toothed (as shown in Figure 4.65).

It is important to note that, regardless of the shape of the belt, for a belt to transmit power there must be a difference in tension on either side of the driver and driven pulleys.

The basic arrangement of a belt drive is shown in Figure 4.66. By similar reasoning to that which we used earlier in relation to a simple gear train, we can conclude that, for a toothed belt drive:

$$\frac{N_B}{N_A} = \frac{T_A}{T_B}$$

Similarly, where the drive uses a flat or V-belt then:

$$\frac{N_B}{N_A} = \frac{d_A}{d_B}$$

Where d_A and d_B are the diameters of the pulley wheels, A and B respectively.

When a belt transmits power, there is a difference in tensions between the tight and slack side due to the friction force between the belt and pulley. The power transmitted is equal to the difference in tensions, T_1 and T_2, and the speed, v, at which the belt is travelling. Hence:

Power transmitted, $P = (T_1 - T_2) \times v$

Chain drives are similar to belt drives but they use a chain based on a series of links (see Figure 4.67a) and, instead of pulleys, toothed sprocket wheels (see Figure 4.67d) are fitted to the driver and driven shafts.

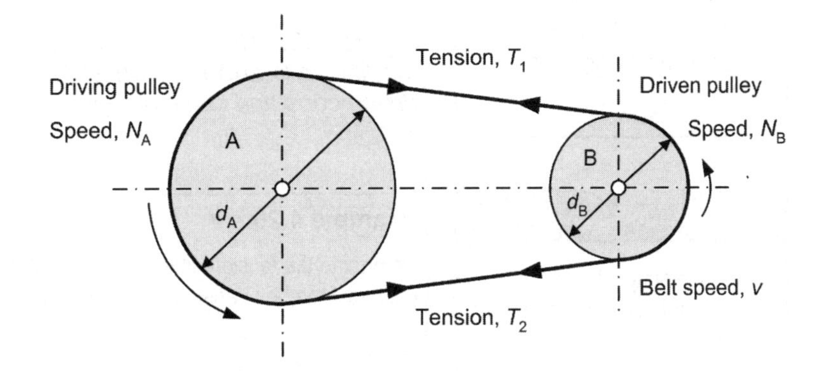

Figure 4.66 *A belt drive*

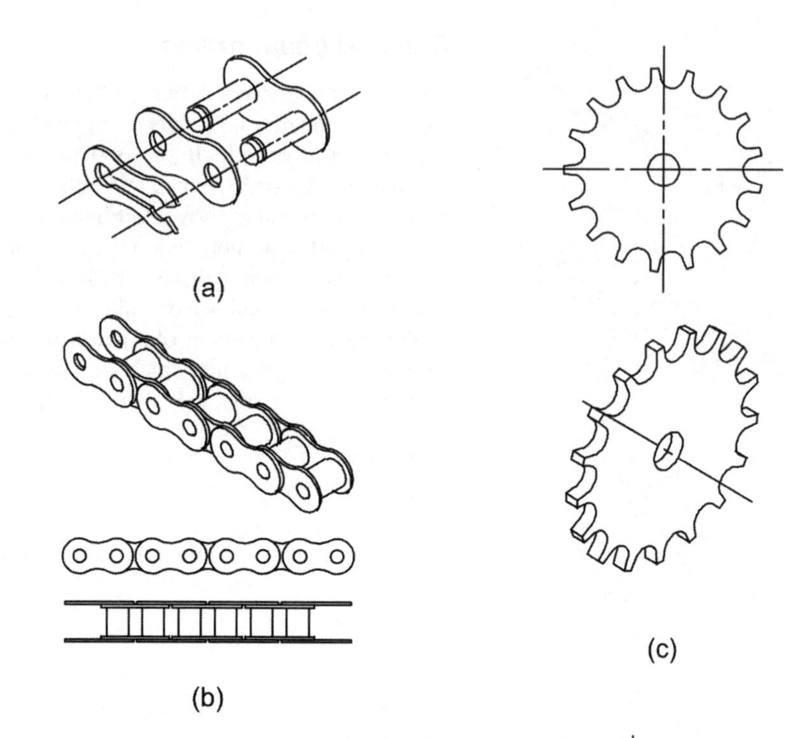

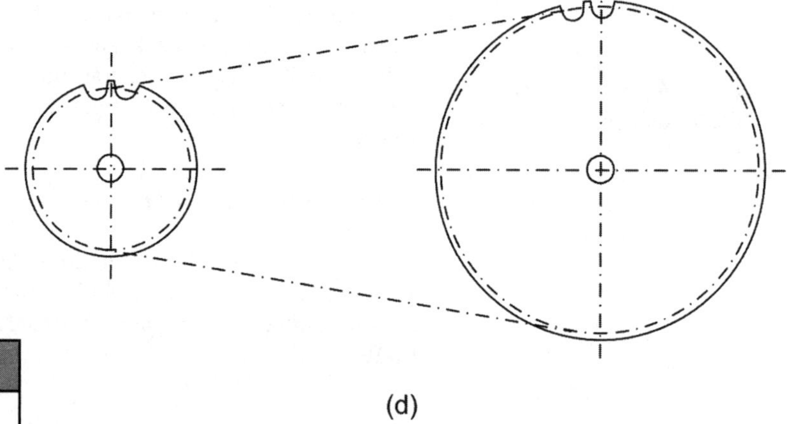

Figure 4.67 *Chain drive components: (a) and (b) show how the links are arranged whilst (c) and (d) show the driving and driven cogs and chain geometry*

Test your knowledge 4.17

Explain, with the aid of a sketch, how a worm drive operates and how it is able to change the line of action of a rotating shaft system.

Test your knowledge 4.18

1. Sketch a simple belt drive arrangement. Label your diagram clearly.

2. Sketch a simple chain drive arrangement. Label your diagram clearly.

Example 4.20

Determine the tension difference in a belt drive moving at a velocity of 0.5 m/s if the belt transmits a power of 1.5 kW.

Re-arranging $P = (T_1 - T_2) \times v$ to make $(T_1 - T_2)$ the subject gives:

$(T_1 - T_2) = P / v = 750 / 0.5 =$ **1.5 kN**

Cranks and linkages

Cranks and linkages are rigid elements that transfer motion from one part of a mechanical system to another. Two linkage mechanisms are shown in Figure 4.69. In Figure 4.69(a), rotary motion is converted into a *reciprocating* (or *linear translational*) motion. A typical example of this type of linkage exists in an internal combustion engine where a piston imparts motion through a crank to a rotating shaft (see Figure 4.70).

Figure 4.69(b) shows a four-bar linkage which produces a rocking motion in the output link. The rocking action is determined by the lengths and position of the links.

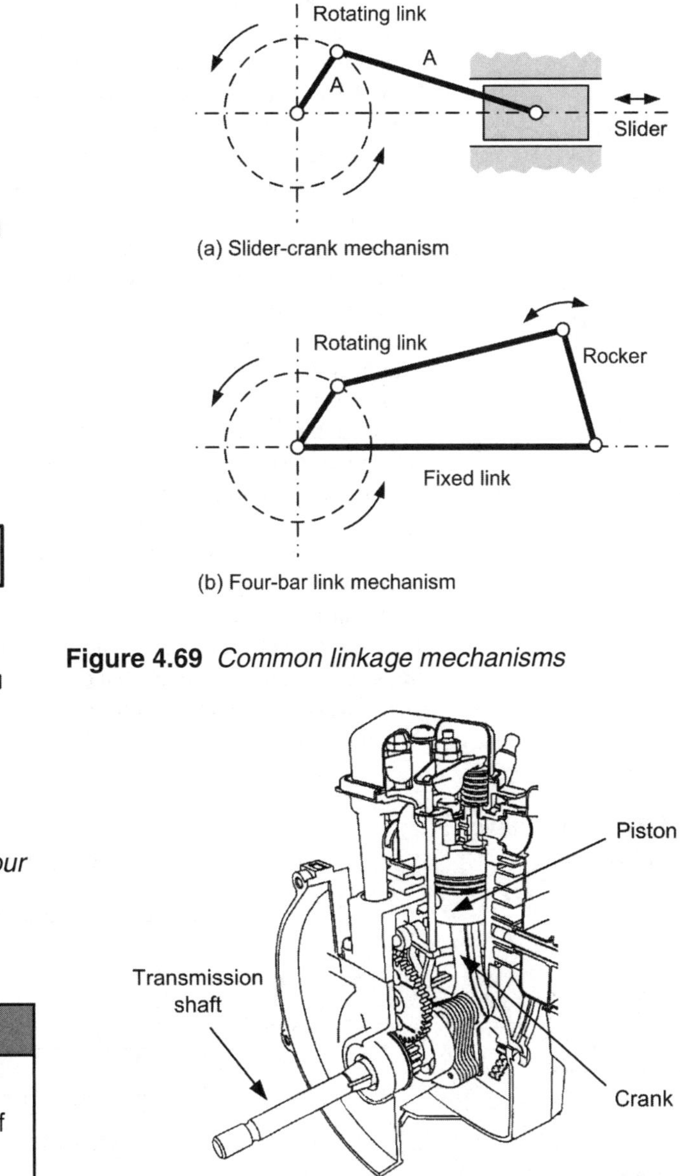

Figure 4.69 *Common linkage mechanisms*

Figure 4.70 *Cutaway view of a simple internal combustion engine*

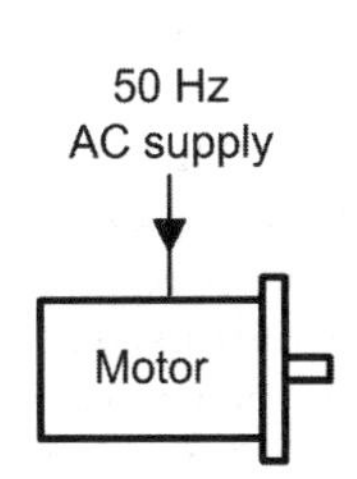

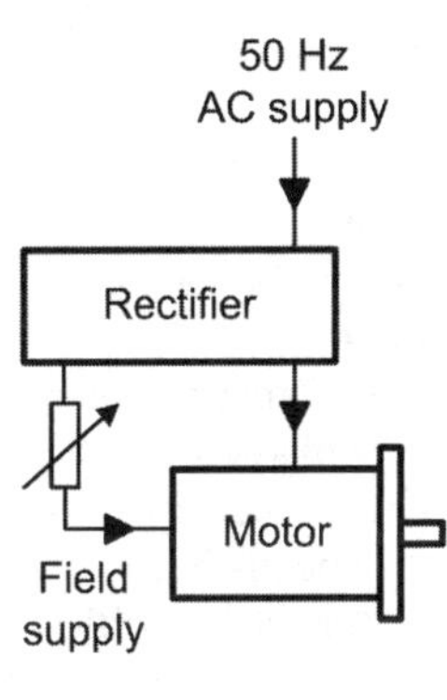

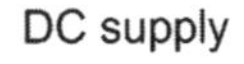

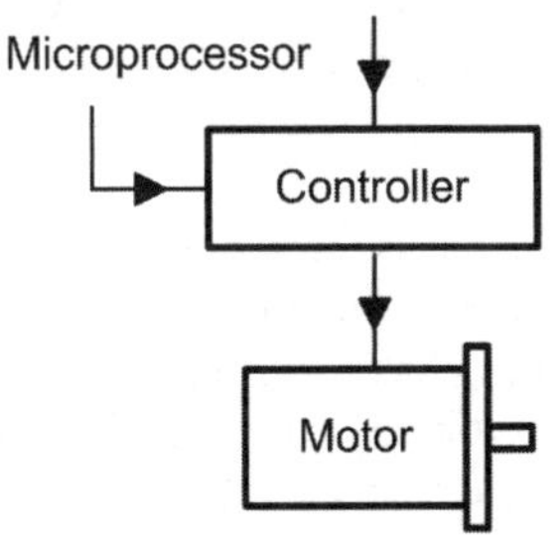

Figure 4.68 *See Test your knowledge 4.19*

Test your knowledge 4.19

Figure 4.68 shows three motor applications. Identify the type of motor used in each of these applications. Give reasons for your answer.

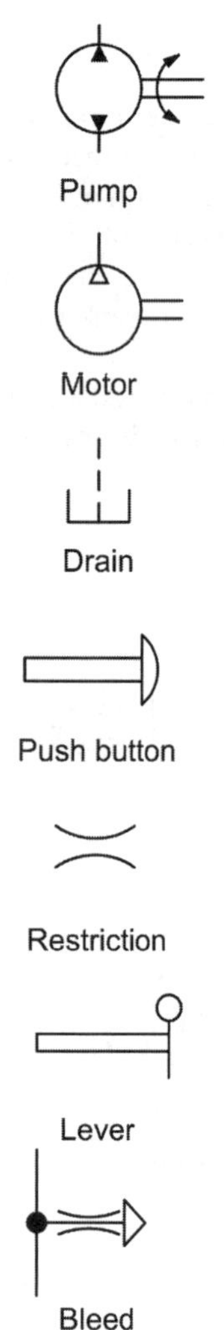

Figure 4.71 *Pneumatic component symbols*

Pneumatic equipment

In a pneumatic system, energy is stored in a potential state in the form of compressed air. Energy to perform work (kinetic energy) is produced when the compressed air is allowed to expand. This expansion occurs in some form of displacement compressor in which a piston moves a crankshaft when driven by pressure exerted on it.

One of the advantages of transmitting energy pneumatically is that energy can be controlled relatively easily using different types of control valve (the DCV that we met earlier in Unit 2). A further advantage is that air can be transmitted through relatively low-cost arrangement of pipes rather than more expensive insulated copper conductors that would be required for an equivalent electrical energy transmission system.

Before going further it would be worth revising some of the symbols and components that we met in Unit 2 (see pages 221 to 236). However, to remind you, a selection of some of the most commonly used symbols are shown in Figure 4.71.

Each DCV is shown by a compound symbol that not only shows the valve action but also shows the actuators that operate the valve. Since a valve can have several positions (each corresponding to a different flow path) , we need to show what happens in each position. We do this by drawing several boxes, one for each position of the valve (as shown in Figure 4.74).

For the box that corresponds to the condition that exists at the start of a machine cycle, we show the connections to the input and output ports of the valve. Note that we don't show these for the other boxes. Inside each box, we show the state of the valve using a 'T' symbol to indicate a blocked (i.e. closed) port and an arrow to show the direction of flow between any open ports. This probably sounds a lot more complicated than it really is!

The symbols for the valve actuators are drawn at each end of the valve boxes. The rule is that each actuator is drawn next to the condition that exists when the actuator is in command of the valve. In Figure 4.73, for example, when the spring has control of the valve the flow is shown in the left-hand box (Position 1) whereas, when the push-button has control of the valve, the flow is shown in the right-hand box (Position 2).

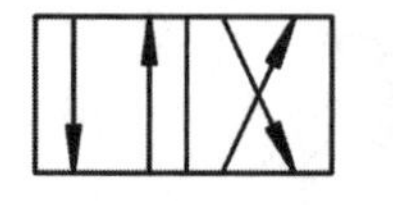

Figure 4.72 *A typical directional control valve (DCV) symbol*

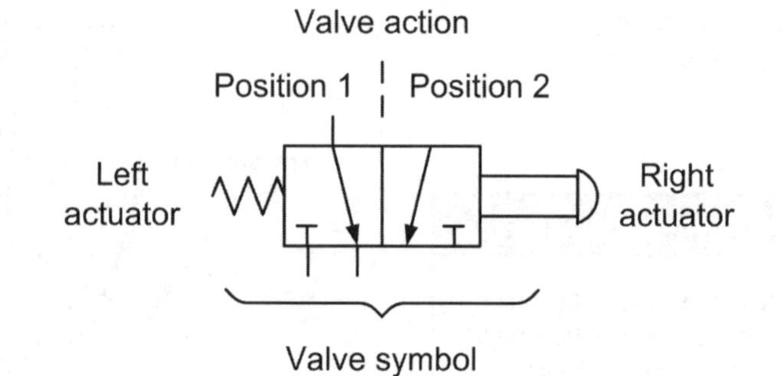

Figure 4.73 *A generic DCV symbol*

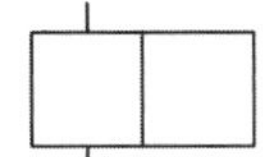

(a) Two-position two-port

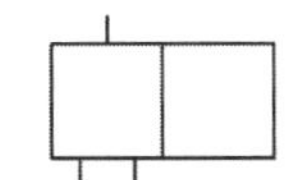

(b) Two-position three-port

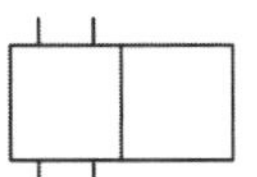

(b) Two-position four-port

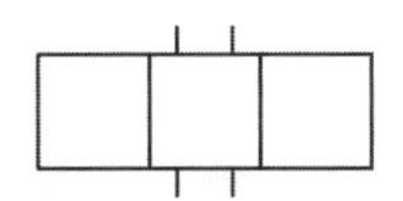

(c) Three-position four-port

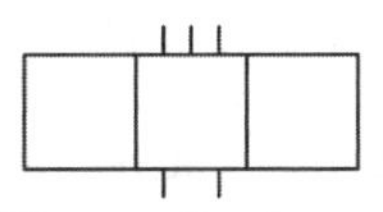

(d) Three-position five-port

Figure 4.74 *Symbols for different types of valve*

Figure 4.75 *A typical solenoid operated two-position, three-port DCV*

Figure 4.76 shows the operation of a two-position, two-port DCV that is open at the start of a cycle and closed at the end of a cycle. Note how the initially open state is shown with the arrow appearing in the left-hand box. Figure 4.77 shows another form of two-position, two-port DCV. This valve is closed at the start of a cycle and becomes open at the end of a cycle.

In order to understand how these valves work it can be useful to compare them with simple electrical switches. The first DCV (Figure 4.76) is similar in action to a normally-open (NO) switch whilst the second (Figure 4.77) is similar to a normally-closed (NC) switch.

A more complicated DCV is shown in Figure 4.78. This shows a two-position, three-port valve. Notice how the direction of flow changes between the two positions.

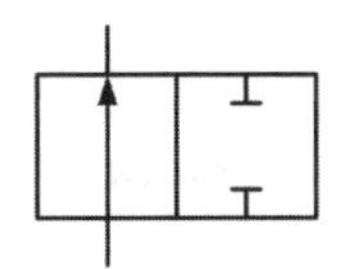

Valve symbol

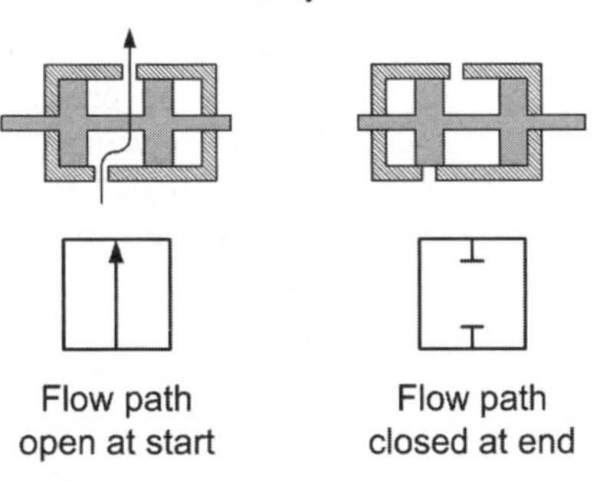

Figure 4.76 *A two-position, two-port DCV, initially open*

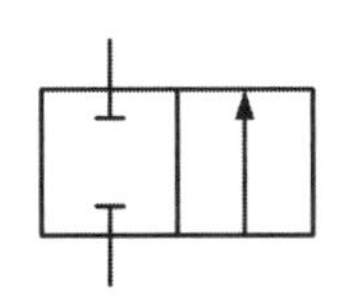

Valve symbol

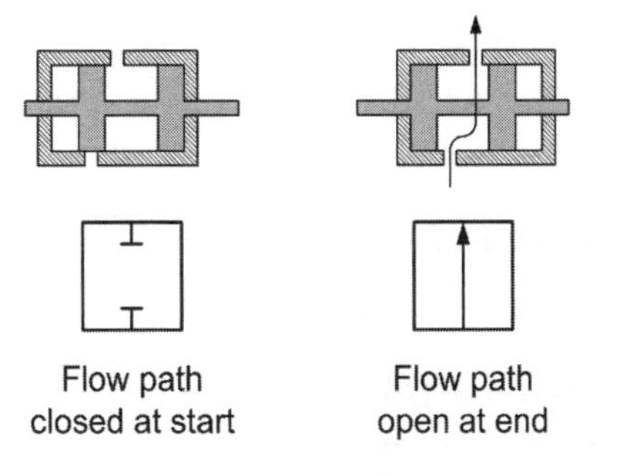

Figure 4.77 *A two-position, two-port DCV, initially closed*

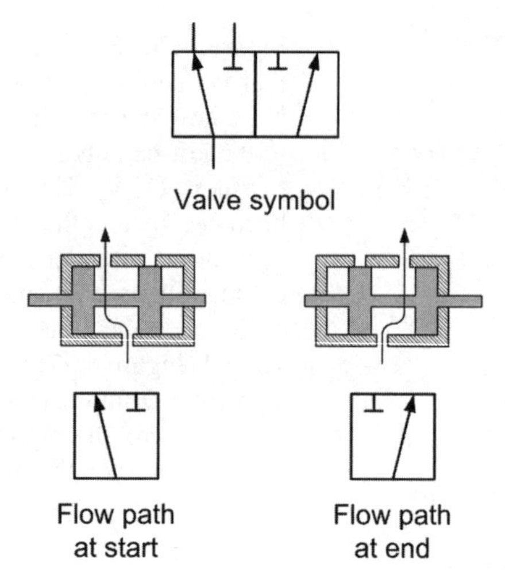

Figure 4.78 *A two-position, three port DCV (see Figure 4.75)*

Figure 4.79 *A pressure relief valve*

Finally, a pressure relief valve is shown in Figure 4.79. Valves of this type are frequently used in pneumatic systems in order to regulate the pressure and to ensure that it does not build up to a dangerous value (which might occur if all of the flow paths in a system should become closed). In the case of the pressure relief valve, the pressure of the incoming air supply acts against a spring. Normally, the pressure force at the input port is less than the force exerted by the spring. In this case, the valve remains closed and there is no flow between the input port and the output port. However, when the pressure force is greater than the force exerted by the spring , the valve will open and air will flow from the input port to the output port.

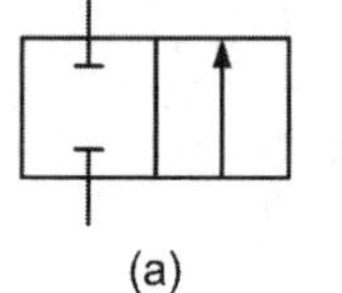

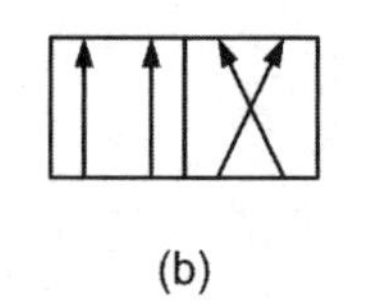

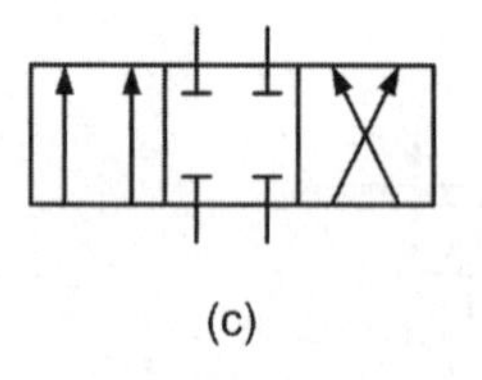

Figure 4.80 *See Test your knowledge 4.20*

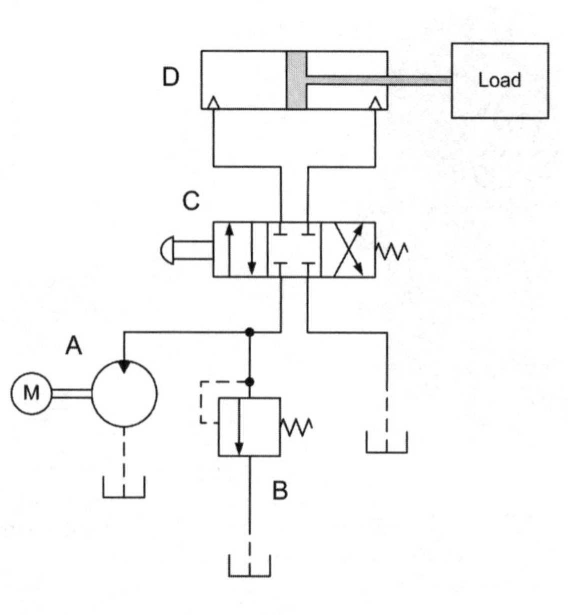

Figure 4.81 *See Activity 4.9*

Test your knowledge 4.20

Identify the three pneumatic components shown in Figure 4.80.

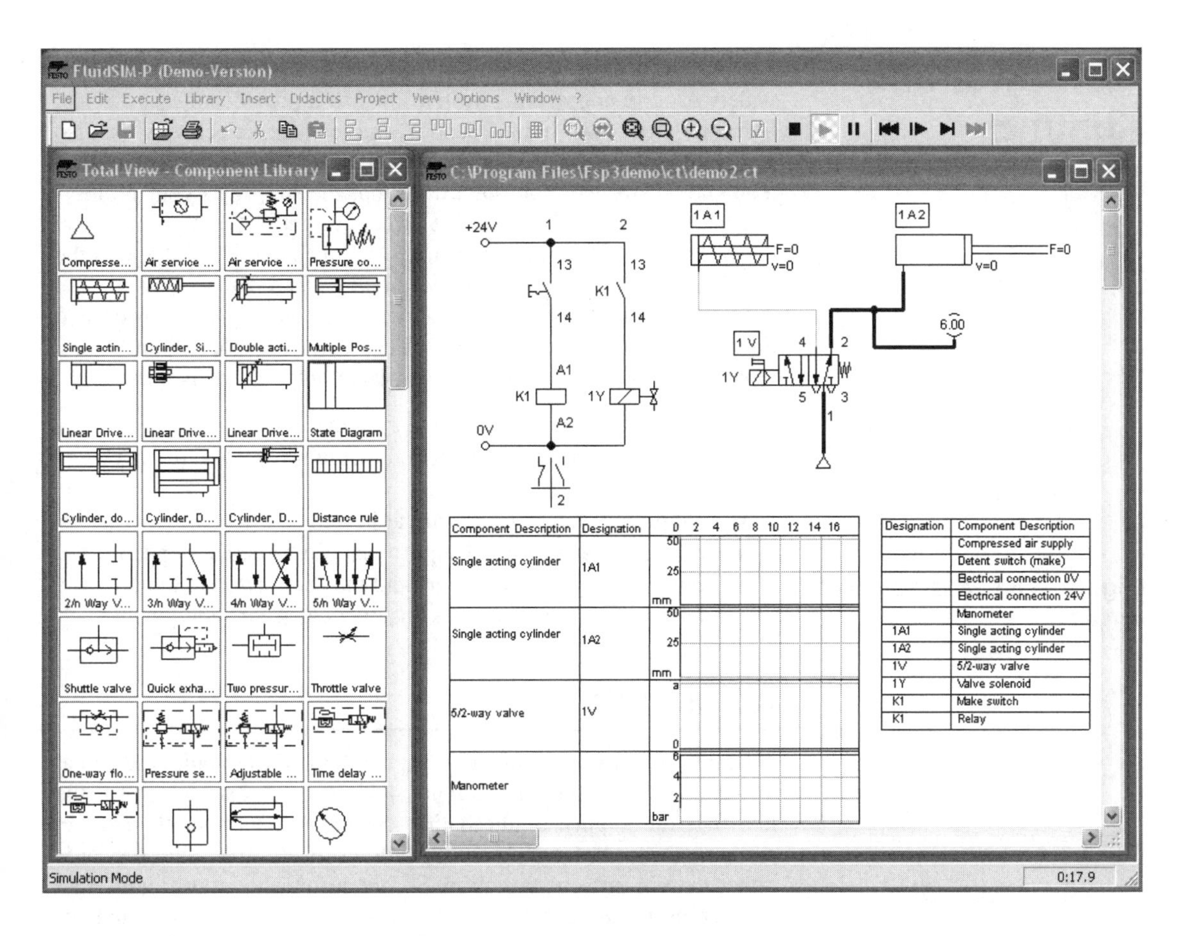

Figure 4.82 *FluidSim software is able to simulate the operation of complex pneumatic and hydraulic systems*

Activity 4.9

Figure 4.81 shows a pneumatic system. Identify components A, B, C and D and explain what the system does and how it operates.

Activity 4.10

Finished parts are accumulating at the end of a conveyer belt. The parts need to be transferred to a second conveyer that will carry them to a final inspection and packaging station. The operator needs to be able to activate and then release a transfer device powered by a pneumatic cylinder.

Design a pneumatic system that will achieve this objective and draw a diagram showing how the various pneumatic components are connected. List the components required to construct the system and write a brief explanation of how the system operates.

4.3 Power and lighting systems

Because of its ease of generation and distribution, electrical energy is widely used in engineering systems. Since it provides the power and light in your own home, school and college, electricity is also something with which you should already be reasonably familiar! You should, for example, be aware that the supply to your home is alternating current (AC) and not direct current (DC). You may also know that the frequency of the supply is 50 Hz and the voltage of the supply is usually somewhere between 220 V and 240 V.

In this third section we shall be investigating the power and lighting systems used in homes, offices and workshops. Some of the typical applications for these *single-phase supplies* are as follows:

- providing power for lighting systems
- providing power for small and medium sized AC motors (see page 297)
- supplying energy for electrical heaters
- delivering an electrical supply that can be easily stepped down (using a *transformer*) and then converted to the direct current (DC) required by consumer electronic equipment.

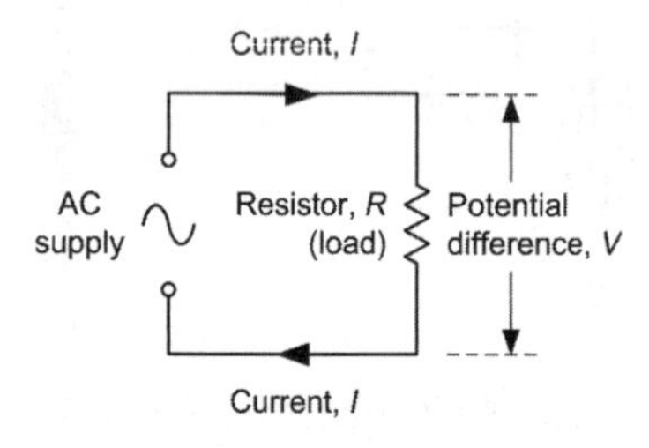

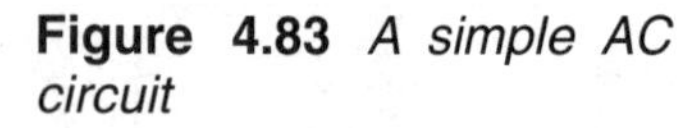
Figure 4.83 *A simple AC circuit*

AC circuits

A simple AC circuit is shown in Figure 4.83. This circuit is identical to the DC circuit shown in Figure 4.41 on page 287 except for the fact that the supply is alternating current (AC) rather than direct current (DC). Alternating current flows backwards and forwards rather than in just one direction. So, the arrow shown in Figure 4.83 just represents a snapshot of the current at a particular instant of time. The difference between AC and DC is shown in Figure 4.84. This shows how the current in the circuit changes with time. Note how cycles of alternating current consists of alternate positive and negative half-cycles.

The frequency of an AC supply is the number of cycles of the current that occurs in a time interval of one second. Frequency is normally expressed in Hertz (Hz). So, for example, in the case of a conventional UK 50 Hz AC supply, the current will reverse fifty times in one second and the time for just one cycle will be a fiftieth of a second (or 20 ms).

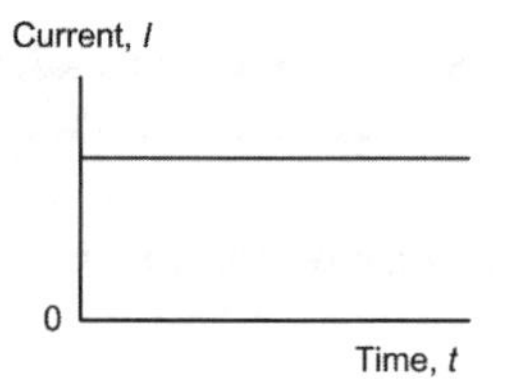

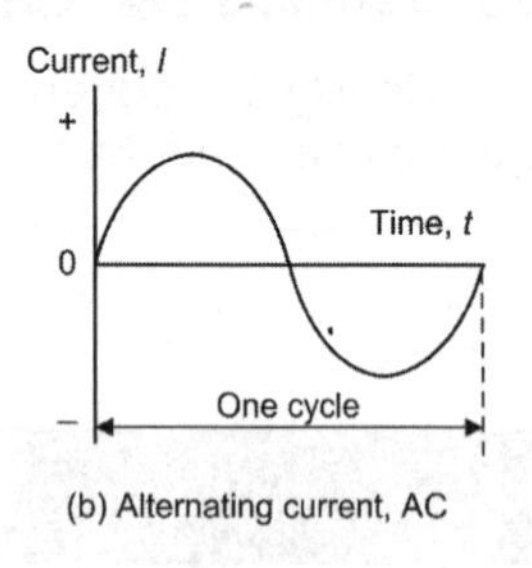

Figure 4.84 *Variation of current with time for (a) DC circuits and (b) AC circuits*

The relationship between the frequency, f, and the time for one complete cycle (the *period*), t, is:

$$f = \frac{1}{t}$$

where f is the frequency in Hertz (Hz) and t is the time in seconds.

Example 4.21

The AC supply in the USA is 60 Hz. What is the periodic time of this supply?

Now $f = 1/t$ so $t = 1/f = 1/60 = 0.01667$ s = **16.67 ms**

Like DC circuits, the voltages and currents in AC circuits are expressed in terms of volts and amps. However, since the *average* value of an alternating current which swings symmetrically above and below zero is zero over one complete cycle, we need to find some other way of expressing the magnitude of the voltages and currents present.

One way of doing this is to use the average value of current or voltage taken over one complete half-cycle (either positive or negative). Alternatively we could use the *peak* value (or *maximum* value) of voltage or current. However, the most useful measure would be that which is the same as the equivalent direct current. This is known as the *root mean square* (*r.m.s.*) or *effective* value of an alternating voltage or current and is the value that would produce the same heat energy in a resistor as a direct voltage or current of the same magnitude.

Since the r.m.s. value of a waveform is very much dependent upon its shape, values are only meaningful when dealing with a waveform of known shape. Where the shape of a waveform is not specified, r.m.s. values are normally assumed to refer to sinusoidal conditions. The relationships between the different ways of expressing voltage and current for a sine wave are shown in Table 4.3.

Table 4.3 *Conversion factors for different ways of specifying sinusoidal current and voltage*

		Wanted quantity:		
		average	*peak*	*r.m.s.*
Given quantity:	*average*	1	1.57	1.11
	peak	0.636	1	0.707
	r.m.s.	0.9	1.414	1

Example 4.22

Determine the peak value of a 230 V r.m.s. AC supply.
To convert from r.m.s. to peak we need to multiply by 1.414. Hence:

$V_{peak} = V_{r.m.s.} \times 1.414 = 230 \times 1.414 =$ **325.22 V**

Test your knowledge 4.21

Explain the difference between AC and DC supplies.

Example 4.23

Determine the r.m.s. value of a current that has a peak value of 27 A.

To convert from peak to r.m.s. we need to multiply by 0.707. Hence:

$I_{r.m.s.} = I_{peak} \times 0.707 = 27 \times 0.707 =$ **19.09 A**

Test your knowledge 4.22

Explain what is meant by the r.m.s. value of sine wave current.

Series and parallel circuits

When more than one load is present in an AC (or DC) circuit, the loads may be connected in series (see Figure 4.85a) or in parallel (see Figure 4.85b) or a combination of both methods.

The equivalent resistance, R_T, of two resistors connected in series (Figure 4.87) is given by:

$$R_T = R_1 + R_2$$

The equivalent resistance, R_T, of two resistors connected in parallel (Figure 4.88) is given by:

$$\frac{1}{R_T} = \frac{1}{R_1} + \frac{1}{R_2}$$

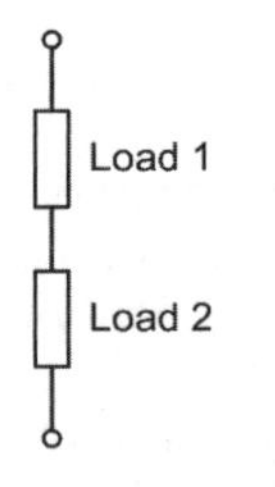

(a) Series connected

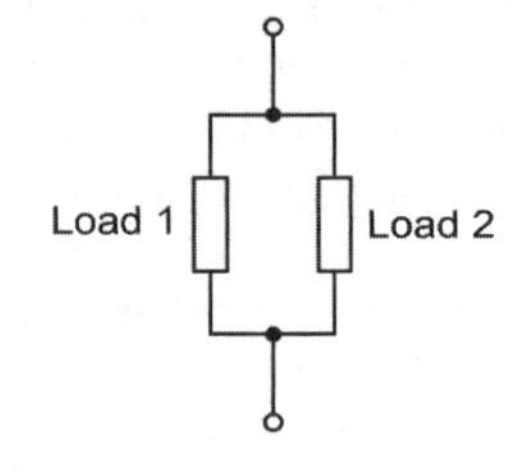

(b) Parallel connected

Figure 4.85 *Series and parallel circuits*

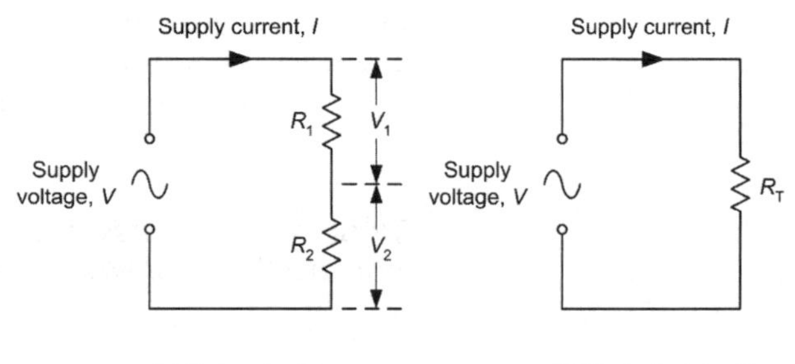

Figure 4.87 *Resistor connected in series*

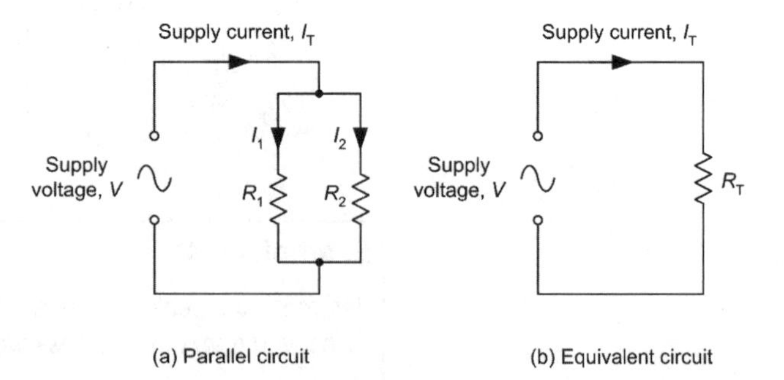

(a) Parallel circuit (b) Equivalent circuit

Figure 4.88 *Resistors connected in parallel*

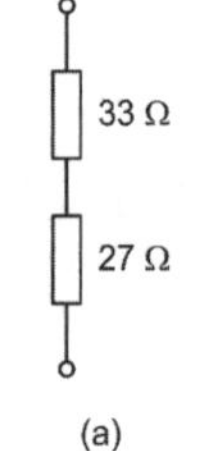

(a)

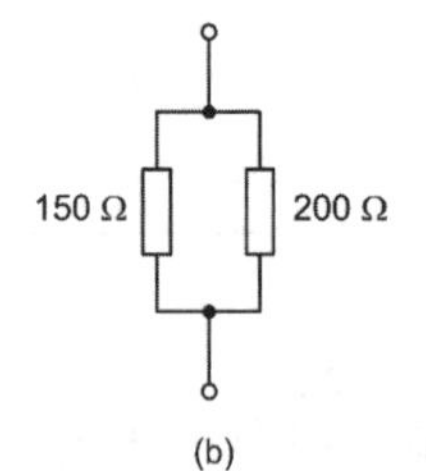

(b)

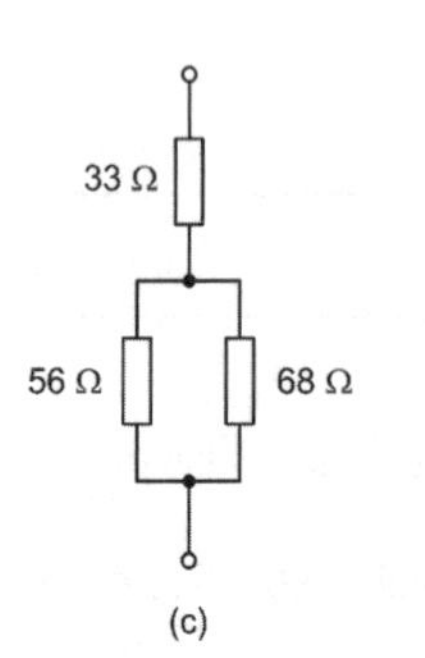

(c)

Figure 4.86 *See Activity 4.11*

Activity 4.11

(a) Determine the equivalent resistance of each of the circuits shown in Figure 4.86.

(b) If each of the circuits shown in Figure 4.86 is connected to a 110 V r.m.s. AC supply, determine the total power supplied to the circuit and the power in each of the loads.

Figure 4.89 *A GLS lamp*

Lighting and power circuit components

General Lighting Service (GLS) lamps

General Lighting Service (GLS) lamps are widely used in industry and in the domestic environment. They are suitable for low-cost uncritical lighting applications and consist of a tungsten filament enclosed in an evacuated glass bulb. Current flowing in the filament causes it to glow and emit visible light. A significant proportion of heat is also produced (only about 5% of the electrical energy supplied to a GLS lamp is turned into useful light energy, the remainder being wasted in the form of heat (see Figure 4.90). Conventional lamps have a limited service life and a typical mean life of between 750 and 1,000 hours is not unusual. The filament of a GLS lamp is coiled in order to allow the gas inside the lamp to circulate freely around it. This helps reduce the filament temperature for a given input power and, as a result, gives a longer working life.

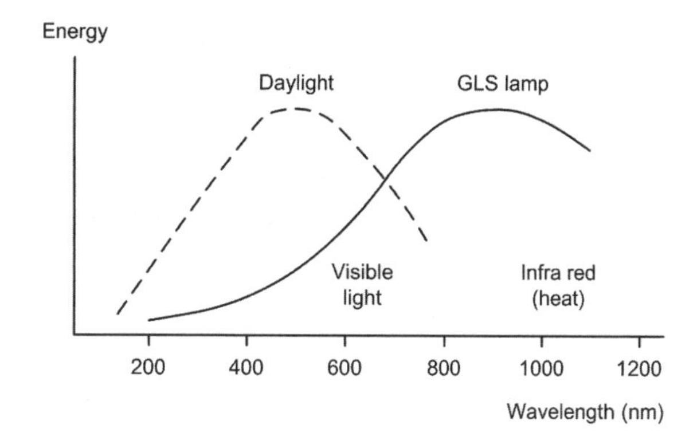

Figure 4.90 *Spectral response of a GLS lamp*

Halogen lamps

Figure 4.91 *A 220 V halogen lamp*

Like GLS lamps, halogen lamps also use a tungsten filament but this is enclosed in a much smaller quartz envelope. Because the envelope is so close to the filament, it would melt if it were made from conventional glass. The envelope is filled with a halogen gas which combines with the tungsten atoms that are evaporated from the filament, allowing them to become redeposit onto the surface of the filament. This helps to improve service life. In addition, the filament can be run hotter which produces more light output per unit of energy than for conventional lamps. It should be noted that this type of lamp runs extremely hot compared with a GLS bulb.

Low-voltage tungsten-halogen technology was developed mainly for automotive and aircraft applications. With extremely compact lamp envelopes and filaments, low-voltage tungsten-halogen lamps can create intense, focused beams.

In order to operate from an AC mains supply, low-voltage halogen lamps require a transformer to produce the low-voltage (usually 12 V) supply, as shown in Figure 4.92. Unfortunately, this transformer adds to system cost, and may potentially cause size, noise, or dimming problems.

Test your knowledge 4.23

Explain why the filament of a GLS lamp is coiled.

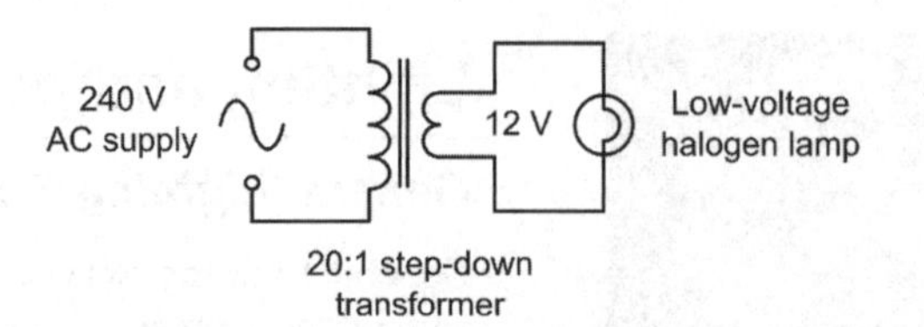

Figure 4.92 *Transformer stepped-down supply for a low-voltage halogen lamp*

Figure 4.93 *A modern fluorescent low-energy replacement for a traditional GLS lamp*

Fluorescent lamps

Fluorescent lamps usually consist of a glass tube filled with argon or krypton gas and containing a few drops of liquid mercury, as shown in Figure 4.94. The interior surface of the glass tube is coated with a fluorescent powder (or *phosphor*) which converts the ultraviolet light produced by the conducting gas inside the tube to visible light. At the end of the tube are the electrodes that alternately serve as the anode and cathode by virtue of the repeated reversal of the alternating supply to which the tube is connected. A starting arrangement (with the starter switch initially closed) is used to heat the cathode and liberate a cloud of electrons. The act of breaking the circuit (when the starter opens) induces a high voltage across the tube produced by the sudden collapse of magnetic flux in the ballast inductor. The excited gas then starts to conduct and visible light is produced. Fluorescent lamps are highly efficient but unsuitable for use where rotating machinery is present due to the strobe action of the light.

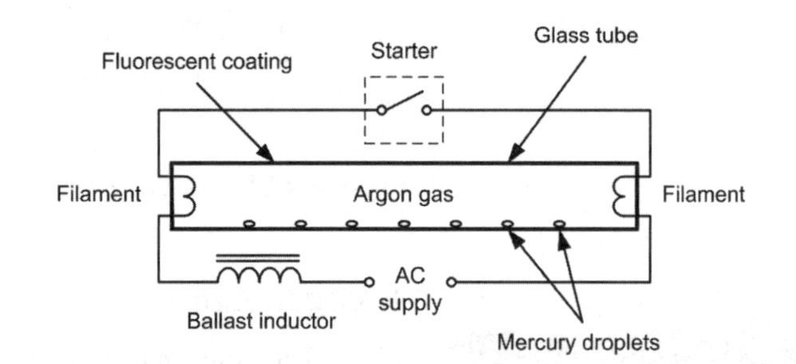

Figure 4.94 *Fluorescent lamp supply*

Cabling, connections and protection

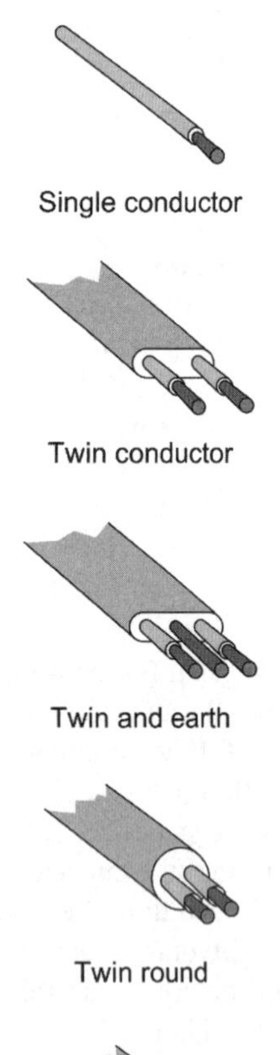

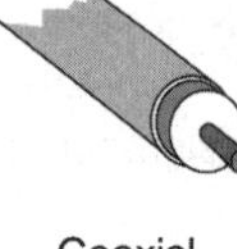

Figure 4.95 *Some different types of cable*

Figure 4.95 shows a variety of different types of cable. Single conductor cable is used extensively for equipment wiring and wiring within junction boxes and consumer units (see Figure 4.96). Semi-rigid flat twin conductor cable is used mainly for lighting circuits (in situations where no separate earth conductor is required) whilst twin and earth is used both for wiring circuits between junction boxes and for ring main power circuits. Flexible twin round cable is used in power leads for appliances that do not require an earth connection (these are often referred to as *double-insulated* appliances). Finally, coaxial cables are used for data and radio frequency (RF) signal applications. This type of cable uses a central conductor surrounded by an earthed screen.

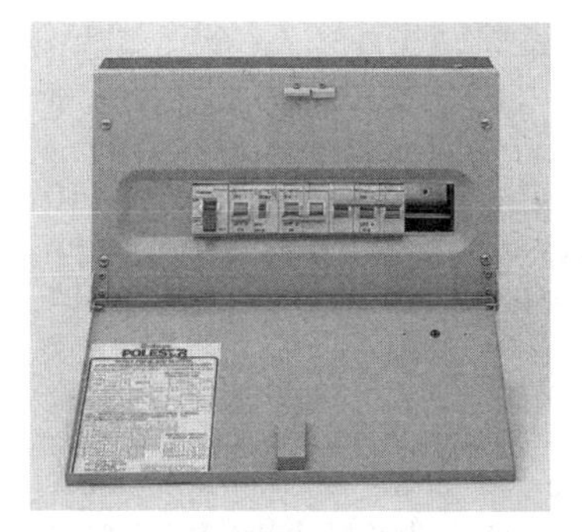

Figure 4.96 *A consumer unit*

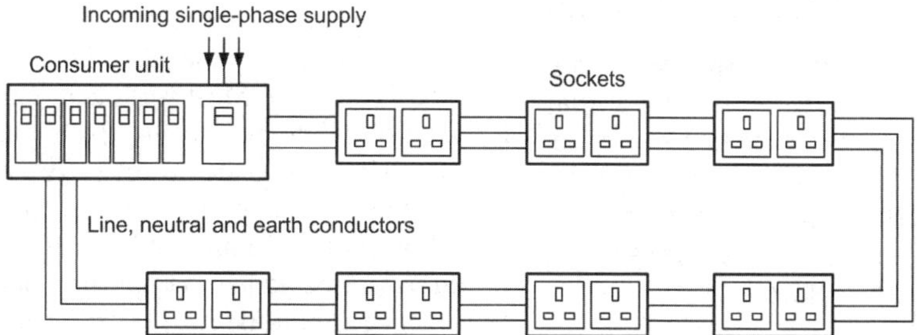

Figure 4.98 *A ring-main circuit*

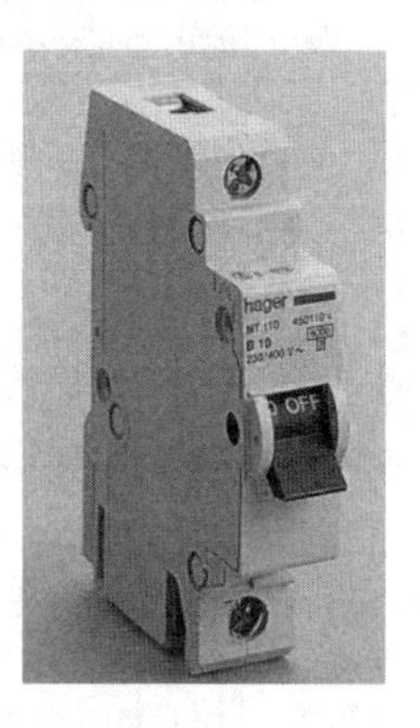

Figure 4.97 *An MCB*

In the UK, mains power in homes and offices is distributed by two main methods; a *ring-main circuit* which distributes power to a number of 13 A sockets, or a *radial circuit* which is used for lighting. Note that it is possible to have several independent ring-main circuits and several independent radial lighting circuits each protected by a *miniature circuit breaker* (MCB) at the consumer unit. The MCB (see Figure 4.97) breaks the circuit whenever there is a demand that is greater than its rated current. However, unlike a cartridge fuse, an MCB can be easily reset once the load or fault has been removed. Typical ratings for an MCB are 6 A for lighting circuits and 32 A for power ring-main circuits. Note that (for obvious reasons) independent lighting and ring-main circuits are normally used for the different floors in a building

A ring-main circuit is shown in Figure 4.98. In this arrangement twin and earth cables are used to carry the line, neutral and earth conductors. A radial lighting circuit is shown in Figure 4.99. The wiring in this circuit is based on twin and earth for the cables that link the junction boxes to the consumer unit and twin only for the wiring to the switches and light fittings.

Test your knowledge 4.24

Sketch the construction of a typical *twin and earth* mains cable.

Test your knowledge 4.25

Explain, briefly, how a fluorescent lamp works.

Test your knowledge 4.26

Sketch a circuit to show how a low-voltage halogen lamp can be connected to an AC mains supply.

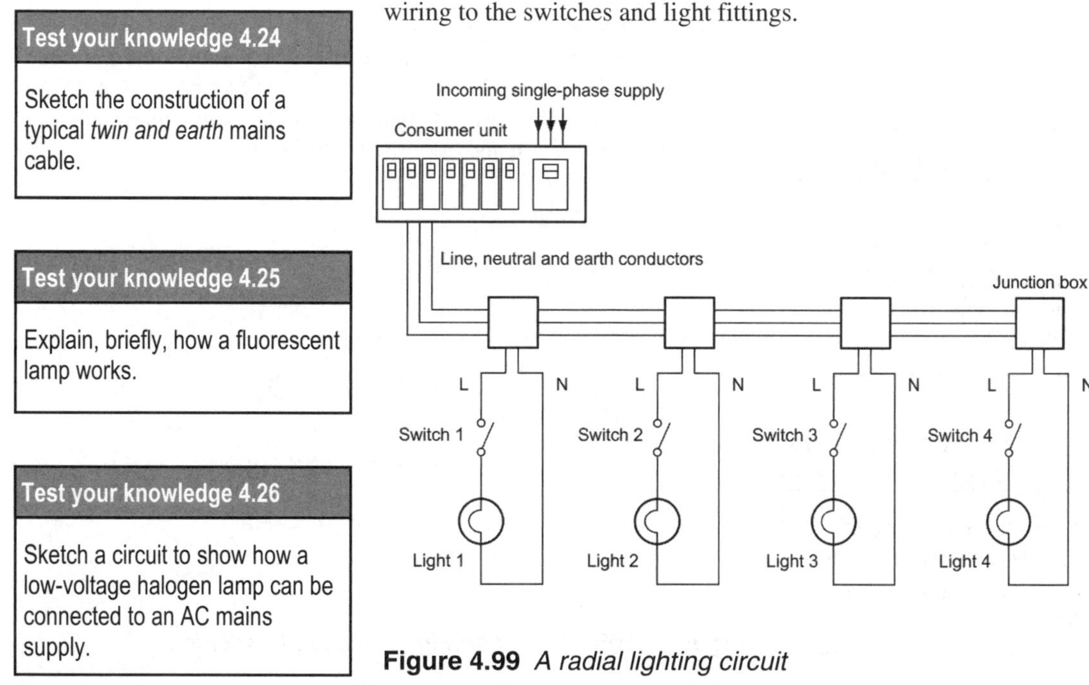

Figure 4.99 *A radial lighting circuit*

Another view

The book's companion website (see page ix) has a spreadsheet tool for solving simple electrical cable problems. You may find this a useful means of checking your results or experimenting with other values!

The physical dimensions of the copper conductors used in electrical wiring (length and cross-sectional area) determine the current carrying capacity and also the load carrying capacity of the circuit (see Table 4.4). Most ring-main circuits can serve areas of up to 110 m^2 using 2.5 mm^2 cable. Note that the advantage of a ring-main circuit is that loads are effectively supplied from both ends of the cabling. By comparison, radial circuits can only serve areas of up to 20 m^2 with 2.5 mm^2 cable or 50 m^2 with 4 mm^2 cable. If necessary *spurs* can be added to both ring-main and radial circuits however this practice should be avoided if possible and all high-power appliances (such as immersion heaters and cookers) should have their own independent radial circuit and MCB.

Table 4.4 *Conductor size and maximum current/load rating*

Conductor size	Maximum current	Maximum load power
1.0 mm^2	10 A	2.4 kW
1.5 mm^2	15 A	3.6 kW
2.5 mm^2	20 A	4.8 kW
4.0 mm^2	25 A	6 kW

If you look carefully at the radial lighting circuit shown in Figure 4.99 you should notice that the light switches are connected to the line conductor and not the neutral conductor. This is important because it removes the potential hazard of one side of the load being 'live' all the time (irrespective of whether the light is switched on or off). This important point is illustrated in Figure 4.100.

Lighting circuits normally use a single-pole single-throw (SPST) switch (as shown in Figures 4.99 and 4.100) however single-pole double-throw (SPDT) switches are used for *two-way lighting circuits* (see Figure 4.101). This type of circuit allows for a light to be switched on and off from two switches that are some distance apart (for example at the top and bottom of a stairway).

Test your knowledge 4.27

Explain the difference between a ring-main and a radial main distribution circuit. Illustrate your answer with appropriate sketches.

Test your knowledge 4.28

Explain why switches should preferably be placed in the line conductor of a mains lighting circuit.

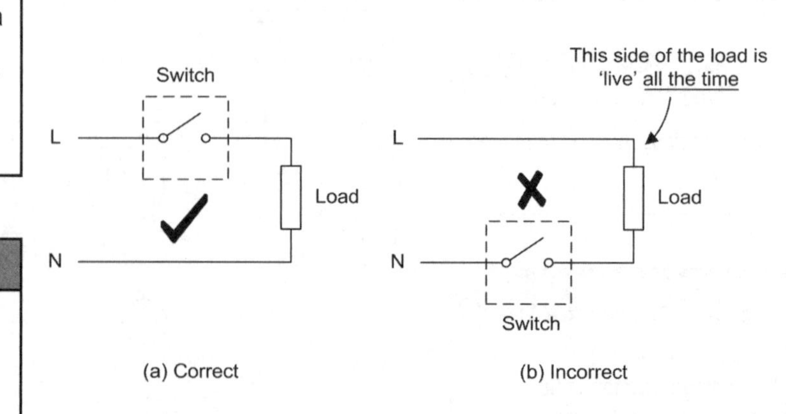

Figure 4.100 *Correct and incorrect load switching*

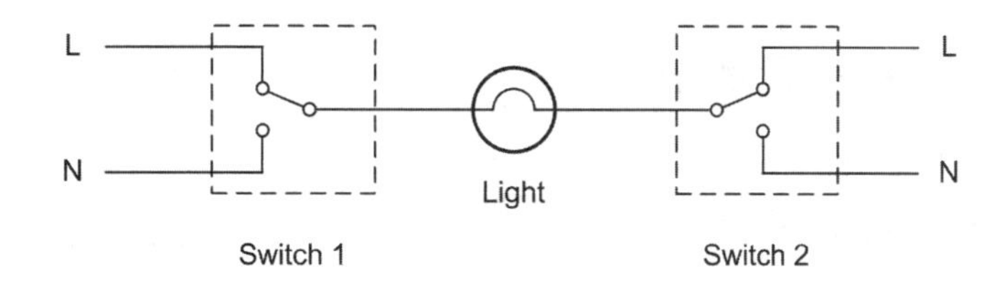

Figure 4.101 *A two-way lighting circuit*

If a person comes into contact with a conductor that is live and at the same time is earthed through the other hand or feet, a current will flow dependent on the resistance of the path through the person's body. We shall consider this situation in greater detail in Section 4.5 when we investigate Health and Safety issues, however it is possible to incorporate protection into an electrical supply that can avoid this situation and prevent the dangerous consequences of receiving an electric shock. This is the function of a *residual current device* (RCD) or *residual current circuit breaker* (RCCB) as shown in Figure 4.102. The device works by sensing the imbalance of current that occurs when current flows between the line conductor and earth. Typical currents at which an RCD will operates are 30 mA or 100 mA (the former giving a higher level of protection). The circuit of a typical RCD is shown in Figure 4.103. Note that the RCD incorporates an in-built test facility.

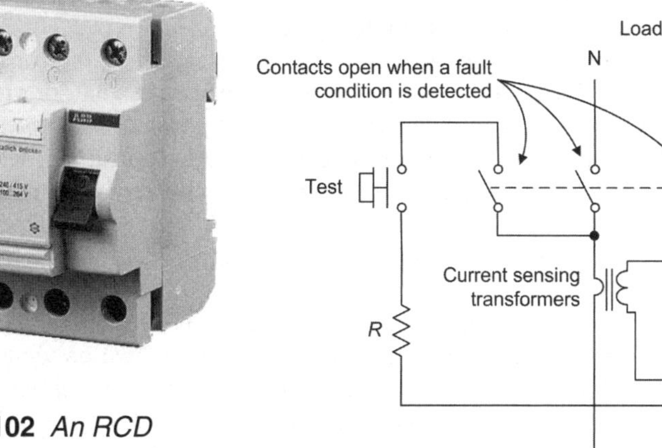

Figure 4.102 *An RCD*

Figure 4.103 *Circuit of an RCD*

Activity 4.12

Investigate the electrical supply distribution in your school or college workshop. Sketch a circuit to show how the electricity is distributed and identify each of the components present (such as mains isolators, MCB, RCD/RCCB, mains outlets, and spurs to any large machines or equipment). Write a brief report explaining your findings.

Test your knowledge 4.29

Explain the function and operation of an RCD.

4.4 Electronics, instrumentation and control

Electronics, and more particularly microelectronics, provides us with a very sophisticated way of monitoring and controlling a wide variety of engineering applications. Examples range from the autopilot systems that can fly an aircraft without intervention to the automatic braking systems that enhance vehicle braking and prevent skidding under adverse road conditions. Despite the obvious differences in what these systems do, they have a number of features in common including the ability to sense what is going on and to apply changes in order to maintain the particular goal of the system (for example, keeping the aircraft in level flight or preventing brakes from 'locking-up').

We begin this section by investigating a wide variety of sensors and transducers before moving on to look at how these sensors are used in real systems. We shall also look at methods of modifying the signals that we derive from sensors and how we interface them with the devices that we use to control a system.

This section also includes a general introduction to control system theory and the use of microprocessors and programmable logic controllers. The section concludes with an introduction to communications, both digital and analogue.

Figure 4.104 *A resistive linear position sensor*

Figure 4.105 *A rotating vane flow sensor*

Figure 4.106 *A liquid level float switch sensor*

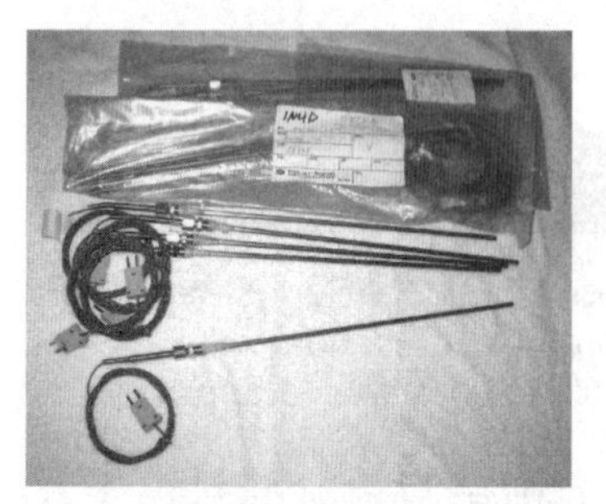

Figure 4.107 *A selection of thermocouple probes*

Sensors and transducers

Sensors and transducers are important components in electronic, instrumentation and control systems.

Transducers are devices that convert energy in the form of sound, light, heat, etc., into an equivalent electrical signal, or vice versa. A *sensor* is a transducer that is used to generate an input signal to a control or measurement system. Before we go further, let's consider a couple of examples that you will already be familiar with. A *loudspeaker* is a device that converts low-frequency electric current into sound. A *thermocouple*, on the other hand, is a transducer that converts temperature into voltage.

It's important to appreciate that transducers are used both as system inputs and system outputs. From the two previous examples, it should be obvious that a loudspeaker is an *output transducer* designed for use in conjunction with an audio system. Whereas, a thermocouple is an *input transducer* which can be used in a temperature control system.

The signal produced by a sensor is an *electrical analogy* of a physical quantity, such as angular position, distance, velocity, acceleration, temperature, pressure, light level, etc. The signals returned from a sensor, together with control inputs from the operator (where appropriate) will subsequently be used to determine the output from the system. The choice of sensor is governed by a number of factors including accuracy, resolution, cost, electrical specification and physical size.

Sensors can be categorised as either *active* or *passive*. An active sensor *generates* a current or voltage output. A passive transducer requires a source of current or voltage and it modifies this in some way (e.g. by virtue of a change in the sensor's resistance). The result may still be a voltage or current *but it is not generated by the sensor on its own.*

Sensors can also be classed as either *digital* or *analogue*. The output of a digital sensor can exist in only two discrete states, either 'on' or 'off', 'low' or 'high', 'logic 1' or 'logic 0', etc. The output of an analogue sensor can take any one of an infinite number of voltage or current levels. It is thus said to be *continuously variable*. A variety of common types of sensor are summarised in Table 4.5.

Table 4.5 *Sensors*

Angular position	
Resistive rotary position sensor	Rotary track potentiometer with linear law produces analogue voltage proportional to angular position.
Optical shaft encoder	Encoded disk interposed between optical transmitter and receiver (infra-red LED and photodiode or photo-transistor).
Differential transformer	Transformer with fixed E-laminations and pivoted I-laminations acting as a moving armature
Angular velocity	
Tachogenerator	Small DC generator with linear output characteristic. Analogue output voltage proportional to shaft speed.
Toothed rotor tachometer	Magnetic pick-up responds to the movement of a toothed ferrous disk. The pulse repetition frequency of the output is proportional to the angular velocity.
Flow	
Rotating vane flow sensor	Turbine rotor driven by fluid. Turbine interrupts infra-red beam. Pulse repetition frequency of output is proportional to flow rate.
Linear position	
Resistive linear position sensor	Linear track potentiometer with linear law produces analogue voltage proportional to linear position. Limited linear range.
Linear variable differential transformer (LVDT)	Miniature transformer with split secondary windings and moving core attached to a plunger. Requires AC excitation and phase-sensitive detector.
Magnetic linear position sensor	Magnetic pick-up responds to movement of a toothed ferrous track. Pulses are counted as the sensor moves along the track.
Light level	
Photocell	Voltage-generating device. The analogue output voltage produced is proportional to light level.
Light dependent resistor (LDR)	An analogue output voltage results from a change of resistance within a cadmium sulphide (CdS) sensing element. Usually connected as part of a potential divider or bridge.
Photodiode	Two-terminal device connected as a current source. An analogue output voltage is developed across a series resistor of appropriate value.
Phototransistor	Three-terminal device connected as a current source. An analogue output voltage is developed across a series resistor of appropriate value.

Table 4.5 *continued*

Liquid level	
Float switch	Simple switch element which operates when a particular level is detected.
Capacitive proximity switch	Switching device which operates when a particular level is detected. Ineffective with some liquids.
Diffuse scan proximity switch	Switching device which operates when a particular level is detected. Ineffective with some liquids.
Pressure	
Microswitch pressure sensor	Microswitch fitted with actuator mechanism and range setting springs. Suitable for high-pressure applications.
Differential pressure vacuum switch	Microswitch with actuator driven by a diaphragm. May be used to sense differential pressure. Alternatively, one chamber may be evacuated and the sensed pressure applied to a second input.
Piezo-resistive pressure sensor	Pressure exerted on diaphragm causes changes of resistance in attached piezo-resistive transducers. Transducers are usually arranged in the form of a four active element bridge which produces an analogue output voltage.
Temperature	
Thermocouple	Junction of wires made from dissimilar metals. A small voltage is generated when a difference of temperature exists between the measuring and reference junctions.
Thermistor	A small resistive device in which the resistance varies with temperature.
Semiconductor temperature sensor	A semiconductor junction which conducts a current that it dependent on temperature.
Vibration	
Electromagnetic vibration sensor	Permanent magnet seismic mass suspended by springs within a cylindrical coil. The frequency and amplitude of the analogue output voltage are respectively proportional to the frequency and amplitude of vibration.

Test your knowledge 4.30

Explain the principle of a magnetic linear position sensor.

Test your knowledge 4.31

State THREE types of sensor that can be used to detect light.

Activity 4.13

Obtain a data sheet for an AD590 semiconductor temperature sensor. Use the data sheet to answer the following questions:

(a) What output current would be produced when the device is used at a temperature of −15°C?

(b) What temperature is being sensed when the output current is 0.32 mA?

Table 4.6 *See Activity 4.14*

Linear position
Linear actuator
Stepper motor actuator
Rotary position
Rotary actuator
Stepper motor
Temperature
Heater
Light
Filament lamp
Light emitting diode

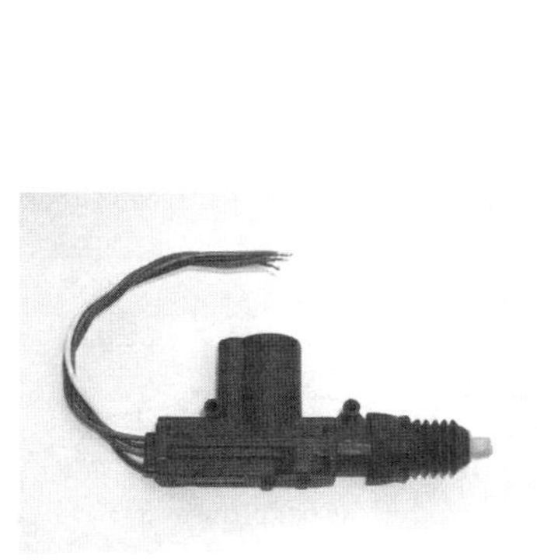

Figure 4.108 *A linear actuator fitted with a DC motor and worm drive*

Activity 4.14

Table 4.6 provides details of some common output transducers. Use the following sentences to complete the table:

(a) Resistive element through which current passes

(b) Stepper motor fitted with a belt drive

(c) Current passing through a filament enclosed in an evacuated glass bulb produces white light

(d) DC motor fitted with a gearbox and rotating mechanism

(e) Current passing through a semiconductor junction produces light at a specific wavelength

(f) DC motor fitted with gearbox, worm drive and a sliding mechanism

(g) Stepper motor driving an output shaft.

Control systems

Control systems are used in many complex machines—particularly those that must operate without human intervention. All control systems comprise a number of elements, components or sub-systems that are connected together in a particular way. The individual elements of a control system interact together to satisfy a particular functional requirement, such as regulating the speed of a motor or operating the control surfaces of an aircraft.

Simple systems (such as a domestic water heater) make use of crude on/off control. With such systems the controlled variable (in this case water temperature) varies continuously above and below the *set point*. This property is known as *hysteresis*.

As an example of a more sophisticated control system, consider the system that operates the elevators of a modern passenger aircraft. When the pilot moves the control column, power is fed to an electric motor that in turn operates a hydraulic actuator. The actuator then moves the elevator control surface to the desired position (the pilot would be unable to exert sufficient force to move the elevator unaided). In order to maintain the desired attitude and altitude, the position of the elevator is detected and fed back to the pitch attitude controller. This continuously monitors the aircraft's attitude and makes fine adjustments to the position of the elevator.

A simplified view of this control system is shown in Figure 4.109. This system has a single input, the desired value (or *set point*) and a single output (the *controlled variable*). In the case of a pitch attitude control system, the desired value would be the hold point (set by the pilot) whereas the controlled variable would be the pitch attitude of the aircraft.

The pitch attitude control system uses three components:

- a *controller* (the pitch computer)
- a final *control element* (the elevator actuator)
- the *controlled process* (the elevator angle).

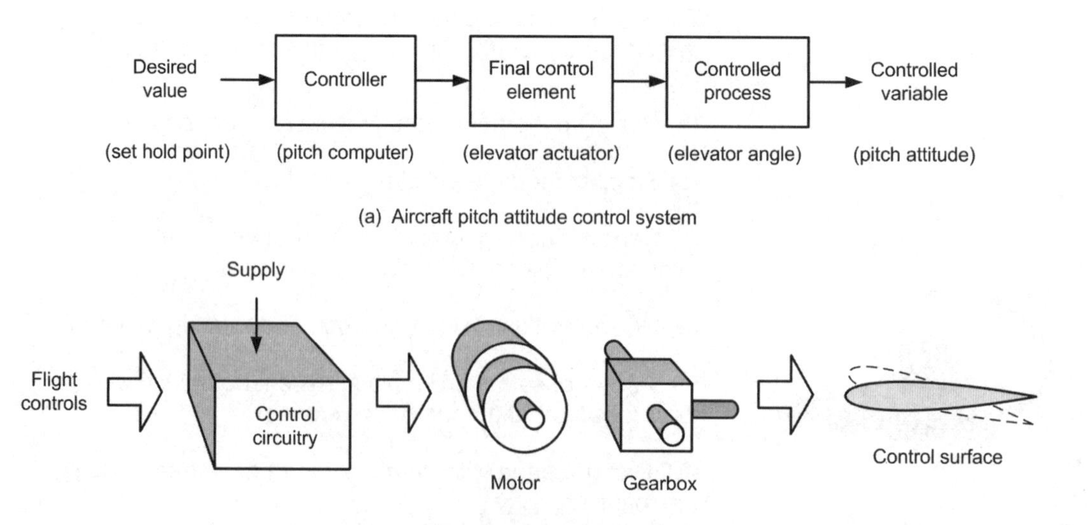

Figure 4.109 *An example of a control system*

Test your knowledge 4.32

A simple hot water system consists of an immersion heater and a thermostat based on a bi-metal strip. The temperature of the water in a tank is regulated by interrupting the electrical supply to the immersion heater.

(a) Explain why this system can be considered *automatic* and how the *feedback* is applied.

(b) Sketch a graph, how the temperature of the water in the tank varies with time.

Control systems are sometimes also referred to as *servomechanisms* or *servo systems*. The important feature of these systems are that they are *automatic* and once set, they are usually capable of operating with minimal human intervention. Furthermore, the input (command) signal used by a servo system is generally very small whereas the output may involve the control or regulation of a very considerable amount of power. For example, the physical power required to operate a high-speed passenger jet's control surfaces greatly exceeds the unaided physical capability of a pilot!

Controlling a system

Controlling a system involves taking into account:

- the desired or input value to the system
- the level of demand (or *loading*) on the output.
- any unwanted variations in the performance of the components of the system.

Different control methods are appropriate to different types of system. The overall control strategy can be based on analogue or digital techniques (or a mixture of the two). At this point it's worth explaining what we mean by these two methods.

Analogue systems

Analogue control involves the use of signals and quantities that are continuously variable. Within analogue control systems, signals are represented by voltage and currents that can take any value between two set limits. Analogue control systems are invariably based on the use of operational amplifiers. These are integrated circuit devices that are capable of performing mathematical operations such as addition, subtraction, multiplication, division, integration and differentiation. Figure 4.110 shows some representative analogue and digital signals.

(a) Analogue signal

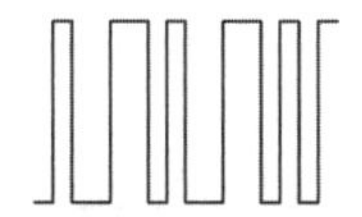

(b) Digital signal

Figure 4.110 *Analogue and digital signals*

Digital systems

Digital control systems use signals and quantities that vary in discrete steps. Values that fall between two adjacent steps must take one or other value as intermediate values are disallowed!

Digital control systems are usually based on digital logic devices or microprocessor-based controllers. Values represented within a digital system are expressed in binary coded form using a number of signal lines. The voltage on each line can be either 'high' (representing logic 1) or 'low' (representing logic 0). The more signal lines the greater the resolution of the system. For example, with just two signal lines it is only possible to represent a number using two binary digits (or 'bits'). Since each bit can be either 0 or 1 it is only possible to represent four different values (00, 01, 10 and 11) using this system. With three signal lines we can represent numbers using three bits and eight different values are possible (000, 001, 010, 011, 100, 101, 110 and 111).

The relationship between the number of bits, n, and the number of different values possible, m, is given by, $m = 2n$. So, in an 8-bit system the number of different discrete states is given by, $m = 28 = 256$.

More complex control systems

More complex control systems may be based on a mixture of both analogue and digital technology and may include electrical, electronic, pneumatic and hydraulic sub-systems. As an example, Figure 4.111 shows the autopilot control system for a modern *fly-by-wire* passenger aircraft.

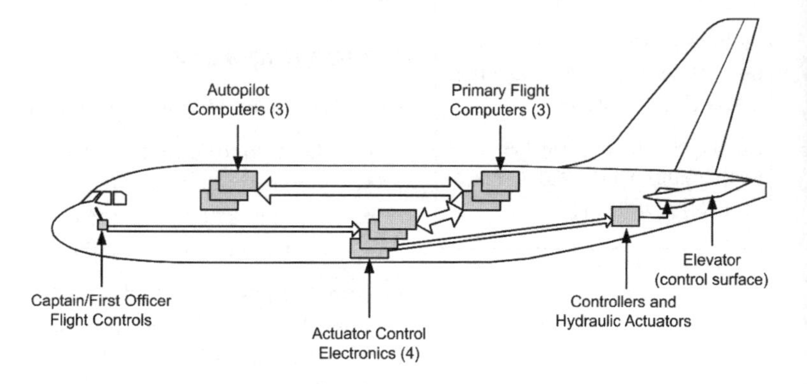

Figure 4.111 *The Airbus A318 autopilot system*

Example 4.24

How many different states can be represented by a 10-bit binary code?

The number of different states is given by 2 raised to the power 10 and 2^{10} = 1024 so 1024 different states will be available.

Example 4.25

A digital control system is required to represent values to a resolution of at least 1%. With how many bits should it operate?

A 1% resolution implies 100 discrete steps. The next highest power of 2 is 128 (2^7) . Thus a 7 (or more) bits will be required.

Signal conditioning

Within a control or instrumentation system, the signals to and from sensors and transducers may need modification or conversion (for example, from digital-to-analogue or analogue-to-digital). Typical signal-conditioning elements are *amplifiers* (to increase signal levels) and *filters* (to remove noise and unwanted signals by restricting the range of frequencies present). Filters can be either low-pass, high-pass, band-pass or band-stop types, as shown in Figure 4.112.

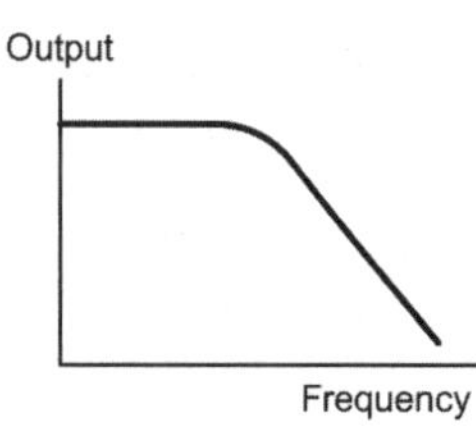

(a) Low-pass filter

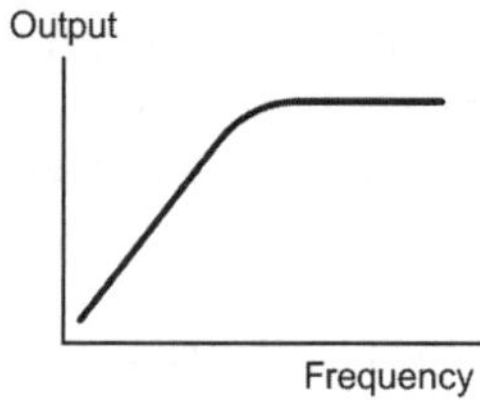

(b) High-pass filter

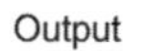

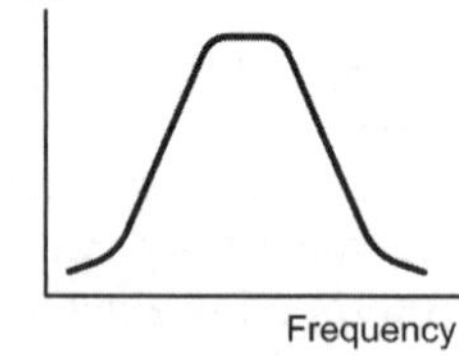

(c) Band-pass filter

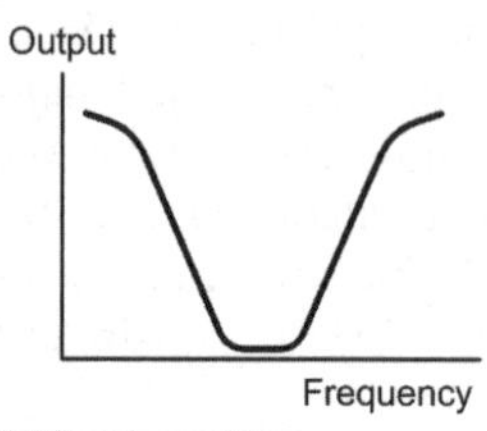

(d) Band-stop filter

Figure 4.112 *The frequency response of various types of filter*

Instrumentation systems

Instrumentation systems are used to perform a wide variety of measurement functions. Many of these systems are automatic, and operate without human intervention, recording data for future analysis. Others provide real-time displays of information used by human operators. The block diagram of a typical instrumentation system is shown in Figure 4.113.

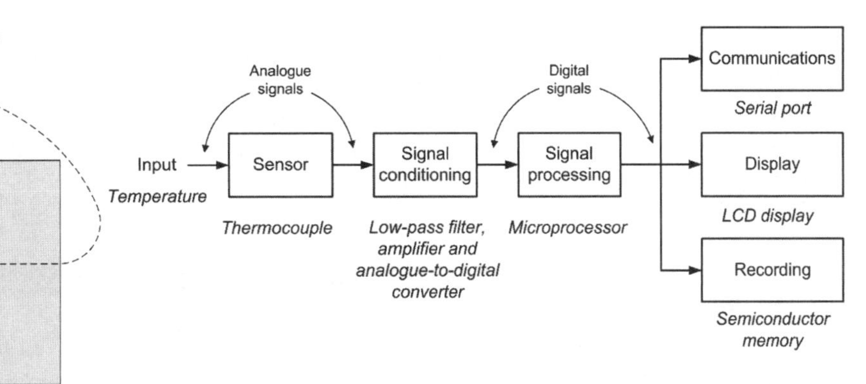

Figure 4.113 *A typical instrumentation system*

Figure 4.114 *Construction of a moving-coil meter*

Figure 4.115 *Moving-coil meters display*

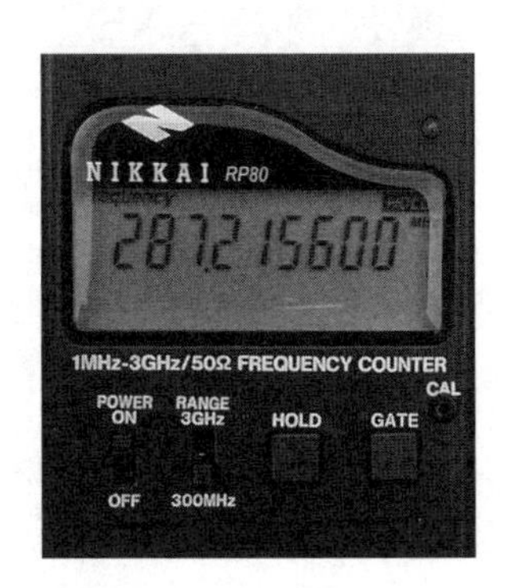

Figure 4.116 *Seven-segment LCD display*

Calibration and reference sources

Measuring systems need to be *calibrated* in order to ensure the accuracy of their indications. This usually involves checking them against a *reference source* that is known to have a higher degree of accuracy. Typical measured parameters and reference sources are listed in Table 4.7.

Table 4.7 *Reference sources and calibration techniques*

Measured parameter	*Reference source/calibration technique*
Temperature	Platinum resistance thermometer connected in a Wheatstone bridge configuration (see Figure 4.122).
Pressure	Capacitance manometer (low pressure), quartz-electrostatic or piezo-resistive pressure sensors (high pressure) calibrated against sub-standards.
Resistance	Standard resistor and Wheatstone bridge (see Figure 4.122).
Voltage	Standard cell and potentiometer (see Figure 4.123). Semiconductor band-gap reference voltage source (less accurate).
Frequency	Quartz crystal-controlled frequency standard (these may themselves be calibrated against an off-air frequency standard checked against an atomic clock).
Time	Timing signal derived from a quartz crystal controlled frequency standard (as above).

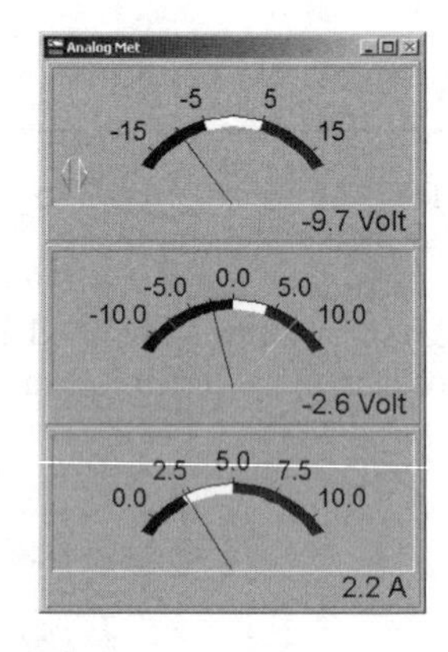

Figure 4.117 *A meter display simulated on a computer screen*

Stand-alone instruments such as chart recorders and X-Y plotters can be rather expensive and so computers are frequently used to simulate the displays produced by these instruments. Another advantage of using a PC in an instrumentation system is that it can easily store vast amounts of data that can be later analysed and/or easily transmitted through a network to other computers.

Figure 4.117 to 4.119 shows some typical simulated (or *virtual*) instrument displays. Unlike conventional display devices, the format of these displays can be very easily changed by simply making changes to software. Figure 4.120 shows how a complete control and instrumentation system can be displayed on the screen of a PC. Note how virtual slider controls are used to alter input values. Finally, Figure 4.121 shows the output of a virtual three-channel chart recorder. The captured data is stored in digital form on disk for future analysis.

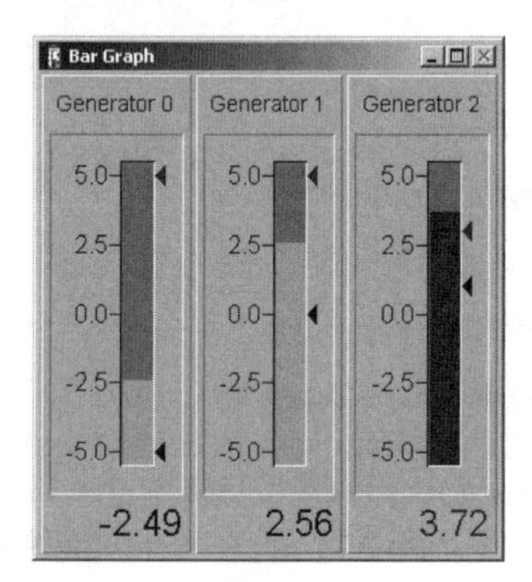

Figure 4.118 *A bar-graph display simulated on a computer screen*

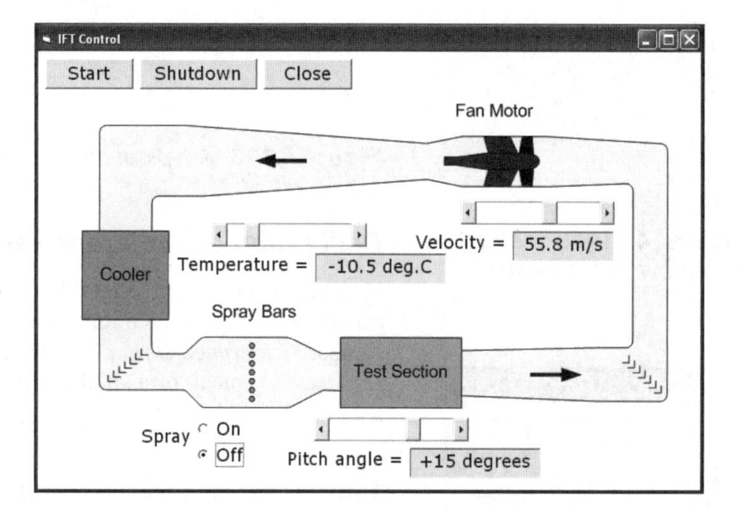

Figure 4.120 *A screen display provided by a PC-based control and instrumentation system*

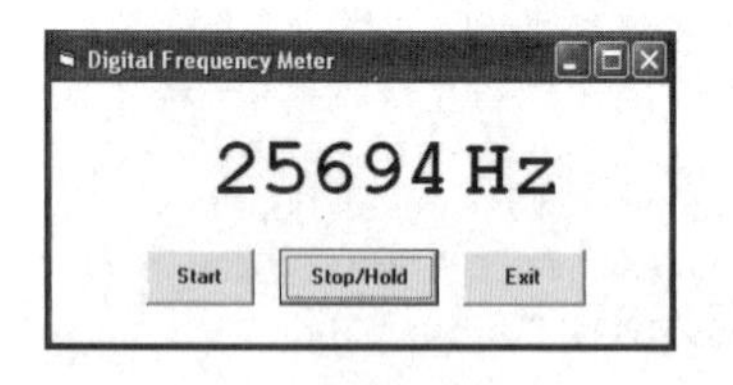

Figure 4.119 *Digital frequency readout on a computer screen*

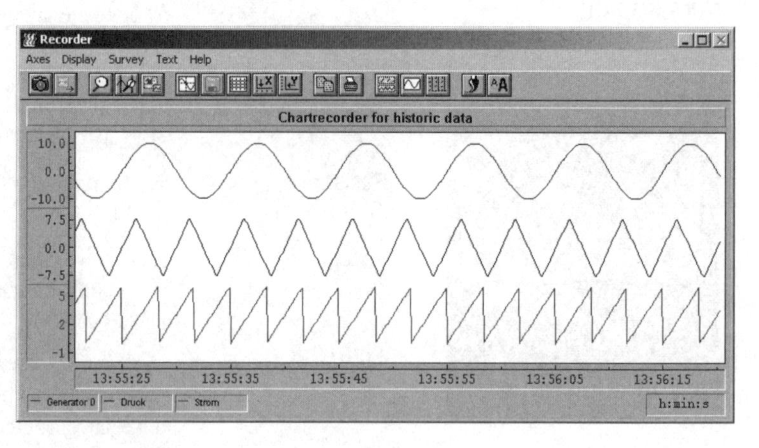

Figure 4.121 *A three-channel chart recorder display simulated on a computer screen*

Test your knowledge 4.33

Explain why a filter may be necessary in a measurement system.

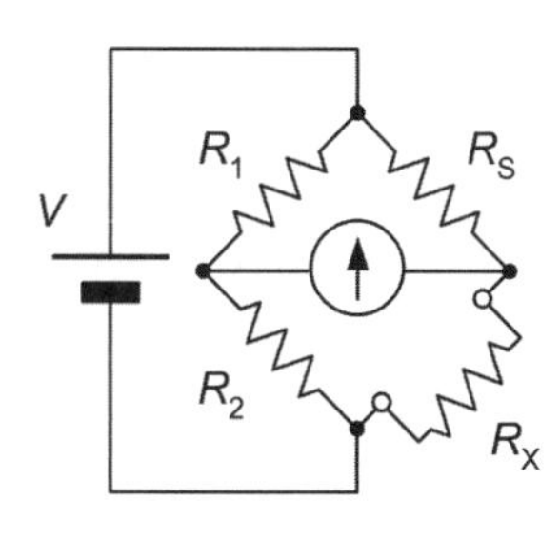

Figure 4.122 *Wheatstone bridge for resistance measurement (the bridge is balanced when no current flows in the galvanometer)*

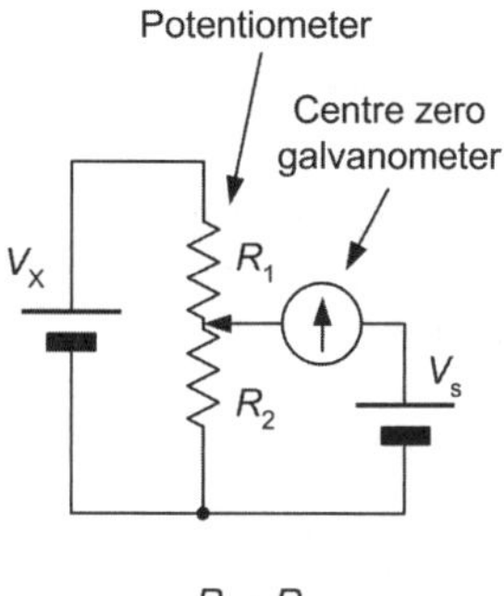

Figure 4.123 *Use of a potentiometer for calibrating a voltage source*

Test your knowledge 4.34

Explain, with the aid of a diagram, how a Wheatstone bridge can be used to check the value of a resistor against a standard.

Activity 4.15

Various methods and devices are used to display the output of different types of instrumentation system, including:

- chart recorders
- X-Y plotters
- seven-segment indicators
- LED bargraphs
- LCD matrix displays.

Investigate each type of display and select the most appropriate type of display for use in each of the following applications:

- indicating the volume of an audio signal being recorded on a tape cassette
- displaying the depth of water beneath a ship's hull
- indicating the deflection of a beam when a load is progressively applied to it
- displaying status message relating to the performance of an engine
- displaying the long-term variation of temperature in a greenhouse.

Give reasons for your answers.

Control system theory

In order to determine the output of a control system corresponding to a particular input we need to determine the overall transfer function of the system. We can do this by first considering the effect of each control element of the system and its contribution to the system overall.

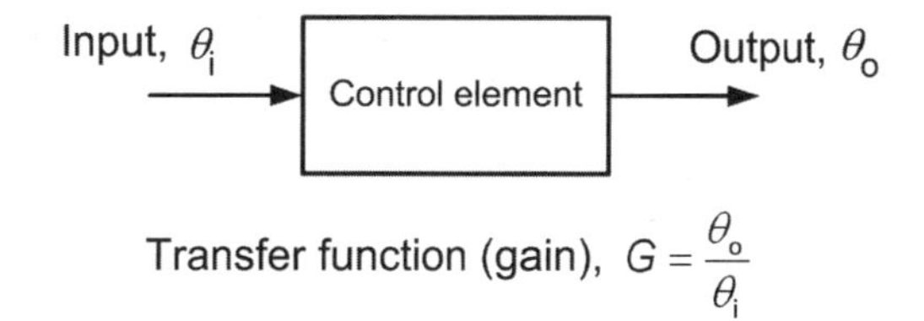

Figure 4.124 *Digital frequency readout on a computer screen*

Figure 4.124 shows a single control element. The input to this control element is θ_i whilst the corresponding output is θ_o. The transfer function, G, for the control is then given by:

$$G = \frac{\text{output}}{\text{input}} = \frac{\theta_o}{\theta_i}$$

For example, consider a linear actuator that produces an output displacement of 2 mm for a change of 1V at the input. The transfer function for the linear actuator would be calculated from:

$$G = \frac{\text{output}}{\text{input}} = \frac{2 \text{ mm}}{0.1 \text{ V}} = 20 \text{ mm/V}$$

Now, suppose that we have three control elements connected in cascade, as shown in Figure 4.125. The combined effect of these three elements would produce a transfer function (or *gain*) of:

$$G = G_1 \times G_2 \times G_3 = \frac{\theta_o}{\theta_i}$$

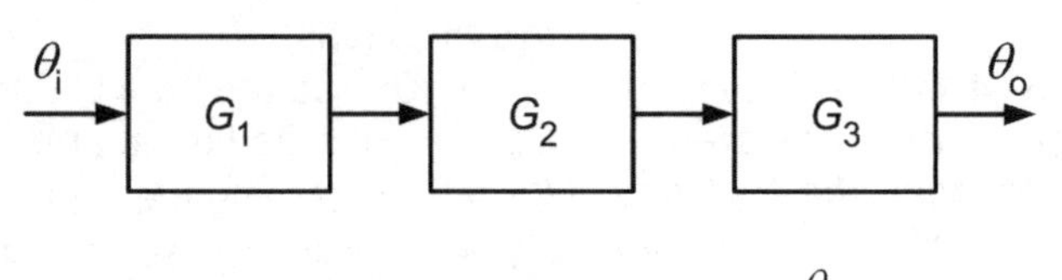

Overall gain, $G_o = G_1G_2G_3 = \frac{\theta_o}{\theta_i}$

Figure 4.125 *Digital frequency readout on a computer screen*

Example 4.26

Figure 4.126 shows the block diagram of a lifting mechanism with the transfer function for each control element shown. Determine the speed at which the load is raised, v, when the input voltage is 150 mV.

The overall transfer function, G, for the lifting mechanism is given by:

$G = G_1 \times G_2 \times G_3 \times G_4 = 40 \times 2.5 \times 0.01 \times 0.35 = 0.35$ m/s/V

The output speed, v, is therefore given by:

$v = 0.35 \times 0.15 =$ **0.0525 m/s**

Test your knowledge 4.35

A tachogenerator has a transfer function of 0.2 V/rev/sec. Determine the speed of the tachogenerator when it is producing an output of 3.4 V.

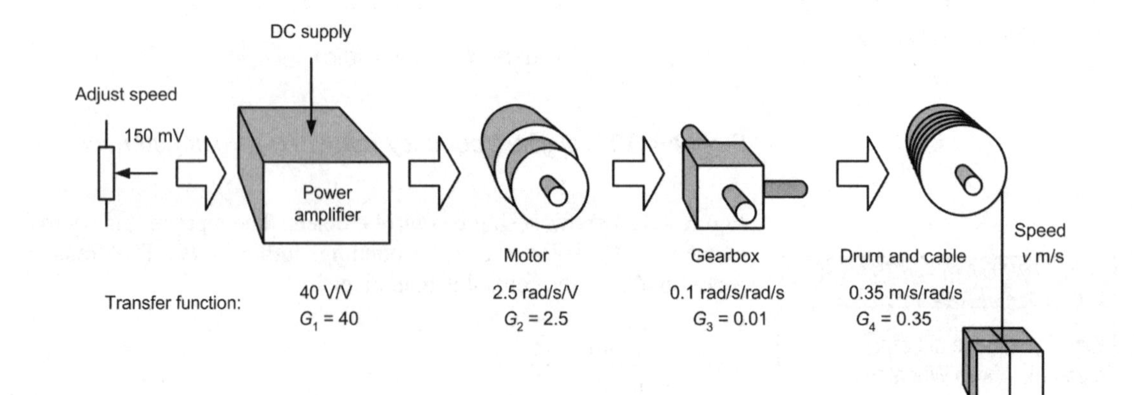

Figure 4.126 *See Example 4.26*

Closed-loop systems

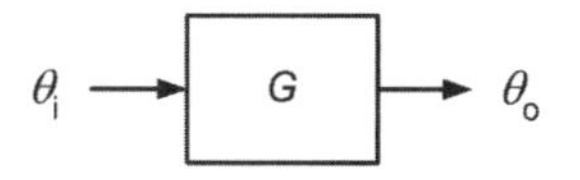

Figure 4.127 *An open-loop system uses no feedback*

The system that we looked at in Example 4.26 has no means of automatically compensating for any variation in load or changes in any of the control elements. Because there is no feedback present, a system of this type is referred to as an *open-loop system* (see Figure 4.127).

Closed-loop systems use *feedback*, as shown in Figure 4.128. A proportion of the output, θ_o, is sensed and fed back to the input where it is compared with the desired value, θ_i. The comparison is performed by a device known as a *comparator* (indicated by the circle in Figure 4.128). If there is any difference between the sensed output value, γ_o, an error signal is produced, ε.

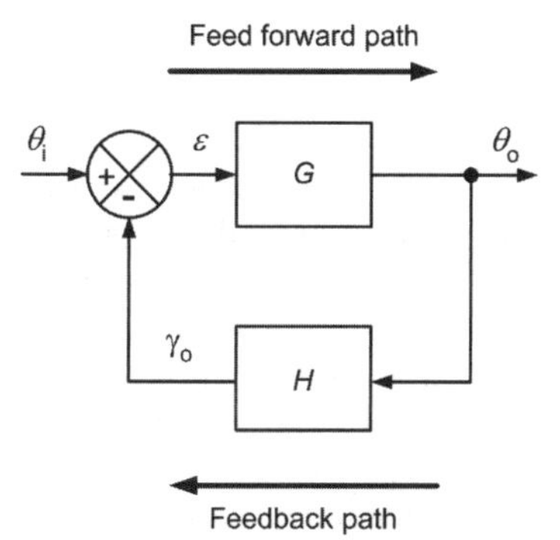

Figure 4.128 *A closed-loop system has feedback*

Using the transfer functions for the forward and feedback paths, we can write down the following:

$$\theta_o = G\,\varepsilon \quad \text{and} \quad \gamma_o = H\,\theta_o$$

We also know that the error, ε, is the difference between θ_i and γ_o, hence:

$$\varepsilon = \theta_i - \gamma_o$$

In order to find out how the system behaves we need to be able to determine its overall transfer function, i.e.:

$$M = \frac{\theta_o}{\theta_i} \quad \text{where } M \text{ is the closed-loop transfer function.}$$

Combining the foregoing equations gives:

$$M = \frac{\theta_o}{\theta_i} = \frac{\theta_o}{\varepsilon + \gamma_o} = \frac{\theta_o}{\varepsilon + H\theta_o} = \frac{G\varepsilon}{\varepsilon + HG\varepsilon} = \frac{G}{1 + HG}$$

This result is very important. Consider what would happen if the value of gain, G, was made very large. The result would be a value of M which become very close to $1/H$. Thus, for large values of G:

$$M \approx \frac{1}{H}$$

This shows that the closed-loop transfer function (closed-loop gain) gain be accurately defined by simply controlling the amount of feedback applied.

Activity 4.16

Determine the value of the closed-loop transfer function, M, for systems in which:

(a) $G = 10$ and $H = 0.1$

(b) $G = 100$ and $H = 0.1$

(c) $G = 1000$ and $H = 0.1$

What do these results suggest?

Test your knowledge 4.36

Explain, with the aid of appropriate diagrams, the differences between closed-loop and open-loop systems.

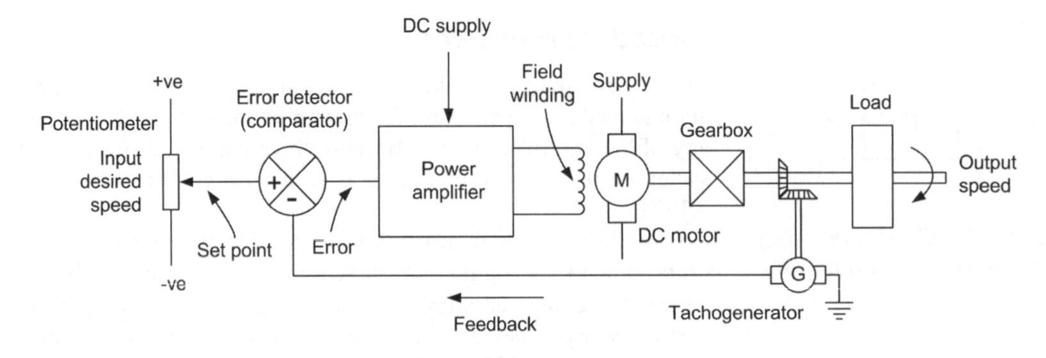

Figure 4.129 *A closed-loop speed control system*

Error-forming devices

Figure 4.129 shows the arrangement of a practical closed-loop speed control system. With the exception of the error-forming device, most of the components used in this system have been described previously in this unit.

The error-forming device (or *summing junction*) can be realised quite easily using an operational amplifier (a single integrated circuit) and a few resistors, as shown in Figure 4.130. Note that the error signal is simply the difference between the input (set point) and the feedback signal. If the two signals are the same then the error signal will be zero (and no adjustment of the output will be necessary).

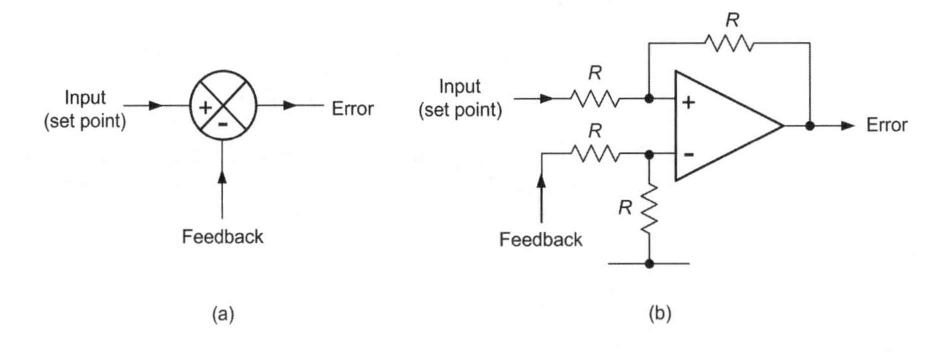

Figure 4.130 *Error-forming devices; (a) symbol and (b) an operational amplifier comparator*

Activity 4.17

A system is to be used to control the rotation of a highly directional aerial system that is fitted to the top of a 15 m steel-guyed mast. The system is to be capable of setting and maintaining the azimuth angle of the aerial system to within ±3°. Design a control system to perform this task and illustrate your answer with a fully labelled block diagram. Identify all of the component parts and explain what they do and why they are needed. State any assumptions that you make.

Closed-loop system response

In a perfect control system the output value would respond immediately to a change in the input. There would be no delay when changing from one value to another and no time taken for the output to settle to its final value. This ideal case is shown in Figure 4.131 (b). In practice, due to friction, inertia, and other real-world constraints, the output of a practical system may take some time to reach its final value or may involve overshooting the desired value and eventually settling back to it, as shown in Figure 4.131(c).

The graph shown in Figure 4.131(c) is known as a *second-order response* curve. This response has two basic components; a growth curve and a damped oscillation. The oscillatory component can be reduced or eliminated by artificially slowing down the response of the system. This is known as *damping*.

Unfortunately, too much damping may cause the system to respond very sluggishly and so a compromise value must often be used. This optimum value of damping will be just sufficient to remove the overshoot and oscillation yet still retain a reasonably fast response. Various conditions of damping are shown in Figure 4.132.

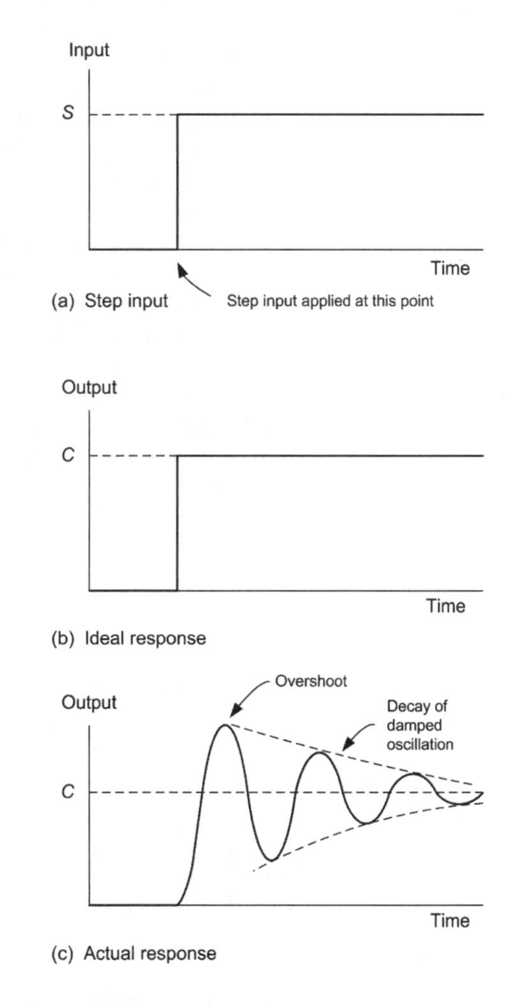

Figure 4.131 *Response of a closed-loop system*

Test your knowledge 4.37

Explain the following terms in relation to the response of a closed-loop system:

(a) Overshoot
(b) Damping
(c) Oscillation.

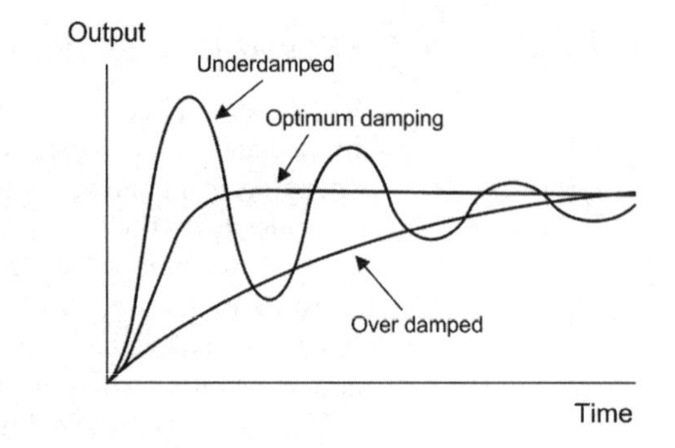

Figure 4.132 *Effect of different amounts of damping*

Microprocessor systems

The information revolution has largely been made possible by developments in electronics and in the manufacture of integrated circuits in particular. Ongoing improvements in manufacturing technology have given us increasingly powerful large-scale integrated circuit chips. Of these, the *microprocessor* (a chip that performs all of the essential functions of a computer) has been arguably the most notable development.

A microprocessor is a single chip of silicon that performs all of the essential functions of a computer *central processor unit* (CPU) on a single silicon chip. *Microprocessor systems* (see Figure 4.133) are found in a huge variety of engineering applications including engine management systems, environmental control systems, domestic appliances, video games, fax machines, photocopiers, etc.

The CPU performs three functions: it controls the system's operation; it performs algebraic and logical operations; and it stores information (or *data*) whilst it is processing. The CPU works in conjunction with other chips, notably those that provide random access memory (RAM), read-only memory (ROM), and input/output (I/O).

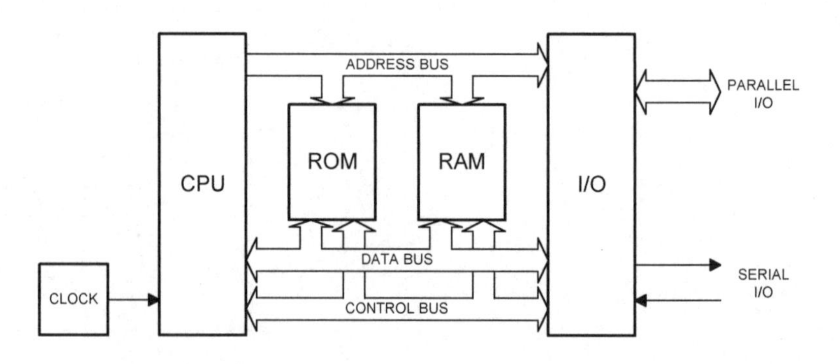

Figure 4.133 *A block schematic diagram of a microprocessor system*

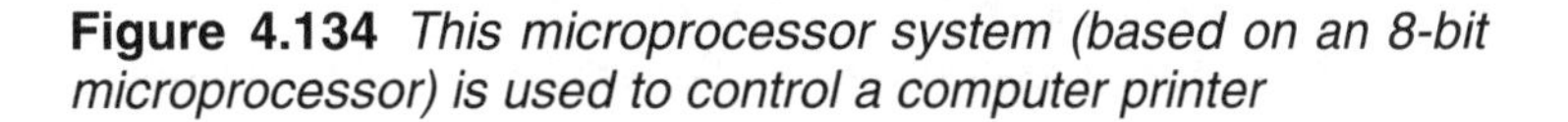

Figure 4.134 *This microprocessor system (based on an 8-bit microprocessor) is used to control a computer printer*

The central processing unit (CPU) is generally the microprocessor chip itself. This device contains the following (see Figure 4.135):

- storage locations (called *registers*) that can be used to hold instructions, data, and addresses during processing
- an *arithmetic logic unit* (ALU) that is able to perform a variety of arithmetic and logical function (such as comparing two numbers)
- a *control unit* which accepts and generates external control signals (such as *read* and *write*) and provides timing signals for the entire system.

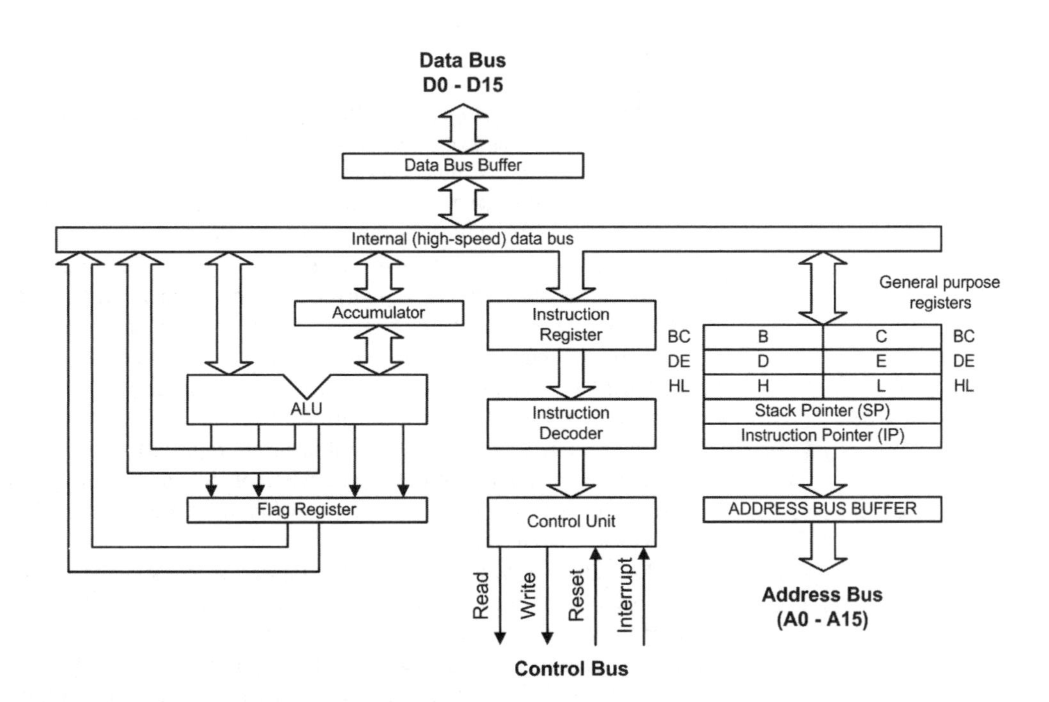

Figure 4.135 *Internal architecture of a typical 8-bit microprocessor*

In order to ensure that all the data flow within the system is orderly, it is necessary to synchronise all of the data transfers using a clock signal. This signal is often generated by a clock circuit (similar to the clock in a digital watch but much faster). To ensure accuracy and stability the clock circuit is usually based on a miniature quartz crystal.

All microprocessors require access to read/write memory in which data (e.g. the results of calculations) can be temporarily stored during processing. Whilst some microprocessors (which are often referred to as *microcontrollers*) contain their own small read/write memory, this is usually provided by means of a semiconductor *random access memory* (RAM), as shown in Figure 4.133.

Microprocessors generally also require more permanent storage for their *control programs* and, where appropriate, operating systems and high-level language interpreters. This is usually provided by means of semiconductor *read-only memory* (ROM), as shown in Figure 4.133.

Test your knowledge 4.38

Briefly explain the function of each of the following in relation to a microprocessor system:

(a) CPU
(b) RAM
(c) ROM
(d) I/O
(e) clock.

To fulfil any useful function, a microprocessor system needs to have connections with the outside world. These are usually supplied by means of one, or more, VLSI devices which may be configured under software control and are therefore said to be *programmable.*

The *input/output* (I/O) devices fall into two general categories; *parallel* (where a *byte* or 8-bits is transferred at a time along eight separate wires), or *serial* (where one *bit* is transferred after another along a single wire).

Programmable logic controllers

Programmable logic controllers (PLCs) are microprocessor systems that are used for controlling a wide variety of automatic processes, from operating an airport baggage handling system to brewing a pint of your favourite lager! PLCs are rugged and modular and they are designed specifically for operating in a *process control* environment.

The control program for a PLC is usually stored in one or more semiconductor memory devices. The program can be entered (or modified) using a simple hand-held programmer, a laptop computer, or downloaded from a local area network (LAN). PLC manufacturers include Allen Bradley, Siemens, and Mitsubishi.

A typical PLC is shown in Figure 4.136. Input/output (I/O) connections are made via the terminal strips on each edge of the unit. A bank of light emitting diode (LED) *status indicators* are used to display the state of each of the I/O lines (i.e. whether they are *on* or *off*) whilst a round multi-pin DIN connector is used to connect the hand-held programmer (see Figure 4.138).

In operation, the PLC's control program repeatedly scans the state of each of its inputs (to determine whether they are *on* or *off*) before using this information to decide on what should happen to the state of each of its output lines.

The PLC usually derives its inputs from switches (such as keypads, buttons, or contacts that make and break) but it can also derive its input from sensors (for example liquid level sensors, temperature sensors, position sensors, motion sensors, etc).

Many programmable logic controllers can also be fitted with modules that allow analogue signals (i.e. signals that can vary

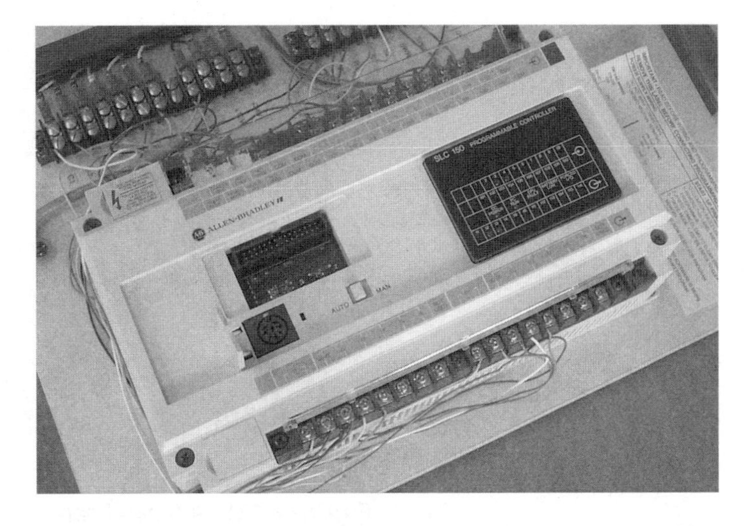

Figure 4.136 *A typical programmable logic controller (PLC)*

continuously in voltage level rather than just be *on* and *off*) to be used both for input and for output. In the former (input) case an *analogue-to-digital converter* (ADC) is required whilst in the latter (output) case a *digital-to-analogue converter* (DAC) is needed (see page 337).

A typical PLC application in which a PLC is controlling a small conveyor belt is shown in Figure 4.137. Note the start and stop buttons and the optical position sensors mounted at the side of the conveyor belt.

Several different methods can be used to program a PLC. These include a simple text-based language using commands based on logic (such as AND, OR, etc) and entered via a keyboard, keypad (see Figure 4.138) and a graphical programming language which is based on *ladder-logic* (see Figure 4.139).

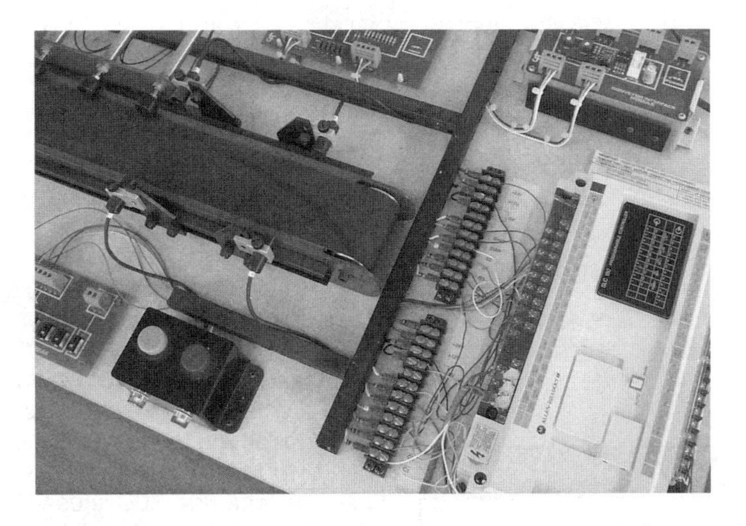

Figure 4.137 *A typical PLC system in which the PLC control a small conveyor belt*

Figure 4.138 *A typical hand-held PLC programmer*

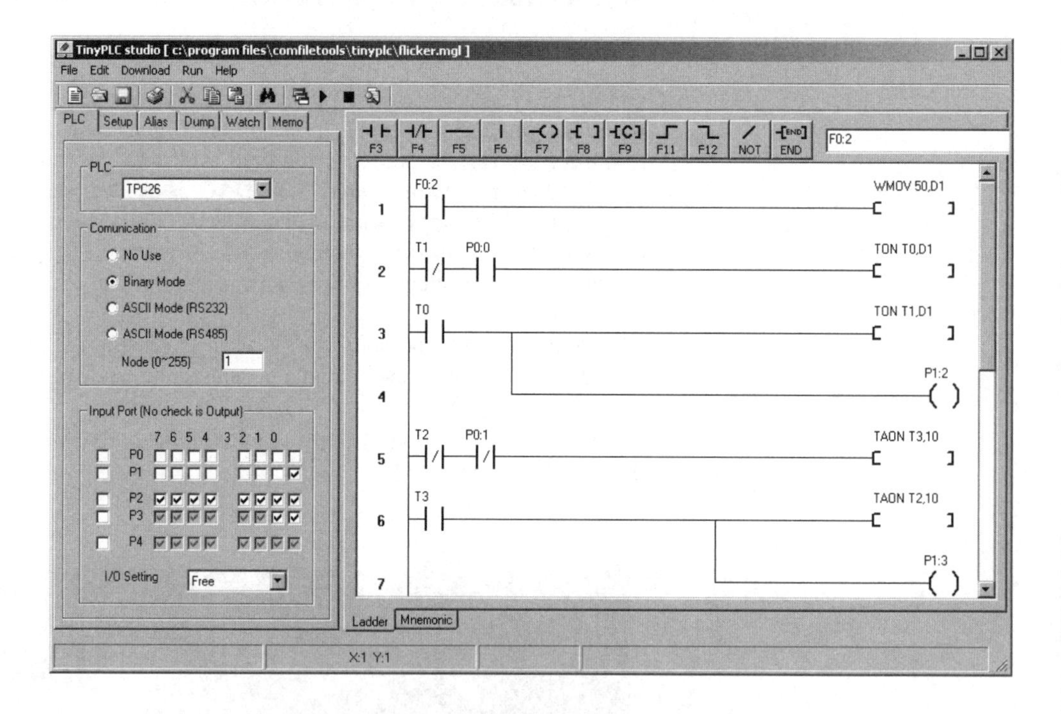

Figure 4.139 *A simple ladder-logic program*

Activity 4.18

Investigate a PLC system in your school or college (if your school does not have such a system they may be able to arrange for you to visit a local engineering manufacturer or a nearby Further Education college that has this equipment). Find out what the PLC is used for and how it is programmed. Write a brief report describing what you have seen and explain, in simple terms, how a PLC works. Illustrate your report with relevant sketches, diagrams and photographs.

Test your knowledge 4.39

Describe TWO typical applications for a programmable logic controller (PLC).

Communications

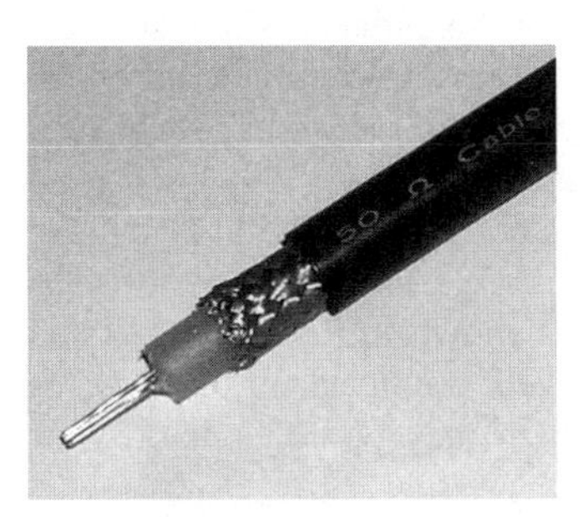

Figure 4.140 *A high-quality coaxial cable*

Whilst many engineering systems operate on a stand-alone basis (and are therefore not interconnected with any other systems) there is an increasing need for systems to share information and data and to communicate with one another. This is particularly true of larger, more complex systems. Good examples of these can be found in the electricity supply industry and in the large-scale manufacturing found in the automotive and aerospace industries.

Communication systems can be based on analogue or digital systems or a combination of both. They can use radio (or wireless), cable (twisted pair or coaxial cables), infra-red, or optical fibres as a medium along which the information can be conveyed. Increasingly, data is being transmitted over digital networks, both private local area networks (LAN) and wide area networks (WAN), and also using Internet.

Analogue to digital conversion

Analogue to digital conversion is necessary when analogue signals (such as audio or video) are required to be processed or transmitted in digital form. The process of conversion (see Figure 4.141a) involves sampling the incoming analogue signal at a fast rate and then converting each sample to a corresponding digital code. The more binary digits (bits) used in the conversion process and the faster the sampling rate, the more accurate will be the digital representation of the analogue signal. In practice, it is necessary to sample an analogue signal at a rate (in terms of samples per seconds) that is at least twice the highest frequency present in the analogue signal. Resolutions of 8-bits can be satisfactory for simple applications (corresponding to 256 different signal voltage levels—see page 323), as many as 10 and 12 bits are used in many practical analogue to digital converters.

Digital to analogue conversion

Digital to analogue conversion is the reverse of analogue to digital conversion and is required whenever an analogue signal is to be produced by a digital system. The process of conversion (Figure 4.141b) involves converting the stream of digital data at regular

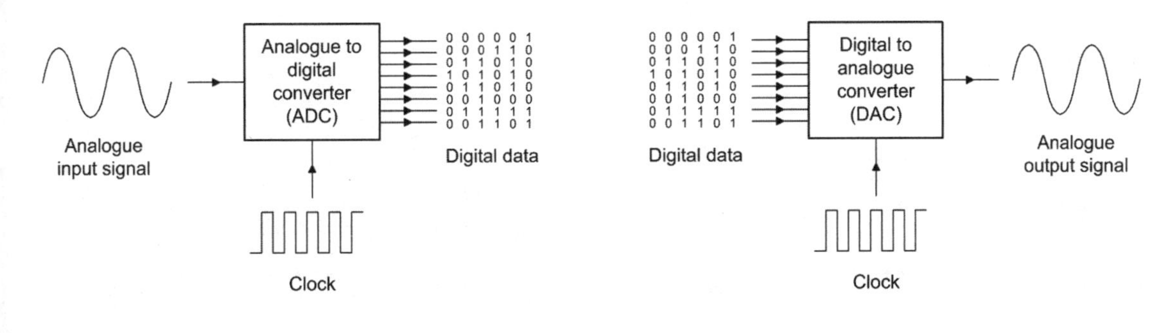

Figure 4.141 *ADC and DAC*

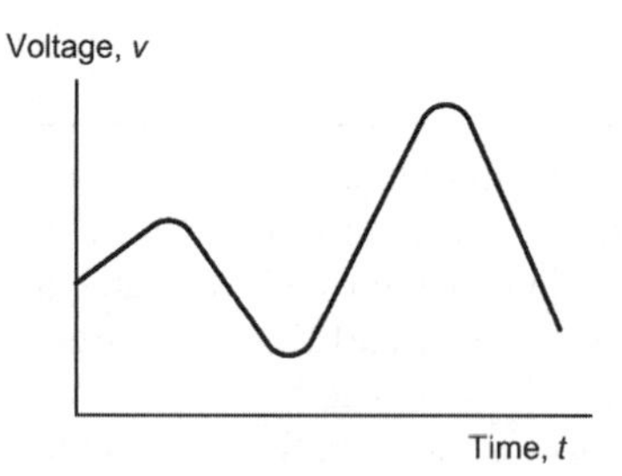

Figure 4.142 *Analogue signal*

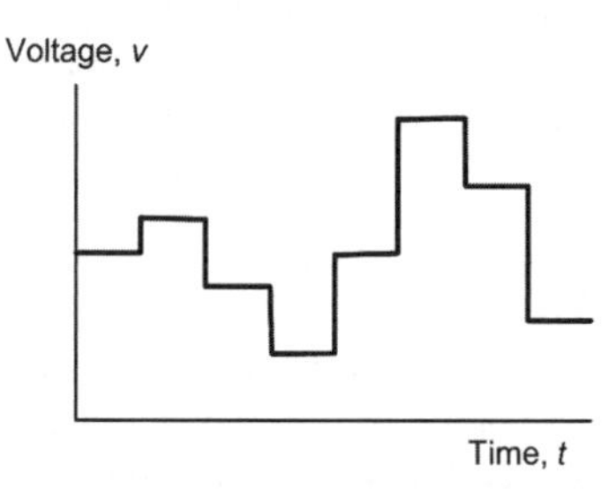

Figure 4.143 *Analogue signal showing quantisation levels*

intervals and then generating an analogue voltage that is equivalent in value to the digital code. Provided that this process is repeated at a fast rate when compared with the highest frequency component present in the analogue signal, the analogue output voltage will be a faithful representation of the digital data.

In both cases it is important to remember that the digital signal can only exist in discrete steps (or *quantisation levels*). This gives rise to an approximation of the true values, as shown in Figures 4.142 and 4.143. The accuracy of the voltage levels produced by the DAC is determined by a voltage reference which may be either internal or external to the DAC chip.

Modulation

In order to transmit signals and data over an appreciable distance using the medium of cables, fibres, infra-red and radio, some modulation and demodulation is required to enable information to be 'carried' using a signal that can be transferred with minimal loss and distortion. This is similar to the process used for broadcasting where speech and music is modulated onto a radio signal at the transmitter, broadcast in the form of an electromagnetic wave, and then demodulated at the receiver.

Figure 4.144(a) shows the process of amplitude modulation (AM) of analogue signals whilst Figure 4.144(b) shows the process of frequency modulation (FM) of analogue signals. Notice that, in either case, the high-frequency carrier is only used for the purpose of transmission and conveys no information (other than the fact that the transmitter is switched on!).

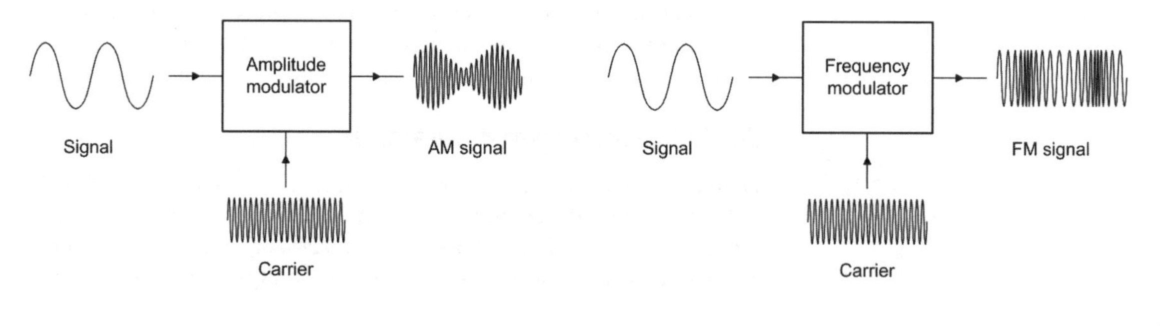

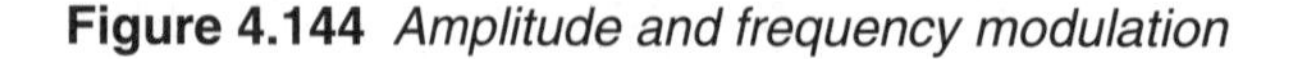

Figure 4.144 *Amplitude and frequency modulation*

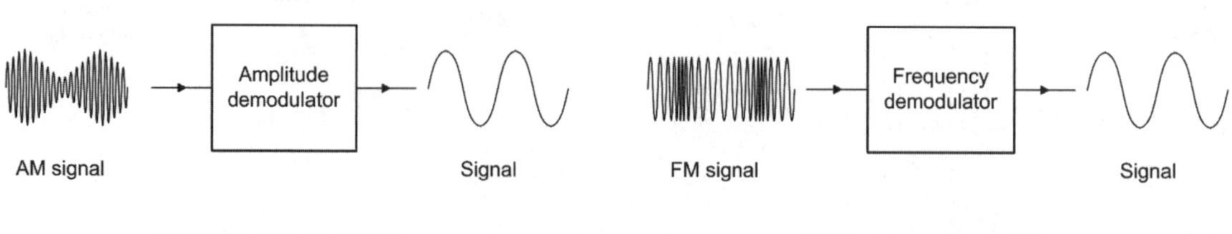

Figure 4.145 *Amplitude and frequency demodulation*

Demodulation

The process of demodulation of analogue signals is shown in Figure 4.145. Figures 4.145(a) and 4.145(b) respectively show the process of amplitude demodulation and frequency demodulation.

Figures 4.146 and 4.147 show a complete radio communication system in which amplitude modulation is used at the transmitter and amplitude demodulation is used at the receiver. The demodulator stage of the receiver is shown in Figure 4.148.

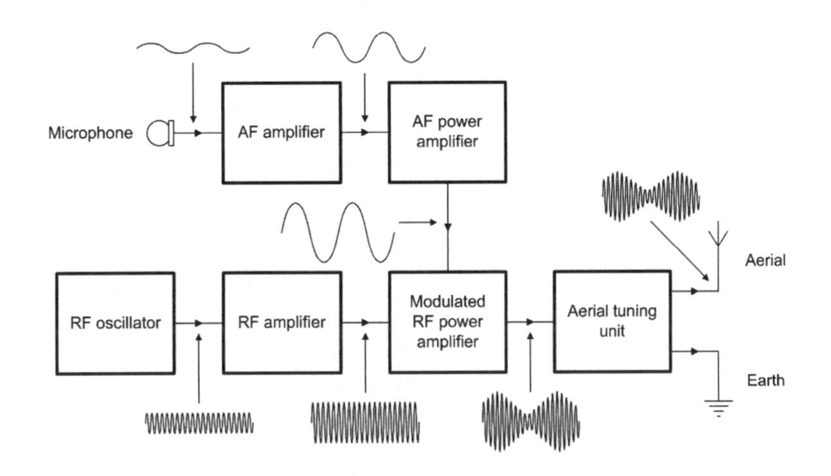

Figure 4.146 *A simple AM radio transmitter*

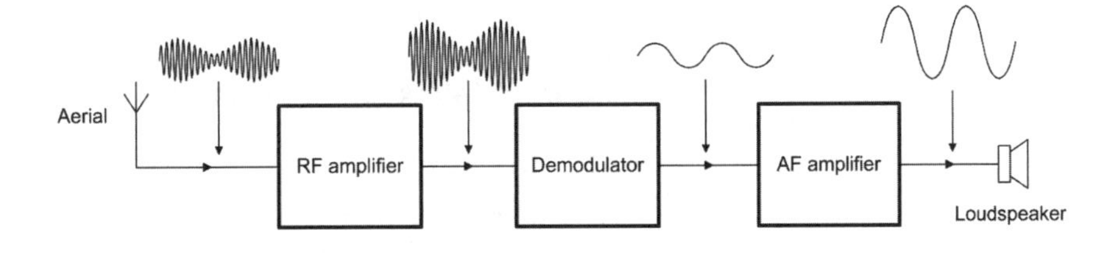

Figure 4.147 *A simple AM radio receiver*

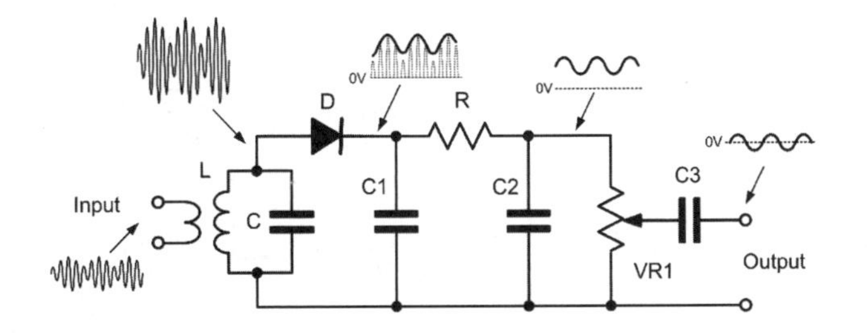

Figure 4.148 *The demodulator stage of the radio receiver*

Test your knowledge 4.40

Explain the terms *modulation* and *demodulation* in relation to a communications system.

MODEMS

In the case of digital signals, the complementary functions of modulation and demodulation are often combined into a single device, a modulator-demodulator, or MODEM. MODEMs are used in a number of applications, such as interfacing a computer to a conventional voice-frequency telephone line (a 'dial-up connection'), a cable, or an optical fibre-based communication system.

Optical fibres

An optical fibre is a long thin strand of very pure glass enclosed in an outer protective jacket. Because the refractive index of the inner layer is larger than that of the outer layer, light travels along the fibre by means of total internal reflection at a speed of around 200 million m/s (approximately 2/3 of the speed of light).

In order to set up an optical data link, the optical fibre must be terminated at each end by means of a transmitter/receiver unit. A simple one-way fibre optic link is shown in Figure 4.149.

The optical transmitter consists of a light emitting diode (LED) coupled directly to the optical fibre. The LED is supplied with pulses of current from a computer interface. The pulses of current produce pulses of light that travel along the fibre until they reach the optical receiver unit.

The optical receiver unit consists of a photodiode (or phototransistor) that passes a relatively large current when illuminated and hardly any current when not. The pulses of current at the transmitting end are thus replicated at the receiving end.

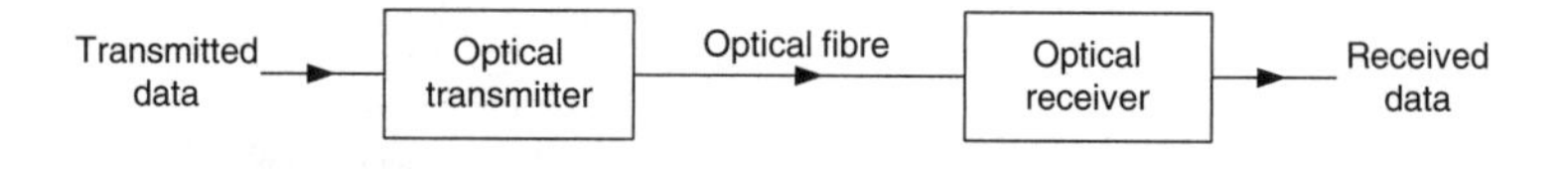

Figure 4.149 *A one-way (unidirectional) optical fibre data link*

A bi-directional optical fibre link can be produced by adding a second fibre and having a transmitter and receiver at each end of the link (Figure 4.150). A simple optical fibre data link of this sort is capable of operating at data rates of up to 500 kb/sec and distances of up to 1 km. More sophisticated equipment (using high-quality low-loss fibres) can work at data rates of up to 250 Mb/s and at distances of more than 10 km.

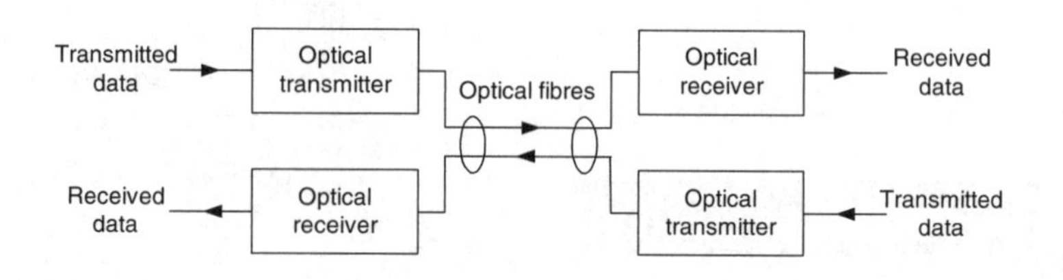

Figure 4.150 *A two-way (bi-directional) optical fibre data link*

Activity 4.19

Obtain a datasheet for an AD558 8-bit DAC. Use the data sheet to answer the following questions:

(a) State the name of the manufacturer of the device.

(b) State the supply voltage range for the device.

(c) State the typical output settling time when the device is used on the 2.65 V range.

(d) Sketch a functional block diagram of the device and label your diagram clearly.

Activity 4.20

There are several different types of optical fibre. Identify at least three different types and explain why they are different. Present your findings in the form of a brief verbal presentation supported by appropriate handouts and visual aids.

Test your knowledge 4.41

Explain how optical fibres can be used to transmit digital data between two computers.

Activity 4.21

Optical fibres can provide a bandwidth that is much greater than that which can be obtained using conventional copper cables. Investigate this statement and write a brief written report explaining why this is.

4.5 Health and Safety

Earlier in Unit 2, we introduced the Health and Safety at Work Act (1974) which makes both the employer and the employee responsible for safety in the workplace. In addition, we mentioned the Control of Substances Hazardous to Health Act (1988) which is designed to safeguard all employees who work with substances hazardous to their health, such as chemicals and toxic dust.

In this section we are going to introduce you to the Health and Safety issues that relate to specific engineering systems. You are expected to be aware of these and how they impact on the design, production and utilisation of the systems to which they relate. This should provide you with a starting point for the unit assessment which will expect you to provide evidence that you have given adequate consideration to issues that relate to Health and Safety.

The Health and Safety at Work Act

The Health and Safety at Work Act (1974) is based on principles that are fundamentally different from any previous health and safety legislation. The underlying reasoning behind the Act was the need to foster a much greater awareness of the problems that surround health and safety matters and, in particular, a much greater involvement of those who are, or who should be, concerned with improvements in occupational health and safety. Consequently, the Act seeks to promote greater personal involvement coupled with the emphasis on individual responsibility and accountability.

You need to be aware that the Health and Safety at Work Act applies to *people*, not to premises. The Act covers all employees in all employment situations. The precise nature of the work is not relevant, neither is its location. The Act also requires employers to take account of the fact that other persons, not just those that are directly employed, may be affected by work activities. The Act also places certain obligations on those who manufacture, design, import or supply articles or materials for use at work to ensure that these can be used in safety and do not constitute a risk to health.

It is the *duty of the employer* to ensure, so far as is reasonably practicable, the health, safety and welfare at work of all the employees, also that all plant and systems are maintained in a manner so that they are safe and without risk to health. The employer is also responsible for:

- the absence of risks in the handling, storage and transport of articles and substances
- instruction, training and supervision to ensure health and safety at work
- the maintenance of the workplace and its environment to be safe and without risk to health
- where appropriate, to provide a statement of general policy with respect to health and safety and to arrangements for safety representatives and safety committees.

It is the *duty of every employee* whilst at work, to take all reasonable care for the health and safety of himself and other persons who may be affected by his acts and omissions. Employees are required to:

- co-operate with the employer to enable the duties placed on him (the employer) to be performed
- have regard of any duty or requirement imposed upon his employer or any other person under any of the statutory provisions
- not interfere with or misuse any thing provided in the interests of health, safety or welfare in the pursuance of any of the relevant statutory provisions.
- In addition to having responsibilities to the employee, the employer also has responsibilities to persons such as the general public.

It is also the *duty of the employer* to:

- conduct his undertakings in such a way so as to ensure that members of the public (i.e. those not in his employ) are not effected or exposed to risks to their health or safety.
- give information about such aspects of the way in which he conducts his undertakings to persons who are not his employees as might affect their health and safety.

It is the duty of *each person* who has control of premises, or access to or from any plant or substance in such premises, to take all reasonable measures to ensure they are safe and without risk.

Identifying potential hazards

When investigating an engineering process or system in relation to Health and Safety factors it is essential to consider the following:

1. The materials and components used.

- What are they and are any of them hazardous?
- What precautions must be observed for handling and storage?
- What precautions should be observed for disposal?
- What protective clothing is required?
- What training is required for those involved with materials handling?

2. The processes used.

- What are they and are any of them hazardous?
- Are any hazardous waste products or by-products produced?
- What tools and equipment are used?
- Are any of these dangerous?
- What training is required for production staff?

3. The energy/fuel supplies used.

- What are they and are any of them hazardous?
- How are the energy or fuel supplies changed/replenished and is this process hazardous?
- What controls are used to regulate and/or protect the supplies (such as pressure relief valves, fuses, RCCB, etc)?
- What training and equipment is required to deal with emergency situations (such as fuel spills, chemical leaks, radiation, etc)?

4. The operation of the system

- What tests are required in order to ensure correct operation of the system?
- What are the consequences of a malfunction of the system?
- How is a system malfunction detected?
- What provision is there for shutting the system down safely in the event of malfunction?
- What provision is there for safely restarting the system?
- What training is required for operators and maintainers?

5. The environment in which the system is used.

- Is the environment constant or is it subject to long-term or short-term change?
- Are there any excessive environmental factors (such as temperature, pressure, humidity, vibration etc.)?

Production and manufacturing processes

From Unit 2, you should be aware that many engineering processes are potentially hazardous and these include activities such as casting, cutting, soldering, welding, etc. In addition, some processes involve the use of hazardous materials and chemicals. Furthermore, even the most basic and straightforward activities can potentially be dangerous if carried out using inappropriate tools, materials, and methods.

In all cases, the correct tools and protective equipment should be used and proper training should be provided. In addition, safety warnings and notices should be prominently placed in the workplace and access to areas where hazardous processes take place should be restricted and carefully controlled so that only appropriately trained personnel can be present. In addition, the storage of hazardous materials (chemicals, radioactive substances, etc) requires special consideration and effective access control.

Processes that are particularly hazardous include:

- casting, forging and grinding
- welding and brazing
- chemical etching
- heat treatment
- use of compressed air.

Operation and maintenance

Hazards are associated with the operation and maintenance of many engineering systems. These include those that use single-phase or three-phase AC mains supplies, compressed air, fluids, gas, or petrochemical fuels as energy sources. They also include those systems that are not in themselves particularly hazardous but which are used in hazardous environments (for example in mining or in the oil and gas industries). Operations such as refuelling a vehicle can be potentially dangerous in the presence of naked flames or if electrical equipment is used nearby. A static discharge, for example, can be sufficient in the presence of petroleum vapour, to cause an explosion. Similar considerations apply to the use of mobile phones or transmitting apparatus in the vicinity of flammable liquids.

Lifting and manual handling

Safe working practices need to be adopted for manual operations (such as lifting, stacking and loading). Personnel need to be trained to perform these tasks and managers should insist that the correct procedures and protective clothing are used at all times. The use of hard hats should be made obligatory wherever overhead work is being carried out and when personnel are working underground or on building sites.

Electrical safety

It should go without saying that, with the exception of portable low-voltage equipment (such as hand-held test equipment) all electrical equipment should be considered potentially dangerous. This applies to *all* mains operated equipment. There are three basic hazards associated with electricity:

- electric shock
- explosion due to the presence of flammable vapours, and
- fire due to overheating of cables and appliances.

Often these hazards result from badly designed installations but all too often the inadequate maintenance of tools and deliberate overloading of circuits results in an accident. The Health and Safety at Work Act 1974 and the Electrical Factories Act Special Regulations must be followed to ensure safe working practice—the observance of these regulations is mandatory.

Before using any electrical equipment it is advisable to carry out a visual check of the condition of the mains lead, plug and connector. Correct earthing of any tools that are not of the double insulated variety is absolutely essential. Good earth continuity is essential for safety since, if a fault should occur within the appliance such that the metal frame became 'live', current would flow through any available path to earth. Since the human body acts as a good conductor, the operator will receive a shock. The amount of current passing through the operator's body would depend on the voltage and the total resistance of the path through the body (which is significantly lower in the presence of dampness, moisture and sweat).

The two ways in which a person may receive a shock is by touching the live conductor of the supply and the neutral conductor at the same time and thus completing the circuit, or by touching the live conductor whilst being in contact with the earth. Because of this, AC mains operated electric power tools used in the workshop can be a potential source of electrical danger if any faults develop. Low-voltage tools (using rechargeable batteries) represent much less of a hazard and there is little risk of electric shock once the equipment is removed from its charging supply. Some modern portable appliances are now supplied double-insulated, whereby an additional plastic body houses the metalwork required for the operation of the power tool and there is no risk of the operator coming into contact with any metal parts.

In all circumstances vigilance and correct maintenance will help to prevent accidents. In the event of a malfunction it is essential to turn-off (or disconnect) the supply of electricity and not to touch any metal parts that may have become 'live'. It is also essential not to make contact with anyone who may have received an electric shock until you are certain that the supply has been disconnected.

Activity 4.22

A long-range radar system uses a high-vacuum cathode ray tube in the display unit and a high-power magnetron in the transmitter. What Health and Safety issues might relate to the manufacture, operation and maintenance of this system. Write a short report explaining your answer.

Activity 4.23

Tractor engines are assembled on a moving assembly line. When each engine is complete it is removed from the assembly line using a hoist and transferred to an engine test cell where the engine is test run before being transferred to a different part of the tractor assembly plant. Fluorescent lighting and compressed air driven power tools are used throughout the plant. What Health and Safety issues might relate to this manufacturing process? Write a short report explaining your answer.

Review questions

1 The cross-section of an aluminium alloy beam is shown in Figure 4.151. If the alloy has a relative density of 3.2 and a length of 1.2 m, determine the weight of the beam.

2 State THREE properties of a force.

3 A force of 2.5 kN acts at angle of 30° to the horizontal. Determine (a) the vertical and (b) the horizontal component of the force.

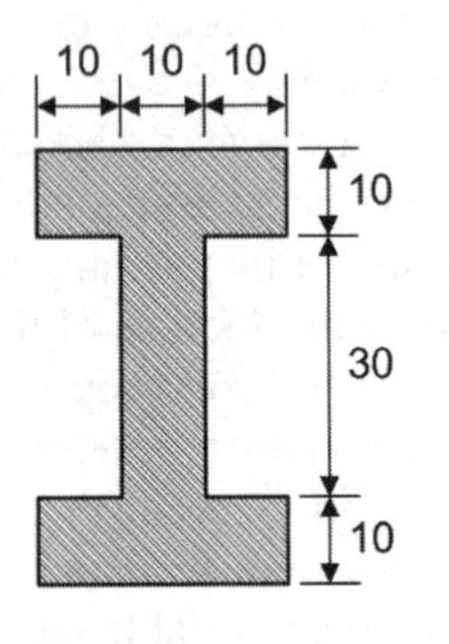

Figure 4.151 *See Question 1*

4 Two forces of 120 N and 160 N act at right angles to one another. Determine the magnitude and direction of the resultant relative top the 160 N force.

5. Determine the magnitude and direction of the resultant of the two forces shown in Figure 4.152.

6 Distinguish, with the aid of simple diagrams, between the following types of force:

(a) tensile force
(b) compressive force
(c) shear force.

7 An oil seal is rated for a maximum pressure of 1.5 kPa. Determine the maximum force that can be applied at right angles to the seal if it has an effective area of 0.005 m^2.

8 Define the terms *stress* and *strain*.

9 State Hooke's Law.

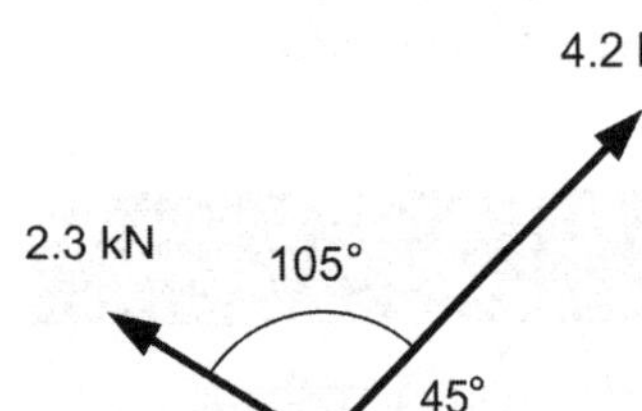

Figure 4.152 *See Question 5*

10 Explain how a tensometer can be used to determine the *elasticity modulus* of a material.

11 Sketch a load/extension graph for an elastic material and label the following points:

(a) elastic limit
(b) yield point
(c) ultimate tensile strength
(d) point of fracture.

12 Sketch, using a common set of axes, typical load/extension graphs for:

(a) hard-drawn brass
(b) cast iron
(b) annealed copper.

Comment, with reasons, on the shape of each graph.

13 Sketch diagrams to show the construction of:

(a) a typical bridge truss
(b) a typical roof truss.

(a)

(b)

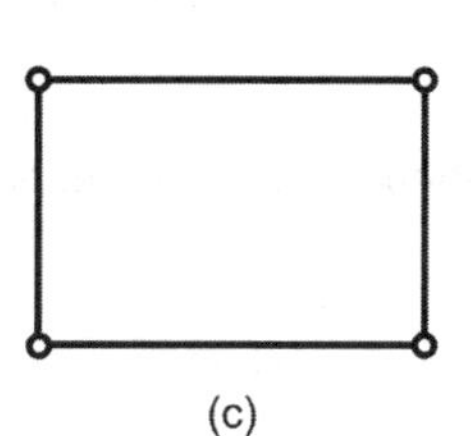

(c)

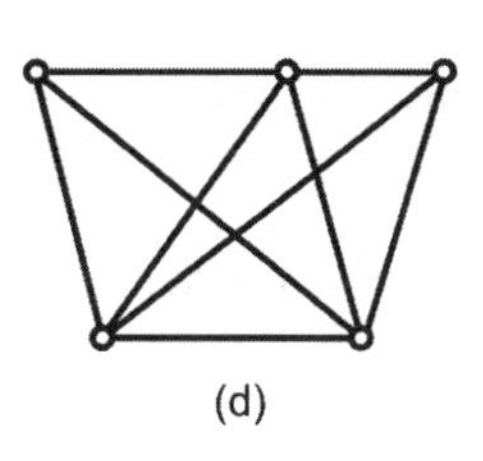

(d)

Figure 4.153 *See Question 14*

14 Four simple pin-jointed frameworks are shown in Figure 4.153. Which of these frameworks is *statically determinate*? Explain your answer.

15 Define the term factor of safety in relation to the design of a framework.

16 A 24 V battery supplies a current of 20 A to the emergency lighting circuit of an aircraft. What power is consumed?

17 In Question 16, what energy is supplied if the lighting circuit operates for 30 minutes?

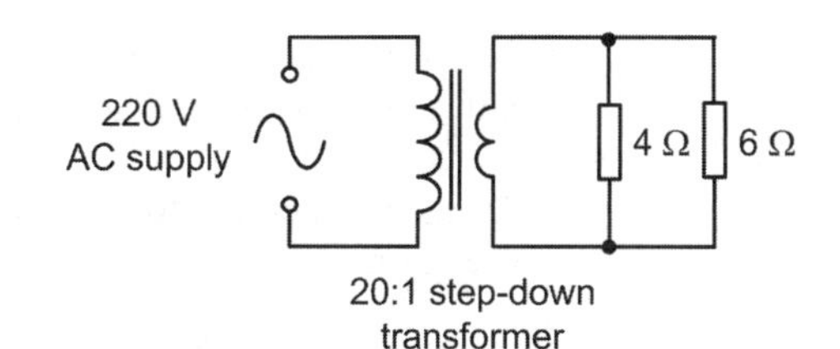

Figure 4.154 *See Question 18*

18 Figure 4.154 shows two loads connected to an AC supply. Determine the power dissipated in each load and the total power supplied.

19 Determine the peak voltage of an AC supply that has an r.m.s. value of 110 V.

20 Identify the types of cable shown in Figure 4.155 and give one example of the use of each type.

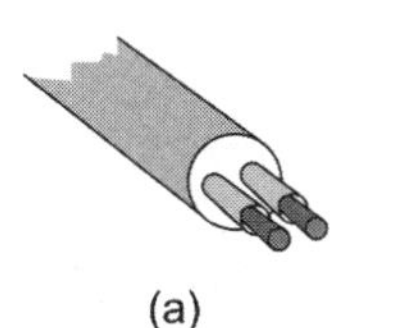

(a)

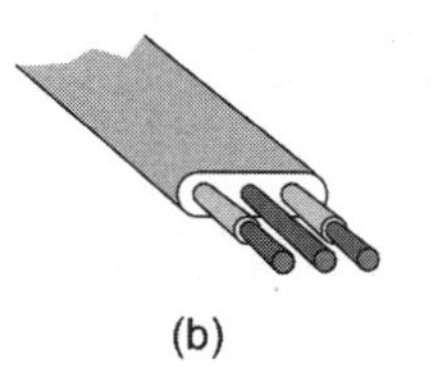

(b)

(c)

Figure 4.155 *See Question 20*

21 Identify the pneumatic symbols shown in Figure 4.156.

22 Explain the function and operation of a pressure relief valve. Describe an application for this component.

23 State TWO advantages and TWO disadvantages of GLS lamps compared with fluorescent lamps.

24 Sketch a circuit that shows how a low-voltage halogen lamp can be operated from a 220 V AC mains supply.

25 Sketch a typical ring-main circuit and explain how this differs from a radial circuit.

26 Explain how a residual current device (RCD) operates.

27 State the typical operating current for an RCD or RCCB.

28 Distinguish between sensors that have digital outputs and those that have analogue outputs. Give one example of each type of sensor.

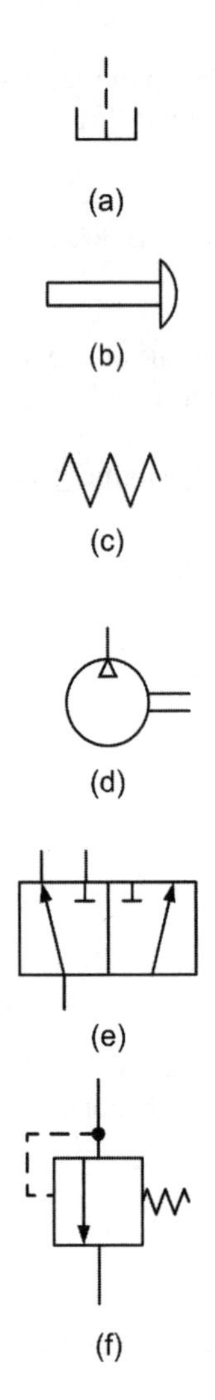

Figure 4.156 *See Question 21*

29 Explain, briefly, the principle of a rotating vane flow sensor.

30 Sketch time related waveforms to show the difference between analogue and digital signals.

31 How many different states can be represented by a 12-bit binary code? Explain your answer.

32 Sketch the simplified block diagram of an instrumentation system. Label your diagram clearly.

33 Explain how reference sources could be used to calibrate an oscilloscope. What types of reference source would be required?

34 Sketch the block diagram of a closed-loop system and explain how the error signal is generated.

35 Sketch a circuit to show how an operational amplifier can be used as an error forming device.

36 Sketch the typical response of a closed-loop system to a suddenly applied load. Label your graph clearly.

37 Sketch the block diagram of a simple microprocessor system. Label your diagram clearly.

38 Explain, in relation to an analogue to digital converter, the terms:
(a) sampling rate
(b) quantisation

39 On what does (a) the resolution and (b) the accuracy of a digital to analogue converter depend?

40 Sketch waveforms to show:
(a) an amplitude modulated (AM) signal
(b) a frequency modulated (FM) signal.

41 Explain the function of a MODEM and describe a typical application for such a device.

42 State FOUR hazards that may exist when an engine is removed from a car.

43 List precautions that need to be taken in order to minimise the risks that you have identified in Question 42.

44 Explain the risks associated with working with mains operated electrical power tools in a damp environment.

45 Explain the advantages of double insulated power tools over tools that may have exposed metal parts.

Unit 5 The engineering environment

Summary

Earlier in Unit 2 you investigated the role of the engineer and how an engineer's role is influenced by many different factors including new technologies, legislation and standards. In this Unit you will further develop your understanding of legislation and standards and how documentation is used to control engineering working practice and product quality. This Unit also explores the environmental impact of engineering and how new technology has affected engineered products and services. The assessment of this unit is based on an in-depth investigation of an engineered product or service (different from the one that you carried out in Unit 2). This investigation will help you put into context what you have learned in this Unit.

5.1 Legislation and documentation in engineering

In Unit 2 we mentioned some of the legislation, regulations and standards that govern the way engineering products are designed, manufactured and maintained. These regulations deal with factors such as:

- electromagnetic compatibility
- radiation emissions
- energy efficiency
- waste disposal
- Certification European (CE).

Engineers need to be aware of the impact of these regulations and how they affect the manufacture, maintenance, and use of the products that they design, manufacture and maintain. In addition, engineers must be aware of and use a variety of documents. These include:

- work procedures
- engineering drawings
- quality manuals
- repair manuals
- operating manuals
- product specifications.

We shall start our study of the engineering environment by taking a brief look at some of the legislation and regulations that we met in Unit 2 before moving on to investigate some of the key documents that engineers use in their everyday working lives.

The EMC Directive

The Electromagnetic Compatibility (EMC) Directive of the European Community (EC) has widespread implications for any engineered product that uses electricity or electronics. The Directive states that products must not emit unwanted electromagnetic pollution which might otherwise cause *interference* to other appliances and services. Equally important is that the Directive also states that products must themselves be immune to a reasonable amount of interference. The standards that are used to demonstrate *compliance* with the Directive are based on a variety of different tests. These tests fall into several classes including:

- *Radiated emissions:* Checks to ensure that the product does not emit unwanted radio signals
- *Conducted emissions:* Checks to ensure the product does not send out unwanted signals along its supply connections and connections to any other apparatus
- *Radiated susceptibility:* Checks that the product can withstand a typical level of electromagnetic pollution
- *Conducted susceptibility:* Checks that the product can withstand a typical level of noise on the power and other connections
- *Electrostatic discharge:* Checks that the product is immune to a reasonable amount of static electricity.

It is important to note that the definitions of the levels above which emissions are defined as "unwanted" (or below which pollution and noise are accepted as being reasonable) are contained in the relevant test standards. Note also that it is not necessary to test every piece of apparatus comprehensively. This is costly and it is usually only necessary to check a sample of the production in order to determine compliance with the appropriate standards. This testing can be carried out 'in-house' (i.e. by the manufacturer) or can use the facilities of an external *testing house*).

Other EC directives

Many other EC directives are applicable to engineering companies and these include:

- *Telecommunications Terminal Equipment* which relates to telecommunications products such as MODEMS, telephones, answering machines, etc.

- *Low Voltage Directive (LVD)* which relates to electrical equipment installation including cables, flexible leads and wiring.
- *Machinery Safety Directive (MSD)* which relates to a wide range of products that comprise a number of linked parts (at least one of which moves), and a source of energy that is other than human. Note that there is a list of exceptions to this directive and this includes tractors, military and police vehicles as well as freight and passenger carrying systems.

CE

Figure 5.1 *The CE mark*

The CE mark

Displaying the CE mark (see Figure 5.1) on a product (or its packaging) is mandatory for most types of product. The CE mark indicates that the product complies with the relevant EC directives. You will probably not be surprised to learn that there is even an EC Directive that governs how the CE mark is used!

It is also worth noting that many products need to demonstrate compliance with more than one EC Directive. Thus, engineering companies need to be fully aware of all of the Directives that could potentially apply to the products that they design and manufacture. Furthermore, since legislation is constantly changing, it is necessary to keep a watchful eye on any new or revised directives.

Test your knowledge 5.1

List THREE EC Directives that might apply to an engineered product.

Documentation

Many different types of document are used in engineering. In this section we shall briefly investigate some of the most important types of document starting with application notes and technical reports.

Application notes and technical reports

Application notes are usually brief notes (often equivalent in extent to a chapter of a book) supplied by manufacturers in order to assist engineers and designers by providing typical examples of the use of engineering components and devices.

An application note can be very useful in providing practical information that can help designers to avoid pitfalls that might occur when using a component or device for the first time. Applications notes often include prototype schematics and layout diagrams.

Technical reports are somewhat similar to application notes but they focus more on the performance specification of engineering components and devices (and the tests that have been carried out on them) than the practical aspects of their use. Technical reports usually include detailed specifications, graphs, charts and tabulated data.

Typical section headings used in application notes and technical reports include:

Summary
A brief overview for busy readers who need to quickly find out what the application note or technical report is about.

Introduction
This sets the context and background and provides a brief description of the process or technology—why it is needed and what is does. It may also include a brief review of alternative methods and solutions.

Main body
A comprehensive description of the process or technology.

Evaluation
A detailed evaluation of the process or technology together with details of tests applied and measured performance specifications. In appropriate cases comparative performance specifications will be provided.

Recommendations
This section provides information on how the process or technology should be implemented or deployed. It may include recommendations for storage or handling together with information relating to Health and Safety.

Conclusions
This section consists of a few concluding remarks.

References
This section provides readers with a list of sources for further information relating to the process or technology, including (where appropriate) relevant standards and legislation.

Application notes explain how something is used in a particular application or how it can be used to solve a particular problem. Application notes are intended as a guide for designers and others

Activity 5.1

Write an application note that explains the use of AA-size NiMh batteries as replacements for the conventional alkaline batteries used in a digital camera. You should carry out some initial research before starting to write your application note but the following are possible headings (these can be combined or expanded if you think it necessary):

- Executive summary
- Introduction
- Basic Requirements for batteries used in digital cameras
- Comparison of three battery types (NiCd, NiMh, and conventional alkaline types)
- Battery life and charging arrangements
- Analysis of costs
- Suitability
- Recommendations
- Reference data (including a list of manufacturers and suppliers).

Test your knowledge 5.2

List the section headings used in a typical application note.

who may be considering using a particular process or technology for the first time. Technical reports, on the other hand, provide information that is more to do with whether a component or device meets a particular specification or how it compares with other solutions. Technical reports are thus more useful when it comes to analysing how a process or technology performs than how it is applied.

Data sheets and data books

Data sheets usually consist of abridged information on a particular engineering component or device. They usually provide maximum and minimum ratings, typical specifications, as well as information on dimensions, packaging and finish. Data sheets are usually supplied free on request from manufacturers and suppliers. Collections of data sheets for similar types of engineering components and devices are often supplied in book form. Often supplementary information is included relating to a complete family of products. An example of a data sheet is shown in Figure 5.4.

Catalogues

Most manufacturers and suppliers provide catalogues that list their full product range. These often include part numbers, illustrations, brief specifications and prices. Whilst catalogues are often extensive documents with many hundreds or thousands of pages, short-form catalogues are usually also available. These usually just list part numbers, brief descriptions and prices but rarely include any illustrations. A brief extract from a short-form catalogue is shown in Figure 5.3.

Catalogues and data sheets are often distributed on compact disks which can provide storage for around 650 Mbytes of computer data. This is equivalent to several thousand pages of A4 text and line diagrams.

Specifications

Written specifications should take the form of a precise and comprehensive description of the product. Specifications should relate not only to the physical characteristics and appearance of a product but also to the performance of a product in terms that can be measured in order to verify its performance.

Since specifications form the basis of a contract between a manufacturer or supplier and a client or customer, they need to be written in terms of what the purchaser requires and in clear, unambiguous words. There are three different types of specification:

- *General specifications:* A detailed written description of the product including its appearance, construction, and materials used.
- *Performance specification:* A list of features of the product that contribute to its ability to meet the needs of the client or end user. For example, output voltage, power, or speed.
- *Standard specification:* Describes the materials and processes (where appropriate) used in the manufacture of the product in terms of relevant standards (e.g. BS 9000).

Test your knowledge 5.3

List THREE different types of specification.

Quality documents

All engineering companies have *Quality Systems* in place to ensure that they produce goods and services of an appropriate quality. These systems are invariably based on documented procedures which often include one or more of the following:

- *Quality Procedures:* A detailed written description of the quality system and the controls that are in place within the company (these often form part of a *Quality Manual* or are documented separately as a *Procedures Manual*).
- *Work Instructions:* A description of a particular operation or task in terms of what must be done, who should do it, when it should be done, and what materials and processes should be used. In many cases, a series of Work Instructions are used to describe each stage in the production or manufacturing process.
- *Test Specifications:* A detailed list of characteristics and features used to verify conformance with the design specification together with details of any measurements that are to be made out and how they should be carried out.

Traceability

Documents are often used to assist in the process of identification and *traceability* of products, components and materials. This is vitally important in critical sectors such as aerospace, nuclear and chemical engineering. Traceability is an essential when it becomes necessary to eliminate the causes of *non-conformance*. Traceability is achieved by coding items and maintaining records that can be updated throughout the working life of a component or record.

Activity 5.2

Refer to the Archer Aerospace data sheet shown in Figure 5.2 and use it to answer the following questions:

1. What three products are described in the data sheet?
2. With what British Standard do these products comply?
3. What units are used to specify the capacity of the batteries?
4. What is the capacity of an HD-120 battery?
6. At what voltage should these batteries be charged?
7. What is the specified temperature range for the batteries?
8. What precautions should be observed when batteries are stored for long periods?
9. What is the weight of an HD-80 battery?
10. What is the effect of operating a battery at temperatures below freezing on battery capacity?

Test your knowledge 5.4

Explain why traceability is important for certain engineered products. Give examples of TWO engineering sectors where traceability is important.

Archer Aerospace

General Purpose Heavy Duty Lead Acid Batteries HD-80, HD-120 and HD-180

Heavy duty sealed, maintenance free lead-acid batteries for use in uninterruptible power supplies, communications equipment, security and other back-up applications. With a service life of 10 years, these batteries comply fully with BS6290 Part 4.

Each battery uses heavy duty anti-corrosive lead-calcium alloy grids separated by a micro-fibre mat which retains the electrolyte. This construction maximises capacity and minimises self-discharge.

Electrical connection is via threaded brass terminal posts. Batteries incorporate low-pressure vents which release gas if internal pressure reaches an unacceptable level due to misuse or an external fault. A connecting bar, nuts and bolts are supplied with each battery. Batteries should be float charged as 2.26V per cell.

HD-80
2V 320Ah

HD-120
6V 160Ah

Technical specification:

Type	Voltage (V)	Capacity (Ah)	Dimensions (mm) Length	Width	Height	Weight (kg)
HD-80	2	320	206	210	240	24
HD-120	6	160	305	210	240	35
HD-180	12	140	595	210	240	65

Notes:

1. Capacities quoted in Ampere-hours are for discharge to a final cell voltage of 1.8V at the 10 hour rate.
2. If a battery is not used for a period of 6 months a recovery top-up charge should be applied.
3. Batteries must always be stored in a fully-charged condition.
4. Batteries must be used and stored in a cool and dry location.
5. Batteries must be kept within the temperature range -15°C to +50°C.

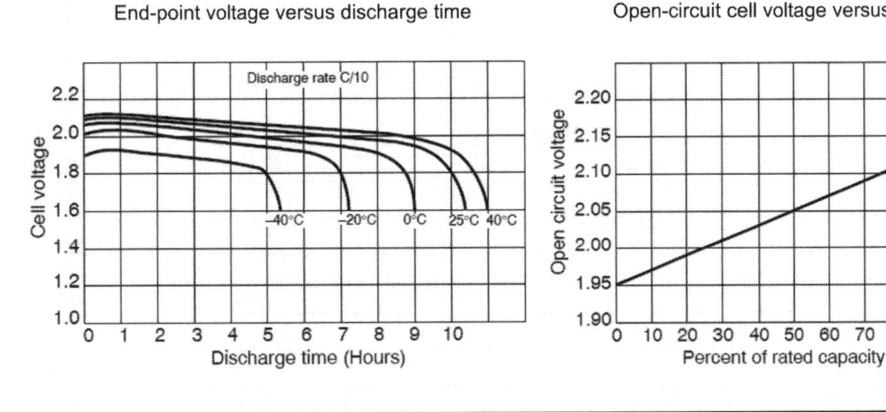

Figure 5.2 *See Activity 5.2*

Diecast Boxes

IP65 Sealed/Painted

A range of high-quality diecast alumnium boxes with an optional grey epoxy paint finish to RAL7001. The lid features an integral synthetic rubber sealing gasket and captive stainless steel fixing screws. Mounting holes and lid fixing screws are outside the seal, giving the enclosure protection to IP 65.

Standard supply multiple = 1

Delivery normally ex-stock

Size							Price each		
L	W	H	T	Finish	Manufacturer's ref:	Stock code	1-9	10-24	25+
90	45	30	3.0	none	1770-1541-21	DB65-01	£4.52	£3.95	£3.50
90	45	30	3.0	grey	1770-1542-21	DB65-01P	£5.40	£4.90	£4.45
110	50	30	4.5	none	1770-1543-22	DB65-02	£5.25	£4.50	£4.15
110	50	30	4.5	grey	1770-1544-22	DB65-02P	£6.42	£5.37	£4.95
125	85	35	5.0	none	1770-1545-23	DB65-03	£6.15	£5.17	£4.71
125	85	35	5.0	grey	1770-1546-23	DB65-03P	£7.10	£6.05	£5.65

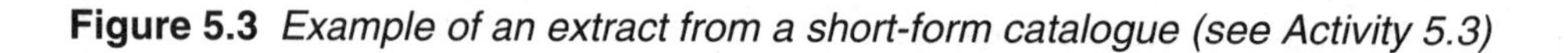

Figure 5.3 *Example of an extract from a short-form catalogue (see Activity 5.3)*

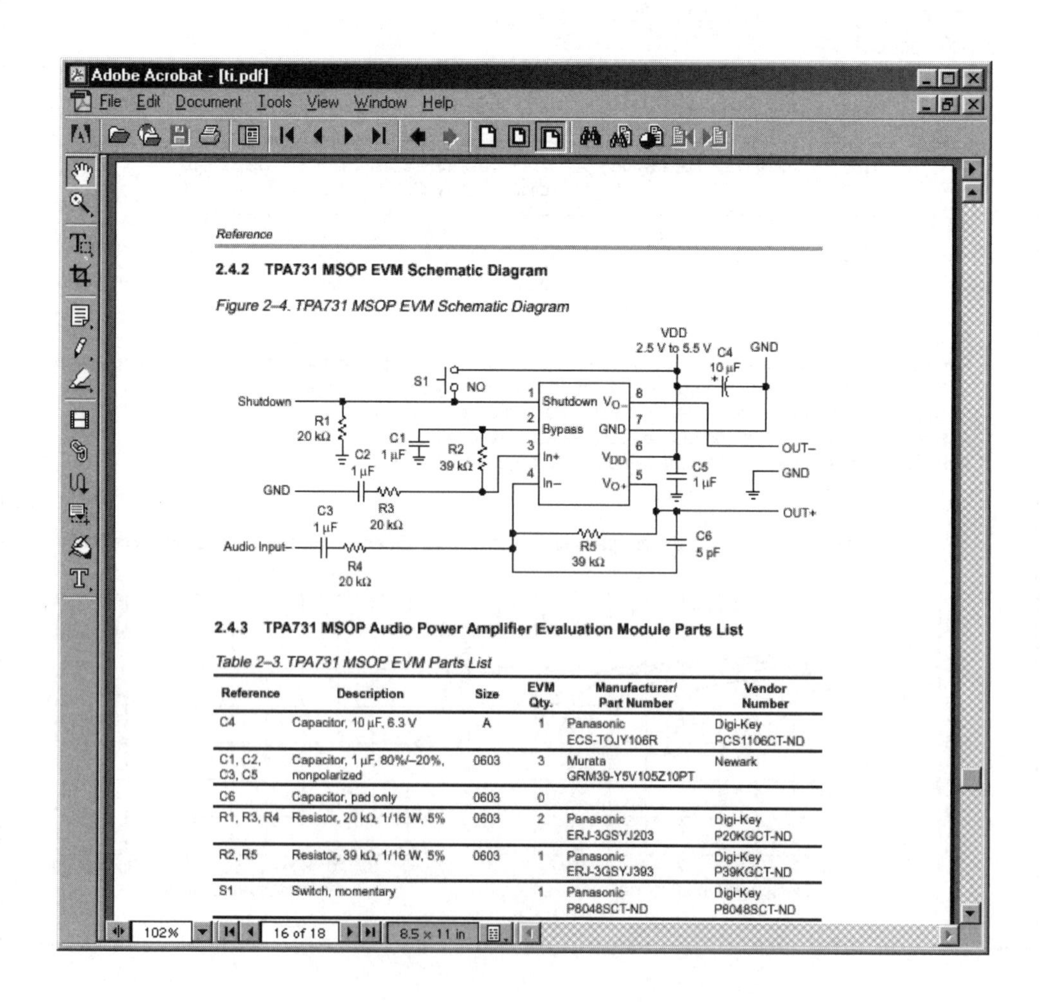

Reference

2.4.2 TPA731 MSOP EVM Schematic Diagram

Figure 2–4. TPA731 MSOP EVM Schematic Diagram

2.4.3 TPA731 MSOP Audio Power Amplifier Evaluation Module Parts List

Table 2–3. TPA731 MSOP EVM Parts List

Reference	Description	Size	EVM Qty.	Manufacturer/ Part Number	Vendor Number
C4	Capacitor, 10 µF, 6.3 V	A	1	Panasonic ECS-TOJY106R	Digi-Key PCS1106CT-ND
C1, C2, C3, C5	Capacitor, 1 µF, 80%/–20%, nonpolarized	0603	3	Murata GRM39-Y5V105Z10PT	Newark
C6	Capacitor, pad only	0603	0		
R1, R3, R4	Resistor, 20 kΩ, 1/16 W, 5%	0603	2	Panasonic ERJ-3GSYJ203	Digi-Key P20KGCT-ND
R2, R5	Resistor, 39 kΩ, 1/16 W, 5%	0603	1	Panasonic ERJ-3GSYJ393	Digi-Key P39KGCT-ND
S1	Switch, momentary		1	Panasonic P8048SCT-ND	Digi-Key P8048SCT-ND

Figure 5.4 *Example of a data sheet published in Adobe Portable Document (PDF) format*

Activity 5.3

Your company requires 15 diecast boxes suitable for enclosing a printed circuit board of thickness 3 mm measuring 80 mm × 35 mm. The tallest component stands 15 mm above the board and a minimum clearance of 5 mm is to be allowed all round the board. The enclosure is to be supplied ready for mounting the printed circuit board and should not need any further finishing other than drilling. Prepare a fax message to Dragon Components (the hardware supplier whose short-form catalogue extract appears in Figure 5.3) giving all the information required to fulfil your order.

Activity 5.4

Visit the Texas Instruments Web site at www.ti.com.

1. Use the search facility to search for information on '74ALS08'.

2. When the device has been located, open the 'Product Folder' to view more information.

3. Click to download the full data sheet for the device (this is provided in the form of an Adobe Acrobat (.PDF) file.

4 View and print the full data sheet.

5. Given that the device employs 'totem-pole' outputs, sketch the circuit that should be used for measuring the switching characteristics of the device and include component values.

6. What are typical values for the 'low state' and 'high state' output voltages?

Present your work in the form of a word-processed 'fact sheet'. Insert your circuit diagram into this document (either by scanning it or cutting and pasting it from a drawing package). Attach your 'fact-sheet' to the printed data sheet.

Manuals

Various types of manual are associated with engineered products including *operating manuals* that are designed to be read by the end-user of the product and *service manuals* or *repair manuals* that are designed to aid the repair and/or the routine maintenance of the product. Manuals are often produced by the company that has manufactured the product but may also be produced by third-party companies and organisations who specialise in producing such information.

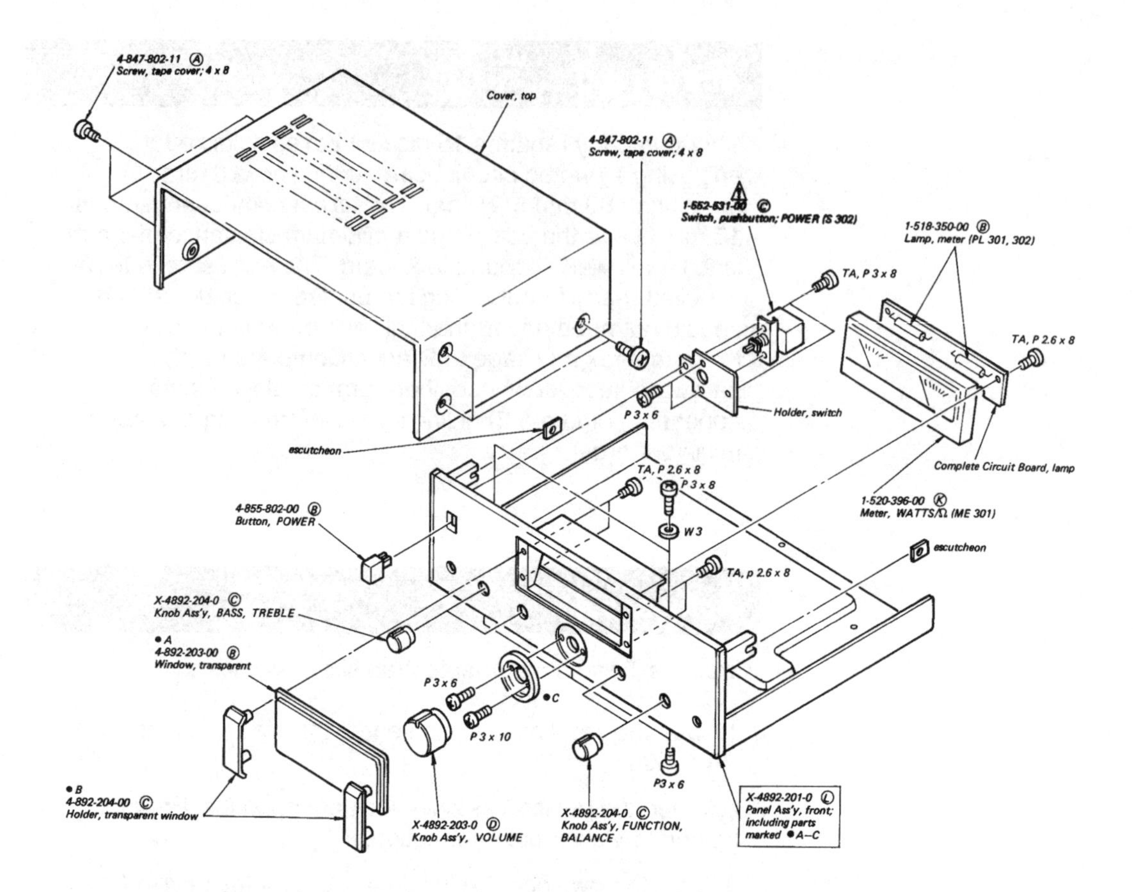

Figure 5.5 *An exploded view of an amplifier taken from a service manual*

Activity 5.5

Obtain a service manual for a car or motorcycle and use it to answer the following questions:

1. Who is the publisher of the service manual?
2. Is the publisher of the service manual the vehicle manufacturer or a third-party company or organisation?
3. What vehicle types and models are covered by the service manual?
4. List the main section headings in the service manual.
5. Does the service manual include specifications for the vehicle?
6. Identify TWO good features of the service manual that could be improved and suggest how each could be usefully changed.

Activity 5.6

Complete the table shown in Figure 5.6 by placing a tick in the column against the most appropriate method for communicating the listed information (tick only one box in each row).

Application	*Application note*	*Data sheet*	*Catalogue*	*Technical report*
Summary of the precautions to be observed when handling a chemical etching fluid				
Cost of diecast boxes supplied in various quantities				
Maximum working temperature for a power transistor				
Recommended printed circuit board layout for an audio amplifier				
Comparison of different types of surface finish for the interior of a domestic microwave oven				
Physical dimensions of a marine radar for fitting to the mast of a small boat				
Description of tests applied to an off-road vehicle				
Performance specification for a satellite TV aerial				

Figure 5.6 *See Activity 5.6*

5.2 The environmental impact of engineering activities

Modern engineering processes and systems are increasingly designed and implemented to minimize environmental affects. Engineering companies must ensure that the negative effects of engineering activities on the natural and built environment are minimized. You need to be able to identify how individual engineering companies seek to do this, such as through:

- design of plant and products which optimizes energy use and minimize pollution

- good practice such as the efficient use of resources and recycling and the use of techniques to improve air and water quality
- management review and corrective action
- relevant legislation and regulations.

Material processing

Many engineering activities involve the processing of materials. Such materials may appear in the product itself or may be used in the manufacturing process. Some of these materials occur naturally and, after extraction from the ground, may require only minimal treatment before being used for some engineering purpose. Examples are timber, copper, iron, silicon, water and air. Other engineering materials need to be manufactured. Examples are steel, brass, plastic, glass, gallium arsenide and ceramic materials. The use of these materials produces effects; some beneficial, some not.

Economic and social effects stem from the regional wealth that is generated by the extraction of the raw material and its subsequent processing or manufacture into useful engineering materials. For example, the extraction of iron-ore in Cleveland and its processing into pure iron and steel has brought great benefit to the Middlesborough region. The work has attracted people to live in the area and the money they earn tends to be spent locally. This benefits trade at the local shops and entertainment centres and local builders must provide more homes and schools and so on. The increased numbers of people produces a growth in local services which includes a wider choice of different amenities, better roads and communications and arguably, in general, a better quality of life.

On the debit side, the extraction of raw materials can leave the landscape untidy. Heaps of slag around coal mines and steelworks together with holes left by disused quarries are not a pretty sight. In recent years much thought and effort has been expended on improving these eyesores.

Slag heaps have been remodelled to become part of golf courses and disused quarries filled with water to become centres for water sports or fishing. Disused mines and quarries can also be used for taking engineering waste in what is known as a landfill operation prior to the appropriate landscaping being undertaken.

Other potential problems can arise from having to transport the raw materials used in engineering processes from place to place. This can have an adverse affect on the environment resulting from noise and pollution.

The effects of waste products

Engineering activities are a major source of *pollutants* causing many types of *pollution*. Air, soil, rivers, lakes and seas are all, somewhere or other, polluted by waste gases, liquids and solids discarded by the engineering industry. Because engineering enterprises tend to be concentrated in and around towns and other built up areas, these tend to be common sources of pollutants.

Electricity is a common source of energy and its generation very often involves the burning of the *fossil fuels*: coal, oil and natural

Figure 5.7 *Many engineering activities produce airborne pollutants*

gas. In so doing, each year, billions of tonnes of carbon dioxide, sulphur dioxide, smoke and toxic metals are released into the air to be distributed by the wind. The release of hot gases and hot liquids also produces another pollutant; heat. Some electricity generating stations use nuclear fuel which produces a highly radioactive solid waste rather than the above gases.

The generation of electricity is by no means the only source of toxic or biologically damaging pollutants. The exhaust gases from motor vehicles, oil refineries, chemical works and industrial furnaces are other problem areas. Also, not all pollutants are graded as *toxic*. For example, plastic and metal scrap dumped on waste tips, slag heaps around mining operations, old quarries, pits and derelict land are all *non-toxic*.

Finally, pollutants can be further defined as *degradable* or *non-degradable*. These terms simply indicate whether the pollutant decomposes or disperses itself with time. For example, smoke is degradable but dumped plastic waste is not.

Carbon dioxide

Carbon dioxide in the air absorbs some of the long-wave radiation emitted by the earth's surface and in so doing is heated. The more carbon dioxide there is in the air, the greater the heating or greenhouse effect. This is suspected as being a major cause of global warming causing average seasonal temperatures to increase. In addition to causing undesirable heating effects, the increased quantity of carbon dioxide in the air, especially around large cities, may lead to people developing respiratory problems.

Oxides of nitrogen

Oxides of nitrogen are produced in most exhaust gases and nitric oxide is prevalent near industrial furnaces. Fortunately, most oxides of nitrogen are soon washed out of the air by rain. But if there is no rain, the air becomes increasingly polluted and unpleasant.

Sulphur dioxide

Sulphur dioxide is produced by the burning of fuels that contain sulphur. Coal is perhaps the major culprit in this respect. High concentrations of this gas cause the air tubes in peoples' lungs to constrict and breathing becomes increasingly difficult. Sulphur dioxide also combines with rain droplets eventually to form sulphuric acid or *acid rain*. This is carried by the winds and can fall many hundreds of miles from the sulphur dioxide source. Acid rain deposits increase the normal weathering effect on buildings and soil, corrode metals and textiles and damage trees and other vegetation.

Smoke

Smoke is caused by the incomplete burning of the fossil fuels. It is a health hazard on its own but even more dangerous if combined with fog. This poisonous combination, called *smog*, was prevalent in the early 1950s. It formed in its highest concentrations around the large cities where many domestic coal fires were then in use. Many deaths were recorded, especially among the elderly and those with respiratory diseases. This led to the first Clean Air Act which prohibited the burning of fuels that caused smoke in areas of high population. So-called *smokeless zones* were established.

Test your knowledge 5.5

Give THREE examples of air pollution caused by engineering activities.

Dust and grit
Dust and grit (or *ash*) are very fine particles of solid material that are formed by combustion and other industrial processes. These are released into the atmosphere where they are dispersed by the wind before falling to the ground. The lighter particles may be held in the air for many hours. They form a mist, which produces a weak, hazy sunshine and less light.

Toxic metals
Toxic metals, such as lead and mercury are released into the air by some engineering processes and especially by motor vehicle exhaust gases. Once again the lead and mercury can be carried over hundreds of miles before falling in rain water to contaminate the soil and the vegetation it grows. Motor vehicles are now encouraged to use lead-free petrol in an attempt to reduce the level of lead pollution.

Ozone
Ozone is a gas that exists naturally in the upper layers of the earth's atmosphere. At that altitude it is one of the earth's great protectors but should it occur at ground level it is linked to pollution. *Stratospheric ozone* shields us from some of the potentially harmful excessive ultra-violet radiation from the sun. In the 1980s it was discovered that emissions of gases from engineering activities were causing a 'hole' in the ozone layer. There is concern that this will increase the risk of skin cancer, eye cataracts, and damage to crops and marine life.

At ground level, sunlight reacts with motor vehicle exhaust gases to produce ozone. Human lungs cannot easily extract oxygen (O_2) from ozone (O_3) so causing breathing difficulties and irritation to the respiratory channels. It can also damage plants.

This ground level or *tropospheric ozone* is a key constituent of what is called photochemical smog or summer smog. In the UK it has increased by about 60% in the last 40 years.

Heat
Heat is a waste product of many engineering activities. A typical example being the dumping of hot coolant water from electricity generating stations into rivers or the sea. This is not so prevalent today as increasingly stringent energy saving measures are applied. However, where it does happen, river and sea temperatures can be raised sufficiently in the region of the heat outlet to destroy natural aquatic life. *Energy efficiency* and the reduction of waste heat is an increasingly important consideration in many engineering processes.

Chemical waste
Chemical waste dumped directly into rivers and the sea, or on to land near water, can cause serious pollution which can wipe out aquatic life in affected areas. There is also the long-term danger that chemicals dumped on soil will soak through the soil into the ground water which we use for drinking purposes and which will therefore require additional purification.

Radioactive waste
Radioactive waste from nuclear power stations or other engineering activities which use radioactive materials poses particular problems.

Figure 5.8 *Waste reprocessing involves transportation of waste products to specialist facilities*

Not only is it extremely dangerous to people—a powerful cause of cancer—its effects do not degrade rapidly with time and remain dangerous for scores of years. Present methods of disposing of radioactive waste, often very contentious however, include their encasement in lead and burial underground or at sea.

Derelict land

Derelict land is an unfortunate effect of some engineering activities. The term derelict land may be taken to mean land so badly damaged that it cannot be used for other purposes without further treatment. This includes disused or abandoned land requiring restoration works to bring it into use or to improve its appearance. Land may be made derelict by mining and quarrying operations, the dumping of waste or by disused factories from by-gone engineering activities.

Activity 5.7

Find out what happens to the domestic waste produced in your locality.

What items of domestic waste are recycled?

Is any of the waste burnt to produce useful heat?

If the waste is transported to another site for processing, where does it go and what processes are used?

What arrangements are there for disposing of hazardous waste?

Present your findings in the form of a brief report. Illustrate your report using a flowchart which shows the sequence of processing that applied to different types of waste.

Test your knowledge 5.6

Explain how damage to the ozone layer has resulted in an increase in the incidence of skin cancer.

Activity 5.8

Write a report based on your investigations into the effects of ONE engineering activity from EACH category selected from the following: *production*, *servicing* and *materials handling*, on the physical environment to include *human*, *natural* and *built*.

Examples of engineering production activities are:

- motor car manufacture,
- steel manufacture,
- coal mining.

Examples of engineering servicing activities are:

- motor car dealership garages,
- local council road maintenance depots,
- maintenance of electricity and gas supplies.

Examples of materials handling activities are:

- container handling terminals,
- moving cargo by rail and road,
- conveying goods on moving belts.

For the engineering activity you have chosen, your report should include a brief description of:

- the environmental effects of the materials used,
- the short-term and long-term environmental effects of any waste products,
- any environmental legislation effects giving specific examples.

Hints:

Make sure that your selected engineering activity gives you the opportunity to produce the necessary amount of evidence to demonstrate your competence and understanding.

You should approach this activity through case studies (e.g. those highlighted by court cases concerning failure to comply with legislation).

Finally, it is important to be clear about the difference between *waste products* and *by-products*. The by-products from one process can be sold as the raw materials for other processes. For example, natural gas is a by-product of oil extraction and a useful fuel used in the generation of electricity. Waste products are those that cannot be sold and may attract costs in their disposal. Nuclear power station waste is a typical example.

Figure 5.9 *See Activity 5.9*

Figure 5.11 *See Activity 5.9*

Figure 5.10 *See Activity 5.9*

Figure 5.12 *See Activity 5.9*

Activity 5.9

1. Identify each of the methods of generating electrical energy shown in Figures 5.9 to 5.12.

2. Explain, briefly, how each method works.

3. State any waste products that may be produced.

4. Which of the methods can be considered to be a 'clean' method of producing electrical energy?

5. State the advantages and disadvantages of these methods of producing electrical energy.

6. What other methods (e.g. based on 'fossil fuels') are currently used to produce electrical energy?

5.3 The application of technology in engineering

The development of technological advances and the development of new materials and manufacturing processes have a considerable impact upon the products and services that engineers create or modify. During the assessment for this Unit you will be required to carry out an investigation of an engineered product or service (this must be different from the one that you carried out previously in Unit 2). During this investigation you must consider (*and provide evidence of*) each of the following:

- *Standards and regulations* (see Section 5.1) which apply to the engineered product or service and how they have influenced the engineering activities (including design, manufacture, production, utilisation, maintenance and eventual disposal or recycling).
- *Types of documentation* (see Section 5.1) that support the engineered product or service during its design, development, manufacture, use and maintenance.
- *Energy efficiency* (see Section 5.2) and how steps are taken to ensure that energy is used efficiently.
- *Environmental impact* (see Section 5.2) and how steps are taken to ensure that the impact of engineering activities is minimised.
- *Technology and techniques* (see Section 5.3) used within the engineered product or service or during its development, manufacture, and maintenance.
- *Evaluation* (see Section 5.4) of the engineered product or service and suggestions for modifications to improve its intended performance.

Later we shall provide you with a sample product investigation however we shall start this section with a brief look at some examples of developments in technology and how they have had an impact on some typical engineered products. We shall start by taking an example of a product that you are very familiar with, a simple domestic radio receiver.

Technology and techniques

Let's begin by looking at how the technology and manufacturing techniques used to produce domestic radio receivers has changed over the last 50 years. Three examples of interior construction, each representative of a different period, are shown in Figures 5.13 to 5.15.

The radio in Figure 5.13 was designed and built in the early 1950's. This receiver uses thermionic valves which are mounted in sockets in a substantial metal chassis. The receiver is built into a veneered wooden cabinet and consumes about 30W of power derived from a 240V AC mains supply. The receiver produces an output power of about 3W from its loudspeaker thus the overall efficiency is a mere 10%. The 27W of wasted power (used mainly to heat and supply the valves with high-voltage DC) is dissipated as heat.

The radio in Figure 5.14 was designed and built in the mid-1960's. By this time transistors had replaced thermionic valves in virtually all domestic receivers. Transistors offered a number of advantages over valves since they required no heater supply, operated from low

Figure 5.13 *Interior construction of an early 1950's radio*

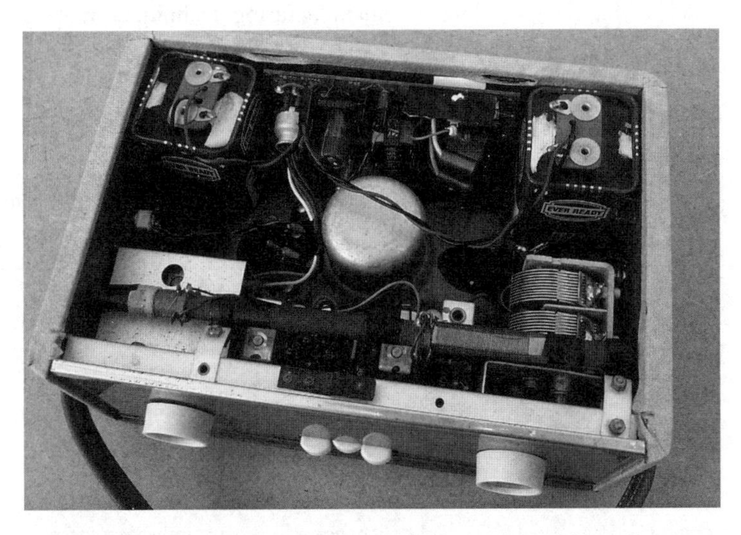

Figure 5.14 *Interior construction of a mid-1960's radio*

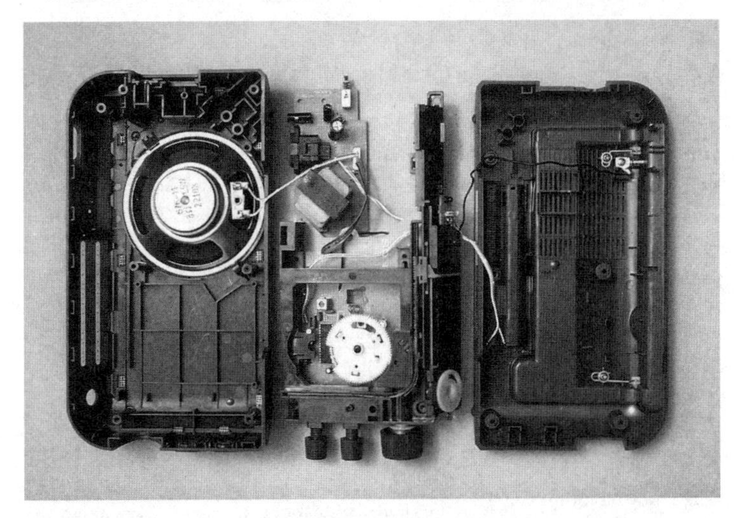

Figure 5.15 *Interior construction of a modern radio*

voltages, and were considerably more reliable. The receiver operated from an 18V DC supply that was derived from two series-connected 9V batteries. The use of internal batteries and a ferrite rod aerial allowed the receiver to be fully portable. The audio output power of the receiver was about 2W with an overall efficiency about 70%. Clearly this represented a considerable improvement on the receivers of a decade earlier!

Figure 5.15 shows the interior construction of a modern radio receiver. This receiver operates from either a 240V AC supply or from internal batteries. Compared with the earlier receivers, this radio operates with considerably fewer components and a single integrated circuit replaces the transistors used in the 1960's receiver. Another feature of the modern receiver is that it provides coverage of the VHF FM band as well as the medium and long AM bands. When used on its internal batteries, the receiver operates from a 6V DC supply derived from four 1.5V LR14 cells connected in series.

Table 5.1 summarises some of the features of the three radio receivers shown in Figures 5.13 to 5.15. Note how the materials and manufacturing techniques have changed considerably over the 50 year period.

Table 5.1 *Comparison of features of the three radio receivers*

Feature	*Early 1950's radio (Figure 5.13)*	*Mid-1960's radio (Figure 5.14)*	*Modern radio (Figure 5.15)*
Enclosure	Veneered wood	Rexine covered wood	ABS
Components	Six thermionic valves	Seven transistors and one diode	One integrated circuit and one diode
Construction	Metal chassis	Internal metal brackets	ABS sub-chassis
Wiring	Point-to-point soldered wires	Two printed circuit boards	Two printed circuit boards
Manufacture	Assembly, wiring and tuned circuit alignment all done by hand	Assembly, wiring and tuned circuit alignment all done by hand	Automated printed circuit board assembly, tuned circuits all pre-aligned
Power source	240V AC	2 x 9V PP9 batteries (18V DC supply)	240V AC or 4 x 1.5V LR14 batteries (6V DC supply)
Power consumption	30W	3W	2W
Frequency coverage	Medium wave and long wave (AM)	Medium wave and long wave (AM)	Medium wave and long wave (AM) and VHF FM

Digital audio broadcasting

It should be apparent from what you have just read that development in the design and manufacture of domestic radio receivers, like many engineered products, have been ongoing during the last 50 years. However, the most significant development, digital audio broadcasting (DAB), has only just appeared. DAB is arguably the biggest revolution in radio since the early days of broadcasting and much of the existing analogue broadcast radio service is likely to be phased out within 10 to 15 years. Not surprisingly, this added a degree of urgency to getting the new digital broadcasting services up and running. Furthermore, the rapidly expanding market for digital receivers has spurred existing radio manufacturers to develop new competitively priced digital models.

The first technology to be used for radio broadcasting was based on the amplitude modulation (AM) of radio signals (see page 338) in the long and medium wave bands. This technology was employed in the first 50, or so, years of radio broadcasting and it was sufficient (in terms of quality) to develop an expanding and enthusiastic radio audience in virtually every country of the world. National coverage can be easily provided by one or more high-power AM transmitters operating in the long and medium wave bands. Furthermore, provided that high-quality audio is not required, the spectrum requirements are relatively undemanding.

FM (frequency modulation) broadcasting first started in the 1950's in an effort to overcome the vagaries of propagation at the lower radio frequencies and also to very significantly reduce problems associated with amplitude noise and propagation disturbances. The FM broadcasting service was essentially a local service, requiring multiple transmitters operating on different frequencies in order to provide national coverage. It was (and still is) very inefficient in terms of its use of radio frequency spectrum.

Conventional AM and FM radio uses signals that are, between certain limits, continuously variable. These analogue signals are prone to various forms of noise and require that special measures (such as compression) are used in the broadcasting process in order to improve signal-to-noise ratios. Ignition and other electrical interference, fading and distortion caused by multi-path reception are further causes of distortion which, whilst much reduced using FM broadcasting techniques, can seriously degrade the quality of the received signal.

Digital radio offers many advantages, including the ability to remain reasonably impervious to the effects of amplitude noise and other signal disturbances as well as the ability to re-use spectrum space (using techniques that are not possible with conventional analogue broadcasting).

The technical specifications for digital audio broadcasting were originally developed by a project team known as 'Eureka 147' and, whereas this is originally a European group (with representatives from a number of interested bodies including broadcasting authorities), it now has members from all over the world).

The World DAB organisation is dedicated to encouraging international co-operation and co-ordination between sound and data broadcasters, network providers, manufacturers, governments and other official bodies in order to gain consensus for the smoother introduction of digital audio broadcasting worldwide.

Conventional terrestrial broadcasting (using transmitter sites all over the UK) is not the only way of receiving DAB. Digital radio can also be received in more than two million homes in the UK via digital satellite broadcasting in the L-band and a set-top box.

Digital radio receivers can also be fitted with pixel matrix displays that will permit the display of graphical symbols and simple pictures. The ability to display scrolling text messages, simple graphics or road maps could be invaluable in many situations.

In DAB, the audio (and data) signals broadcast in a digital radio system must be converted to digital format before they can be broadcast. At the receiving end, the digital information recovered from the received radio wave must be converted back into analogue form. Digital modulation techniques must be employed at the transmitter whilst digital demodulation must be employed at the receiver.

A number of digital radio programmes and services are conveyed in one contiguous block of radio frequencies. This band of frequencies is known as the *multiplex*. The multiplex allowing broadcasters to group together a number of programmes and additional data services that are all transmitted within a frequency channel. Each multiplex can carry a mixture of stereo and mono broadcasts and data services.

Since there is a trade-off between bandwidth and audio quality (in terms of the highest audio frequency that can be transmitted) the fewer the number of services carried, the higher the audio quality that can be allocated to each service. Multiplexing allows digital broadcasters to exploit the trade-off between bandwidth and audio quality and they thus have a high degree of flexibility in how they use their allocated bandwidth.

A DAB multiplex is made up of 2,300,000 bits (*bi*nary digi*ts*) that are used for carrying audio, data and error correction. Of these, approximately 1,200,000 bits are used for the audio and data services. In order to ensure efficiency, a different number of bits may be allocated to each service.

The seven multiplexes in the UK have been allocated as follows:

- one multiplex for BBC national radio
- one multiplex for national commercial radio (awarded to Digital One)
- five multiplexes for local radio in England and national stations in Scotland, Wales and Northern Ireland (BBC and commercial).

The UK Government has allocated the seven multiplexes to Band III (the range of frequencies formerly occupied by 405-line ITV signals) in the frequency range from 217.5 MHz to 230.0 MHz. The BBC national multiplex is located at 225.648 MHz. Each multiplex can carry a mixture of stereo and mono broadcasts and data services, the number of each is determined by the audio quality required but services can be varied throughout the day according to programme schedules.

Apart from the obvious improvement in sound quality resulting from an improvement in noise reduction, digital radio offers a number of other significant advantages over conventional analogue-based broadcasting technologies. These include:

- *Enhanced services* Digital radio makes far more efficient use of the available bandwidth than do conventional broadcasting technologies. This allows broadcasters to provide additional services as well as providing a wider range of programmes. Most UK listeners will, for example, be able to receive more than 12 national and 12 regional and local services broadcast both by the BBC and by commercial broadcasting companies. Additional data-related services will include the provision of news, traffic information, sports results as well as selective programme content which will offer listeners more choice and customizable options.
- *Reduced interference* Provided that received signal levels are above a certain minimum threshold, digital radio signals remain virtually impervious to noise and amplitude borne interference. Common problems associated with the propagation of radio waves (such as fading on HF and aircraft reflection on VHF) have negligible effect on digital signals.
- *Simplified tuning requirements* Digital radio receivers (particularly those designed for portable and in-car use) require minimal re-tuning when the receiver is moved from place to place. The user can simply select a particular programme without having to alter the tuning to locate the most favourable transmitter source.
- *Data displays* Digital radio receivers will be able to accept data as well as conventional audio signals. This will allow them to display scrolling text and graphical information using a built-in LCD display. It will, for example, be possible to display news headlines, sports result and half-time scores as well as programme schedules, details of artists, etc.
- *Improved quality and reduced distortion* With digital radio there is no need for the audio compression that has to be applied to signals broadcast using conventional analogue broadcasting technology. The removal of audio compression is instrumental in producing audio signals which very closely resemble the original source material. A comparison that is sometimes made when attempting to describe the improvement in signal quality is the difference that exists between vinyl records and compact disks. In reality, and because of the vagaries of the transmission medium employed, the improvement in quality between digital and analogue radio is *much* greater!

DAB technology

Digital audio broadcasting (DAB) is not just one technology but the combination of several technologies, notably:

- *MUSICAM* MUSICAM is a digital *compression* system that is used to reduce the vast amount of digital information required to be broadcast to a manageable amount. It does this by eliminating from the broadcast channel signals that will not be perceived by the listener (e.g. very quiet sounds that are masked by other louder sounds).

- *COFDM* COFDM (or *coded orthogonal frequency division multiplexing*) ensures that signals are received reliably, even in environments normally prone to interference. Using a precise mathematical transformation, the MUSICAM signal is split across 1,536 different carrier frequencies, and also divided in time. This process ensures that even if some of the carrier frequencies are affected by interference, or the signal disturbed for a short period of time, the receiver is still able to recover the original sound. The interference that disturbs conventional FM radio signals (e.g. multi-path reception and aircraft reflection) is significantly reduced by applying COFDM technology. COFDM also allows spectrum re-use so that the same broadcast frequency can be used across the entire country.
- *SFN* In digital radio broadcasting, the same frequency block of spectrum is effectively re-used throughout the service area in what is known as a *single frequency network* (SFN). Within the SFN, all the transmitters use the same frequency to broadcast the same digital radio signal. Hence there is no need to re-tune a receiver that is moved from place to place. Once selected, the required programme remains tuned.
- *TPEG* TPEG is a set of traffic information protocols. The Transport Protocol Experts Group (TPEG) was commissioned by the European Broadcasting Union's EBU Broadcast Management Committee to develop a new protocol for traffic and travel information for use in the multimedia broadcasting environment. TPEG technology is expected to revolutionise traffic information services by providing a personal travel service that allows a receiver to only use the traffic news that applies to the area in which it is currently located. This digital service will provide traffic information on demand and in much greater detail than ever previously possible. In addition to providing up to the minute information such as delays and accidents, the data can be translated into a wide range of formats including text or graphical display and voice synthesised. In future it is also possible that data could be available in several different languages. TPEG also offers the possibility of selecting information that is relevant to an individual motorist. Not only can it help in the selection of routes to avoid delays caused by traffic congestion or road works but it can also suggest alternative routes. Integration of digital radio receivers using TPEG with vehicle navigation systems (such as those based on RDS and Navstar satellites) is a further possibility that we are likely to see in the near future.

Each service component within a DAB signal carries an audio or data part of a service. Information that links the various components of a particular service is carried in the *fast information channel* (FIC).

Broadcasting organisations can link a number of different *service components* together to produce a complete service. One component must be defined as the *main component*, the others are called *secondary components*. For example, the BBC's principal radio services (Radios 1, 2, 3, 4 and 5 Live) will be transmitted as main service components. Each of the principal services may have secondary components. For example, Radio 5 Live Plus could be introduced into the multiplex to provide additional coverage of a major sports event. This provides broadcasting organisations with considerable flexibility in determining how their output should be made available to their listening audience.

Another view

The development of broadcast radio receivers over the last 50 years makes an interesting case study of how successive advances in different technologies have had an impact on the design and manufacture of an engineered product with which we are all familiar. Can you think of other products that have developed along similar lines? It shouldn't be too difficult!

Activity 5.10

Use the Internet or other information sources to obtain information and specifications on at least three currently available DAB receivers (these are manufactured and/or distributed in the UK by companies like Arcam, Clarion, Cymbol, Grundig, Kenwood, Pioneer, Sony, Tag McLaren, and Technics). Present your finding in the form of a series of word processed data.

Test your knowledge 5.7

What does each of the following abbreviations stand for?

(a) DAB
(b) AM
(c) FM
(d) SFN
(e) COFDM
(f) TPEG
(g) RDS
(h) FIC.

Activity 5.11

Write a brief word-processed article for your local paper suitable for non-technical readers explaining the advantages of DAB.

Test your knowledge 5.8

Explain why VHF FM broadcasting provides better quality reception than conventional AM broadcasting. Give at least *two* reasons.

Activity 5.12

Use the Internet to obtain information on the DAB services that are currently available from the BBC. Explain how these services improve on services based on conventional analogue broadcasting technology. Present your findings in the form of a brief word-processed fact sheet.

Test your knowledge 5.9

Some DAB receivers provide coverage of signals in L-band (with a typical tuning range of 1,452 MHz to 1,492 MHz) as well as Band III (with a typical tuning range of 174 MHz to 240 MHz). Explain why this is.

Activity 5.13

Future digital radio receivers are likely to be very different from conventional radio receivers. Explain why this is and suggest ways in which this new technology will provide additional services of specific benefit to the motorist. Present your findings in the form of a brief class presentation with appropriate visual aids.

Figure 5.16 *A modern GPS receiver designed for use in a vehicle*

Global Positioning System

Another example of technological development is that of the Global Positioning System (or GPS). GPS is a collection of satellites owned by the U.S. Government that provides highly accurate, worldwide positioning and navigation information, 24 hours a day. It is made up of 24 NAVSTAR GPS satellites that orbit 12,000 miles above the Earth, constantly transmitting the precise time and their position in space. GPS receivers on (or near) the surface of the earth, listen in on the information received from three to twelve satellites and, from that, determine the precise location of the receiver, as well as how fast and in what direction it is moving.

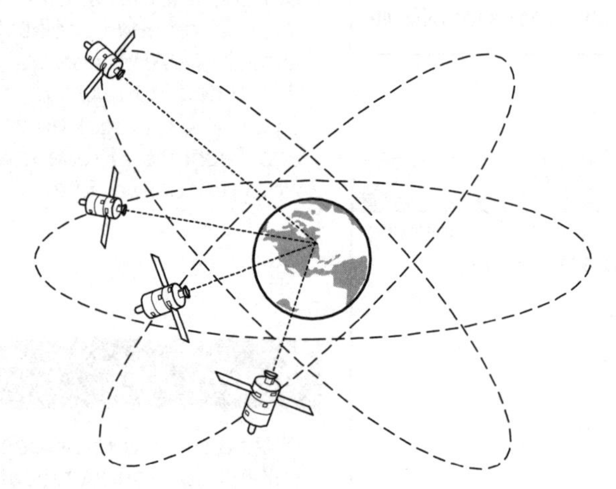

Figure 5.17 *GPS system using triangulation—a minimum of three satellites is required to determine position*

GPS uses the triangulation of signals from the satellites to determine locations on Earth (see Figure 5.17). GPS satellites know their location in space and receivers can determine their distance from a satellite by using the travel time of a radio message from the satellite to the receiver. After calculating its relative position to at least three or four satellites, a GPS receiver can determine its position using triangulation. GPS satellites have four highly accurate atomic clocks on board. They also have a database (sometimes referred to as an 'almanac') of the current and expected positions for all of the satellites that is frequently updated from Earth. That way when a GPS receiver locates one satellite, it can download all satellite location information, and find the remaining needed satellites much more quickly.

Over the last several years, an increasing variety of affordable GPS receivers have been released for the average consumer. As the technology has improved, many additional features are added to these units, while the price and size continue to decrease.

Some of the more specialized GPS receivers currently available include:

- hand-held GPS receivers that have background maps
- GPS receivers fitted to cars and lorries (with integral databases of maps and road information)
- GPS receivers for large and small boats (including those integrated with other navigational equipment)
- aircraft GPS receivers with built-in airport information
- GPS receivers that combine with Internet access and/or e-mail into one unit.

GPS products have been developed for use for many commercial applications including surveying, map making, tracking systems, navigation, construction and mapping natural resources. The recreational use of GPS includes sea fishing, hiking, skiing and mountain walking.

Activity 5.14

Obtain the data sheets and/or technical specification for at least two different types of low-cost GPS receiver. Identify the market for these receivers and write a brief article for your local newspaper explaining what GPS can do for ordinary readers.

Activity 5.15

Visit a local car showroom and investigate the GPS systems that may be supplied with the latest models. Write a brief article for a car enthusiasts' magazine explaining, in simple terms, how GPS works and how it can benefit the car driver.

Sample product investigation

This sample product investigation is designed to help you to carry out your own product investigation. The sample product investigation was carried out by James Taylor and Debbie Morris who are two students at North Downs College.

James and Debbie have decided to base their product investigation on a local company, Howard Associates. This company manufactures a range of filters and oscillators based on quartz crystals, piezoelectric resonators, surface acoustic wave (SAW) devices, and conventional tuned circuit components. The particular products that James and Debbie have chosen to investigate are those that are based on quartz crystals. The two students have arranged to visit the company in order to interview Nick Jones, the Senior Design Engineer.

James and Debbie are aware that they must take into account each of the following during their investigation (see Page 366):

- *Standards and regulations*
- *Types of documentation*
- *Energy efficiency*
- *Environmental impact*
- *Technology and techniques*
- *Evaluation of the engineered product.*

Prior to their visit to the company, James and Debbie decide to carry out some research into quartz crystal-based products, including filters and oscillators. A visit to North Downs College library provides some basic information on the manufacture and use of quartz crystals including the following:

"Quartz is a crystalline material that is based on both silicon and oxygen (silicon dioxide). The quartz crystals used in electronic circuits usually consist of a thin slice of quartz onto the opposite faces of which film electrodes of gold or silver are deposited. A fine supporting and connecting wire is soldered at a nodal point on each electrode and the complete assembly is enclosed in an evacuated glass or metal envelope. Lead-out pins or wires facilitate connection with external circuitry.

Whilst quartz crystals occur quite naturally, they can also be manufactured to ensure consistency both in terms of physical properties and supply. The growing of quartz crystals simply involves dissolving quartz from small chips and allowing the quartz to grow on prepared seeds. These encompass a batch process that requires about 21 days to result in the desired size crystals. Approximately one half day is required to bring the charge and the equipment to operating temperature.

The quartz chips are dissolved in sodium hydroxide solution during which temperatures are maintained above the critical temperature of the solution. The growth process of the quartz is controlled by a two-zone temperature system such that the higher temperature exists in the dissolving zone and the lower in the growth zone.

In the actual manufacturing process, the quartz chips (or nutrients) are placed in the bottom of a long vertical steel autoclave that is specifically designed to withstand very high temperatures and pressures (much like the barrel of a large gun).

The quartz wafers, sliced along the basal plane, are suspended in the top zone of the vessel of seeds. Carefully controlled amounts of de-ionized water and sodium hydroxide pellets are added and the vessel is sealed. External heat is then applied to achieve the two isothermal zones.

Special insulation and a careful controlled pattern of heat application are important to obtain the proper results; the small quartz chips, the raw material, are dissolved by the caustic solution. This solution is then carried by convection currents to the cooler growing zone where it becomes supersaturated and growth begins on the seed plates.

Of paramount importance is temperature control, because the temperature affects the dissolution rate of the nutrient, the deposition of quartz on the seeds and the convective transfer of nutrients

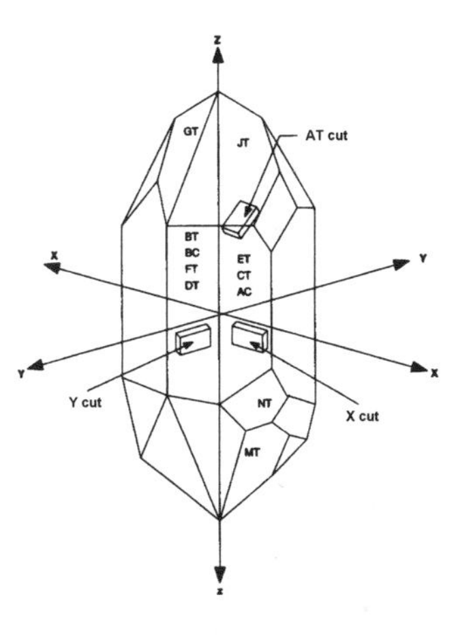

Figure 5.18 *Quartz crystal showing cutting planes*

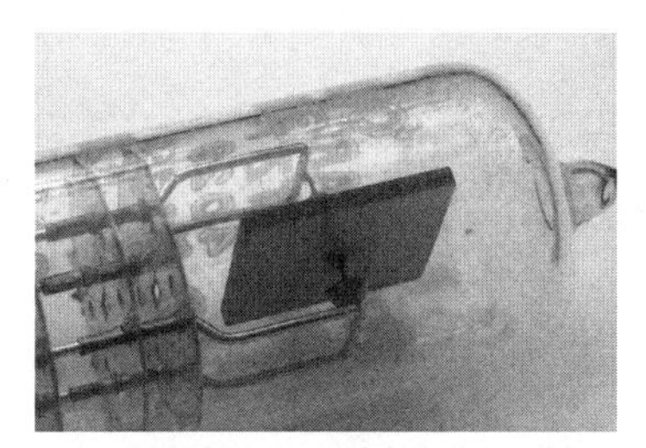

Figure 5.19 *A low-frequency quartz crystal mounted in an evacuated glass envelope*

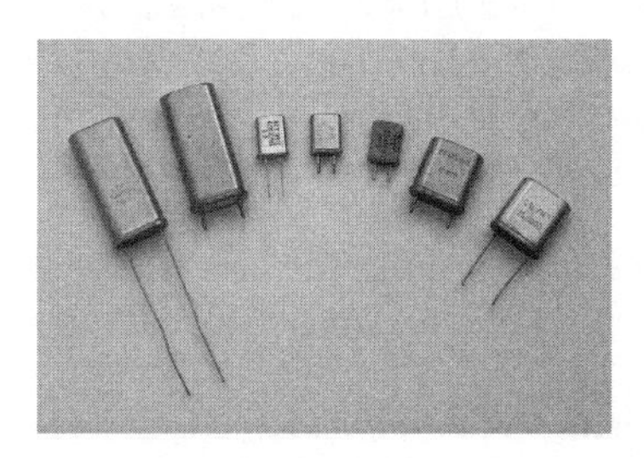

Figure 5.20 *A selection of typical quartz crystal units*

between zones. For example, a high temperature differential between the zones leads to rapid growth, but faults tend to occur if the temperature difference is too high."

James and Debbie learn that crystals can be cut in different ways and the way that they are cut affects the way that the crystal element vibrates when a small electric charge is applied across its opposite faces (the *piezoelectric effect*):

"The basic element of an electronic quartz crystal is a small slice of quartz with parallel sides. This can be cut from a complete crystal in a variety of different ways, as shown in Figure 5.18.

The angle at which the quartz slice is cut from the quartz blank is actually very important as it is instrumental in determining the modes of vibration that can be induced in it. The angle of cut also affects the temperature stability of the crystal. For many applications, the AT cut is used."

From another reference source they find that:

"It is possible to excite vibrations in the quartz element that are mechanical overtones of a component's fundamental frequency. Overtone modes may be excited on the 3rd, 5th, 7th, 9th harmonic of AT cut ('thickness shear' plates). The properties for a given crystal unit and given order of overtones are quite different from those of its fundamental frequency or other orders of overtones. Consequently, no reliance should be placed on the behaviour of a crystal unit at any frequency other than that for which it was designed.

Crystals, and their associated oscillator circuits tend to fall into one of two main classes, fundamental and overtone. Crystals manufactured for fundamental operation are designed to oscillate at their basic resonant frequency whereas those intended for overtone operation oscillate at, or very near, an integral multiple of their fundamental resonant frequency.

Generally, the third overtone is preferred although fifth, seventh, and even ninth overtone devices are available. At high frequencies, crystals become extremely thin and are consequently more difficult and more expensive to manufacture. Thus fundamental crystals are normally used at frequencies up to about 20 MHz, beyond this, overtone units are usually employed. A typical expanded impedance/frequency characteristic for a quartz crystal is shown in Figure 5.21. This shows overtone resonances as well as some (unwanted) spurious resonance.

The frequency of a crystal is usually specified in MHz for units of 1.0 MHz and above and kHz below 1.0 MHz. The fundamental or overtone mode required should be stated and the following table indicates the normal frequency limits for each mode:

- *Fundamental mode (up to 155MHz)*
- *3rd overtone (20 to 75MHz)*
- *5th overtone (50 to 150MHz)*
- *7th overtone (125 to 175MHz)*
- *9th overtone (150 to 225MHz)."*

A further reference provides the following information on quartz crystal manufacturing and encapsulation processes:

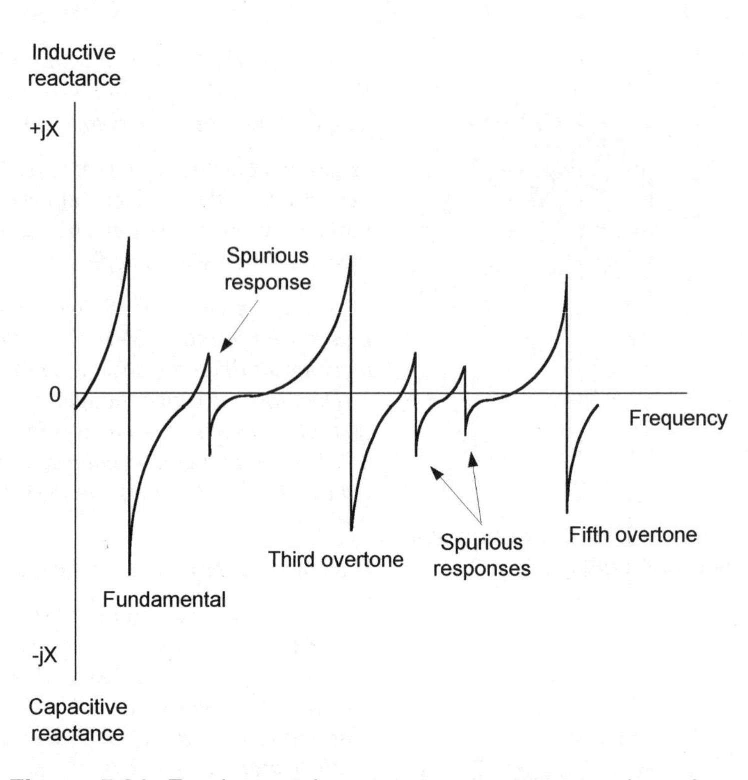

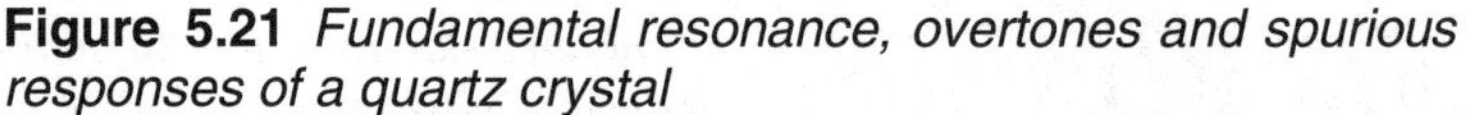
Figure 5.21 *Fundamental resonance, overtones and spurious responses of a quartz crystal*

"The size of the package in which a quartz crystal is encapsulated is often important in circuit board design. If a quartz resonator is to be supplied in the smallest possible enclosure there is usually a trade off against many other parameters in particular ESR, temperature coefficient, drive level and spurious response.

Modern resistance weld enclosures provide a good compromise between cost and quality and the MIL spec HC49 enclosure is the most widely used. There are enclosures of glass, metal and plastic with glass used for either low frequency units or where minimum aging rates are required and plastic for very low cost including the enclosing of ceramic resonators.

Ceramic surface mount crystal packages are increasingly used in volume applications and provide low profile units and specifications to replace many metal case types.

There are three methods used to seal crystal units in metal cases; solder seal, resistance weld and cold weld, each method of sealing the encapsulation has slightly different characteristics, as follows:

- *Cold weld: Case sealed using pressure to weld the metal faces, almost no internal pollution, excellent aging, high quality, very good seal, expensive.*
- *Resistance weld: Case sealed using heating effect of electric current through steel case, low internal pollution, very good aging rate, good quality, good seal, reasonable cost. This is the most widely used method of sealing for general purpose crystals.*

Figure 5.22 *A selection of quartz crystal filters*

- *Solder seal: Case sealed using pre-tinned faces and flow soldering or hand soldering to achieve a seal, the case is then further filled with an inert gas and totally sealed using solder. Very poor aging, lowest cost, low quality, high internal pollution and initial thermal shock."*

Having gathered information on the technology used in the manufacture and packaging of quartz crystals, James and Debbie continue their library research by looking for information on crystal filters and crystal oscillators. Their search for information on crystal filters reveals the following:

"Whilst single crystals are used as the frequency determining element in crystal controlled oscillators, multiple arrangements of crystals can be used to produce frequency selective filters. Indeed, by careful selection of resonant frequencies it is possible to produce a network of crystals that has near-perfect characteristics for use as a filter (i.e. very high attenuation in the stop-band coupled with very low loss in the pass-band).

A typical crystal filter frequency response (showing the pass-band and stop-band regions) is shown in Figure 5.23. The stop-band attenuation for this filter is approximately 60 dB whilst the pass-band ripple is less than 3 dB. The shape factor for the filter is usually specified in terms of the pass-band bandwidth (A) and the bandwidth (B) corresponding to a particular value of attenuation. Hence:

Shape factor = (Bandwidth A/Bandwidth B)

It is worth noting that the spurious responses associated with quartz crystal filters can be made less significant by using conventional tuned circuit matching transformers.

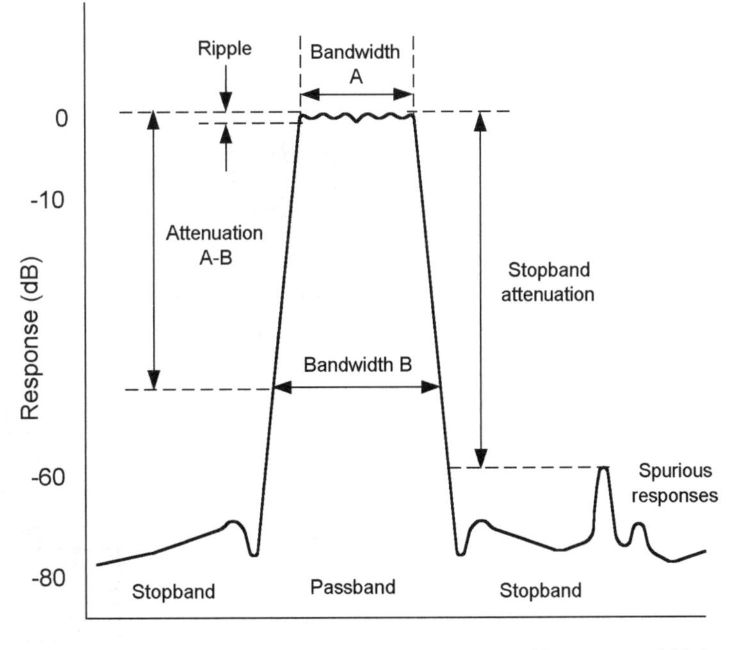

Figure 5.23 *Crystal filter response curve*

A typical 10.7MHz crystal filter for use in an HF communications receiver would have the following specification:

- *Centre frequency:* *10.7 MHz*
- *Passband bandwidth:* *3.75 kHz at –6 dB*
- *Stopband bandwidth:* *12.5 kHz at –60 dB*
- *Ripple:* *2 dB (max)*
- *Insertion loss:* *3 dB (max)*
- *Input/output impedance:* *1.5 kΩ."*

Their search for information on crystal oscillators produces the following:

Figure 5.24 *A crystal oscillator module on a surface mounted PCB*

"Quartz crystals are widely used in both RF and digital systems in order to provide an accurate and stable frequency source (see Figure 5.24). Three common crystal oscillator circuit configurations are shown in Figure 5.25. For simplicity, a junction-gate FET has been used as the active device in each of these arrangements and, where a bipolar transistor is employed, additional bias components are required.

The Pierce oscillator of Figure 5.25(a) is a fundamental oscillator that does not require a tuned circuit connected in the drain (or collector) load. Feedback is via the internal capacitance of the FET and an inductive load is normally employed.

The Colpitts oscillator of Figure 5.25(b) is another fundamental mode arrangement and the inductor is included to provide a low-resistance DC path for the source current whilst exhibiting a relatively high impedance at the operating frequency.

The Miller oscillator shown in Figure 5.25(c) is suitable for either fundamental or overtone operation depending upon the frequency at which the tuned circuit drain load is resonant.

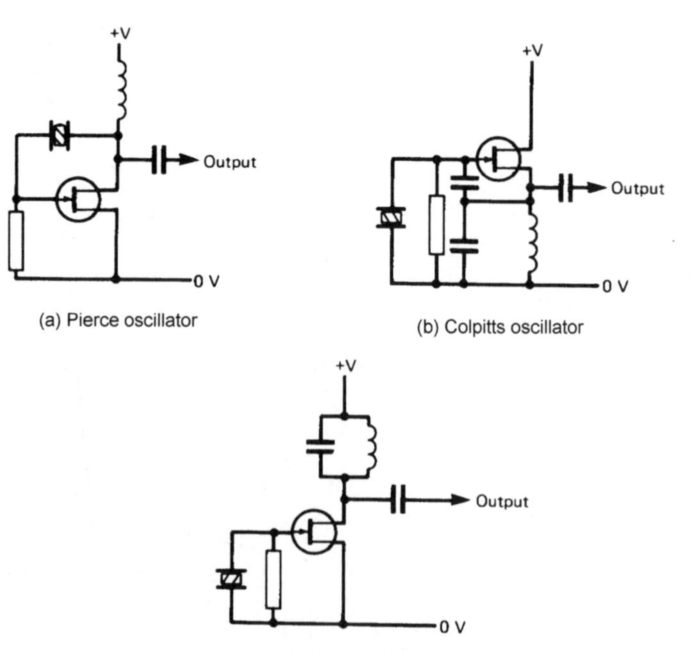

Figure 5.25 *Crystal oscillator circuits*

A typical 10 MHz crystal frequency standard (comprising an oscillator stage and buffer amplifier) would have the following specification:

- *Output frequency:* *10 MHz*
- *Frequency stability:* *± 20 parts per million*
- *Output purity:* *–60 dB total harmonic content*
- *Output voltage:* *1V r.m.s. into 50 Ω*
- *Temperature stability:* *± 50 parts per million"*

Having obtained information from their college library, James and Debbie next carry out an Internet search to locate information from manufacturers' websites. The results of this search produces the following (all of which will be listed in their written report):

C-MAC MicroTechnology
Quartz crystals, filters and other frequency products
http://www.cmac.com

Precision Devices UK (Hy-Q)
Quartz crystals, crystal filters, standard oscillators, etc
http://www.hy-q.co.uk

Euroquartz Group
Quartz crystals, oscillators, filters, etc
http://www.euroquartz.co.uk

ECM Electronics Group
Quartz crystals, oscillators and filters
http://www.ecmelectronics.co.uk

Total Frequency Control Ltd
Quartz crystals and accessories
http://www.tfc.co.uk

Golledge Electronics Ltd
Crystals and components for frequency control
http://www.golledge.com

AEL Crystals
Crystals, clock oscillators and modules
http://www.aelcrystals.co.uk

QuartsLab Marketing Ltd
Quartz crystals for amateur radio, PMR and general applications
http://www.quartslab.demon.co.uk

ESKA Crystals
Quartz crystals and clock oscillators
http://www.eska.dk

Corning Frequency Control
Technical paper on the history and physics of quartz crystals:
http://www.corningfrequency.com/piezo/papers/qcao.html

Armed with this information, James and Debbie were ready to visit their chosen company. In order to prepare for their meeting with Nick Jones, the Senior Design Engineer at Howard Associates, Debbie produced the checklist of questions shown in Figure 5.26.

VISIT RECORD FOR HOWARD ASSOCIATES

Interviewers: James Taylor and Debbie Morris, North Downs College

Debs, can I have a copy of this?

Person interviewed: Nick Jones, Senior Design Engineer at Howard Associates

Date of visit: 17th March 2006

Please explain your job role? Responsible for leading the design team who design new products and modify existing products. Also works with clients who need special products developed.

Who do you report to? Martin Richards - Production Manager

Who reports to you? Steve Wilson, Dave Hart, Anna Parker - Design Engineers

How long have you had this job? 4 years

Worked for British Aeropsace before joining the company.

What qualifications do you have? Degree in Electronic Engineering, IEng.

What does this mean?

What products are you responsible for? Design and prototype/testing of filters, oscillators, and tuned(?) components.

What technology/techniques are used in the products? Quartz crystals, surface acoutsic waves(?), hemertic seals, surface mounting, high-Q ceramics

What standards do you work to? BS9000, EN45001, EN54002 (Test labs), ISO 9001 quality system, CENELEC(?)

What documents are used to support the product? Data sheets and drawings; Quality Procedures (QPs), Work Instructions (WIs), Test Specifications (TSs)

Didn't we get a copy of one of these?

What software or CAD tools are used? CAD for filter design, PCB design, and 2D CAD.

Why no 3D CAD??

How is energy efficiency taken into account? Temperature controlled soldering stations and energy controlled flow soldering plante. Filters use only passive components and don't use any power. Oscillators use low-power devices (milliwatts only).

What is the environmental impact? Soldering stations have fume extraction and use solders with low lead content. Oscillators and filters are fully screened to prevent EMC problems.

Where does waste go?

What is the most difficult part of your job? Contract review with clients.

Wasn't this because some contracts didn't have enough detail?

What is the most interesting part of your job? Meeting new clients. Solving problems.

Debs, Nick said we could speak to Martin Richards about energy aspects of the production plant - when shall we follow this up?

Figure 5.26 *James and Debbie's Visit Record*

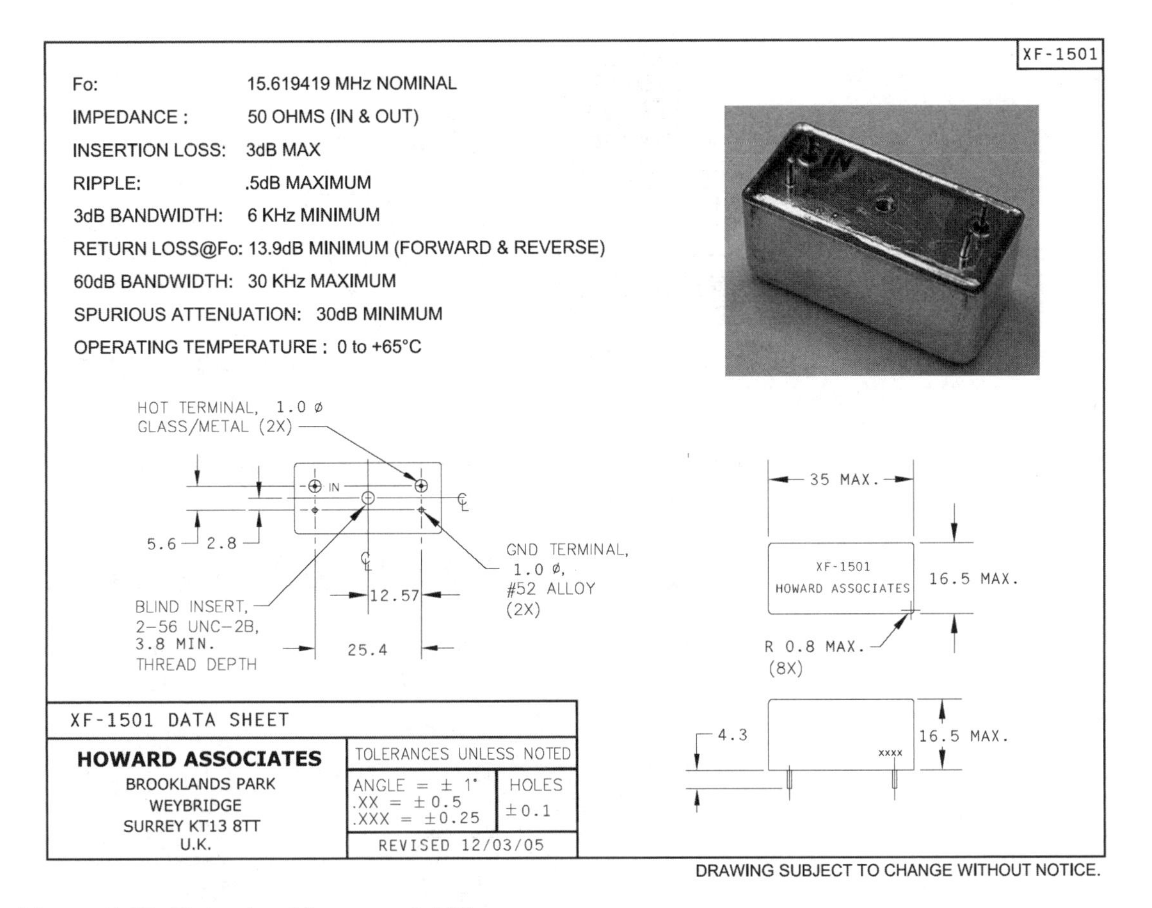

Figure 5.27 *Data sheet for a crystal filter*

Figure 5.28 *The crystal oscillator module used in James's investigation*

Figure 5.29 *The crystal filter used in Debbie's investigation*

The two students' visit to Howard Associates (and their meeting with Nick Jones) enabled them to find out how the company operates as well as providing plenty of information that they were able to use in their product investigations.

During the interview, it was agreed that Debbie would ask the questions and James would make notes (see Figure 5.26). In addition to these notes, Nick Jones provided the two students with copies of a number of documents, including some sample work instructions, a product catalogue, data sheets, and specifications. Nick also demonstrated some of the CAD software used within the company and showed James and Debbie some of the processes used in design, manufacturing and testing the company's products.

Following their visit to Howard Associates, James and Debbie decided to concentrate on different products for their own personal investigations. James chose to base his investigation on crystal oscillators (Figure 5.28) whilst Debbie chose to concentrate on crystal filters (Figure 5.29). This meant that the two students could share in the initial research and visit preparation but each could prepare an individual product investigation.

Figures 5.30 to 5.32 show extracts from some of the documents that Nick Jones provided the students with whilst Figure 5.33 shows the results of a typical CAD analysis of a prototype filter. James and Debbie incorporated extracts from several of these documents into their written reports.

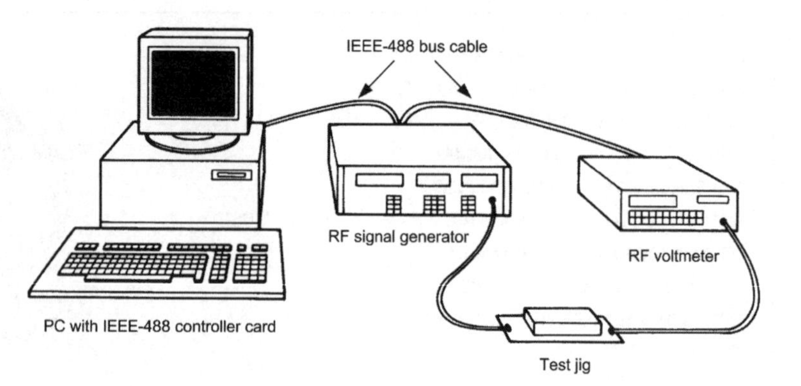

Figure 5.30 *Test jig for measuring filter specifications used in Debbie's investigation report*

Product Search

Type: Select 1
- Bandpass
- Bandstop
- Highpass
- Lowpass
- Diplexer

Frequency: Select at least 1
- 0 - 1 MHz
- 1 - 100 MHz
- 101 - 500 MHz
- 501 - 999 MHz
- 1GHz -

Technology: Select at least 1
- Ceramic
- Crystal
- LC/Microwave
- Monolithic
- Saw
- Tunable

GO!

Figure 5.31 *On-line product search incorporated into Debbie's investigation (this screen shows the initial filter selection dialogue)*

C:\CBA\QC.EXE

```
Performance data for oscillator ref: HXO

Performance measured at: 22 deg.C
Maximum frequency:   5.016176E+07  Hz
Minimum frequency:   5.015923E+07  Hz
Mean frequency:      5.016055E+07  Hz
Frequency drift:     2532  Hz
```

Figure 5.32 *Automatic test equipment report for an L-C oscillator undergoing a long-term frequency drift test (used in James's report)*

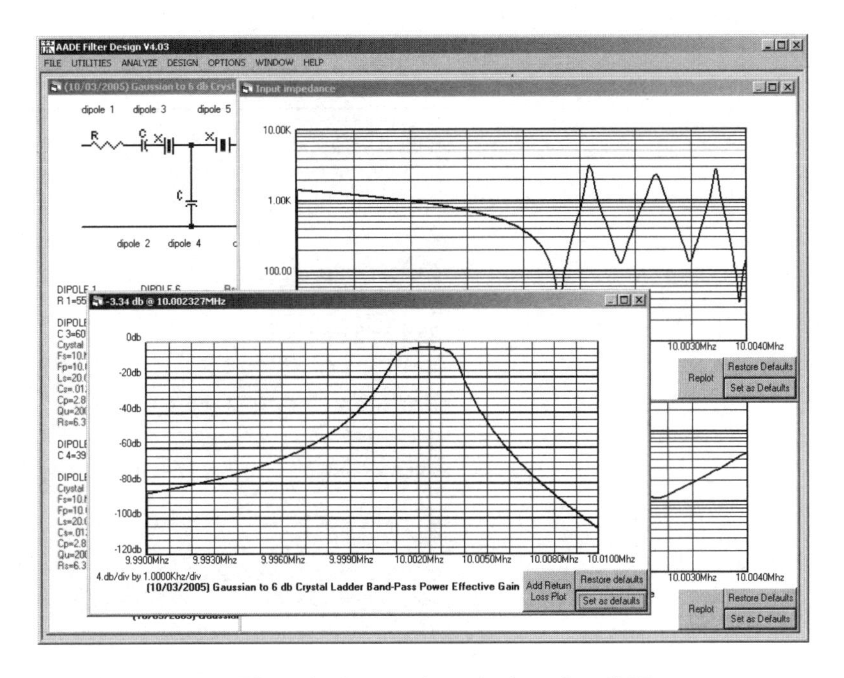

Figure 5.33 *Filter design and analysis using CAD*

5.4 Evaluation and modification

When you have completed your product investigation you will be expected to carry out a detailed evaluation of your product or service in order to determine whether it is fit for purpose. To do this you will need to make an assessment of the functions and specifications for the product or service and decide whether these fully meet the needs of the client or customer. You will also be expected to make suggestions for modifications, where appropriate, that will improve the performance of the engineered product or service. Whilst your suggested modifications can be presented in the form of a written report, diagrams, sketches or drawings, you will probably wish to make use of a combination of these methods. It is also important that your modifications are correctly reasoned and you will need to show evidence of this in your answer. In order to provide you with an idea of what is required, we will next show you how James Taylor (one of the students that we met in the last section) tackled this task in relation to the product that he had chosen to investigate.

Sample product modification

James Taylor is a student at North Down College and his chosen product investigation was a crystal oscillator module produced by a local engineering company, Howard Associates. The oscillator uses a quartz crystal frequency-determining element and a buffer amplifier to increase the output level so that the output signal can deliver to a 50 Ω load. The quoted specification for the crystal oscillator is shown in Table 5.2.

When James visited Howard Associates for a second time, he met Martin Richards the Production Manager. Martin had mentioned a particular problem with the 10 MHz crystal oscillator module that James was basing his investigation on. Martin said that a number of

Table 5.2 *Crystal oscillator specification*

Frequency:	10 MHz (sine wave)
Frequency stability:	+/– 20 parts per million
Temperature stability:	+/– 50 parts per million
Second harmonic:	–28 dB
Third and higher harmonics:	At least –40 dB
Output voltage:	1 V r.m.s. into 50 Ω
Supply voltage:	4.75 V to 5.25 V
Supply current:	40 mA

customers had reported that the output of the oscillator was insufficiently pure and that the levels of the harmonic components present were unacceptably high. Martin suggested that this was something that James might like to follow up and report back with his findings. When James questioned Martin about what level of harmonic would be acceptable, Martin said that he felt that the harmonic content should be not less than –60 dB relative to the desired output (see Figure 5.34).

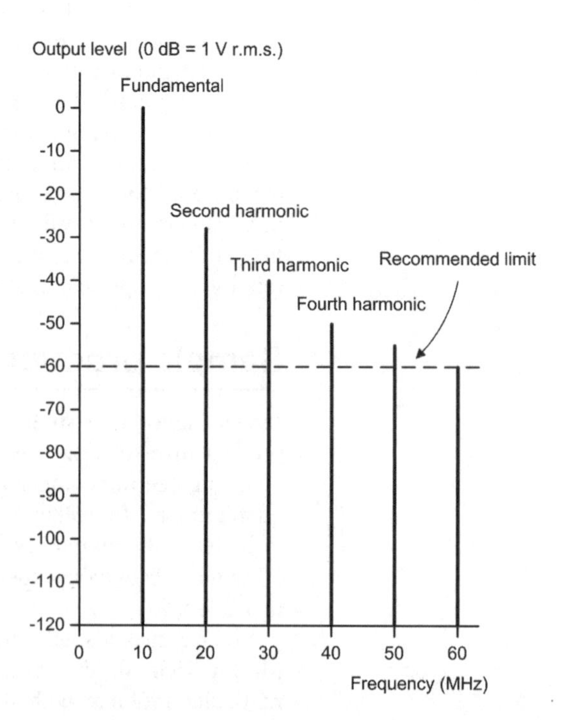

Figure 5.34 *Frequency spectrum of the oscillator's output*

James decided to investigate the problem that Martin had identified and attempt to find a solution. His tutor at North Downs College suggested that he could design a filter that would remove some of the harmonic components. At North Downs College students have access to several electronic CAD packages including the student version of Tina Pro so, after carrying out some further research in the North Downs College library, James decided to investigate the performance of a modified *m*-derived filter. After some initial experimentation using Tina Pro to simulate the filter, he arrived at the design shown in Figure 5.35. His filter was designed to that it provides a sharp notch at the second harmonic frequency (20 MHz) and progressively increasing attenuation at the frequencies of the higher order harmonics. This filter was then constructed in the laboratory and tested in conjunction with one of the prototype crystal oscillators (note that it was necessary to increase the gain of the internal buffer amplifier to compensate for the slight loss of the filter at 10 MHz). In order to make the filter adjustable (and thus compensate for any minor variations in 'real' components), a miniature 10 pF pre-set trimmer was used for the capacitor, C1, instead of a 5 pF fixed value component. The three inductors were wound on miniature formers with adjustable ferrite cores and then trimmed to the required values by means of an LCR component bridge.

The response of the modified oscillator was then measured and later confirmed by Nick Jones at Howard Associates. The modified frequency spectrum of the oscillator output is shown in Figure 5.36 and its specification shown in Table 5.3. James included full details of these modifications (as well as the rationale for making them) in his product evaluation.

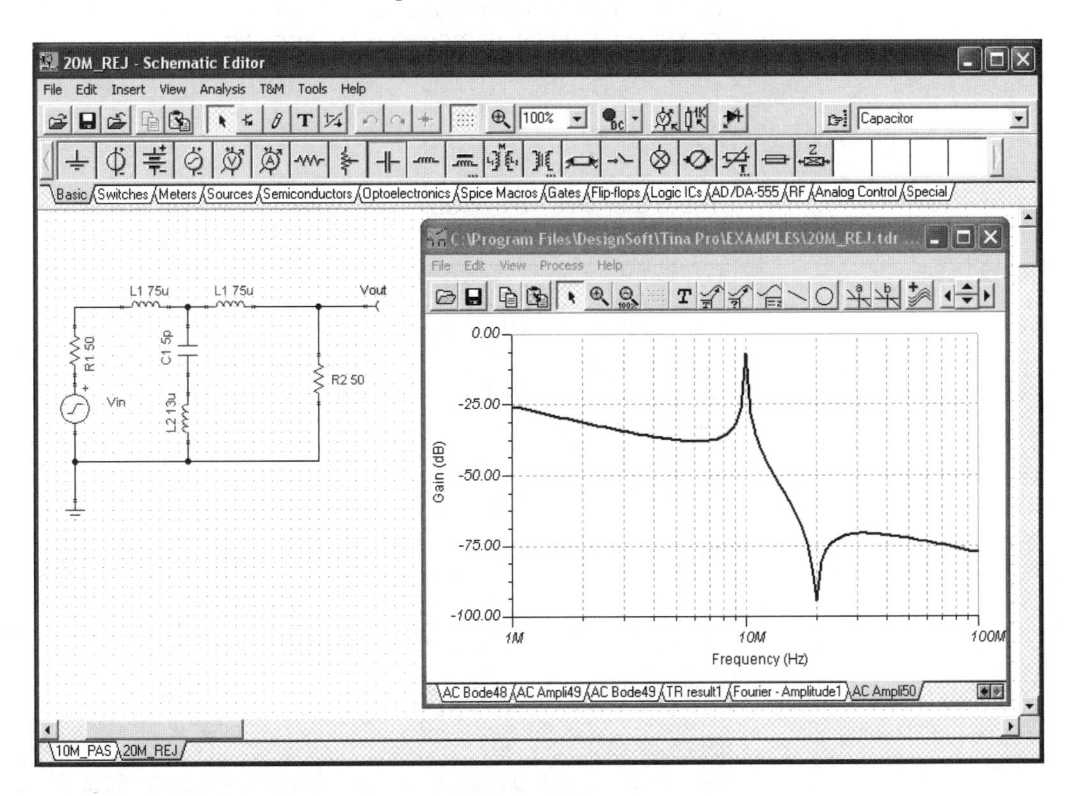

Figure 5.35 *Using Tina Pro to design the output filter*

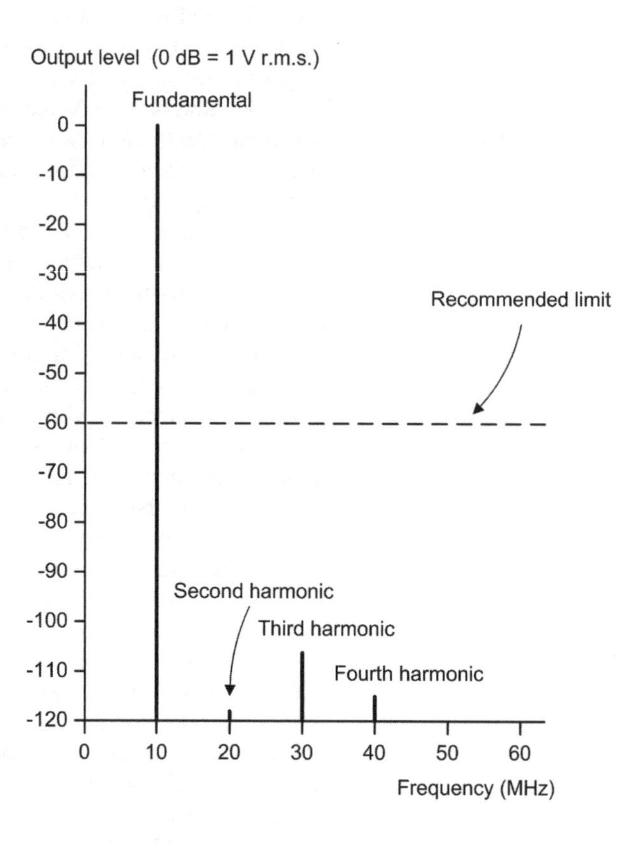

Figure 5.36 *Frequency spectrum of the modified oscillator showing a significant reduction in harmonic content*

Table 5.3 *Modified crystal oscillator specification*

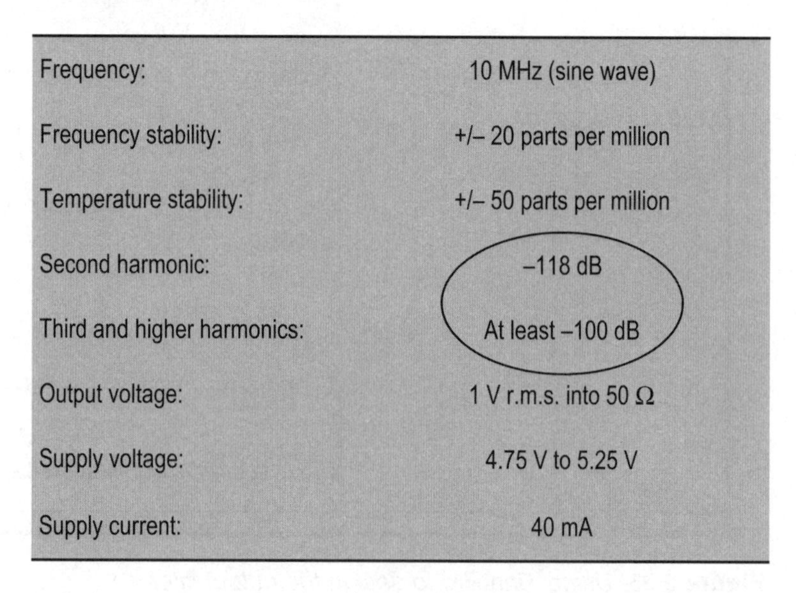

Frequency:	10 MHz (sine wave)
Frequency stability:	+/– 20 parts per million
Temperature stability:	+/– 50 parts per million
Second harmonic:	–118 dB
Third and higher harmonics:	At least –100 dB
Output voltage:	1 V r.m.s. into 50 Ω
Supply voltage:	4.75 V to 5.25 V
Supply current:	40 mA

Review questions

1 List THREE tests that might be required to ensure that an engineered product complies with the EMC Directive.

2 A MODEM is to be connected between a computer and a telephone line. The MODEM is to be powered from the 240V AC mains supply. State TWO EC directive that apply to the MODEM.

3 Explain the purpose of:
(a) An application note
(b) A short-form catalogue
(c) A technical report.

4 Give ONE example of where each of the documents in Question 3 might be used.

5 Sketch the CE Mark and explain what it signifies.

6 Describe TWO documents that might form part of a Quality System used by a typical engineering company.

Figure 5.37 *Temperature controlled soldering iron with extraction facilities*

7 Explain why *traceability* is important in the case of components and spare parts used in a civil aircraft.

8 A data sheet is to be produced for a low-voltage halogen lamp. List FOUR essential items that should appear in the data sheet.

9 Explain why *energy efficiency* is important in the design, manufacture and utilisation of an engineered product.

10 Give TWO examples of where energy efficiency can be improved in the case of a domestic central heating system.

11 Explain the difference between degradable and non-degradable pollutants. Give TWO examples of each type of pollutant.

12 Describe TWO methods of disposing of nuclear waste.

Figure 5.38 *Satellite MODEM for global digital communications*

13 Figure 5.37 shows a temperature controlled soldering iron being used to replace surface mounted components on a printed circuit board. Explain why the soldering iron is fitted with extraction facilities.

14 Figure 5.38 shows a satellite MODEM which permits connection to the Internet anywhere in the world. List THREE technologies that are used in this product and explain how they are used.

15 Explain why nickel cadmium batteries must not be disposed of with normal domestic waste.

16 Give TWO examples of how engineering activities have affected the landscape in parts of the UK.

Some possibilities for case studies and/or product investigations

Here are a few ideas for products that are suitable for use as case studies (or for your own product investigation). In all cases you are strongly advised to check the suitability of your chosen engineered product or service with your tutor. He or she will be able to advise on suitability and may also be able to suggest alternatives. Before you make a final choice of your engineered product or service it is essential to ensure that you have adequate material available to address each of the factors listed on page 366. You should also ensure that you have sufficient resources available to carry out a thorough evaluation of the engineered product or service and that you are in a position to suggest modifications that can be made to it. Access to a manufacturer or supplier could also be important.

A flat screen LCD display

Flat screen LCD displays that are rapidly replacing computer and television displays that have conventionally been based on cathode ray tubes (CRT). Investigate the design, manufacture, and utilisation of an LCD display, taking into account each of the factors listed on page 366. Also consider ways in which an LCD display could be improved (for example in respect of pan and tilt mounting systems).

A dual-fuel vehicle

Several car manufacturers have produced designs for dual-fuel vehicles and at least one manufacturer is currently selling these vehicles in the UK market. Investigate the design, manufacture, and utilisation of a dual-fuel vehicle, taking into account each of the factors listed on page 366. Also consider ways in which a dual-fuel vehicle could be improved (for example in respect of instrument panels and driver displays).

A charger for Ni-MH and/or Ni-Cd batteries

Nickel Metal Hydride and Nickel Cadmium batteries are widely used in portable equipment such as digital cameras and other equipment that requires a high-capacity DC supply. Various types of charger are available for use with these batteries including those that cope with different types of cell and those that provide fast charging facilities. Investigate the design, manufacture, and utilisation of an Ni-MH and Ni-Cd battery charger, taking into account each of the factors listed on page 366. Also consider ways in which the charger could be improved (for example with a built-in discharge facility or a battery test function).

A compact loudspeaker system

Compact loudspeaker systems (or ‘bookshelf speakers’) are frequently used with smaller domestic hi-fi systems. Various techniques are used to improve the sound quality and power handling capacity of small loudspeakers including the use of ‘long-throw’ loudspeaker units. Investigate the design, manufacture, and utilisation of a compact loudspeaker system, taking into account each of the factors listed on page 366. Also consider ways in which a compact loudspeaker could be improved (for example in the design of the enclosure or the use of separate high and low-frequency drivers).

Unit 6 Applied design, planning and prototyping

Summary

This unit is the culmination of your GCE Engineering studies and it builds on the skills that you gained in Unit 3: *Principles of Design, Planning and Prototyping*. Based on a detailed client brief, you will be required to undertake a project leading to the manufacture of an engineered product or the design of an engineering service. You will produce a working prototype that proves the design principles and demonstrates how your product will look and function. You will need to demonstrate a clear understanding of why you have selected particular materials, components and processes, making appropriate references to the underpinning scientific and/or mathematical principles. You must also carry out a detailed evaluation of your product or engineering service. In order to help you carry out your product design, development, manufacture and evaluation we have, as with the previous unit, included a typical example showing how a student at our mythical North Downs College tackled the assessment of this unit.

6.1 Research

Before you make a start on the design and development of your engineered product or service it is vitally important to carry out some initial research and also to ensure that you have a thorough understanding of your client's needs. You need to resist the temptation to rush into the design and manufacture of your product. Instead, you should ensure that you devote sufficient time to clarifying the problem and collecting together information from a variety of different sources.

Client brief

The starting point for this unit is the client brief. This must be a clear description of the problem that you will solve. In particular, the brief will include the identification of a problem, a statement of needs and sufficient information to place the problem into context. In discussion with your client (or your tutor), you will need to ensure that the information included in the client brief is simple, focused and concise, and that you *fully* understand it!

In addition, you will need to identify and confirm each of the key features of the product or service with your client and carry out an analysis of their needs. At this stage you should avoid stating a solution even though this may be obvious or your client may be able to offer a number of suggestions. At this stage it is important to keep an open mind as to what is required to solve the problem. It is also important that the client brief is significantly different to the one that you were given in Unit 3: *Principles of Design, Planning and Prototyping*.

What is an engineered product?

In the context of this unit (and its assessment) an 'engineered product or service' can be taken as meaning:

- a physical product, for example an item of mechanical, electrical, electronic or fluidic equipment
- a service product, for example, the delivery of an engineering service such as the routine testing, inspection and maintenance of portable electrical equipment
- a 'system' product, for example, remote monitoring of the performance of the engine of a rally car using appropriate hardware, software and telecommunications equipment.

Choosing a project

There are a number of factors that you should think about when choosing your project. These include:

- The original client brief from which you must design and develop an appropriate solution (the 'product' or 'service'). Your primary aim must be that of satisfying your client's needs—not that of producing a particular product!
- You should keep in mind that the project that you undertake is the vehicle through which you will be assessed. For this reason it is essential that you maximise the potential for gaining marks against the *full* range of requirements stated in the assessment criteria.
- Your project should not be over-ambitious or overdemanding in terms of the knowledge, skills, materials, resources or time available for the unit.
- In practice, projects leading to 'service' products are likely to be more challenging because of difficulties in the research and product implementation (broadly equivalent to 'manufacturing') phases of the project.
- You should choose a project that will extend your understanding of a subject that already interests you, and that you wish to further develop. For example, if you are interested in motor vehicle technology you might like to consider a project that will allow you to explore a particular aspect of the design of performance cars.

Another view

When you study this unit and work on your assessed project it can be extremely useful to take into account what you previously learned in Unit 3. In particular, feedback and comments from your tutor can be invaluable in helping you to improve your marks in this unit. A good starting point would be to look critically at the work you did in Unit 3, asking yourself what worked well and what could be improved.

Finally, it is worth remembering that the best projects are those that arise out of a real requirement, possibly one identified by a customer, or through an industrial or commercial contact.

Resources

You will need to have access to appropriate workshop facilities in order to manufacture the engineering product or service. During the course of design, development and manufacture of the prototype, you should support and justify your decisions using relevant scientific and mathematical principles. Witness statements and photographic evidence can be used to provide evidence of safe, correct and competent use of appropriate tools, equipment, techniques and processes.

Your portfolio

Your portfolio should contain evidence of your work carried out in the design, development and manufacture of the engineered product or service. You should submit your work on A4 paper apart from the engineering drawings that must be submitted on A3 paper. You must also include photographic evidence of your final prototype. Your portfolio must include evidence of:

- appropriate research and the development of a technical specification
- generation of at least THREE alternative design ideas and their development into a final design solution using appropriate current industry standards and conventions
- discussions with other engineers (your peer group) on your initial design solutions
- planning for production
- prototype manufacture
- testing, evaluation and suggestions for modifications to improve the performance of the engineered product
- relevant health and safety issues.

Contact with industry

Contact with industry, either in person (based on visits and interviews) or as a result of research and analysis, is critical in being able to demonstrate that your project is realistic, both in terms of the original client brief and your chosen solution. Industrial contact will usually be much more effective if it is at a personal level but you may also find it useful to make contact in a more general way. For example, through requests for data sheets, catalogues, specifications or other technical information relevant to your project. Most engineering companies will be very willing to assist with enquiries from students but they will appreciate you explaining precisely what it is that you are seeking from them and why you need it!

Most schools and colleges have good links with local industry and your tutor may well be able to supply you with named contacts in local companies that would be willing to help you. However, in all cases it is important to check with your tutor before making contact. He or she may be able to suggest how (and to whom) an approach should be made and this will usually lead to a more successful outcome for you!

You will need to collect and record information that will assist you in developing a realistic technical specification. You may also need to carry out market research to establish the design details preferred by your target market. You should also analyse existing products to determine for example, how they function, how they are constructed, what materials and processes have been used in their manufacture, and also how much they cost. All of this information should be incorporated into your portfolio.

6.2 Technical specification

You will need to develop a technical design specification, containing key points identified from the client brief and from the research you carried out. Your specification should be as comprehensive as possible and should contain measurable parameters that your prototype can be evaluated against. You should develop your specification in consultation with your client and it should typically contain information on:

- function of the product
- user requirements
- performance requirements
- material and component requirements
- quality and safety issues
- required conformance standards
- scale of production and cost.

Activity 6.1

Archer Aerospace have asked you to assist the company with the development of a portable battery tester. The tester is to be used in conjunction with the batteries described in the data sheet shown in Figure 5.2 on page 355. Develop an outline technical specification for the portable battery tester and identify any outstanding items that would need to be clarified with the client.

Activity 6.2

Write a letter to Graham Milnes, the Sales Engineer of a local company that manufactures off-road vehicles. Let him know that you are a GCE Engineering student and ask him to supply you with full technical specifications of each off-road vehicle that is in current production. Also ask if it would be possible to arrange for you to take a photograph of one of the vehicles. Don't forget to thank him for responding to your letter!

6.3 Generation of alternative ideas and their development

From your technical design specification you should produce at least THREE alternative design solutions that offer different proposals for solutions to the product requirements. Each different design solution should be realistic and match the requirements set down in the specification. You should include accompanying notes that review and evaluate each design solution for its fitness for purpose.

When developing your design solutions you should consider the following:

- the selection of appropriate materials from the information gathered in your research
- how you will manufacture the product with the facilities available to you (note that you should also consider how the product would be manufactured on a larger scale)
- ergonomics (including effectiveness and ease of use)
- health and safety
- relevant regulations, codes of practice and standards.

Final design solution

Your final design solution should bring together the most suitable sub-systems or part designs taken from your initial design ideas and develop them into a workable solution that fully matches the specification. Development must show change and how the initial ideas have moved on in response to feedback and evaluation. Modelling and testing using computer software and/or hardware will take place during development.

Where numerical values are included as part of the proposed solution, you should show how you applied your scientific and mathematical understanding to arrive at your outcome. Where materials are selected for use because of their properties, you should show any relevant scientific and mathematical data you used in selecting the material.

6.4 Formative evaluation

At the design and development stage of your project, your progress must be reviewed and evaluated by a team of 'engineers' (drawn from your class or group) who should be asked to provide objective feedback on how the design solution matches the specification and whether the intended design is likely to succeed. You must use this peer-group evaluation in the development of your final design solution.

Activity 6.3

The portable battery tester that Archer Aerospace have asked you to design (see Activity 6.1) will require an enclosure. Given that the portable battery tester is to be manufactured in medium volume, suggest THREE possible materials for use in the manufacture of the enclosure and select the most appropriate. Give reasons for your choice.

6.5 Planning for production

You will need to produce a plan for production that considers all the manufacturing processes that would be involved in the manufacture of your product on a commercial scale. You must include proposed timings for particular tasks as well as points at which quality control checks need to be carried out. You should provide a full explanation of conformance quality checks. Your planning must also include a proposed costing for the manufacture of the product (including any assumptions made).

Your plan for production should include a record and explanation of the appropriate regulations, standards and documentation. It is important to be aware that a plan for production is a forward-looking document which considers each stage in the manufacture, testing, and adjustment of the product. It is not a haphazard list of processes or a retrospective diary of events.

The prototype

You must produce a working prototype that fully matches your final design solution. A prototype is a first attempt at a representative working product. Its primary function is to prove the design principles and to demonstrate how the product will look and function.

In your prototype production, you should demonstrate a clear understanding of why you have selected particular materials, components and processes referring to any scientific and/or mathematical principles you used.

You should ensure that your final outcome relates fully to all of the features you have specified in your design solution, for example material, construction, finish. During manufacture, you should demonstrate high-level manufacturing skills that show precision and attention to detail. You must also demonstrate a high level of safety awareness when working with machinery, tools and equipment.

When you have completed your prototype, you must test and evaluate its performance against what you set out to achieve in your specification. You should devise appropriate tests for each of the measurable points of the specification and check that your prototype matches the quality of performance specified.

These tests should be performed under real working conditions and ideally will include potential users, who should comment on their findings. Your test results, and those of the user group, can be used to evaluate your prototype objectively against the specification.

Activity 6.4

Production plans can be usefully summarised using simple flow diagrams (see page 182). Refer to appropriate references and use them to construct a flow diagram for the manufacture of a printed circuit board using photo-sensitive coated copper laminate board and a 1:1 photographic master of the track layout.

Enclosures

You will need to strike a balance between the amount of effort applied to the 'technology' content of your project and the effort directed to the use of materials to produce an enclosure in which to hold the parts and components used in your product. In most cases, the enclosure will be a relatively minor part of your product and will play little part in its overall functionality. Hence it is inappropriate to to spend a great deal of time developing and manufacturing the enclosure. It is also important to avoid using circuits, components or mechanisms that are not well understood or have not been adapted or personalised in any way.

Your work should reflect your knowledge and understanding related to the technology involved in your product or service and evidence of this should be incorporated into your portfolio. Design and development of the enclosure should be of secondary importance. As a general guide, you should aim to divide your efforts in a ratio of about 70:30 in favour of the technological content of your work. You should aim to produce a sound and correctly-reasoned solution to the client's design brief rather than a product that looks wonderful but does not actually satisfy the identified need!

6.6 Final evaluation

As a result of testing and evaluation, ideas for possible modifications will arise and these should be included in this section in sufficient detail to explain how their use will improve the design and performance of your product in the future. Your final evaluation should relate to the original client brief and the degree to which you have been able to satisfy your client's needs. You might also like to suggest alternative applications for your product or service as well as any further work that could be usefully carried out in order to enhance the product.

Figure 6.1 *Example of an exterior view of a project*

Photographic evidence

You will need to incorporate photographic evidence in your portfolio. This is most easily achieved using a digital camera. Your school or college will probably be able to supply this but you may also wish to use your own camera to produce images that show your work at various stages of the project. It is important to note that photographs that only show the external appearance of your product will be insufficient to fully satisfy the unit's assessment requirements.

In order to provide adequate photographic evidence, the following evidence will normally be required:

- general external view of the product (showing controls, indicators and displays)
- front and/or rear view of the product (where appropriate)
- internal view of the product (with covers or enclosure removed)
- close-up views of particular features (e.g. printed circuit boards, mechanisms, drives, etc).

Figure 6.2 *Example of an interior view showing manufacturing details and quality of production*

Once again, it is important to ensure that you provide sufficient evidence to fully satisfy the assessment requirements for the unit.

Sample product development

James Taylor is a student at North Downs College. His chosen project involved the development of an engineered product or system that will solve a particular problem identified by a local garage, car dealership and service agent, Portmore Motors.

The client brief

Portmore Motors (the 'client') asked James to find a way of extending the coverage of their existing wireless local area network (LAN) to include a PC that is installed in the Service Manager's Office. Unfortunately, the coverage of the wireless LAN (specified by the manufacturer) is only 30 metres but the line-of-sight (LOS) distance between the wireless access point (WAP) and the PC in the Service Manager's Office is 100 m.

The client brief, therefore, was to extend the wireless LAN in order that the Service Manager can access the company's client records and database and also to allow him to place orders for spare parts using the Internet. In order to clarify the client brief (and before attempting to develop a technical specification) James arranged to carry out an initial visit to Portmore Motors in order to talk to the Service Manager, Greg Smith. James was able to view the existing network and assess the layout of the site and, using sketches and rough measurements, he was later able to produce the drawing shown in Figure 6.3. This drawing provided a useful

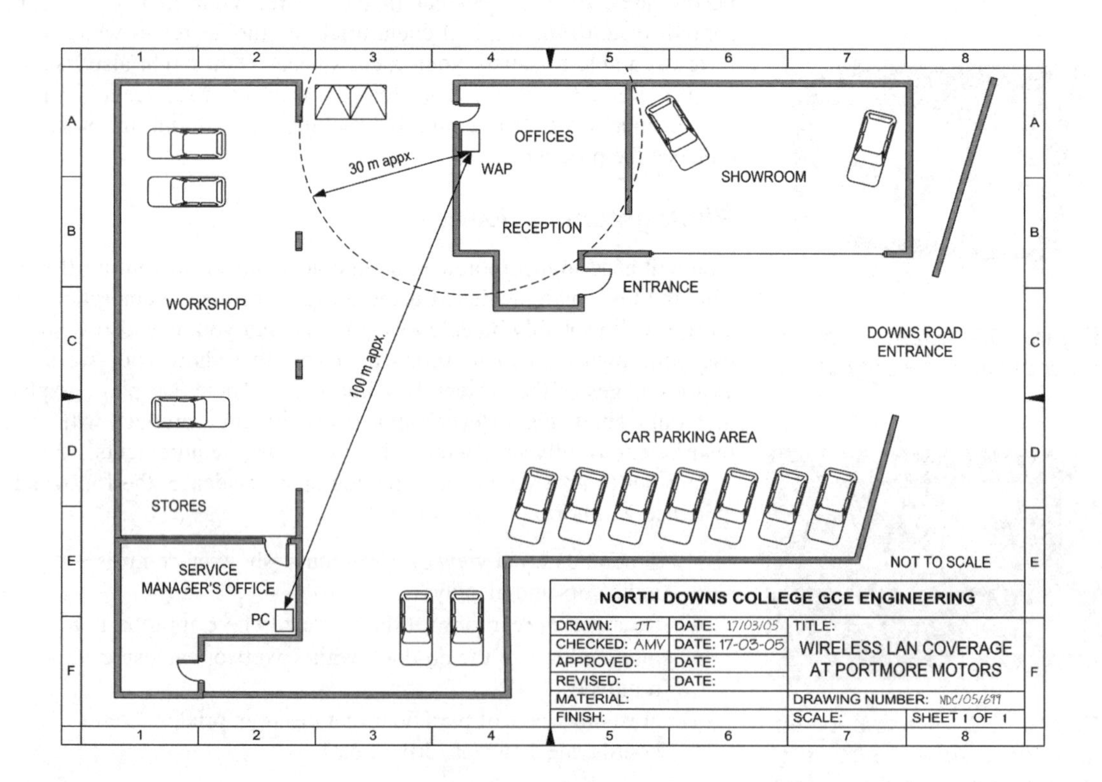

Figure 6.3 *Portmore Motors and the location of the two horn antennas*

clarification of the client brief and was included in his portfolio. James also took along his digital camera and obtained some photographs of the existing wireless LAN, including the wireless adapters (shown in Figure 6.4) and the wireless access point (WAP) and cable MODEM (shown in Figure 6.5).

James also spent some time discussing alternative solutions (for example, those based on cables and fixed LAN access points) but these were felt to be unsuitable on the basis of the likely cost and lack of flexibility. The company had found the wireless network to be very satisfactory in the office and showroom areas but unfortunately it was inaccessible from the workshop and the Service Manager's Office.

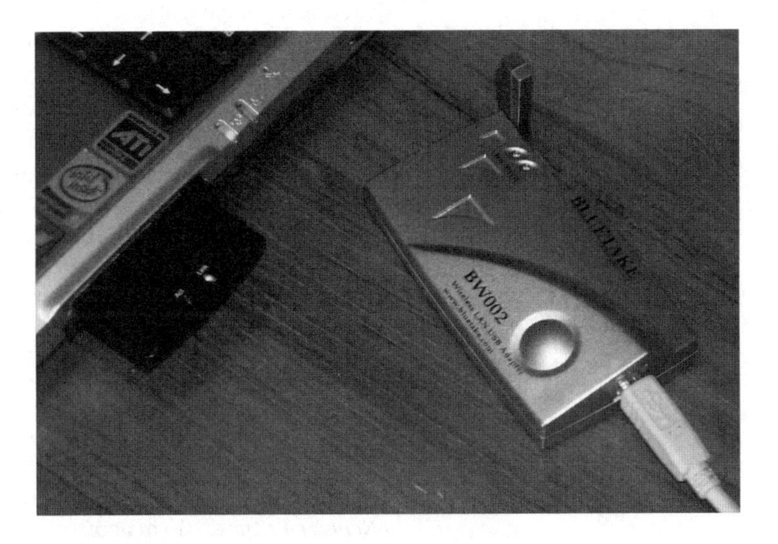

Figure 6.4 *Wireless network adapters in use at Portmore Motors: the PCMCIA adapter (left) has a low-gain internal antenna whilst the USB wireless adapter (right) has a small quarter-wave vertical antenna*

Figure 6.5 *Cable MODEM and wireless network adapter*

Initial research

Following his initial visit to Portmore Motors, James decided to carry out some research using the North Downs College Library and the Internet. James searched specifically for information on wireless networks and on ways in which the range (effective distance) of this equipment could be improved,

James found that the easiest way of increasing the range of the wireless LAN would be to make improvements to the antenna systems used. By replacing one of the low-gain omni-directional antennas with a directional antenna he would be able to increase the coverage by adding 'gain' into the system (the gain being associated with the directivity of the antenna compared with an omni-directional component).

James also found that the required minimum four-fold increase in power gain would be equivalent to a power gain of 6 dB. James was also pleased to find that this increase in gain could be achieved by simply replacing the current antennas and that no additional power supply or amplifying equipment would be required. James kept a record of the information sources that he used so that he could add these to his portfolio.

The technical specification

James was now in a position to formulate a full draft technical specification for the project:

Project aim:	Extension of existing wireless LAN to provide access from the Service Manager's Office
Product:	Directional antenna system
Required range:	100 m approx. (current range = 30 m)
Antenna gain:	12 dBi to 14 dBi (min. additional gain = 2 × 6 dB)
Frequency:	2.450 GHz
Data rate:	100 Kbps min.
Connector:	N-type female coaxial connector
Impedance:	50 Ω
Weight:	less than 600 g
Size:	less than 300 mm square
Standards:	IEEE 802.11b, IEEE 802.11g
Features:	Rugged, weatherproof, simple mounting, easily adjustable
Production:	2 units

He included this specification in his portfolio and returned to it later to confirm that the specification was being met. The specification became particularly important at the evaluation stage.

Generation of alternative ideas

Having completed the technical specification and having arrived at a means of solving the problem that Portmore Motors had identified in their client brief, James was confronted with the problem of arriving at a suitable antenna design. This required him to carry out some focused library and Internet research from which he was able to identify the five different antennas (or 'candidate solutions') shown in Figure 6.6.

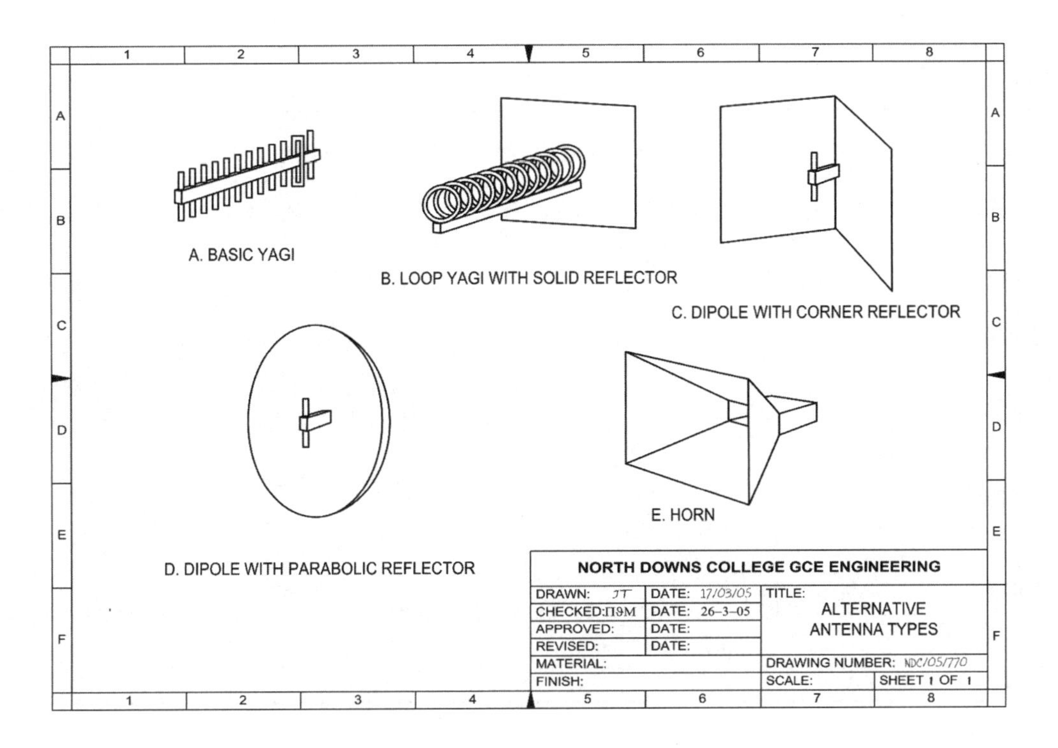

Figure 6.6 *Alternative antenna types*

In order to evaluate these potential solutions, James decided to make use of an evaluation matrix (see Figure 6.7). He chose six different features to take into consideration and arrived at an overall 'suitability rating' for each of them. This result of the evaluation was that the horn antenna was found to be most suitable followed closely by the corner reflector and basic Yagi antennas (both of which were slightly more complex). In order to confirm his choice, James consulted several technical references, including the *VHF/UHF Manual* by George Jessop, from which he found that " .... *horns are an attractive form of antenna, particularly for use at the higher microwave frequencies. They are fundamentally broadband devices which show a virtually perfect match over a wide range of frequencies. They are simple to design, tolerant of dimensional inaccuracies in construction, and they need no adjustment.*" James incorporated this quotation and gave a full reference to its origin in his portfolio.

Final design solution

James's final design solution was the horn antenna shown in Figure 6.8. His next problem was that of determining the dimensions of the antenna and the specification of suitable materials and production processes.

A further Internet search provided James with formulae for calculating the dimensions of the horn aperture and launcher, and he was also able to locate some CAD software that simplified the design process (see Figure 6.9). James compared the calculated values with those from the CAD software and found that they were in close agreement. James included these calculations and screen dumps in his portfolio.

Antenna:	A.	B.	C.	D.	E.
Type:	Basic Yagi	Loop Yagi	Corner reflector	Parabolic reflector	Horn
Gain (typical):	12 dBi to 15 dBi	15 dBi to 18 dBi	12 dBi to 15 dBi	15 dBi to 30 dBi	10 dBi to 20 dBi
Construction:	Fairly simple	Complex	Fairly simple	Complex	Simple
Size:	Medium	Medium	Large	Large	Medium
Mounting:	Simple	Fairly simple	Fairly simple	Could be difficult	Fairly simple
Signal feed:	Fairly simple	Could be difficult	Fairly simple	Must be at focal point	Easy
Materials:	Aluminium or brass rods	Brass or copper strip	Tinplate or aluminium sheet	Aluminium sheet	Tinplate or brass sheet
Suitability:	Possible?	Too complex	Possible?	Too complex	Good

Figure 6.7 *James's evaluation matrix was used to evaluate possible design solutions and included in his portfolio*

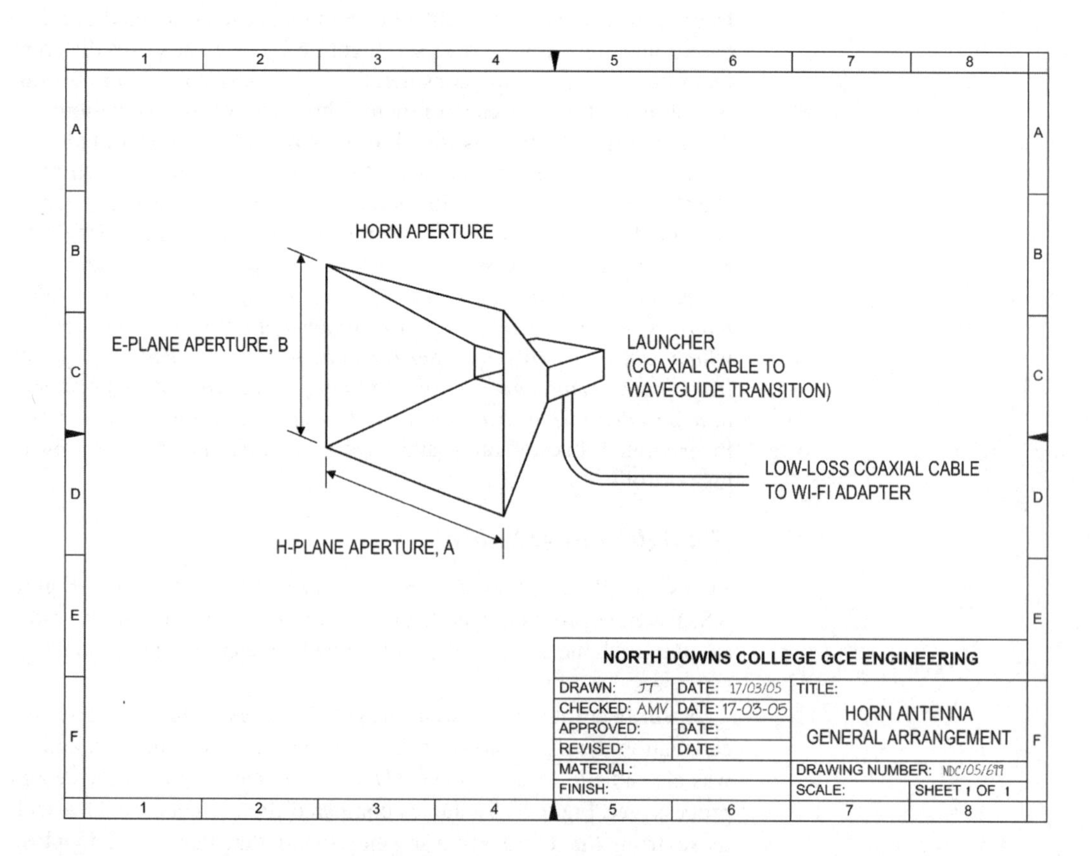

Figure 6.8 *James's final design solution. This general arrangement drawing was included in his portfolio.*

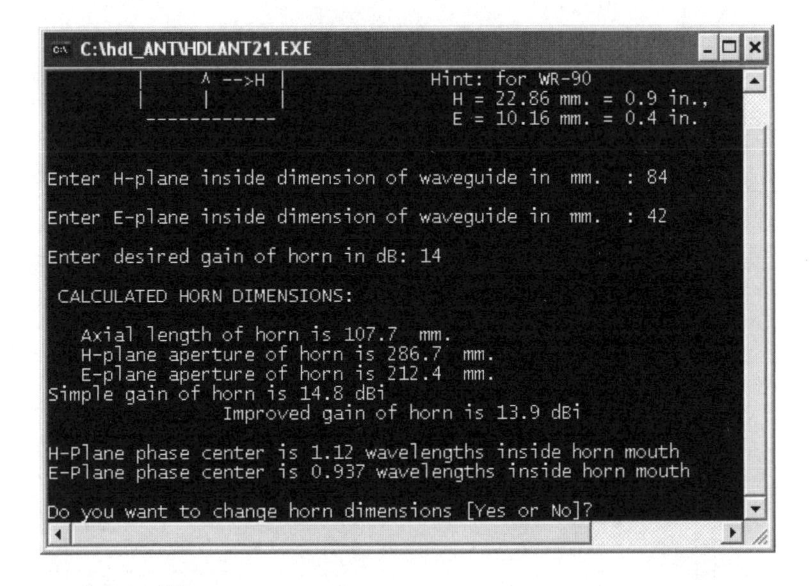

Figure 6.9 *The horn antenna design software used by James. This screen dump (as well as the check calculations) were included in his portfolio.*

Formative evaluation

Having arrived at a design solution, James was invited to present his work to the rest of the class in a 'review and evaluation' session which had been arranged by his tutor. At this session, James first outlined the client brief and described his initial visit to Portmore Motors. He explained the technical specification and the process by which he had arrived at it. James then gave a brief presentation on the different types of antenna that he had evaluated and his reasons for choosing the horn antenna for his product development.

The class asked a number of questions and made several useful comments (including the suggestion of using double sided copper laminate as a material from which to construct a prototype antenna). James's tutor took notes of the discussion and gave these to James for further reference and for inclusion in his portfolio. One member of the class recalled a series of articles that had appeared in the computer press showing how it was possible to construct a directional antenna for a wireless LAN using a tin can. He promised to provide James with copies of the relevant articles so that James could refer to these in his portfolio.

Planning for production

James was now ready to put his ideas into practice and plan the production of his prototype. First, however, he needed to give some thought as to how he would actually construct the horn and the launcher and join them together. Figure 6.10 shows the different techniques that James considered for constructing the horn aperture. By using a further evaluation matrix, James arrived at method D (fully folded construction) but he also decided to build an early prototype using method A with copper laminate cut and soldered along all four seams (as suggested by his classmates).

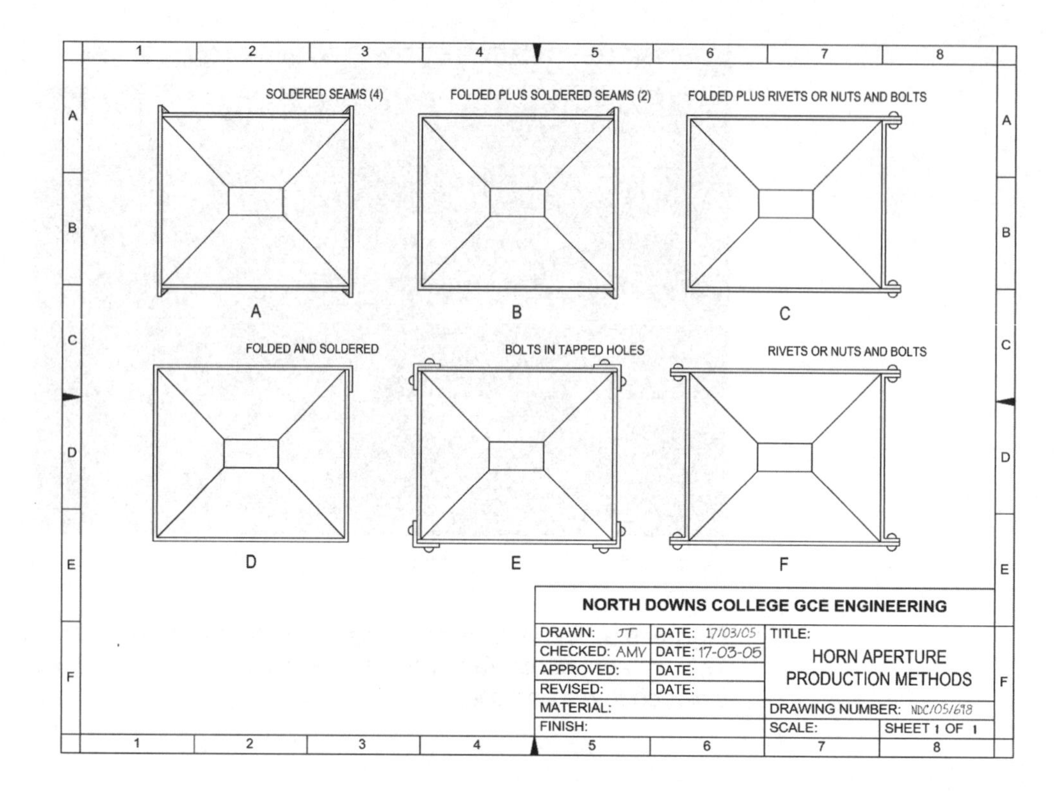

Figure 6.10 *James considered six different methods of constructing the horn antenna. By using a further evaluation matrix he decided to use method D (fully folded construction).*

Figure 6.11 *Early prototype horn antenna using method A*

James's early prototype antenna is shown in Figure 6.11. This was tested in order to 'prove' the design before moving on to production of the first two antennas using folded construction (as shown in Figure 6.12) and the launcher (shown in Figure 6.13). James incorporated all of these drawings and photographs into his portfolio.

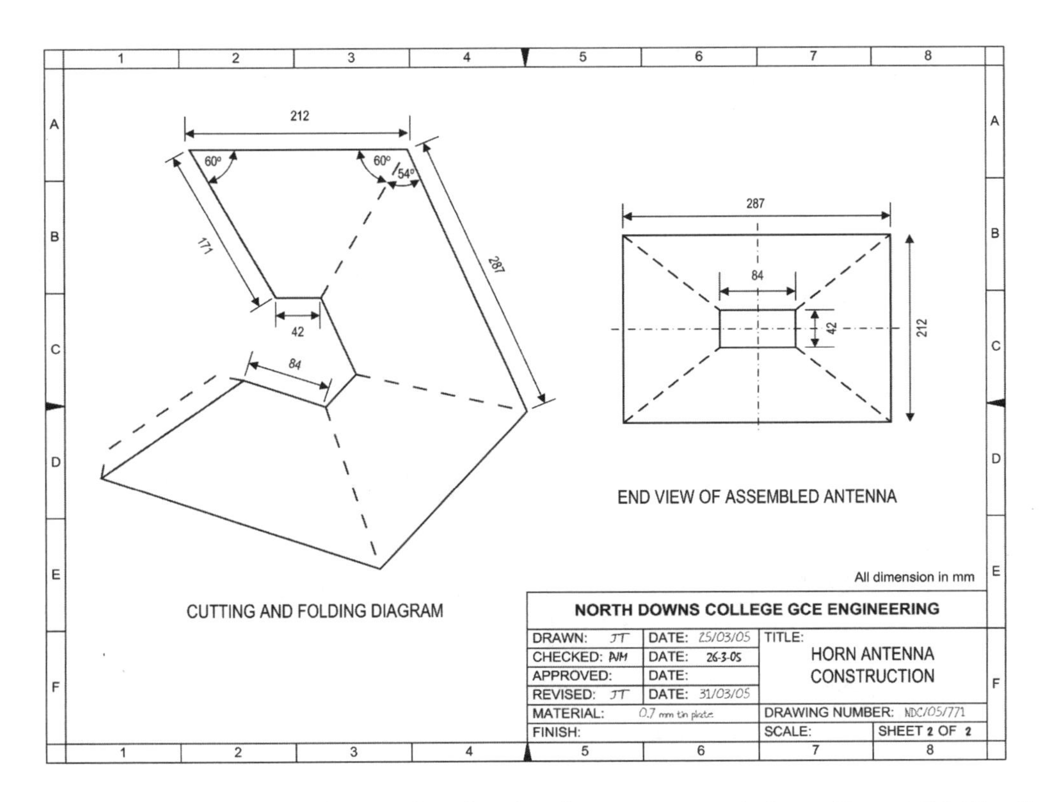

Figure 6.12 *Horn antenna construction showing the arrangements for cutting and folding the brass or tinplate sheet*

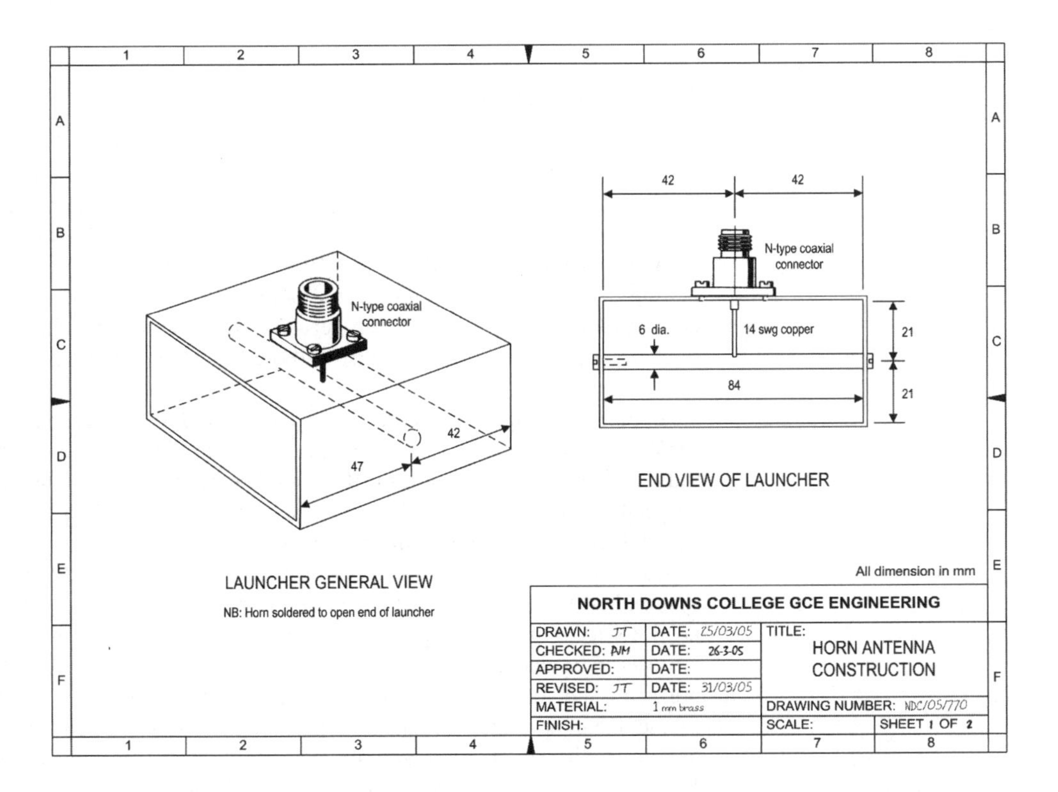

Figure 6.13 *Horn antenna construction showing the launcher (coaxial cable to waveguide transition). The launcher is to be soldered to the base of the horn aperture.*

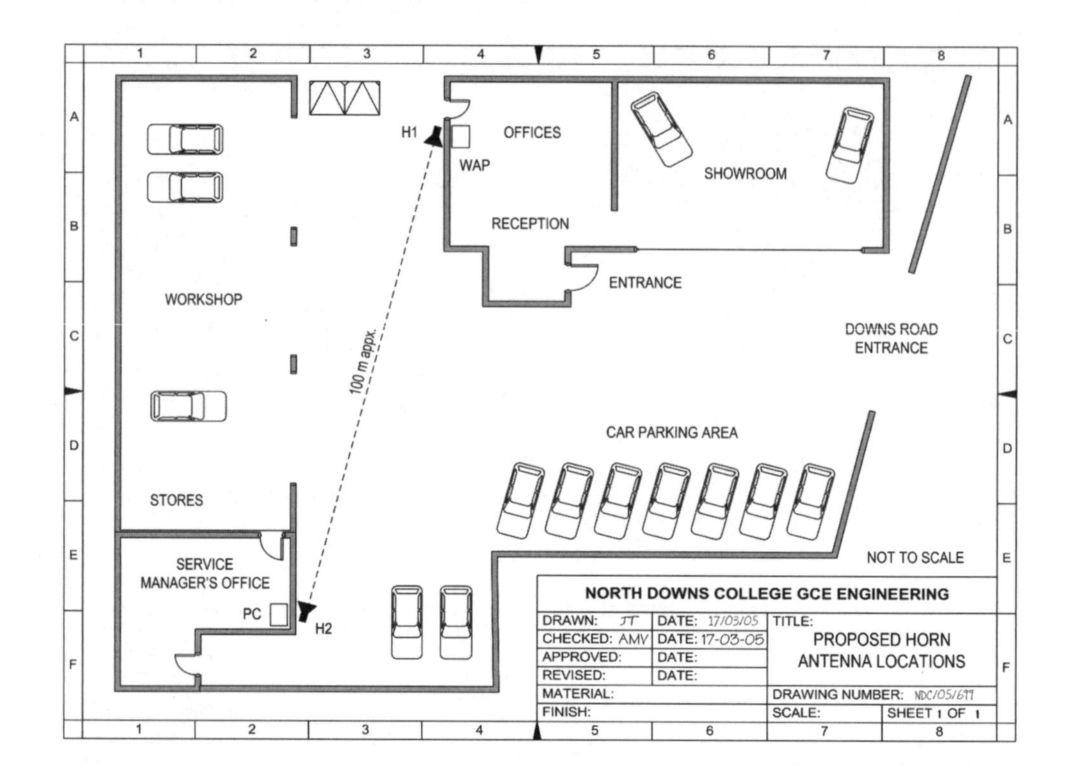

Figure 6.14 *The location of the two horn antennas (H1 and H2) at the Portmore Motors site*

Final evaluation

Having completed the production and assembly of his two horn antennas, James was ready to carry out a full site test using the wireless LAN at Portmore Motors (see Figure 6.14). He obtained two low-loss coaxial connectors to link the horn antennas to the wireless network adapter (H2) and to one of the two antenna ports provided on the wireless access point (H1). James used the wireless LAN configuration utility to set up the network (see Figure 6.15) and also to check the signal strength and overall quality of the link (see Figure 6.16).

The indicated signal strength was '100%' and the signal quality '95%' (showing some errors in data transmission). The data rate indicated reached a maximum of about 100 Kbps (in-line with the original technical specification). Greg Smith was invited to test the network connection and report any problems to James.

After a period of approximately two weeks, Greg reported that the system had worked without fault but he expressed concern about the need to seal the horn aperture against the ingress of water and airborne dust and dirt. James included this valuable feedback in his portfolio and modified his original design by fitting perspex covers to the front apertures of the two horn antennas. He included full details in the 'evaluation and modification' section of his report. James also included several photographs of the completed system including close-ups of the antennas showing their construction and how they were mounted. This completed his portfolio.

Figure 6.15 *Wireless network configuration screen. James included this as part of his portfolio evidence.*

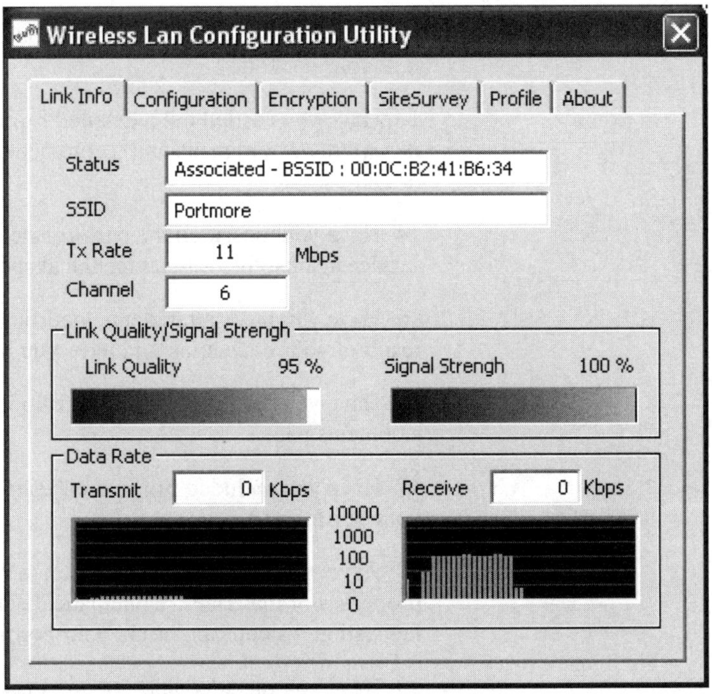

Figure 6.16 *Wireless network performance information. James used this screen as evidence that his horn antenna had complied with its original technical specification.*

Project checklist

The following checklist is designed to help you address each of the requirements and thus maximise your potential for earning marks in the assessment of this unit:

1 Have you clearly identified your client?

2 Do you have a full client brief?

3 Is the client brief clear and do you fully understand it?

4 Have you carried out appropriate research and investigation in order to clarify the problem, the context, and the client's needs?

5 Have you included details of your research and investigation in your portfolio?

6 Have you evaluated at least THREE alternative design ideas?

7 Have you formulated a detailed technical specification?

8 Does your technical specification include measurable parameters?

9 Have you arrived at a final design solution and included full details of it in your portfolio?

10 Have you presented your client brief and alternative design solutions to a peer group review?

11 Have you included details of this review (and any suggestions or changes made as a result of it) in your portfolio?

12 Have you produced a detailed production plan and included details of it in your portfolio?

13 Have you carried out a detailed formative evaluation and have you incorporated comments and recommendations received from your client or end-users?

14 Have you measured the performance of your product or engineered service against the original technical specification?

15 Have you incorporated any modifications or improvements as a result of your evaluation and have you included details in your portfolio?

16 Have you organised your portfolio in a logical sequence and included a contents page?

17 Have you included full references to your information sources in your portfolio?

18 Have you referred to relevant underpinning scientific principles, properties of materials, mathematical calculations, standards and legislation (as appropriate) in your portfolio?

19 Have you included appropriate drawings and photographic evidence in your portfolio?

20 Have you included acknowledgments in your portfolio?

Index

2D CAD 140, 143, 144
3D CAD 142, 145, 146
3D rendered view 147

A9CAD 143, 145
AC circuits 310
AC motor 297
ADC 335, 337
ALU 333
AM 338, 339, 369
AT cut 377
Abbreviations 211
Absolute dimensioning 214
Absolute permittivity 36
Acceptance phase 155, 169
Acid rain 361
Acrylics 15
Action 262
Address bus 332
Adherend 66
Adhesion 66, 68
Adhesive 19, 66, 79
Admiralty brass 7
Adobe Portable Document Format 356
Aeronautical engineering 128
Aerospace engineering 128
Allowable working stress 286
Alloy 2
Alloying 45
Alternative ideas 395, 400
Aluminium 9
Amendment 156
Ammeter 239
Amplifier 241, 324
Amplitude modulation 338, 369
Analogue signal 323, 338
Analogue system 323
Analogue to digital converter 335, 337
Angle grinder 118
Angle plate 104
Angular position sensor 319
Angular velocity 319
Annealing 43
Anode 47
Antenna 401, 404
Application note 351
Araldite 19
Arithmetic logic unit 333
Armature 295
Asbestos at Work Regulations 168
Ash 362
Atomic bonds 23
Atomic clock 325
Audit 156
AutoCAD 143
AutoSketch 143
Autoclave 87
Automatic system 323
Automatic test equipment 384
Automotive engineering 128
Auxiliary view 199, 201
Average value 311

BMP 144
BS 2197 204
BS 5070 204
BS 8888 181, 183, 203, 206, 211, 212
BS ENO ISO 5457 204
BS ISO 128-20 203
BS ISO 128-21 204
BS ISO 129-1 204
PS PP 8888 181, 183, 211, 212
BSI 163
BSI Kitemark 163
Bag-moulding 85, 86
Band-pass filter 324
Band-stop filter 324
Bandwidth 379
Bar-graph display 326
Battery 237, 287, 390
Belt drive 303
Binary coded decimal 323
Binary digit 323, 337
Bits 323, 337
Block diagrams 181
Blow moulding 82
Bolt 60, 61
Bolted connection 72
Bonding 66
Bonds 23
Bookshelf speaker 390
Boring 111, 113

Boring tool 112
Bowls 81
Brainstorming 251
Brass 7
Brass alloys 8
Brazing 58
Bridge 387
Bridge truss 279
British Standards 163, 181
Brittleness 29
Bronze 8
Bronze alloys 9
Bulk modulus 28
Bus 332
Business plan 248, 249
Butadiene-styrene 18
Butt joint 63
Button die 96
Butyl 18
Buyer 154

CAD 140, 141, 143, 144, 178, 385, 387, 401
CAD/CAM 140, 147, 148
CAE 140, 147, 150
CAM 140, 141, 147
CE mark 161, 163, 351
CFRP 85
CNC 148, 149, 150
COFDM 372
COSHH 168, 341
CPU 332, 333
CRT 390
Cables 314
Calender 80, 81
Calibration 325
Camera view 145
Candidate solutions 254, 255, 400
Capacitive proximity switch 320
Capacitor 237
Capacity planning 136
Carbon dioxide 361
Carbon fibre reinforced plastic 85
Carbon steel 5, 6, 7
Carrier 338
Casting 75
Castle nut 62
Catalogue 353, 356
Cathode 47
Cathode ray tube 390
Cathodic protection 51
Cell 237, 287
Cellulose acetate 15
Cellulose plastics 15
Central processing unit 332, 333
Centre drill 108
Centre lathe 106, 107, 108
Ceramic coating 51
Ceramics 2, 3, 10, 11, 84
Chain dimensioning 214
Chain drive 303
Chamfer 113
Chamfering tool 112
Charge carriers 287
Chart recorder 326
Chartered Engineer 132
Charts 189
Chemical attack 46
Chemical blacking 120
Chemical engineering 128
Chemical waste 362
Chemically deposited coating 51
Chuck 101, 109
Circuit 237, 240
Circuit diagram 183, 184, 185
Civil engineering 128
Clamped connection 70, 72
Classes of material 2
Clean Air Act 165
Clearance angle 89, 112
Client brief 250, 254, 391, 398
Closed-loop system 329, 330, 331
Coating 51, 119
Coaxial cable 314, 337, 405
Codes of Practice 168
Coefficient of linear expansion 39
Coercive force 36
Cogs 303
Cohesion 67
Cold pressing 76
Cold weld 378
Cold working 44
Colpitts oscillator 380
Communications 337
Communications engineering 128
Commutator 295, 296
Comparator 330
Competent body 161
Composite material 14, 16, 17
Composite materials processing 83
Composites 3, 83
Composition cell 50
Compound gear 302
Compound slide 106, 112
Compound wound motor 296
Compression 22, 279, 371
Compression moulding 81, 82
Compressive force 23, 271, 272
Compressive strain 273
Compressive strength 22
Compressive stress 273

Computer aided design 140
Computer aided engineering 140, 147
Computer aided manufacture 140, 147
Computer integrated manufacturing 147
Concentration cell 50
Conclusions 256, 257
Conditioning devices 234
Conditioning equipment 233
Conducted emissions 162, 350
Conducted susceptibility 350
Conduction 40
Conductivity 31, 32
Conductor 31, 314, 316, 316
Conical surfaces 113
Constraints 139
Construction line 196, 198
Consumer unit 315
Contract Manager 154, 155, 156
Contract life-cycle 154
Contract of Employment Act 159
Contracts 154, 156
Contracts of employment 159
Control 318
Control bus 332
Control element 322
Control of Substances Hazardous to Health 168, 341
Control program 334
Control system 150, 322, 324, 327
Control theory 327
Control unit 333
Controlled process 322
Controlled variable 322
Controller 322, 334
Conventional flow of current 287
Conventions 212
Copper 6
Corrosion 46, 52
Corrosion fatigue 53
Cost estimate 155, 156
Counterboring 105, 106
Countersink 108
Countersinking 103, 106
Couple 294
Crank 305
Cranks 305
Creep 30
Crimped joints 70
Crimping 72
Crimping tool 72
Cross filing 92, 93
Crystal filter 379, 383
Crystal frequency standard 381
Crystal oscillator 380, 383, 385, 388
Crystal structure 76
Crystals 42
Current 287
Cutting plane 217
Cutting screw threads 95
Cutting tools 89
Cyanoacrylate 19

DAB 369, 370, 371, 372
DAC 335, 337
DC 287
DC motor 293, 295, 301, 321
DC power supply 177
DCV 225, 227, 229, 306, 307, 308
DXF 144
Damping 331, 332
Data 189, 337
Data book 353
Data bus 332
Data link 340
Data sheet 177, 353, 356, 383
Decay 331
Deep drawing 76, 77
Degradable pollutant 361
Delivery phase 155
Demodulation 338, 339
Demodulator 339
Density 37, 263, 264
Deposit of Poisonous Wastes Act 165
Derelict land 363
Design 133, 250, 391
Design brief 175, 250, 254
Design criteria 255
Design engineer 135, 139
Design portfolio 254
Design review panel 252
Design solution 395
Design specification 175, 250, 255
Design strategy 244
DesignCAD 143
Detail drawing 186, 187
Deterioration 46
Development 133
Development process 134
Dezincification 7
Diagram 181, 182, 183, 189, 221
Diamagnetic materials 34
Die casting 75
Die holder 96, 97
Die moulding 85
Dielectric 36
Differential pressure vacuum switch 320
Diffuse scan proximity switch 320
Digital audio broadcasting 369
Digital signal 323
Digital storage oscilloscope 152

Digital system 323
Digital to analogue converter 335, 337
Dimensioned drawing 202
Dimensioning 214, 215, 216
Diodes 239
Dipole 34
Dipping 120, 121
Direct chemical attack 46
Direct current 287
Direct stress 24
Directional control valve 223, 306
Discharge 350
Disciplinary procedure 159
Dislocation 43
Dismissal 160
Display 326
Dispute 156
Documentation 349, 351, 366
Doping 33
Double-insulated appliance 314
Draw filing 92, 93
Drawing 181, 186, 187, 192, 193, 196, 204
Drawing conventions 212
Drawing scale 205
Drawing sheet 190, 191
Drill 98, 99, 100
Drill chuck 101
Drill tang 101
Drilling 98, 102, 103
Drills 98
Driven pulley 303
Driver 300
Driving pulley 303
Drop forging 75
Drunken thread 96
Dual-fuel vehicle 390
Ductility 29
Duralumin 45
Dust 362

E.m.f. 287
EC directive 350
EEA 162
EMC 165
EMC Directive 161, 162, 165, 350
EU 130
EU Directive 165
Economy 130
Effective value 311
Efficiency 293, 362
Elastic limit 275
Elastic materials 274
Elasticity 28, 29, 274
Elasticity modulus 274, 275, 278
Elastomers 2, 18
Electric circuit 240, 287
Electric drill 98
Electric shock 345
Electrical and electronic engineering 128
Electrical safety 344
Electrical schematic 237
Electricity at Work Regulations 168
Electrochemical attack 46
Electrochemical series 49
Electrolyte 46
Electromagnetic compatibility 162
Electromagnetic vibration sensor 320
Electromechanical systems 287, 293
Electronic circuit 237
Electronic symbols 238
Electronics 318
Electroplating 46, 117
Electrostatic discharge 350
Elevation 196
Elongation 275
Emissions 350
Employment 159
Employment Protection Act 159
Employment Rights Act 159
Employment tribunal 159
Emulsifier 91
Enclosures 397
Encumber 156
End view 196
Energy 288, 289, 291
Energy converter 222, 224
Energy efficiency 362, 366
Engineer's file 92
Engineered product 392
Engineering 128, 130
Engineering Council 132
Engineering Technician 132
Engineering drawing 181, 192, 203
Engineering roles 132
Engineering sectors 128
Engineering stress 24
Environmental Protection Act 165
Environmental impact 359
Environmental legislation 165
Epoxy resin 19
Equilibrium 262, 270
Error-forming device 330
Etching 116
European Community 161
European Economic Area 162
European Union 130
European standards 161
Evaluation 169, 180, 255, 385, 395, 403, 406
Evaluation matrix 402
Expansion 38, 39

Exploded view 186, 188, 189, 358
Exterior view 397
Extruder 81
Extrusion 76, 77, 80, 82

FIC 372
FM 338, 339, 369
Face-plate 110, 111
Facilities management 136
Facing tool 112
Factor of safety 286
Fast Information Channel 372
FastCAD 143
Fatigue 30
Fault-finding 182
Ferric chloride 74
Ferromagnetic materials 34
Ferrous metals 5, 7
Fibre 16, 86
Fibre mats 17
Field winding 295, 296
Filament 313
File 92
File card 94
Files 91, 93
Filing 91
Filler 12, 121
Filler rod 59
Filter 234, 324, 387
Final control element 322
Final design solution 395, 401
Final evaluation 406
Financial manager 156
First-angle projection 196, 197, 198
Fitness for purpose 169
Fitter's vice 88
Flat belt 303
Fleming's left-hand rule 293
Flow diagrams 182
Flow sensor 318, 319
Flowchart 182
Fluid control valve 232
Fluid power diagram 223, 224
Fluid power schematic 221
FluidSim 309
Fluidized bed dipping 120
Fluorescent lamp 314
Flux 35, 56, 78, 294
Fly-by-wire 324
Flywheel 142
Follower 300
Force 262, 265, 266, 267, 297
Force diagrams 264
Force/extension curve 276
Forging 77
Formal warning 159
Formative evaluation 252, 395, 403
Four arcs method 195
Four-bar linkage 305
Four-jaw chuck 109
Fracture stress 25
Fracture toughness 30
Framed structures 279
Framework 279, 280, 281
Frenkel defect 44
Frequency 310
Frequency modulation 338, 369
Frequency spectrum 385, 388
Frequency standard 325
Friction welding 59
Functional manager 156
Fundamental mode 377
Fuse 237
Fusion welding 53, 54, 79

GA drawing 186, 187
GLS lamps 313
GPS 374, 375
GRP 13, 16
GUI 143
Gain 327, 328
Galvanic cell 47, 50
Gantt charts 247, 248
Gas welding 53
Gauge specimens 275
Gear train 300, 301
General arrangement drawing 186
General lighting service lamps 313
General specification 353
Generating ideas 400
Germanium 33
Glass reinforced fibre 16
Glass reinforced plastics 13
Global Positioning System 374
Glue 19
Grain 42
Grain boundary 42
Graphical user interface 143
Graphs 189
Gravitational force 262
Gravity die casting 75
Grey cast iron 5
Grinding 118
Grit 362
Gross misconduct 160
Groups of paint 122
Guards 167

Hacksaw blade 89, 95
Hacksaws 94

Half-double scenario 249
Halogen lamps 313
Hand lay-up 85
Hand tap 97
Handouts 256
Hard soldering 57
Hardening 45
Hardness 29
Harmonic 385, 388
Harmonized standards 161
Hatching 218, 219
Hazards 167, 343
Headstock 107
Health and Safety 73, 166, 341
Health and Safety at Work Act 165, 166, 341, 342
Heat 38, 41, 362
Heat energy 288
Heat exchanger 234
Heat treatment 42
Helical spring 147
High carbon steel 5
High-pass filter 324
High-temperature corrosion 52
Hoist 229, 292
Holes 34
Hooke's law 25, 26, 274
Hot dip galvanizing 120
Hot gas welding 59, 60
Hot plate welding 59
Hot working 75
Hydraulic circuit 185
Hysteresis 35, 36, 322

I/O 332, 334
ISO 203
Ideas 251, 400
Idler 302
Incorporated Engineer 132
Indexing logic 298, 299
Induction motor 297
Inductor 237
Industry 393
Initial research 400
Injection moulding 83
Input transducer 318
Input/output 334
Instrumentation 318
Instrumentation system 325
Insulators 31
Integrated circuits 239
Interference 162, 371
Intergranular corrosion 52
Interior view 397
Intermittent flow 136
Internal architecture 333
Internal combustion engine 305
International Standards Organization 203
Intrinsic function 162
Investigation 254
Investment appraisal 135
Ion 47
Isometric drawing 193, 194
Isometric view 145, 146, 197

JPEG 144
Jack-screws 109
Joining 53, 79
Joint 63, 64, 67, 68, 69

Keys 97, 98
Kilowatt-hour 289
Kinetic energy 288, 291
Kitemark 163, 164
Knurling tool 112

LAN 334, 337, 398
LCD display 325, 371, 390
LCR bridge 387
LDR 319
LED 334, 340
LOS 398
LVD 351
LVDT 319
Ladder logic 335, 336
Laminated plastics 13
Lamps 313
Lap joint 63, 67
Lathe 106, 107, 108
Launcher 405
Lay-up 85, 86
Layout diagram 184
Leader lines 209, 210
Left-hand rule 293
Legislation 154, 349
Letters 211
Lifting 344
Light dependent resistor 319
Light emitting diode 340
Light energy 288
Light fittings 315
Light level sensor 319
Lighting circuit 315
Line flow 136
Line types 207, 208
Line-of-sight 398
Linear actuator 227, 228, 321
Linear expansion 39, 40
Linear friction welding 59
Linear position sensor 319

Linear translational motion 305
Linear variable differential transducer 319
Lines 206, 209
Linkages 305
Links 303
Liquid level float switch 318
Liquid level sensor 320
Liquid plastisol dipping 121
Load 275, 316
Load/extension graph 275
Loading 323
Local area network 334, 337
Local economy 130
Locking nut 62
Loop 294
Loudspeaker 318, 390
Low Voltage Directive 161, 351
Low carbon steel 5
Low-pass filter 324
Lubricator 234

M-derived filter 387
MCB 315, 316
MDSolids 283, 284
MIG welding 55
MODEM 340, 350, 389, 399
MSD 351
MSDS 73
MUSICAM 371
Machine vice 104
Machinery Safety Directive 351
Machining 78
Magnetic dipole 34
Magnetic field 294
Magnetic flux 35, 294
Magnetic flux density 35
Magnetic linear position sensor 319
Maintenance 167
Malleability 29
Management of Health and Safety at Work Act 168
Manometer 325
Manual 357
Manual handling 344
Manufacturing 255
Manufacturing processes 343
Market research 137, 176
Marketing 137, 138
Martensite 45
Mass 262
MatSdata 21
Material Safety Data Sheets 73
Material processing 360
Material removal 88
Materials 1, 2, 22
Materials Safety Data Sheet 74
Materials processing 75
Matrix 16, 84
Matrix board 71
Matrix materials 86
Maximum value 311
Mechanical engineering 128
Mechanical work 291
Mechanism 305
Megnetization 34
Metals 2, 84
Method of joints 281
Microcontroller 150, 151, 334
Microprocessor 333
Microprocessor system 332
Microswitch 320
Mild steel 7
Milestone 155, 156
Miller oscillator 380
Milling machine 148
Mind mapping 251
Miniature circuit breaker 315
Mining 363
Modification 169, 180, 385
Modulation 338, 369
Modulator-demodulator 340
Modulus of elasticity 28
Modulus of rigidity 274
Moment 294
Motor 295, 297
Motor vehicle engineering 128
Motors 293
Moulding 81
Moving coil meter 325
Multiplex 370

NCB 163
NRV 232
National Standards Body 163
National economy 130
Natural materials 4
Naval brass 7
Navstar 372, 374
Needs analysis 175
Network 337
New Approach 161
Ni-Cd battery 390
Ni-MH battery 390
Nitrogen 361
Non-conformance 354
Non-degradable pollutant 361
Non-ferrous metal 6
Non-metals 10, 79
Non-return valve 232, 233
Non-toxic pollutant 361

Notified body 161
Numbers 211
Nut 60, 61, 62
Nylon 15

OLE 144
Object linking and embedding 144
Oblique cutting 90
Oblique drawing 193
Off-hand grinding machine 118
Off-the-wall ideas 251
Ohm's law 287
Oil blueing 120
On-call contract 156
On-line product search 384
Open-loop system 329
Operating manual 357
Operational amplifier 330
Optical fibre 337, 340
Optical shaft encoder 319
Organic coating 51
Orthogonal cutting 90
Orthographic drawing 196
Orthographic view 145, 146
Oscillator 380
Oscilloscope 153
Output transducer 318
Overshoot 331
Overtone 377, 378
Oxidation reaction 47
Oxide layer 51
Oxides of nitrogen 361
Oxidizing 120
Oxy-acetylene welding 53
Ozone 362

PCB 70
PCB track diagram 184
PCFCV 231
PDF 356
PLC 334, 335
PLC programmer 336
PMC 85
PNG 144
PP 7307 181, 203, 204, 237
PP 7308 181, 203, 204
PRV 231
PSI 164
PTFE 2, 15, 120
PVC 2, 15
PVC coating 121
Paint 122
Paint groups 122
Painting 121
Paper sizes 204, 205
Parallel I/O 334
Parallel circuits 312
Parallelogram of forces 267
Paramagnetic materials 34
Parison 82
Passive metal 51
Performance specification 353
Period 310
Permanent elongation 275
Permeability 35
Permittivity 36
Personal protection 167
Petroleum engineering 128
Phosphating 51
Phosphor 314
Phosphor-bronze 9
Photoconduction 34
Photodiode 319
Photographic evidence 397
Phototransistor 319, 340
Pierce oscillator 380
Piezo-electric pressure sensor 320
Piezoelectric effect 377
Pigment 121
Pillar drill 99
Pinning 92, 94
Pipeline 236
Piping diagram 185
Piston 305
Pitting 52
Plain washer 62
Plan view 196
Planning 243, 245, 391, 396, 403
Planning phase 155
Plastic coating 120
Plasticity 29
Plastics 2, 13, 14
Plumbing circuit 185
Pneumatic equipment 36, 306
Pneumatic hoist 229
Pneumatic symbols 306
Point defect 44
Poisson's ratio 31
Polishing 119
Pollutant 360, 361
Pollution 359, 360, 362
Pollution of Rivers Act 165
Polyester 15, 18
Polymer matrix composites 85
Polymer processing 79, 80, 81, 82
Polymerization 12
Polymers 2, 84
Polypropylene 15
Polystyrene 15
Polytetrafluoroethylene 2

Polythene 15
Polyurethane 18
Polyvinyl chloride 2, 15
Portfolio 254, 393
Position sensor 319
Post audit 156
Potential energy 288, 291
Potentiometer 327
Power 288, 289, 303
Power supply 177, 181, 182
Precipitation hardening 45
Presentation 254, 256
Presentation drawings 255
Pressure 271
Pressure compensated flow control valve 231, 232
Pressure relief valve 230, 231
Pressure sensor 320
Primer 121
Printed circuit board 70, 71, 116, 177
Process annealing 44
Process control 334
Process management 136
Processing materials 75
Procurement phase 155
Product 392
Product development 133
Product search 384
Production 134
Production constraints 252
Production engineer 134, 135, 139
Production planning 396, 403
Production processes 343
Products 173, 175
Program 334, 336
Programmable logic controller 334, 335
Programme timing 134
Project 156, 392
Project Manager 154, 155
Project Team 154
Project phase 250
Project planning 243, 245
Project presentation 254
Project schedule 156
Project scope 156
Project-based production 136
Properties of a force 265
Properties of materials 22, 84
Protection 167
Protective coating 51
Protective oxide layer 51
Prototype 177, 178, 396
Prototype manufacture 253
Prototyping 391
Proximity switch 320
Pulley 303
Purchaser 154
Putty 121

Quality documents 354
Quality of life 360
Quality procedures 354
Quality systems 354
Quantisation 3, 338
Quarrying 363
Quartz 376
Quartz crystal 325, 377, 378, 379, 380
Quenching 45

RAM 332, 334
RCCB 317
RCD 317
RDS 372
RIDDOR 168
ROM 332, 334
Radial circuit 315
Radiated emissions 162, 350
Radiated susceptibility 162, 350
Radio 366, 367, 368
Radio receiver 339
Radioactive waste 361, 362
Radius forming tool 112
Rake angle 89, 90, 91, 112
Random access memory 332, 334
Reactant 48
Reaction 48, 262
Read 333
Read-only memory 332, 334
Reaming 114
Reciprocating motion 305
Recommendations 256, 257
Recrystallization 44
Redox series 49
Reduction reaction 48
Redundant member 280
Reference axes 146
Reference source 325
References 256, 257
Refractoriness 12
Regional economy 130
Register 333
Registered Firm 164
Reinforced concrete 3
Reinforced plastics 13
Reinforced polyester 18
Reinforcing fibres 86
Relative density 264
Remanence 36
Removal of material 88
Render mode 145

Rendered drawing 142
Rendering 146
Repair manual 357
Reporting lines 135
Reporting of Injuries, Diseases and Dangerous Occurrences Regulations 168
Reports 256
Research 252, 254, 400
Residual current circuit breaker 317
Residual current device 317
Resistance 33, 240, 287
Resistance weld 378
Resistivity 31, 32
Resistor 237
Resonance 378
Resource plan 154
Resources 393
Responsive materials 19
Resultant 265
Rigidity 274
Ring main circuit 315
Risk assessment 73
Rivet 64
Riveted joint 62, 63, 64
Riveting 65
Roles of engineers 132
Roof truss 279
Root mean square 311
Rotating vane flow sensor 318, 319
Rotor 295, 297, 299
Rubber 2, 18

SFN 372
SI 165
SMA 19
SPDT switch 316
SPST switch 316
Sacrificial anode 51
Safety 286, 341, 342, 343, 344
Safety education 167
Sales 137, 138
Sales engineer 139
Satellite MODEM 389
Sawing 94
Scale 205
Scarf joint 67
Scatter diagram 189
Schedule 156
Schematic 185
Schematic diagrams 221
Schottky defect 44
Screw 60
Screw extruder 80, 81
Screw thread 203
Screw thread cutting 95
Screwed fastening 59, 61
Second-order response 331
Section 217
Sectional view 216
Sectioning 215, 220
Self-locking nut 62
Seller 154
Semiconductors 33
Sensors 319
Sequence valve 230
Serial I/O 334
Series circuits 312
Series wound motor 295, 296
Service manual 357, 358
Servo system 323
Servomechanism 323
Set of a saw blade 95
Set point 322
Shaft encoder 319
Shape factor 379
Shape-memory alloys 19
Shear force 23, 271, 272
Shear modulus 28
Shear strain 273, 274
Shear stress 272, 273
Shock hazard 345
Short-form catalogue 356
Shunt wound motor 295, 296
Shuttle valve 232
Signal 323, 337, 338
Signal conditioning 324
Silica 3
Silicon 33
Simulation 151, 152
Single frequency network 372
Single-phase supply 310
Sintering 76, 77
Slider-crank mechanism 305
Slideways 106
Smart materials 19
Smog 361
Smoke 361
Soft soldering 55, 78
Solder seal 379
Soldered joint 55, 56, 57, 69
Soldering 55, 57, 78
Soldering flux 56
Soldering iron 389
Solenoid 294, 295
Solution development 178
Solvent 121
Spanners 97, 98
Speaker 390
Specific heat capacity 38, 39
Specific weight 37

Specification 175, 179, 250, 353, 394, 400
Speed control system 330
Splice joint 63
Spot facing 106
Spot-facing 105
Spray-up 85
Spring washer 62
Squirrel cage rotor 297
Stable framework 280
Standard 161, 164
Standard cell 325
Standard resistor 325
Standard specification 353
Starter 314
Static structural systems 261
Statically determinate 280
Statically indeterminate 280
Stator 297, 299
Status indicator 334
Statutory Instrument 165
Steel 5, 7
Stepper motor 298
Stepper motor controller 299
Stiffness 274
Straight grinder 118
Straight shank drill 100
Strain 24, 271, 272, 273
Strain energy 291
Strainer 234
Stratospheric ozone 362
Strength 25, 26
Stress 24, 25, 271, 272, 273
Stress cell 50
Stress corrosion 52
Strip board 71, 177
Structures 279
Suds 91
Sulphur dioxide 361
Summing junction 330
Super glue 19
Superconductivity 32
Supplier 154
Surfacing 111, 113
Susceptibility 350
Swarf 102
Sweating 57
Switch 239, 316
Symbols 211, 223, 238
Synchronous motor 297
System 322, 327, 329, 331

TIG welding 55
TPEG 372
Tab washer 62
Tachogenerator 319
Tachometer 319
Tailstock 107, 108, 114, 188
Tang 101
Tap wrench 96, 97
Taper shank drill 100, 101
Taper washer 62
Tapping 114
Tapping size hole 95
Taps 97
Technical report 256, 351
Technical specification 394, 400
Technician 132
Techniques 366
Technology 140, 366
Teflon 15, 120
Telecommunications Terminal Equipment Directive 350
Telecommunications engineering 128
Temperature resistivity coefficient 32
Temperature sensor 320, 320
Tempering 45
Tensile force 23, 271
Tensile fracture stress 27
Tensile strain 273
Tensile strength 22, 278
Tensile stress 273
Tensile testing 275
Tension 22, 279
Tensometer 275
Tesla 35
Test jig 384
Test specifications 354
Testing 179
Thermal conduction 40
Thermal conductivity 38, 41
Thermal expansion 38, 40
Thermistor 320
Thermocouple 320
Thermocouple probe 318
Thermoplastics 2, 13, 15, 58
Thermosets 12
Thermosetting plastics 2, 12
Thinner 121
Third-angle projection 196, 199
Threading 114
Threads 96, 203
Three-jaw chuck 110
Tin bronze 8
Tin-bronze 9
Tina Pro 387
Tinning 55
Title block 205
Toothed belt 303
Toothed rotor tachometer 319
Top coat 121

Torque 294
Toughness 30
Toxic metals 362
Toxic pollutant 361
Traceability 354
Transducer 318
Transfer function 327, 328
Transformation temperature 19
Transformer 237, 314
Transistor amplifier 241
Transistors 239
Transmission 305
Transmitter 339
Treaty of Rome 161
Triangle of forces 268, 269
Triangulation 374
Tribunal 159
Tropospheric ozone 362
True stress 24
Truss 279, 280
Tufnol 13, 116
Tungsten 77
TurboCAD 143
Turning tool 112
Turret 112
Twin and earth 314
Twist drill 100, 101, 102, 103
Two-pack paints 122
Two-roll mill 80
Two-way lighting 240, 316, 317

Ultimate strength 26
Ultimate tensile strength 27
Ultimate tensile stress 286
Ultra-violet radiation 362
Ultrasonic welding 58
Undercoat 121
Undercutting tool 112
Unit of electricity 289
Unstable framework 280

V-belt 303
VLSI 334
Vacuum switch 320
Vacuum-bagging process 86
Value analysis 134
Value engineering 134
Valve control methods 225, 226
Vehicle 121
Vibration sensor 320
Vibration welding 59
Vice 88
Virtual display 151
Virtual instrument 152, 153, 326
Visit record 382
Voltage 287
Voltmeter 239

WAN 337
WAP 398, 399
Washer 62
Waste products 360
Waste reprocessing 363
Water trap 234
Waveguide 405
Wedge angle 89
Weight 37, 262
Welding 53, 54, 58, 59, 79
Wetting agent 78
What-if analysis 249
Wheatstone bridge 325, 327
Wide area network 337
Wire wrapping 70
Wireframe 145, 146
Wireless LAN 398
Wireless LAN adapter 399
Wireless access point 398
Wireless network 407
Work 288, 291
Work instructions 354
Work-holding 102, 104, 105, 106, 108
Work-in-process 135
Working drawings 255
Work plane 146
Working practices 167
Worm drive 302
Write 333
Written warning 159, 160

X-Y plotter 326

Yield point 275
Yield strength 278
Yield stress 25, 26, 286
Young's modulus 28, 274, 278

Zinc 7